REFINING PROCESSES HANDBOOK

REFINING PROCESSES HANDBOOK

Surinder Parkash

AMSTERDAM BOSTON HEIDELBERG LONDON
NEW YORK OXFORD PARIS SAN DIEGO
SAN FRANCISCO SINGAPORE SYDNEY TOKYO
Gulf Professional Publishing is an imprint of Elsevier

Gulf Professional Publishing is an imprint of Elsevier.

Recognizing the importance of preserving what has been written, Elsevier prints its books on acid-free paper whenever possible.

Library of Congress Cataloging-in-Publication Data
A catalog record for this book is available from the Library of Congress.

ISBN-13: 978-0-7506-7721-9
ISBN-10: 0-7506-7721-X

British Library Cataloguing-in-Publication Data
A catalogue record for this book is available from the British Library.

The publisher offers special discounts on bulk orders of this book.
For information, please contact:

Manager of Special Sales
Elsevier
200 Wheeler Road
Burlington, MA 01803
Tel: 781-313-4700
Fax: 781-313-4882

For information on all Gulf Professional publications available, contact our World Wide Web home page at: http://www.gulfpp.com

10 9 8 7 6 5 4 3 2

Printed in the United States of America

Cover Photo by Mieko Mahi, energyimages.com

To My Wife

RITA

Contents

Preface

Petroleum refineries have grown rapidly in complexity and so, too, the refinery operations. However, the published information on the refinery processes and operation is scant and mostly confined to licensor's data, which reveal little beyond what is absolutely necessary for process sale, even when these processes have been in operation for a number of years and in many refineries. This book is an overview of the processes and operations concerned with refining of crude oil into products. The streams coming from processing units are not finished products; they must be blended to yield finished products. The refining operations presented here are those concerned with blending products in an optimum manner with the twin objectives of meeting product demand and maximizing refinery profit. The objective here is to provide basic instructions in refinery practices employing the methods and language of the industry.

Presented in the book are refinery processes, such as crude desalting and atmospheric and vacuum distillation; gasoline manufacturing processes, such as catalytic reforming, catalytic cracking, alkylation, and isomerization; hydrodesulfurization processes for naphtha, kerosene, diesel, and reduced crude; conversion processes such as distillate and resid hydrocracking; resid conversion processes such as delayed coking, visbreaking, solvent deasphalting, and bitumen manufacture; pollution control processes such as sulfur manufacture, sulfur plant tail gas treatment, and stack gas desulfurization. Also presented here are operations performed in refinery off-site facilities, such as product storage and blending, refinery steam and fuel systems, refinery boiler feedwater treatment, and wastewater treatment.

The process details include process flowsheets, process description, chemistry involved, detailed operating conditions, process yields and utilities. Among the refinery operations and practices presented are product blending, refinery inventory forecasts, spreadsheet and LP modeling of refineries, and methods for pricing crude oil, petroleum products, and intermediate stocks.

It must be recognized, however, that many variants of the same process are found in the industry, and the operating conditions can be quite diverse, depending on the type of catalyst used and feedstock processed. We have insufficient space for bibliographic comparison and evaluations of identical basic processes from different licensors. The data presented here represent typical industrial operations practiced in refineries today. Where no mention is made of recent contributions to the literature, no slight is intended. The few references quoted are those where an industrial practice is known to have originated.

Another important subject presented in this volume is concerned with the operation of joint ownership refineries. Building a grassroots refinery requires large capital investment. It is feasible for two companies to own and operate a refinery as if it were build of two independent refineries. Each company may operate its share of the refinery virtually independent of other; that is, each company may bring in its own feedstock and produce product slate independent of the other with no need to build separate product storage facilities for the two companies.

The basic rules of operations of joint ownership refineries is discussed in this book. A typical pro-forma processing agreement between the participants is presented in the Appendix of this book. This covers detailed procedures for refinery production planning, product allocation, inventory management, and allocation of refinery operating cost to participants. Product allocation is the split of total refinery production among the participants on the basis of the feedstock processed by each. Keeping in view that the participants do not process identical feedstocks or produce identical product grades, product allocation for establishing the ownership of stock, must be done at the end of every month. This is a complex exercise and a detailed procedure for this is presented in a separate chapter.

The methods for preparing inventory forecasts and tracking refinery operating expenses in a joint ownership refinery scenario are presented as well. Even though such practices—product allocation, inventory and ullage allocation, operating costs allocation—exist in refining industry, there is no known literature examining them.

CHAPTER BREAKDOWN

Chapter 1 covers atmospheric and vacuum distillation and crude desalting. Chapter 2 covers the refinery hydrotreating processes: naphtha hydrotreating, kerosene hydrotreating, gas oil hydrodesulfurization and atmospheric resid desulfurization. Chapter 3 presents the distillate hydrocracking, mild hydrocracking, and resid hydrocracking processes.

Chapter 4 covers gasoline manufacturing processes: catalytic reforming, alkylation, isomerization, catalytic cracking, and MTBE manufacture. Chapter 5 looks at the manufacture of hydrogen for hydrotreating and hydrocracking process and its recovery from some of the hydrogen-bearing streams coming from these units. Chapter 6 presents refinery residuum processing units, on delayed coking, visbreaking, solvent deasphalting, and bitumen blowing.

Chapter 7 examines treating processes for catalytic cracker light and heavy naphthas and kerosene-type jet fuels. Chapter 8 presents sulfur manufacture and pollution control processes, such as sulfur plant, sulfur tail gas treatment, and stack gas desulfurization.

Chapter 9 examines the refinery water system. This includes treatment of cooling and boiler feedwater, the refinery's oily wastewater, and stripping the refinery's sour water.

Chapter 10 looks at the off-site and utility systems of a refinery. The topics include the tankage requirements for product export and product blending; batch and in-line product blending systems; refinery flare system, including principals of flare system design; the refinery steam system; and liquid and gaseous fuel systems.

Chapter 11 describes the procedures for product blending. Chapter 12 presents the procedure for preparing a refinery material balance using a spreadsheet program. Chapter 13 describes the general principles of building a refinery LP model. Chapter 14 discusses the mechanism of pricing petroleum products, including intermediate streams and products. Chapter 15 describes the concept of a definitive operating plan for the refinery during an operating period.

Chapter 16 shows the methodology behind product allocation in joint-ownership refineries. Chapter 17 explains methods of estimating available tankage capacity as a part of an inventory forecast system in both single- and joint-ownership refineries. Chapter 18 explains how these inventory forecasts are prepared for planning shipment of product in both single-ownership and joint-ownership refineries. Chapter 19 presents procedures for estimating the operating costs of the refinery and, in case of joint-ownership refineries, the allocation of refinery operating costs to the participants.

An appendix explains the organizational structure of joint-ownership refineries and presents an example of a processing agreement among the participants required for operating such a refinery.

We hope this book will serve as a useful tool for both practicing engineers concerned with refinery operational planning as well as for academics.

CHAPTER ONE

Refinery Distillation

Crude oil as produced in the oil field is a complex mixture of hydrocarbons ranging from methane to asphalt, with varying proportions of paraffins, naphthenes, and aromatics. The objective of crude distillation is to fractionate crude oil into light-end hydrocarbons (C_1–C_4), naphtha/gasoline, kerosene, diesel, and atmospheric resid. Some of these broad cuts can be marketed directly, while others require further processing in refinery downstream units to make them saleable.

The first processing step in the refinery, after desalting the crude, is separation of crude into a number of fractions by distillation. The distillation is carried out at a pressure slightly above atmospheric. This is necessary for the following considerations:

1. To raise the boiling point of the light-end carbons so that refinery cooling water can be used to condense some of the C_3 and C_4 in the overhead condenser.
2. To place the uncondensed gas under sufficient pressure to allow it to flow to the next piece of processing equipment.
3. To allow for pressure drop in the column.

Crude oil is preheated in exchangers and finally vaporized in a fired furnace until approximately the required overhead and sidestream products are vaporized. The furnace effluent is flashed into the crude column flash zone, where the vapor and liquid separate. The liquid leaving the flash zone still contains some distillate components, which are recovered by steam stripping. After steam stripping, the bottom product, also known as *reduced crude*, is discharged from the tower. The bottom temperature is limited to 700–750°F to prevent cracking.

The atmospheric resid is fed to a furnace, heated to 730–770°F and next to a vacuum tower operated at a minimum practical vacuum (80–110 mm Hg). The operating conditions are dictated by cracking and product quality required. The objectives of vacuum distillation is generally to separate vacuum gas oil (VGO) from reduced crude. The VGO may become feedstock for FCCU or hydrocracker units or used to make lube base stocks. Depending on the end use, there may be one or more sidestreams. The bottom stream from the vacuum distillation unit may be used to produce bitumen or used for fuel oil production after mixing it with small amounts of cutter stocks (in the diesel/kerosene range).

If the crude contains very high percentages of light-ends, a flash drum or a prefractionator with an overhead condensing system is added ahead of atmospheric tower. The prefractionator is designed to recover most of the light-ends and a part of the light naphtha. The bottom stream from prefractionator becomes feed to atmospheric tower.

PROCESS VARIABLES

The following variables are important in the design of crude columns:

1. The nature of the crude—water content, metal content, and heat stability. The heat stability of the crude limits the temperature to which crude can be heated in the furnace without incipient cracking.
2. Flash zone operating conditions—flash zone temperature is limited by advent of cracking; flash zone pressure is set by fixing the reflux drum pressure and adding to it to the line and tower pressure drop.
3. Overflash is the vaporization of crude over and above the crude overhead and sidestream products. Overflash is generally kept in the range of 3–6 LV% (LV = Liquid Volume). Overflash prevents coking of wash section plates and carryover of coke to the bottom sidestream and ensures a better fractionation between the bottom sidestream and the tower bottom by providing reflux to plates between the lowest sidestream and the flash zone. A larger overflash also consumes larger utilities; therefore, overflash is kept to a minimum value consistent with the quality requirement of the bottom sidestream.
4. In steam stripping, the bottom stripping steam is used to recover the light components from the bottom liquid. In the flash zone of an atmospheric distillation column, approximately 50–60% of crude is vaporized. The unvaporized crude travels down the stripping section of the column containing four to six plates and is stripped of any low boiling-point distillates still contained in the reduced

crude by superheated steam. The steam rate used is approximately 5–10 lbs/bbl of stripped product.[1] The flash point of the stripped stream can be adjusted by varying the stripping steam rate.

5. Fractionation is the difference between the 5% ASTM curve of a heavy cut and the 95% point on the ASTM curve of a lighter cut of two adjacent side products. A positive difference is called a *gap*,[2] and a negative difference is called an *overlap*.

The design procedures used for atmospheric and vacuum distillation are mostly empirical, as crude oil is made of a very large number of hydrocarbons, from methane to asphaltic pitch. The basic data required, refinery crude distillation column, and a brief overview of the design procedures follow.

TRUE BOILING POINT CURVE

The composition of any crude oil sample is approximated by a true boiling point (TBP) curve. The method used is basically a batch distillation operation, using a large number of stages, usually greater than 60, and high reflux to distillate ratio (greater than 5). The temperature at any point on the temperature-volumetric yield curve represents the true boiling point of the hydrocarbon material present at the given volume percent point distilled. TBP distillation curves are generally run only on the crude and not on petroleum products. Typical TBP curves of crude and products are shown in Figures 1-1 and 1-2.

ASTM DISTILLATION

For petroleum products, a more rapid distillation procedure is used. This is procedure, developed by the American Society for Testing and Materials (ASTM), employs a batch distillation procedure with no trays or reflux between the still pot and the condenser.[3] The only reflux available is that generated by heat losses from the condenser.

EQUILIBRIUM FLASH VAPORIZATION

In this procedure,[4] the feed material is heated as it flows continuously through a heating coil. Vapor formed travels along in the tube with the remaining liquid until separation is permitted in a vapor separator or

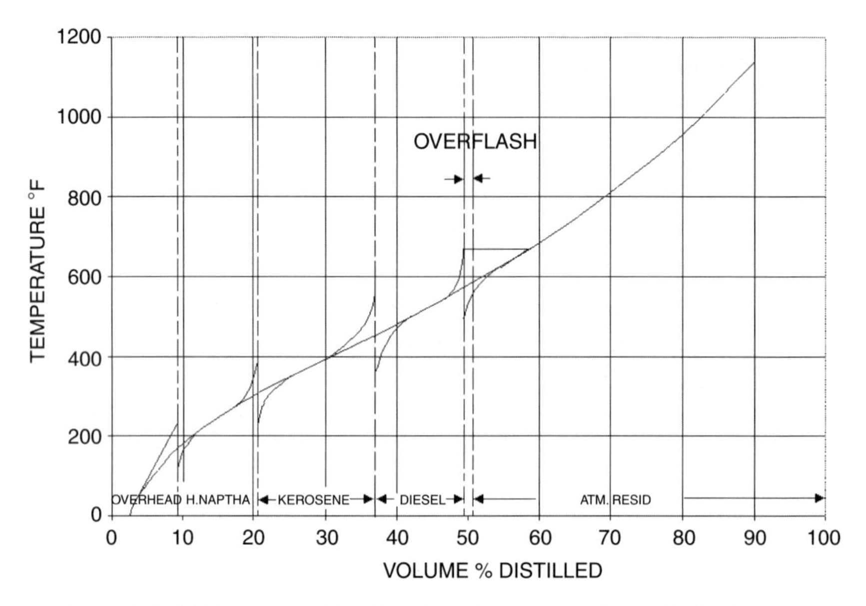

Figure 1-1. TBP curves of feed and products atmosphere distillation tower.

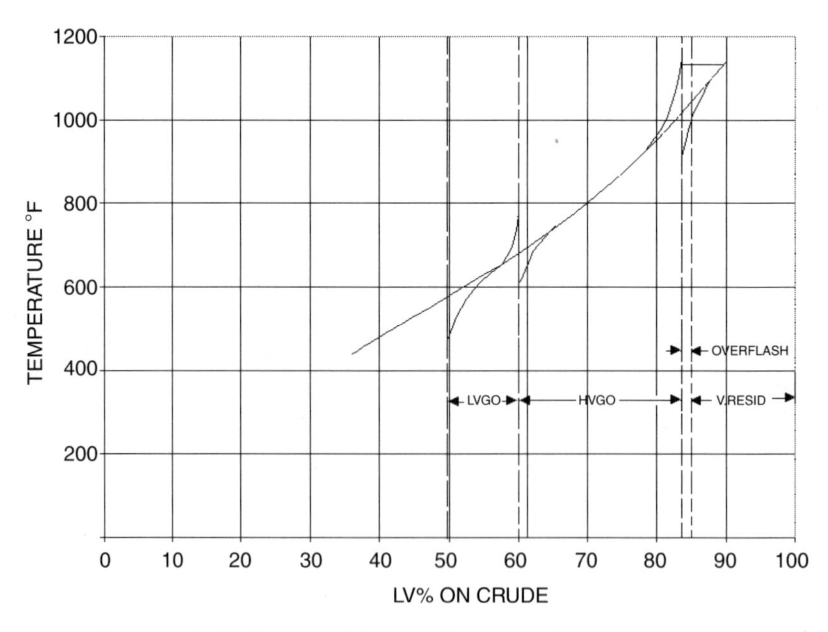

Figure 1-2. TBP curve of feed and products for vacuum tower.

vaporizer. By conducting the operation at various outlet temperatures, a curve of percent vaporized vs. temperature may be plotted. Also, this distillation can be run at a pressure above atmospheric as well as under vacuum. Equilibrium flash vaporization (EFV) curves are run chiefly on crude oil or reduced crude samples being evaluated for vacuum column feed.

CRUDE ASSAY

The complete and definitive analysis of a crude oil is called *crude assay*. This is more detailed than a crude TBP curve. A complete crude assay contains some of the following data:

1. Whole crude salt, gravity, viscosity, sulfur, light-end carbons, and the pour point.
2. A TBP curve and a mid-volume plot of gravity, viscosity, sulfur, and the like.
3. Light-end carbons analysis up to C_8 or C_9.
4. Properties of fractions (naphthas, kerosenes, diesels, heavy diesels, vacuum gas oils, and resids). The properties required include yield as volume percent, gravity, sulfur, viscosity, octane number, diesel index, flash point, fire point, freeze point, smoke point, and pour point.
5. Properties of the lube distillates if the crude is suitable for manufacture of lubes.
6. Detailed studies of fractions for various properties and suitability for various end uses.

PROCESS DESIGN OF A CRUDE DISTILLATION TOWER

A very brief overview of the design steps involved follows:

1. Prepare TBP distillation and equilibrium flash vaporization curves of the crude to be processed. Several methods are available for converting TBP data to EFV curves.
2. Using crude assay data, construct TBP curves for all products except gas and reduced crude. These are then converted to ASTM and EFV curves by Edmister,[5] 'Maxwell,'[6] or computer methods.
3. Prepare material balance of the crude distillation column, on both volume and weight bases, showing crude input and product output.

Also plot the physical properties, such as cut range on TBP and LV%, mid vol% vs. SG, molecular weight, mean average boiling point, and enthalpy curves for crude and various products.

4. Fractionation requirements are considered next. Ideal fractionation is the difference between the 5% and 95% points on ASTM distillation curves obtained from ideal TBP curves of adjacent heavier and lighter cuts. Having fixed the gaps as the design parameter, the ideal gap is converted into an actual gap. The difference between the ideal gap and actual gap required is deviation. Deviation is directly correlated with (number of plates × reflux).
5. The deviation or gap can be correlated with an F factor,[7] which is the product of number of plates between two adjacent side draws offstream and internal reflux ratio. *Internal reflux* is defined as volume of liquid (at 60°F) of the hot reflux below the draw offplate of the lighter product divided by the volume of liquid products (at 60°F) except gas, lighter than the adjacent heavier products. This implies that the reflux ratio and the number of plates are interchangeable for a given fractionation, which holds quite accurately for the degree of fractionation generally desired and the number of plates (5–10) and reflux ratios (1–5) normally used. The procedure is made clear by Example 1-1.

NUMBER OF TRAYS

Most atmospheric towers have 25–35 trays between the flash zone and tower top. The number of trays in various sections of the tower depends on the properties of cuts desired from the crude column, as shown in Table 1-1.

The allowable pressure drop for trays is approximately 0.1–0.2 psi, per tray. Generally, a pressure drop of 5 psi is allowed between

Table 1-1
Number of Trays between Side Draws in Crude Distillation Unit

SEPARATION	NUMBER OF TRAYS
NAPHTHA–KEROSENE	8–9
KEROSENE–LIGHT DIESEL	9–11
LIGHT DIESEL–ATM RESID	8–11
FLASH ZONE TO FIRST DRAW TRAY	4–5
STEAM STRIPPER SECTION	4–6

Table 1-2
Typical Separation Obtainable in Atmospheric and Vacuum Towers

SEPARATION	(5–95) GAP°F
NAPHTHA–KEROSENE	12°F GAP
KEROSENE–LIGHT DIESEL	62°F OVERLAP
LIGHT DIESEL–HEAVY DIESEL	169°F OVERLAP
HEAVY DIESEL–VGO	70°F OVERLAP
VGO–VACUUM BOTTOMS	70°F OVERLAP

OVERLAP IS A GAP WITH A NEGATIVE SIGN.

the flash zone and the tower top. Flash zone pressure is set as the sum of reflux drum pressure and combined pressure drop across condenser and trays above the flash zone. A pressure drop of 5 psi between the flash zone and furnace outlet is generally allowed.

FLASH ZONE CONDITIONS

The reflux drum pressure is estimated first. This is the bubble point pressure of the top product at the maximum cooling water temperature. The flash zone pressure is then equal to reflux drum pressure plus pressure drop in the condenser overhead lines plus the pressure drop in the trays.

Before fixing the flash zone temperature, the bottom stripping steam quantity and overflash are fixed. The volume percentage of strip-out on crude is calculated using available correlations.[8] If D is the sum of all distillate streams, V is percent of vaporization in the flash zone, OF is overflash, and ST is strip out, then

$$V = D + OF - ST$$

From the flash curve of the crude, the temperature at which this vaporization is achieved at flash zone pressure is determined. This temperature should not exceed the maximum permissible temperature. If it does, the quantity of overflash and stripping steam are changed until a permissible temperature is obtained.

The temperature at which a crude oil begin to undergo thermal decomposition varies from crude to crude, depending on its composition (naphthenic, paraffinic, or aromatic base) and the trace metals present in the crude. Decomposition temperature can be determined only by actual test runs. For most paraffinic and naphthenic crudes, it is in the range of 650–670°F.

COLUMN OVERHEAD TEMPERATURE

The column top temperature is equal to the dew point of the overhead vapor. This corresponds to the 100% point on the EFV curve of the top product at its partial pressure calculated on the top tray.

A trial and error procedure is used to determine the temperature:

1. The temperature of reflux drum is fixed, keeping in view the maximum temperature of the cooling medium (water or air).
2. Estimate a tower overhead temperature, assuming steam does not condense at that temperature.
3. Run a heat balance around top of tower to determine the heat to be removed by pumpback reflux. Calculate the quantity of pumpback reflux.
4. Calculate the partial pressure of the distillate and reflux in the overhead vapor. Adjust the 100% point temperature on the distillate atmospheric flash vaporization curve to the partial pressure.
5. Repeat these steps until the calculated temperature is equal to the one estimated.
6. Calculate the partial pressure of steam in the overhead vapor. If the vapor pressure of steam at the overhead temperature is greater than the partial pressure of steam, then the assumption that steam does not condense is correct. If not, it is necessary to assume a quantity of steam condensing and repeat all steps until the partial pressure of steam in the overhead vapor is equal to the vapor pressure of water at overhead temperature. Also, in this case, it is necessary to provide sidestream water draw-off facilities.
7. To calculate overhead gas and distillate quantities, make a component analysis of total tower overhead stream consisting of overhead gas, overhead distillate, pumpback reflux, and steam. Next make a flash calculation on total overhead vapor at the distillate drum pressure and temperature.
8. The overhead condenser duty is determined by making an enthalpy balance around the top of the tower.

BOTTOM STRIPPING

To determine the amount of liquid to be vaporized by the stripping steam in the bottom of the tower, it is necessary to construct the flash curve of this liquid (called the *initial bottoms*). The flash curve of the reduced crude can be constructed from the flash curve of the whole crude.[9] It is assumed that the initial bottom is flashed in the presence of stripping steam at the pressure existing on top of the stripping plate and at the exit temperature of liquid from this plate.

Approximately 50–60% of the crude is vaporized in the flash zone of the atmospheric tower. The unvaporized crude travels down the stripping section of the tower, containing four to six plates, and is stripped of any remaining low-boiling distillates by superheated steam at 600°F. The steam rate used is approximately 5–10 lb/bbl of stripped product. The flash point of the stripped product can be adjusted by varying stripping steam rate.

SIDESTREAM STRIPPER

Distillate products (kerosene and diesel) are withdrawn from the column as sidestream and usually contain material from adjacent cuts. Thus, the kerosene cut may contain some naphtha and the light diesel cut may contain some kerosene-range boiling material. These side cuts are steam stripped using superheated steam, in small sidestream stripper columns, containing four to six plates, where lower-boiling hydrocarbons are stripped out and the flash point of the product adjusted to the requirements.

REFLUX

In normal distillation columns, heat is added to the column from a reboiler and removed in an overhead condenser. A part of the distillate condensed in overhead condenser is returned to the column as reflux to aid fractionation. This approach is not feasible in crude distillation because the overhead temperature is too low for recovery of heat. Also the vapor and liquid flows in column increase markedly from bottom to top, requiring a very large-diameter tower. To recover the maximum heat and have uniform vapor and liquid loads in the column, intermediate refluxes are withdrawn, they exchange heat with incoming crude oil before entering the furnace and are returned to the plate above in the column (Figure 1-3).

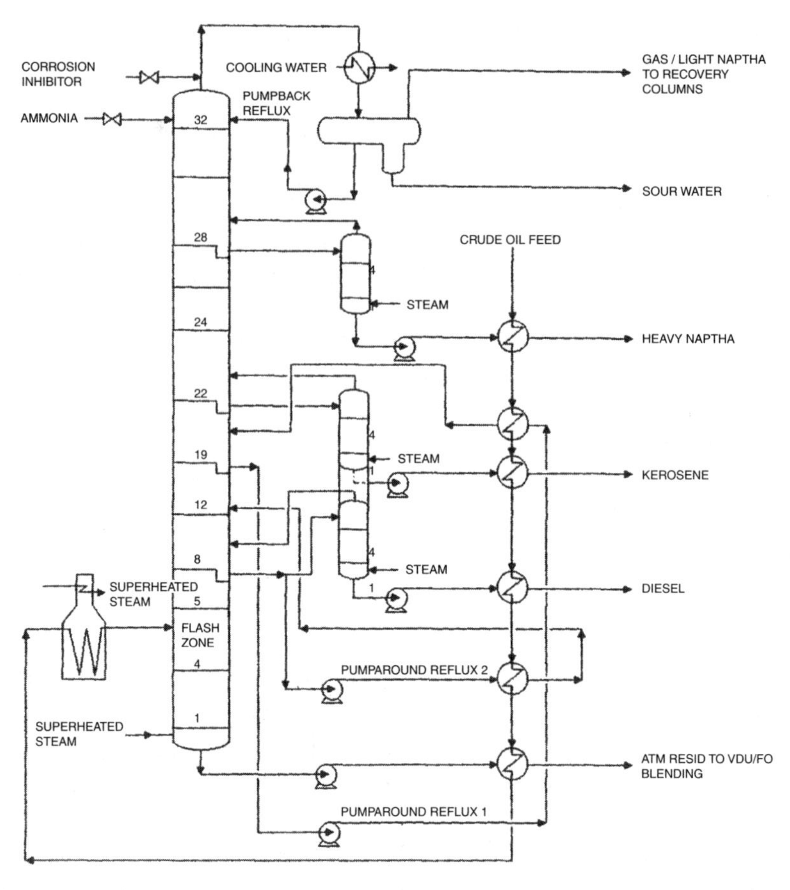

Figure 1-3. Atmospheric crude column with pumpback and pumparound reflux.

SIDESTREAM TEMPERATURE

The flash curve of the product stream is determined first. This product is completely vaporized below the sidestream draw-off plate. Therefore, the 100% point of the flash curve is used. To determine the partial pressure of the product plus reflux vapor, both of which are of same composition, the lighter vapors are considered inert.

$$\begin{array}{l}\text{Partial pressure}\\ \text{of side stream}\end{array} = \frac{(\text{moles of sidestream} + \text{moles of reflux})}{(\text{total moles of vapor below plates})} \times \text{total pressure}$$

EXAMPLE 1-1

The 95% point of heavy naphtha is 315°F and the 5% ASTM distillation point of kerosene is 370°F. The flash point of kerosene is 127.2°F. Calculate the deviation from actual fractionation between heavy naphtha and kerosene for the steam-stripped kerosene fraction and the number of plates and reflux required for separation.

$$\text{Ideal gap} = 370 - 315, \text{ or } 55°\text{F}$$

The actual 5% ASTM distillation point of a fraction can be correlated from its flash point (known), by following relation:

$$\text{Flash point (°F)} = 0.77 \times (\text{ASTM 5\% point, °F}) - 150$$

The actual 5% point on the ASTM distillation curve of kerosene, by this correlation, equals 360°F, which is 10° less than ideal. Since kerosene is to be steam stripped, 95% of heavy naphtha will be 325°F. Therefore,

$$\text{Actual gap} = (360 - 325), \text{ or } 35°\text{F}$$

$$\text{Deviation from ideal fractionation} = (55 - 35), \text{ or } 20°\text{F}$$

From the Packie's correlation, an F factor of 11.5 is required.

CHARACTERIZATION OF UNIT FRACTIONATION

In commercial atmospheric and vacuum units, the distillation is not perfect. For example, a kerosene fraction with a TBP cut of 300–400°F will have material (referred to as *tails*) that boils below 300°F and other material that boils above 400°F. Because of these tails, the yield of the required product must be reduced to stay within the desired product quality limits.

The size and shape of the tails of each product depends on the characteristics of the unit from which it was produced. The factors affecting the fractionation are the number of trays between the product draw trays, tray efficiency, reflux ratio, operating pressure, and boiling ranges of the products.

Several approaches are possible to characterize fractionation in an operating unit. One approach is to characterize the light tail at the front end of a stream in terms of two factors:

V_1 is the volume boiling below the cut point, expressed as LV% of crude.

T_f is the temperature difference between the cut point and the TBP initial boiling point (1 LV% distilled) of the stream.

Consider the TBP distillation of products from an atmospheric distillation column (Figure 1-1). The front-end tail of kerosene (TBP cut 300–400) contains 1.5% material on crude boiling below 300°F (see Table 1-3); therefore, $V_f = 1.5$.

The initial boiling point of kerosene cut (1 LV% distilled) is 240°F and the temperature difference between the cut point (300°F) and IBP is 60°F; therefore, $T_F = 60$.

The shape of the front tail can be developed using these two parameters on a probability plot. Having established these parameters, the same values are used for the front end tails of kerosenes on this unit for different cut-point temperatures (e.g., for different flash-point kerosenes).

A similar approach is used for back end tail; in the preceding example, the lighter heavy straight-run (HSR) naphtha cut is before kerosene. The

Table 1-3
Front and Back Tail Characterization of a Typical Atmospheric Crude Unit

	FRONT END TAIL		BACK END TAIL	
STREAM	V_F **LV%**	$T_F \Delta T$	V_B **LV%**	$T_B \Delta T$
C_4	—	—	0.0	0.0
LSR	—	—	1.0	35.0
HSR	1.0	40.0	1.5	50.0
KEROSENE	1.5	60.0	2.0	65.0
LIGHT DIESEL	2.0	70.0	3.5	120.0
RESID	3.5	160.0	—	—

NOTE:
KEROSENE V_F = HSR V_B.
LIGHT DIESEL V_F = KEROSENE V_B.
RESID V_F = LIGHT DIESEL V_B.

volume of HSR material boiling above the kerosene cut point of 300°F must be 1.5 LV% (on crude), equal to the front end tail volume on kerosene. Let us call it V_B; therefore,

$$V_F = V_B = 1.5\% \text{ (LV on crude)}$$

The HSR end point (99% LV distilled) is 250°F and the cut point is 300°F. Therefore,

$$T_B \text{ for HSR} = 300 - 250 = 50°\text{F}$$

The shape of the back end tail can be estimated using a probability paper. Similarly the shape of front and back end tails for all cuts on vacuum units can also be determined (Table 1-4).

Having established these parameters, the same values are used, for example, for all kerosene cuts on this unit at different front end cut temperatures. This is an excellent approximation, provided the changes in cut point and boiling range are not too large.

Having established the appropriate unit fractionation parameters, the individual product distillations can be established based on selected TBP cut temperatures. These are defined by the points where the produced yield cuts the crude TBP curve. For example, referring to Figure 1-1, the yield of a product lighter than kerosene is 20.4 LV%; hence, the kerosene

Table 1-4
Front and Back Tail Characterization of a Typical Vacuum Unit

	FRONT END TAIL		BACK END TAIL	
STREAM	V_F **LV%**	$T_F \Delta T$	V_B **LV%**	$T_B \Delta T$
WET GAS OIL	—	—	—	—
DRY GAS OIL	—	—	1.0	32.0
HEAVY DIESEL	1.0	60.0	2.2	108.0
VACUUM RESID	2.2	100.0	—	—

NOTE:
HEAVY DIESEL V_F = DRY GAS OIL V_B.
VACUUM RESID V_F = HEAVY DIESEL V_B.
RESID V_F = LIGHT DIESEL V_B.

initial cut point is 300°F where the crude volume percent distilled is 20.4. The kerosene back end TBP cut point is 448°F where the crude volume percent distilled is 36.8, giving the required kerosene yield of 16.4 LV% on the crude.

The product volume and product qualities can be determined by breaking the distillation into narrow cuts, called *pseudocomponents*, and blending the qualities of these using the properties of the narrow cuts from the crude assay data.

GENERAL PROPERTIES OF PETROLEUM FRACTIONS

Most petroleum distillates, especially those from the atmospheric distillation, are usually defined in term of their ASTM boiling ranges. The following general class of distillates is obtained from petroleum: liquefied petroleum gas, naphtha, kerosene, diesel, vacuum gas oil, and residual fuel oil.

DISTILLATES

Liquefied Petroleum Gas

The gases obtained from crude oil distillation are ethane, propane, and *n*-butane isobutene. These products cannot be produced directly from the crude distillation and require high-pressure distillation of overhead gases from the crude column. C_3 and C_4 particularly are recovered and sold as liquefied petroleum gas (LPG), while C_1 and C_2 are generally used as refinery fuel.

Naphtha

C_5-400°F ASTM cut is generally termed *naphtha*. There are many grades and boiling ranges of naphtha. Many refineries produce 400°F end-point naphtha as an overhead distillate from the crude column, then fractionate it as required in separate facilities. Naphtha is used as feedstock for petrochemicals either by thermal cracking to olefins or by reforming and extraction of aromatics. Also some naphtha is used in the manufacture of gasoline by a catalytic reforming process.

Kerosene

The most important use of kerosene is as aviation turbine fuel. This product has the most stringent specifications, which must be met to ensure the safety standards of the various categories of aircraft. The most important specifications are the flash and freeze points of this fuel. The initial boiling point (IBP) is adjusted to meet the minimum flash requirements of approximately 100°F. The final boiling point (FBP) is adjusted to meet the maximum freeze point requirement of the jet fuel grade, approximately −52°F. A full-range kerosene may have an ASTM boiling range between 310 and 550°F. Basic civil jet fuels are

1. Jet A, a kerosene-type fuel having a maximum freeze point of −40°C. Jet A-type fuel is used by mainly domestic airlines of various countries, where a higher freeze point imposes no operating limitations.
2. Jet A-1, a kerosene-type fuel identical with Jet A but with a maximum freeze point of −47°C. This type of fuel is used by most international airlines. Jet A and Jet A-1 generally have a flash point of 38°C.
3. Jet B is a wide-cut gasoline-type fuel with a maximum freeze point of −50 to −58°C. The fuel is of a wider cut, comprising heavy naphtha and kerosene, and is meant mainly for military aircraft.

A limited number of additives are permitted in aviation turbine fuels. The type and concentration of all additives are closely controlled by appropriate fuel specifications. The following aviation turbine fuel additives are in current use:

- *Antioxidants*. Its use is mandatory in fuels produced by a hydrotreating process, to prevent formation of hydrogen peroxide, which can cause rapid deterioration of nitrile rubber fuel system components.
- *Static dissipators*, also known as *antistatic additives* or *electrical conductivity improvers*. Its use is mandatory to increase the electrical conductivity of the fuel, which in turn promotes a rapid relaxation of any static charge build-up during the movement of fuel.
- *Fuel system icing inhibitor* (FSII). The main purpose of FSII is to prevent fuel system blockage by ice formation from water precipitated from fuels in flight. Because of the biocidal nature of this additive, it is very effective in reducing microbiological contamination problems in aircraft tanks and ground fuel handling facilities.

As most commercial aircrafts are provided with fuel filter heaters, they have no requirement for the anti-icing properties of this additive. FSII is therefore not usually permitted in civil specifications, its use is confined mainly to military fuels.

- *Corrosion inhibitor/lubricity improver.* Its use is optional to protect storage tanks and pipelines from corrosion and improve the lubricating properties of the fuel.

Diesel

Diesel grades have an ASTM end point of 650–700°F. Diesel fuel is a blend of light and heavy distillates and has an ASTM boiling range of approximately 350–675°F. Marine diesels are a little heavier, having an ASTM boiling end point approximately 775°F. The most important specifications of diesel fuels are cetane number, sulfur, and pour or cloud point. Cetane number is related to the burning quality of the fuel in an engine. The permissible sulfur content of diesel is being lowered worldwide due to the environmental pollution concerns resulting from combustion of this fuel. Pour point or cloud point of diesel is related to the storage and handling properties of diesel and depends on the climatic conditions in which the fuel is being used.

Vacuum Gas Oil

Vacuum gas oil is the distillate boiling between 700 and 1000°F. This is not a saleable product and is used as feed to secondary processing units, such as fluid catalytic cracking units, and hydrocrackers, for conversion to light and middle distillates.

Residual Fuel Oil

Hydrocarbon material boiling above 1000°F is not distillable and consists mostly of resins and asphaltenes. This is blended with cutter stock, usually kerosene and diesel, to meet the viscosity and sulfur specifications of various fuel oil grades.

VACUUM DISTILLATION PRODUCTS

In an atmospheric distillation tower, the maximum flash zone temperature without cracking is 700–800°F. The atmospheric residuum, commonly known as *reduced crude*, contains a large volume of distillable oils that can be recovered by vacuum distillation at the maximum permissible flash zone temperature. The TBP cut point between vacuum gas oil and vacuum resid is approximately 1075–1125°F. The cut point is generally optimized, depending on the objectives of the vacuum distillation, into asphalt operation and pitch operation.

Asphalt Operation

Given the specification (penetration) of the asphalt to be produced, the corresponding residuum yield can be determined from the crude assay data. The total distillate yield is determined by subtracting asphalt yield from the total vacuum column feed. In case a number of lubricating oil distillates is to be produced, the distillation range of each has to be specified, and the corresponding yields can be determined from the crude assay data. Lube cuts are produced as sidestreams from the vacuum column.

In asphalt operation, some gas oil must remain in the pitch to provide the proper degree of plasticity. The gravity of an asphalt stream is usually between 5 and 8° API. Not all crudes can be used to make asphalt. Experimental data for asphalt operation are necessary to relate asphalt penetration to residual volume. The penetration range between 85 and 10, are possible and the units are generally designed to produce more than one grade of asphalt.

The principal criteria for producing lube oil fractions are viscosity, color, and rejection to residuum the heavy impurities and metals. These oils are further refined by solvent extraction, dewaxing, and other types of finishing treatment, such as hydrotreating. Vacuum towers for the manufacture of lubricating oils are designed to provide same relative degree of fractionation between streams as in the atmospheric tower. Sidestreams are stripped in the external towers to control front-end properties. The number of trays between the draw trays is set arbitrarily. Generally, three to five trays are used between draws. Sieve trays are more popular for vacuum column service.

Pitch Operation

The objective in this case is to produce maximum distillate and minimum pitch, which is used for fuel oil blending. In this case, the TBP cut point between the distillate and pitch has to be set by unit design, generally around 1100°F. From the crude assay data, the total distillate yield from the crude up to the cut point is known; deducting the total distillates yield in the atmospheric column, the total yield of vacuum distillate can be estimated. The light vacuum distillate yield is set at approximately 30% of the total vacuum gas oil, to facilitate heat recovery at two levels of heat.

The unit design has to specify the amount of overflash, depending on the purity of the heavy vacuum gas oil (HVGO) required. If the color requirements or level of metal contaminants is not severe, 1–2 vol% (volume %) of vacuum feed is taken as overflash.

Vacuum column design calculation is similar to atmospheric column design with some differences in technique as follows:

- A material balance is made for vacuum feed vs. the products—vacuum bottoms, sidestream products (vacuum gas oils), and overhead condensable hydrocarbons. The assumed quantity of noncondensables is not carried in the material balance nor considered in the flash zone calculations but must be estimated for vacuum ejector calculations.
- The construction of flash vaporization curve (AFVC, atmospheric flush vaporization curve) of the reduced crude, feed to vacuum distillation unit is done in the same manner as for the whole crude.

ATMOSPHERIC DISTILLATION UNIT

In Figure 1-4, the crude oil received from off-site storage tanks through booster pumps is pumped by charge pump P-101 and preheated in parallel trains of preheat exchangers with hot intermediate streams and products. A small quantity of water and demulsifier chemicals are added before preheating. The hot crude is mixed with washwater and fed to electric desalters V-106 A and B to reduce the salt content by an electric desalting process. The water phase, containing most of the dissolved salts contained in the crude, separates out. The desalted crude is dosed with an NaOH solution to a fixed chloride content. The desalted crude is further heated through two parallel trains of heat exchangers and fed to preflash tower V-117. The preflash tower overhead vapor is cooled by exchanging heat with crude oil and condensed in an overhead drum V-118. Part of this liquid naphtha is used as reflux in the column; the rest of the liquid and the vapor from the drum are sent to the naphtha processing unit.

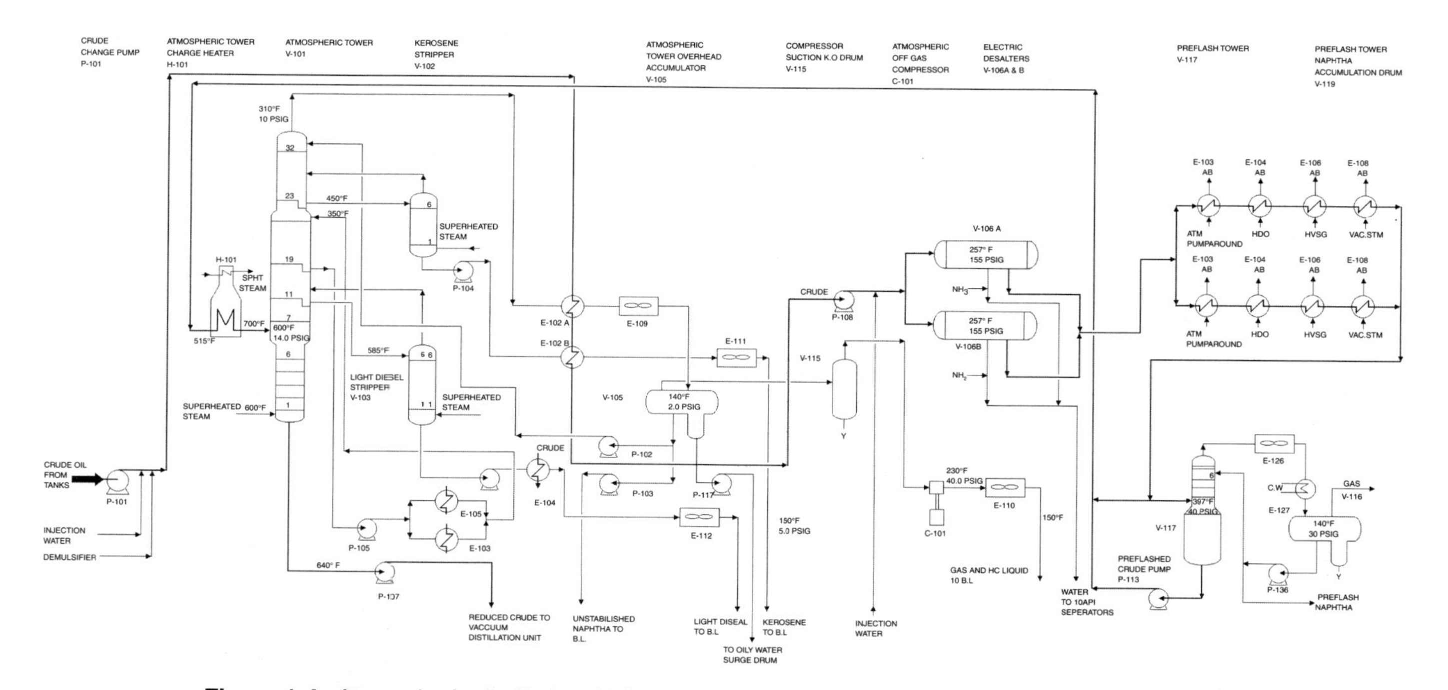

Figure 1-4. Atmospheric distillation. K.O. = knockout; C.W. = cooling water; B.L. = battery limits.

The crude from the bottom of the preflash tower is pumped through the heat exchangers, recovering heat from vacuum tower bottoms and sidestream HVGO, and sent to fired atmospheric heater H-101. The crude is partially vaporized in the fired heater before entering the flash zone of the atmospheric tower V-101. Superheated stripping steam is introduced through the bottom of the column.

The tower overhead vapor is cooled by exchanging heat with crude oil, condensed in air cooler E-109, and routed to overhead product accumulator V-105. The overhead gases from this accumulator are compressed in compressor C-101 to about 40 psig pressure and sent to the refinery gas recovery system. The condensed naphtha in the accumulator is separated from water. A part of this naphtha is sent back to the column by reflux pump P-102 and the rest is withdrawn as an intermediate product for processing in naphtha fractionation unit.

Kerosene and light diesel cuts are withdrawn as sidestreams from the atmospheric distillation tower. These are steam stripped in steam strippers V-102 and V-103, respectively. The kerosene and light diesel product streams exchange heat with crude oil feed in the crude preheat train and finally cooled in air fin coolers E-106 and E-111 and sent to storage.

The hot atmospheric bottoms or reduced crude, at approximately 660°F, is transferred by P-107 to vacuum tower heater H-102.

VACUUM DISTILLATION UNIT

The reduced crude from vacuum heater H-102 enters the flash zone of vacuum tower V-104. The column operates under vacuum by means of an ejector/condenser system to achieve the required separation between the heavy components at lower temperature. Some gaseous hydrocarbons are produced due to cracking of the feed in vacuum heater H-102. This sour gas is burnt in atmospheric tower heater H-101 while the condensate water is routed to desalter feed water surge drum V-106.

The tower is provided with a cold recycle (quench) to lower the bottom temperature and avoid coking. A superheated stripping stream is introduced at the bottom of the tower. The heavy diesel product is drawn as a sidestream and exchanges heat with crude oil in the preheat train. It is partly used as top and intermediate reflux to the column, and the balance is sent to storage after cooling in E-121 and E-113.

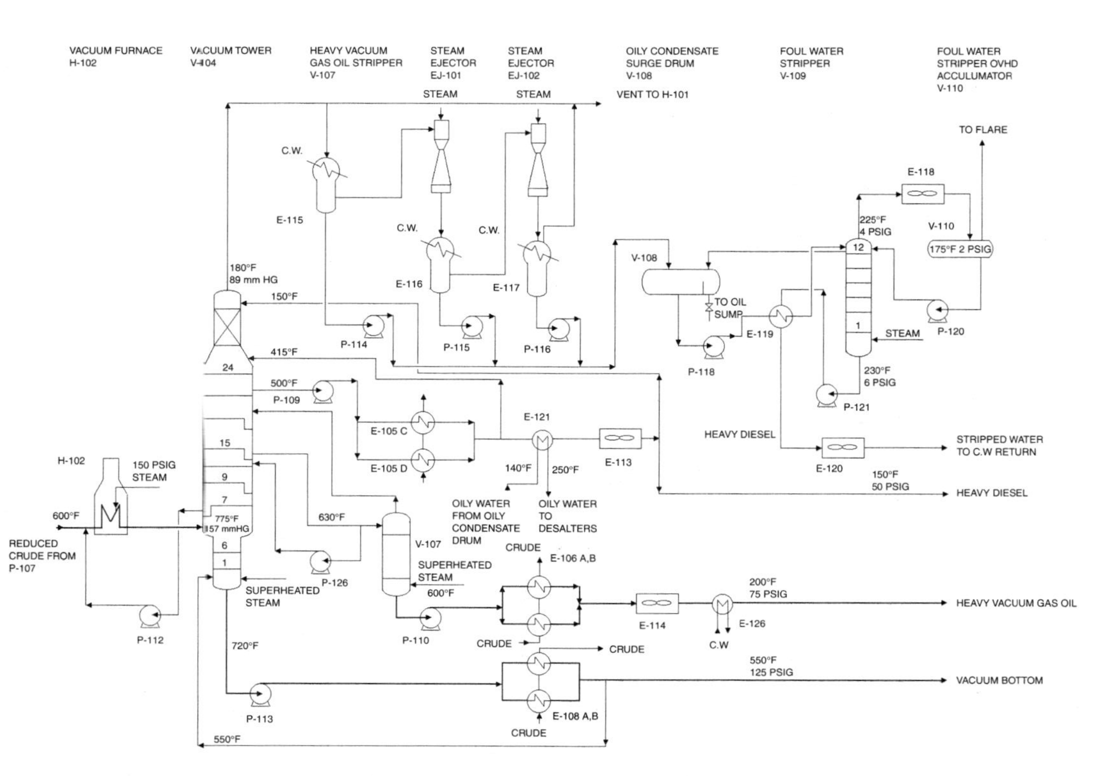

Figure 1-5. Vacuum distillation unit. C.W. = cooling water; OVDH = overhead.

Vacuum gas oil drawn as bottom sidestream is stripped in steam stripper V-107 and cooled by exchanging heat with crude in the preheat train and finally in air cooler E 114 and trim cooler E-126 before being sent to off-site storage tanks.

The bottom product, the vacuum residue, exchanges heat with crude coming from the preflash tower bottoms and in the preheat train before being sent to off-site storage.

To control corrosion, a 3% ammonia solution and inhibitor is injected into the top of preflash, atmospheric, and vacuum towers.

Foul water is generated in the overhead accumulator drum of atmospheric distillation column and in the ejector/condenser system of the vacuum distillation column. All foul water streams are combined in oily condensate surge drum V-108. From V-108, the oily water is transferred by P-118 to foul water stripper V-109. Superheated steam is admitted at the bottom of this 12-plate tower for stripping H_2S and NH_3 from the foul water.

The overhead gases are cooled and condensed in air fin cooler E-118. Noncondensable gases are routed to the flare header. Condensed and concentrated H_2S/NH_3 liquid is returned to the column as total reflux. The hot stripped water from the column bottom is partly recycled to desalters and the rest is to a water treatment plant.

The typical operating conditions for an atmospheric and vacuum distillation towers are shown in Tables 1-5 through 1-7.

CRUDE DESALTING

Crude desalting is the first processing step in a refinery (see Figure 1-6). The objectives of crude desalting are the removal of salts and solids and the formation water from unrefined crude oil before the crude is introduced in the crude distillation unit of the refinery.

Salt in the crude oil is in the form of dissolved or suspended salt crystals in water emulsified with the crude oil. The basic process of desalting is to wash the salt from crude oil with water. Problems occur in efficient and economical water/oil mixing, water wetting of suspended solids, and separation of oil from wash water. The separation of oil and washwater is affected by the gravity, viscosity, and pH of the crude oil and the ratio of water/crude used for washing.

An important function of the desalting process is the removal of suspended solids from the crude oil. These are usually very fine sand and soil particles, iron oxides, and iron sulfide particles from pipelines,

Table 1-5
Atmospheric Tower Operating Conditions

OPERATING PARAMETER	UNITS	
TEMPERATURES	°F	
TRANSFER LINE		660
FLASH ZONE		657
TOWER TOP		359
KEROSENE DRAW-OFF		469
PUMPAROUND DRAW OFF		548
PUMPAROUND RETURN		345
LIGHT DIESEL DRAW OFF		603
TOWER BOTTOM		648
PRESSURE	psig	
REFLUX DRUM		2.0
TOWER TOP		10.3
FLASH ZONE		14.7
REFLUX RATIO, REFLUX/LIQUID DIST.		0.6
STRIPPING STEAM		
TO ATMOSPHERIC TOWER	lbs/bbl RESID	5.5
TO KEROSENE STRIPPER	lbs/bbl RESID	5.9
TO DIESEL STRIPPER	lbs/bbl RESID	2.1
ATMOSPHERIC HEATER		
PROCESS FLUID CONDITIONS		
TEMPERATURE IN	°F	453
TEMPERATURE OUT	°F	660
PRESSURE DROP	psi	138
TUBE SKIN TEMPERATURE (AVG)	°F	735
STACK GAS TEMPERATURE	°F	725
FRACTIONATION EFFICIENCY		
95%–5% ASTM DISTRIBUTION GAP		
ATMOSPHERIC NAPHTHA-KEROSENE		GAP + 10
KEROSENE-LIGHT DIESEL		GAP − 36

NOTE: BASIS 154000 BPSD KUWAIT CRUDE RUN.

tanks or tankers, and other contaminants picked up in transit or from processing.

Until recently, the criteria for desalting crude oil was 10 lb salt/1000 bbl (expressed as NaCl), but due to more stringent requirements of some downstream processes, desalting is now done at the much lower level of 1.0 lb/1000 bbl or lower. Reduced equipment fouling and corrosion

Table 1-6
Vacuum Tower Operating Conditions

OPERATING PARAMETER	UNITS	
TEMPERATURES	°F	
TRANSFER LINE		740
FLASH ZONE		711
TOWER TOP		307
HEAVY DIESEL DRAW-OFF		447
TOP REFLUX TEMPERATURE		121
HVGO DRAW-OFF		613
TOWER BOTTOM		670
PRESSURE	mmHg	
TOWER TOP		64
FLASH ZONE		125
TOP REFLUX RATIO; REFLUX/FEED		0.15
HOT REFLUX RATIO; REFLUX/FEED		0.97
WASH OIL RATIO; WASH OIL/FEED		0.14
BOTTOM QUENCH OIL RATIO; QUENCH/FEED		0.24
STRIPPING STEAM		
TO VACUUM TOWER	lbs/bbl RESID	8.0
TO HVGO STRIPPER	lbs/bbl RESID	4.6
VACUUM HEATER		
PROCESS FLUID CONDITIONS		
TEMPERATURE IN	°F	645
TEMPERATURE OUT	°F	736
PRESSURE DROP	psi	73
TUBE SKIN TEMPERATURE (AVG)	°F	850
STACK GAS TEMPERATURE	°F	845
FRACTIONATION EFFICIENCY		
95%–5% ASTM DISTRIBUTION GAP		
LIGHT DIESEL–HEAVY DIESEL		GAP – 145
HEAVY DIESEL-HVGO		GAP + 25

NOTE: BASIS 154000 BPSD KUWAIT CRUDE RUN.

Table 1-7
Atmospheric and Vacuum Crude Distillation Utility Consumption

UTILITY	UNITS	CONSUMPTION
ELECTRICITY	kWhr	8.7
FUEL	mmBtu	0.6
STEAM	mmBtu	0.09
COOLING WATER	MIG*	0.31
DISTILLED WATER	MIG*	0.02

**THOUSAND IMPERIAL GALLONS.*

NOTE: THE UTILITY CONSUMPTIONS (PER TON FEED) ARE FOR AN INTEGRATED CRUDE AND VACUUM UNIT.

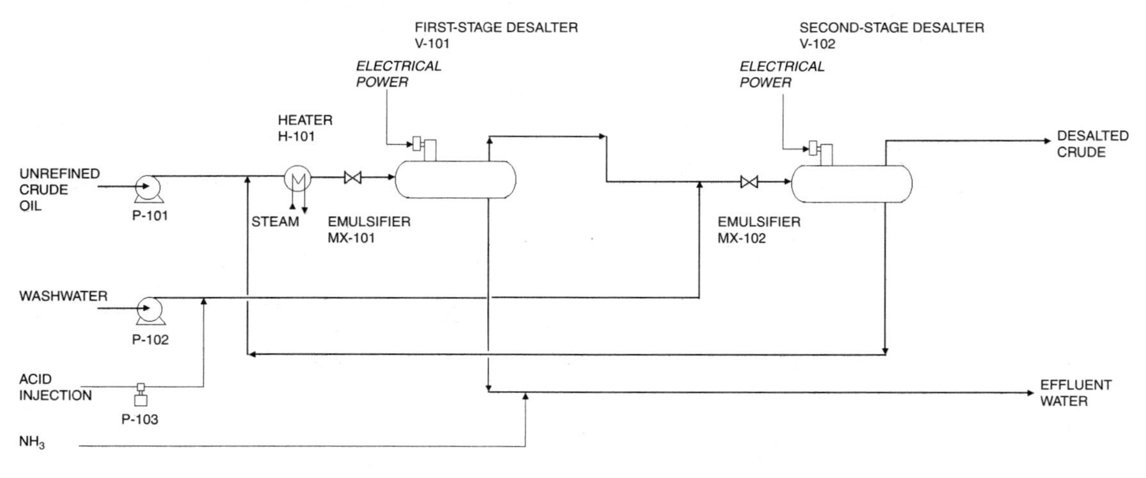

Figure 1-6. Two-stage desalter.

and longer catalyst life in downstream processing units provide justification for this additional treatment.

Desalting is carried out by emulsifying the crude oil with 3 to 10 vol% (volume %) water at a temperature of 200–300°F. Both the ratio of water to oil and the operating temperature are functions of the gravity of the crude oil. Typical operating conditions are given in Table 1-8.

The salts are dissolved in the washwater and oil and water phases are separated in a settling vessel either by adding chemicals to assist in breaking up the emulsion or by the application of an electrostatic field to coalesce the droplets of saltwater more rapidly (see Table 1-9). Either an AC or DC field may be used (see Table 1-10) and potentials of 16,000–35,000 V are used to promote coalescence. Efficiencies up to 90–95% water removal are achieved in a single stage and up to 99% in a two-stage desalting process.

Heavy naphthenic crudes form more stable emulsions than most other crudes, and desalters usually operate at lower efficiency when handling them. The crude oil densities are close to density of water, and temperatures above 280°F are required.

It is necessary to adjust the pH of the brine to obtain a value of 7 or less. If the pH of the brine exceeds 7, emulsions are formed because of the presence of sodium naphthenate and sodium sulfide. For most crude oils, it is desirable to keep the pH below 8. Better dehydration is obtained in electrical desalters when they are operated at a pH of 6. The pH is controlled by the addition of acid to the incoming or recycle water.

Makeup water is added to the second stage of a two-stage desalter. The quantity is 4–5% on crude oil volume. For very heavy crude oil ($API < 15$), gas oil is added as a diluent to the second stage to obtain more efficient separation. The gas oil is recovered in the crude column and recycled to the desalter. Frequently, the washwater used is from the vacuum crude unit barometric condenser or other refinery sources containing phenols. The phenols are preferentially soluble in crude oil, thus reducing the phenol content of the water sent to the refinery wastewater handling system.

Suspended solids are another major cause of water-in-oil emulsions. Wetting agents are frequently added to improve the water wetting of solids and reduce oil carry under in the desalters. Oxyalkylated phenols and sulfates are the most frequently used wetting agents.

Table 1-8
Washwater Requirements of Desalters

CRUDE API	WASHWATER, VOL%	TEMPERATURE, °F
$\text{API} > 40$	3–4	240–260
$30 < \text{API} < 40$	4–7	260–280
$\text{API} < 30$	7–10	280–300

Table 1-9
Operating Conditions

PARAMETER	UNITS	VALUE
CRUDE TO DESALTER*	bpsd	98000
WATER TO DESALTER	gpm	145
WATER TO CRUDE RATIO	%	5
DEMULSIFIER INJECTION	ppmw	10–15
PRESSURE		
CRUDE TO DESALTER	psig	125
DELTA P MIXING VALVE	psig	20
TEMPERATURE		
CRUDE TO DESALTER	°F	270
WATER TO DESALTER	°F	265
CRUDE FROM DESALTER	°F	260
ANALYSIS RESULTS		
CRUDE INLET SALT	lb/1000 bbl	3.94
CRUDE INLET SALT	ppmw	12.87
CRUDE OUTLET SALT	ppmw	1.2
OUTLET BS&W	% MASS	0.05
WATER		
INLET SALT CONTENT	ppm	100
OUTLET SALT CONTENT	ppm	310
INLET OIL CONTENT	ppm	7
OUTLET OIL CONTENT	ppm	10
pH INLET		6.5
OUTLET pH		6.5
OUTLET pH AFTER NH3 INJECTION		7

** 30.4 API CRUDE.*

NOTE: BASIS 98000 BPSD CRUDE.

Table 1-10
Utility Consumption

UTILITY	UNITS	CONSUMPTION
ELECTRICITY	kWhr	0.014–0.070
WATER	GALLONS	10–18

NOTE: PER TON FEED.

NOTES

1. W. L. Nelson. *Oil and Gas Journal*, (March 2, 1944; July 21, 1945; May 12, 1945).
2. J. W. Packie. "Distillation Equipment in Oil Refining Industry." *AIChE Transctions* 37(1941), pp. 51–78.
3. *Standard Test Method for Distillation of Petroleum Products.* ASTM Standards D-86 and IP 123/84.
4. J. B. Maxwell. *Data Book on Hydrocarbons*. Princeton, NJ: Van Nostrand, 1968. W. C. Edmister. *Applied Hydrocarbons Thermodynamics*. Houston: Gulf Publishing, 1961. W. L. Nelson. *Petroleum Refinery Engineering*. New York: McGraw-Hill, 1958.
5. W. C. Edmister. *Applied Hydrocarbons Thermodynamics*. Houston, Gulf Publishing, 1961.
6. Maxwell, *Data Book on Hydrocarbons*.
7. Packie, "Distillation Equipment in Oil Refining Industry."
8. R. N. Watkins. *Petroleum Refinery Distillation*. Houston, Gulf Publishing, 1981.
9. Maxwell, *Data Book on Hydrocarbons*.

CHAPTER TWO

Distillate Hydrotreating

Hydrotreating processes aim at the removal of impurities such as sulfur and nitrogen from distillate fuels—naphtha, kerosene, and diesel—by treating the feed with hydrogen at elevated temperature and pressure in the presence of a catalyst. Hydrotreating has been extended in recent years to atmospheric resids to reduce the sulfur and metal content of resids for producing low-sulfur fuel oils. The operating conditions of treatment are a function of type of feed and the desulfurization levels desired in the treated product. The feed types considered here are

Naphtha.
Kerosene.
Gas oils.
Atmospheric resids or reduced crudes.

The principal impurities to be removed are

Sulfur.
Nitrogen.
Oxygen.
Olefins.
Metals.

The basic reactions involved are outlined in Figure 2-1.

Sulfur

The sulfur-containing compounds are mainly mercaptans, sulfides, disulfides, polysulfides, and thiophenes. The thiophenes are more difficult to eliminate than most other types of sulfur.

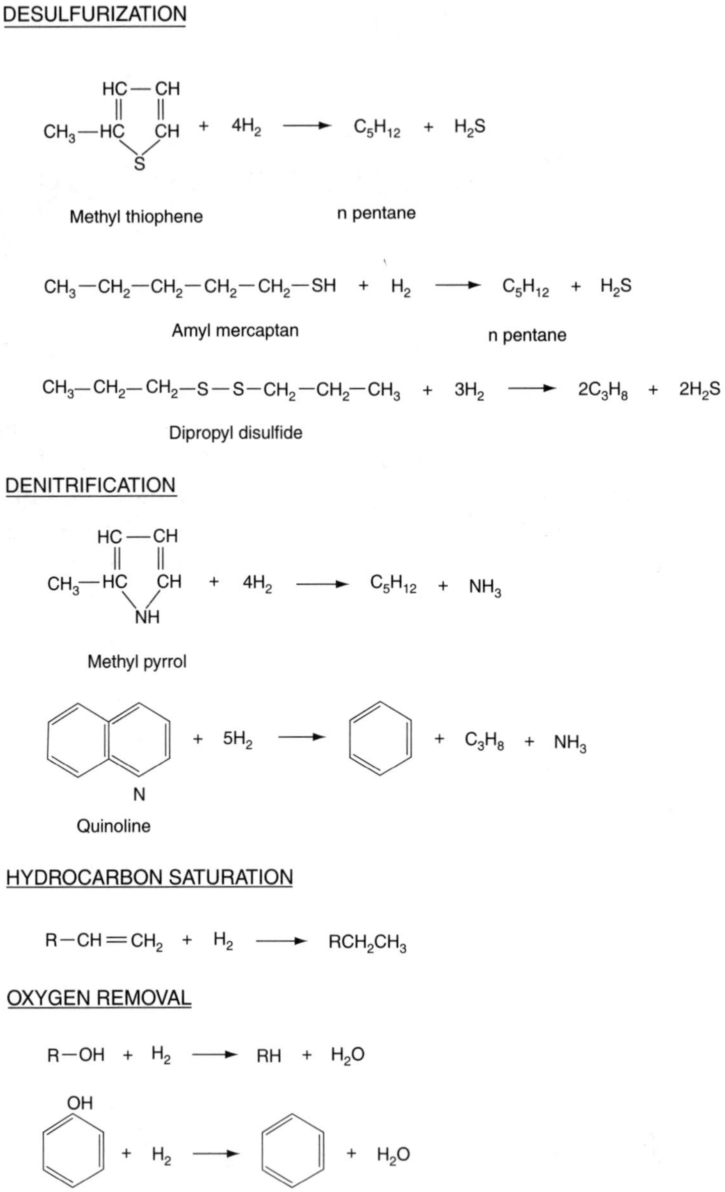

DESULFURIZATION

$$CH_3{-}HC\overset{\text{HC}=\text{CH}}{}CH\ (\text{ring with S}) + 4H_2 \longrightarrow C_5H_{12} + H_2S$$

Methyl thiophene — n pentane

$$CH_3{-}CH_2{-}CH_2{-}CH_2{-}CH_2{-}SH + H_2 \longrightarrow C_5H_{12} + H_2S$$

Amyl mercaptan — n pentane

$$CH_3{-}CH_2{-}CH_2{-}S{-}S{-}CH_2{-}CH_2{-}CH_3 + 3H_2 \longrightarrow 2C_3H_8 + 2H_2S$$

Dipropyl disulfide

DENITRIFICATION

$$CH_3{-}HC\overset{\text{HC}=\text{CH}}{}CH\ (\text{ring with NH}) + 4H_2 \longrightarrow C_5H_{12} + NH_3$$

Methyl pyrrol

$$\text{Quinoline} + 5H_2 \longrightarrow \text{Benzene} + C_3H_8 + NH_3$$

N

Quinoline

HYDROCARBON SATURATION

$$R{-}CH{=}CH_2 + H_2 \longrightarrow RCH_2CH_3$$

OXYGEN REMOVAL

$$R{-}OH + H_2 \longrightarrow RH + H_2O$$

$$\text{Phenol (OH)} + H_2 \longrightarrow \text{Benzene} + H_2O$$

Figure 2-1. Basic reactions.

Nitrogen

The nitrogen compounds inhibit the acidic function of the catalyst considerably. These are transformed into ammonia by reaction with hydrogen.

Oxygen

The oxygen dissolved or present in the form of compounds such as phenols or peroxides are eliminated in the form of water after reacting with hydrogen.

Olefins

The olefinic hydrocarbons at high temperature can cause formation of coke deposits on the catalyst or in the furnaces. These are easily transformed into stable paraffinic hydrocarbons. Such reactions are highly exothermic. Straight run feeds from the crude unit usually contain no olefins. If, however, the feed contains a significant amount of olefins, a liquid quench stream is used in the reactor to control the reactor outlet temperature within the design operating range.

Metals

The metals contained in the naphtha feed are arsenic, lead, and to a lesser degree copper and nickel, which damage the reforming catalyst permanently. Vacuum gas oils and resid feeds can contain a significant amount of vanadium and nickel. During the hydrotreating process, the compounds that contain these metals are destroyed and the metals get deposited on the hydrotreating catalyst.

OPERATING VARIABLES

The principal variables for hydrodesulfurization (HDS) reactions are temperature, the total reactor pressure and partial pressure (PPH_2) of hydrogen, the hydrogen recycle rate, and the space velocity (VVH).

Temperature

The HDS reactions are favored by an increase in temperature, but at the same time, high temperature causes coking reactions, diminishing the activity of the catalyst. The desulfurization reactions are exothermic and the heat of reaction is approximately 22–30 Btu/mole hydrogen. It is necessary to find a compromise between the reaction rate and the overall catalyst life. The operating temperature (start of run/end of run) is approximately 625–698°F according to the nature of the charge. During the course of a run, the temperature of the catalyst is gradually raised to compensate for the fall in activity due to coke deposits until the maximum permissible temperature limit (EOR) for the HDS catalyst is reached. At this stage, the catalyst must be regenerated or discarded.

Pressure

The increase in partial pressure of hydrogen increases the HDS rate and diminishes the coke deposits on the catalyst, thereby reducing the catalyst fouling rate and increasing the catalyst life. Also, many unstable compounds are converted to stable compounds. Operation at higher pressure increases the hydrodesulfurization rate because of higher hydrogen partial pressure in the reactor, requiring a smaller quantity of catalyst for a given desulfurization service. In an operating unit, higher-pressure operation can increase the feed throughput of the unit while maintaining the given desulfurization rate.

Space Velocity

The liquid hourly space velocity (LHSV) is defined as

$$\text{LHSV} = \frac{\text{per hour feed rate of the charge } (\text{ft}^3/\text{hr})}{\text{volume of the catalyst bed } (\text{in ft}^3)}$$

Hydrodesulfurization reactions are favored by a reduction in VVH. The rate of desulfurization is a function of (PPH_2/VVH) or the ratio of partial pressure of hydrogen in the reactor to liquid hourly space velocity. For a given desulfurization rate (at constant temperature), the ratio PPH_2/VVH is fixed. Fixing the total reactor pressure automatically fixes the partial pressure and the required hydrogen recycle rate. In general, the total reactor pressure is fixed from the available hydrogen pressure, the hydrogen partial pressure, and other variables such as VVH are adjusted until these fall within the acceptable limits.

Recycle Rate

In an HDS process, the hydrogen separated in a high-pressure (HP) separator drum is recycled to the reactor via a recycle compressor and furnace. This stream joins the incoming fresh feed, which has been heated in the feed furnace. The recycle rate is the ratio of volume of hydrogen at 1 atm and 15°C to the volume of fresh liquid feed at 15°C.

MAKEUP HYDROGEN

While the hydrogen recycling assures the requisite partial pressure of hydrogen in the reactor, makeup hydrogen is required to replace the hydrogen consumed in HDS reactions. The hydrogen is lost through solution losses in the liquid phase and purges for regulating unit pressure and maintaining recycle gas purity.

PURGE RATIO

A purge ratio is the ratio of the volume of hydrogen in the purged gas to the volume of hydrogen in the makeup gas. Purging is required to prevent the buildup of inert gases and light hydrocarbons in the recycle gas. The quantity of purge directly influences the purity of the hydrogen in the recycle gas. For low-sulfur feeds such as naphtha, the purge ratio required is small. For heavy, high-sulfur feeds the purge ratio required to maintain the purity of recycled hydrogen is quite high. Typical purge values used are shown in Table 2-1.

Table 2-1
Purge Requirements of HDS Processes

PROCESS	HYDROGEN IN PURGE/ HYDROGEN IN MAKEUP GAS
NAPHTHA HDS	10%
KEROSENE HDS	15%
DIESEL HDS	20%
VGO HDS	30%

CATALYSTS

The catalyst consists of two parts, the catalyst support and the active elements. The support consists of solid substances with high porosity and able to withstand the temperature, pressure, and the environment encountered in HDS reactors. The support utilized by the HDS catalyst is alumina in the form of balls or extrudates. The active elements are the metals deposited on the support in form of oxides. Before operation, the catalyst is sulfided in order to moderate activity.

The principal types of catalyst used in the HDS service are

1. Cobalt (molybdenum on alumina support). This is the general catalyst for HDS service.
2. Nickel (molybdenum on alumina support). This is used especially for denitrification.
3. Cobalt (molybdenum and other metals on neutral support). The neutral support prevents the polymerization of the olefins. This catalyst is employed in conjunction with other catalysts for olefinic feeds.

NAPHTHA HYDRODESULFURIZATION PROCESS

Naphtha is hydrodesulfurized to make the feed suitable for subsequent treatment; catalytic reforming to improve octane or steam reforming for hydrogen production. In the first case, this takes the name *pretreatment*. The process consists of treating the feed with hydrogen in the presence of a catalyst under suitable operating conditions to remove feed contaminants (see Figure 2-2).

The naphtha feed enters the unit through charge pump P-101. It is mixed with hydrogen gas coming from a cat reforming unit or hydrogen plant. The feed/hydrogen mix is next heated in succession through heat exchange with reactor effluent in E-101 and fired heater H-101. The heated feed/hydrogen mix next enters the desulfurization reactor V-101 at the top. The desulfurization reactions take place over a Co-Mo on alumina catalyst contained in the reactor. The reactor effluent contains the desulfurized naphtha, excess hydrogen, H_2S, and light end elements formed as a result of reaction of sulfur in the feed with hydrogen in the presence of the catalyst.

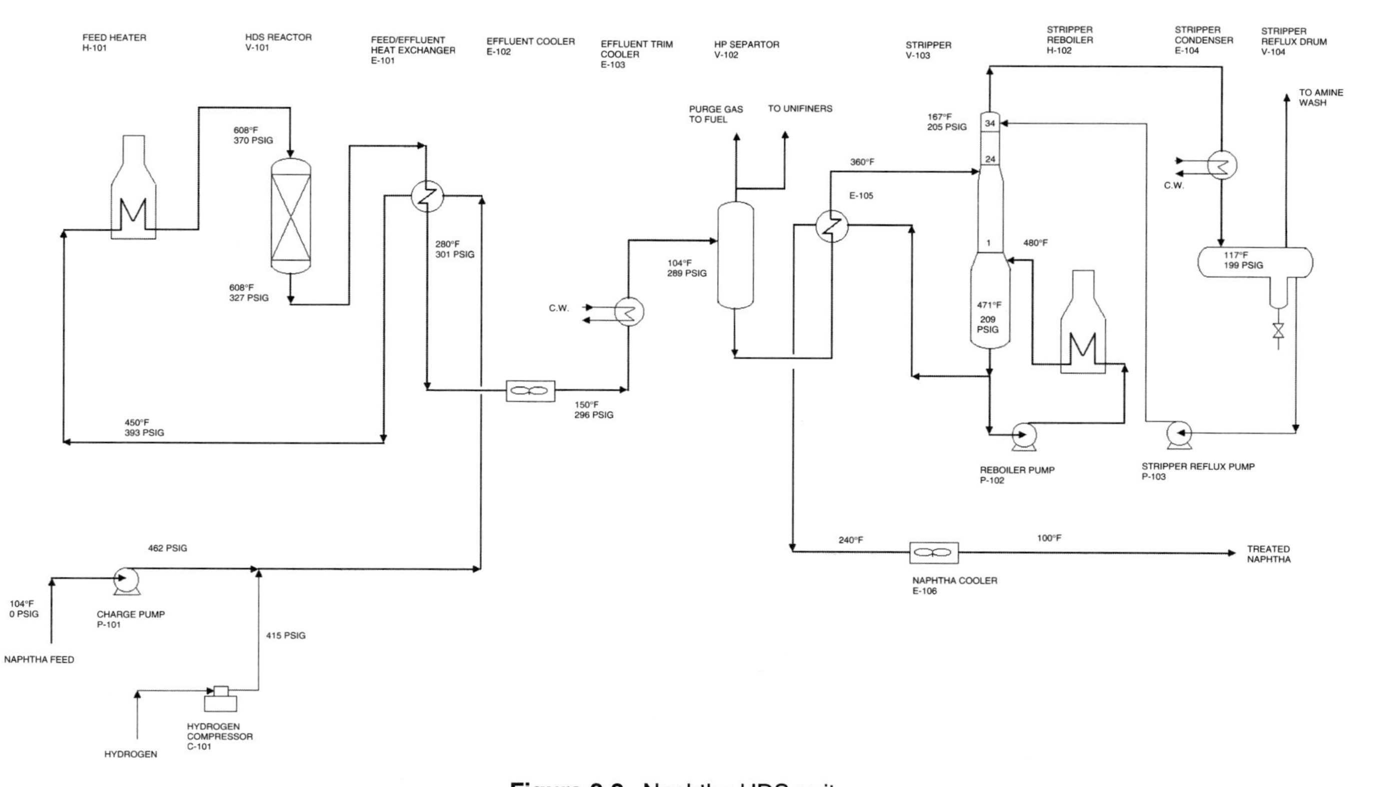

Figure 2-2. Naphtha HDS unit.

The reactor effluent is cooled and partially condensed through feed/effluent heat exchanger E-101, an air cooler E-102, and a trim cooler E-103, before flowing into high-pressure separator V-102. The separation between the vapor and liquid phases occurs in this separator drum. The vapor from this drum containing H_2S, light hydrocarbons formed as a result of desulfurization reactions, and excess hydrogen are purged and sent out of unit's battery limit.

The liquid from V-102 is preheated through a heat exchange with the stripper bottom in heat exchanger E-105 before charging to the stripper column V-103, where the dissolved hydrogen and H_2S contained in the HP separator liquid is removed as overhead product. The gross overhead product from V-103 is partially condensed through heat exchanger E-104. The uncondensed vapor, containing most of the separated H_2S, is sent to an amine unit for H_2S recovery. The liquid is refluxed back to the column through pump P-103.

The column is reboiled with a fired heater H-102. The bottom recirculation is provided by P-102. The stripper bottom product is cooled by heat exchange with incoming feed in E-105 and next in E-106 to 100°F before going out of unit's battery limits.

Operating conditions of a naphtha HDS unit for preparing cat reformer feed are shown in Table 2-2. Corresponding feed and product properties, unit yields, and utility consumption are shown in Tables 2-3 to 2-5.

Table 2-2
Naphtha HDS Operating Conditions

OPERATING PARAMETERS	UNITS	
REACTOR INLET TEMPERATURE		
SOR	°F	608
EOR	°F	698
TOTAL PRESSURE AT SEPARATOR DRUM	psia	303
HYDROGEN PARTIAL PRESSURE AT REACTOR OUTLET	psia	160
LIQUID HOURLY SPACE VELOCITY (LHSV)	hr^{-1}	4.00
HYDROGEN CONSUMPTION	scf/bbl	45

CATALYST: CO-MO ON ALUMINA SUPPORT
TYPICAL COMPOSITION AND PROPERTIES:
CO = 2.2%, MO = 12.0%
SURFACE AREA = 225 m^2/gm
PORE VOLUME = 0.45 cm^3/gm
CRUSH STRENGTH = 30 kg

Table 2-3
Naphtha HDS Feed (Sulfur-Run) and Product Properties

QUALITY	UNITS	
SULFUR GRAVITY		0.734
TBP DISTILLATION		
IBP	°F	194
10%	°F	203
30%	°F	221
50%	°F	239
70%	°F	257
90%	°F	275
FBP	°F	284
HYDROCARBON TYPE		
PARAFFINS	VOL%	69
NAPHTHENES	VOL%	20
AROMATICS	VOL%	11
TOTAL SULFUR	Wt%	0.015
MERCAPTAN		
SULFUR	Wt%	0.008
TOTAL NITROGEN	ppmw	1
MOLECULAR WEIGHT		111
PRODUCT SULFUR	ppmw	0.5

IBP = INITIAL BOILING POINT; FBP = FINAL BOILING POINT.

Table 2-4
Naphtha HDS Unit Yields

	YIELD WEIGHT FRACTION
FEED	
NAPHTHA FEED	1.0000
HYDROGEN	0.0080
TOTAL FEED	1.0080
PRODUCTS	
ACID GAS	0.0012
H_2 RICH GAS	0.0110
LPG RICH GAS	0.0058
HYDROTREATED NAPHTHA	0.9900
TOTAL PRODUCT	1.0080

Table 2-5
Naphtha HDS Unit Utility Consumption per Ton Feed

UTILITY	UNITS	VALUE
FUEL GAS	mmBtu	0.4330
STEAM	mmBtu	0.3680
POWER	kWhr	10.0000
COOLING WATER	mig	1.0600
DISTILLED WATER	mig	0.0025

KEROSENE HYDROTREATING

The objective of kerosene hydrotreating is to upgrade raw kerosene distillate to produce specification products suitable for marketing as kerosene and jet fuel. Sulfur and mercaptans in the raw kerosene cuts coming from the crude distillation unit can cause corrosion problems in aircraft engines and fuel handling and storage facilities. Nitrogen in the raw kerosene feed from some crude oils can cause color stability problems in the product. For aviation turbine fuels (ATF), the ASTM distillation, flash point, and freeze point of the hydrotreated kerosene cut has to be rigorously controlled to meet the stringent requirements. This is done by distillation in a series of columns to remove gases, light ends, and heavy kerosene fractions. The upgrading is achieved by treating hydrogen in the presence of a catalyst, where sulfur and nitrogen compounds are converted into hydrogen sulfide and ammonia.

Because of the very stringent product specifications, the ATF product can have only straight run kerosene or hydrotreated blend components. Another important property of aviation turbine fuel is its smoke point, which in turn is a function of the aromatic type hydrocarbons in the cut. Higher aromatic content yields lower smoke point kerosene cuts, which may not meet the aviation turbine fuel specification. Depending on the severity of hydrotreating, the smoke point of the kerosene may be improved by saturation of aromatics to corresponding naphthenes (see Figure 2-3).

Kerosene feed from storage is pumped via charge pump P-101 and preheated in effluent/feed exchanger E-103, followed by final heating in fired heater H-101. The effluent from H-101 next joins the recycle

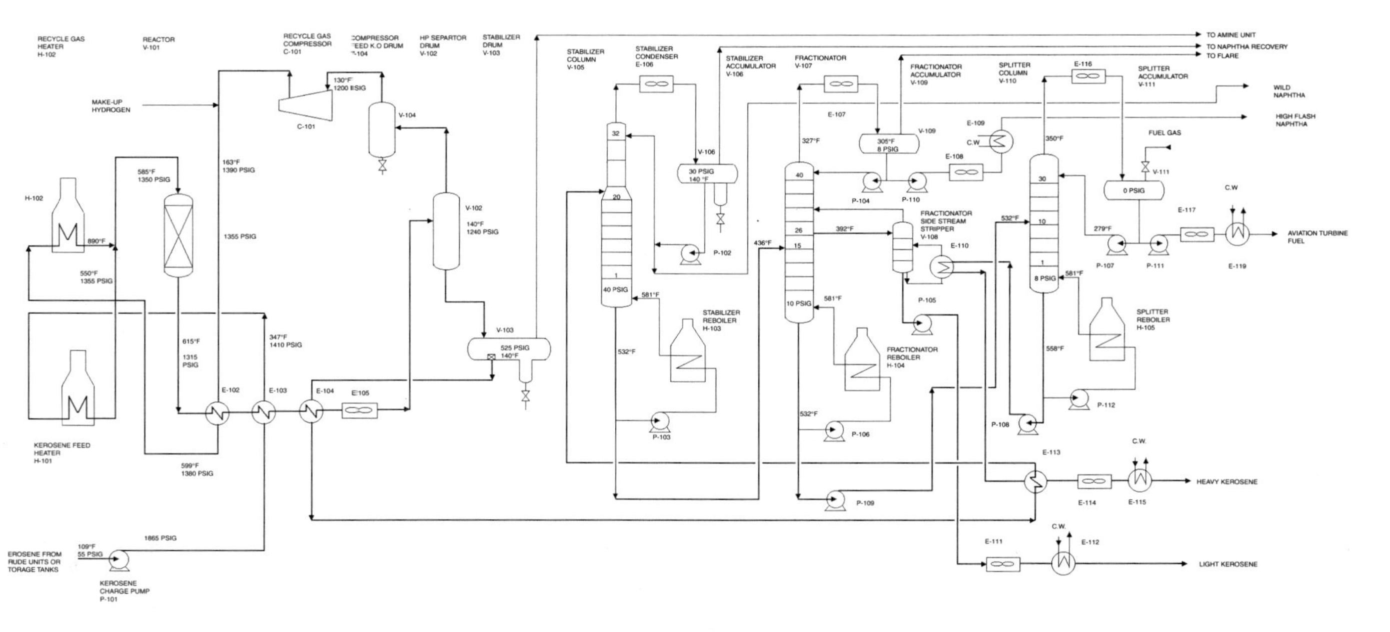

Figure 2-3. Kerosene HDS unit. C.W. = cooling water.

hydrogen coming from compressor C-101 and is heated successively in feed/effluent exchanger E-102 and fired heater H-102. The heated kerosene feed and hydrogen mix stream next flow through reactor V-101, loaded with a Co-Mo or Mo-Ni catalyst. Hydrodesulfurization and hydrodenitrification reactions take place in the reactor. These reactions are exothermic. The reactor effluent is cooled in the effluent/feed exchangers E-102, E-103, and E-104 by exchanging heat with incoming kerosene feed and hydrogen. The effluent is next cooled in air cooler E-105 before being flashed in high pressure separator drum V-102 at 140°F.

The hydrogen-rich gas from the separator is compressed and recycled to the reactor section by centrifugal compressor C-101. Recycled hydrogen gas is preheated in effluent/hydrogen exchanger E-102. It is further heated in fired heater H-102 and joins the hydrocarbon feed to reactor V-101.

The hydrocarbon liquid from the separator drum is depressurized into flash drum V-104. The flash gas is sent to the amine unit for H_2S removal before being sent to refinery fuel system. The liquid from the flash drum is sent to a stabilizer column V-105. The stabilizer overhead vapor is partially condensed in air cooler E-106 and flows into accumulator V-106. A part of the accumulator liquid naphtha is returned to the column as reflux, the rest is withdrawn as wild naphtha.

The stabilizer bottom product is sent to fractionator column V-107, where a high flash naphtha cut is taken as overhead product. Light kerosene base stock is withdrawn from the fractionator as a sidestream. It passes through kerosene side stripper V-108 to adjust its flash point and cooled in E-111 and E-112 before sending to storage. The stabilizer column is reboiled by fired heater H-103.

Fractionator bottoms flow to splitter column V-110, where aviation turbine kerosene is withdrawn as an overhead product. Antioxidant is injected into the ATK product before it is finally sent to storage. The splitter column is heated in a forced recirculation-type reboiler, the heat provided by the H-106 fired heater.

The bottom product is pumped through air cooler E-114 and water trim cooler E-115 to storage as heavy kerosene. This product is used as a blend stock for diesel or as a cutter for various fuel oil grades.

The operating conditions of a kerosene hydrotreating unit are shown in Table 2-6. The corresponding feed and product properties, unit yields, and utility consumption are shown in Tables 2-7 to 2-10.

Table 2-6
Kerosene HDS Operating Conditions

OPERATING PARAMETERS	UNITS	
TEMPERATURE (W.A.B.T.)		
SOR	°F	600
EOR	°F	698
REACTOR ΔT	°F	30
REACTOR ΔP	psi	35
TOTAL REACTOR PRESSURE	psig	1350
HYDROGEN PARTIAL PRESSURE	psia	1105
RECYCLE RATIO	scf/bbl	3072
HYDROGEN CONSUMPTION	scf/bbl	555
% DESULFURIZATION	Wt%	99.6
% DENITRIFICATION	Wt%	98.0
SEPARATOR DRUM		
PRESSURE	psig	1240
TEMPERATURE	°F	140
RECYCLE GAS COMPRESSOR		
SUCTION PRESSURE	psig	1200
DISCHARGE PRESSURE	psig	1390
DISCHARGE TEMPERATURE	°F	161

W.A.B.T. = WEIGHTED AVERAGE BED TEMPERATURE.

GAS OIL HYDRODESULFURIZATION

Gas oil hydrodesulfurization is designed to reduce the sulfur and other impurities (e.g., nitrogen) present in the raw gas oil cuts. The feed to the unit may be a straight run diesel cut from the crude distillation unit or secondary units such as FCCU (light cycle gas oil) or delayed coker. The feed from these secondary units may contain significant amount of olefinic hydrocarbons, which must be converted to saturates in the diesel hydrotreating unit to improve the storage stability of these products.

The primary improvement in product quality is with respect to sulfur and conradson carbon. The raw diesel cut from most Middle Eastern crudes, for example, may contain as much as 1–2% sulfur. Because of atmospheric pollution concerns, the sulfur content of saleable diesel grades allowable in most countries has fallen very rapidly. Until a few years ago 1.0% sulfur was acceptable. But, because of very rapid increase in the number of vehicles using diesel worldwide, the pollution level has

Table 2-7
Kerosene HDS Unit Feed and Product Properties

PROPERTY	FEED	NAPHTHA	STABILIZER BOTTOMS	HIGH FLASH NAPHTHA	LIGHT KERO	ATK	HEAVY KERO
ANILINE POINT, °F	143		151.5				156.5
API GRAVITY	46.44	58.89	45.40	56.02	47.93	44.82	41.17
DENSITY	0.7952	0.7432	0.7999	0.7546	0.7886	0.8025	0.8195
AROMATICS VOL%	22	10.3		12.3	18.2	19.1	19.9
CLOUD POINT, °F							2
CORROSION, Ag STRIP					ZERO	ZERO	
ASTM DISTILLATION °F							
(IBP)	192	124	318	256	322	384	452
5 VOL%	306	178	344	264	338	396	464
10	324	202	354	268	342	400	472
20	346	232	368	272	346	404	478
30	362	256	386	276	350	406	484
50	396	292	424	280	362	412	500
70	440	316	468	286	382	422	518
90	504	348	520	296	418	440	544
95	526	360	536	302	436	448	556
(FBP)	556	394	562	326	470	470	576
FLASH POINT °F			138		140	184	226
FREEZE POINT °C					−59	−54	
POUR POINT °F							ZERO
SMOKE POINT mm	24		25		27	26	25
SULFUR PPMW	4500	1900	17	6.4	3	3.6	41.1
VIS KIN. @ 122°F	1.1		1.31	0.66	0.99	1.32	2.2

IBP = INITIAL BOILING POINT; FBP = FINAL BOILING POINT.

Table 2-8
Kerosene HDS Unit Overall Yields

STREAM	WEIGHT FRACTION
FEED	
KEROSENE FEED	1.0000
H_2 GAS	0.0137
TOTAL FEED	1.0137
PRODUCTS	
GAS FROM UNIFINER	0.0109
HP GAS	0.0060
ACID GAS	0.0018
NAPHTHA	0.1568
ATK	0.7582
HEAVY KEROSENE	0.0800
TOTAL PRODUCT	1.0137

Table 2-9
Kerosene HDS Unit Utility Consumption per Ton Feed

UTILITY	UNITS	CONSUMPTION
FUEL	mmBtu	1.7
POWER	kWhr	15
STEAM	mmBtu	0.03
COOLING WATER	MIG	0.48
DISTILLED WATER	MIG	0.0034

risen exponentially. To curb this increasing atmospheric pollution, sulfur specifications in most developed countries have fallen to 0.005 wt% or lower. The average sulfur decline is often pushed by vehicle and engine manufacturers aiming at lower particulate emissions.

In Figure 2-4, the diesel feed is pumped by charge pump P-101 to effluent/feed heat exchangers E-103 and E-101, then it joins the hot recycle hydrogen stream before entering reactor V-101 loaded with the desulfurization catalyst (Co-Mo or Ni-Mo on alumina type). The recycled gas from compressor discharge C-101 is heated in effluent/feed heat exchanger E-102, next in fired heater H-101 then mixed with the hydrocarbon feed before going to HDS reactor V-101.

In the reactor, hydrodesulfurization reactions take place, in which sulfur and nitrogen attached to hydrocarbon molecules are separated

Table 2-10
Typical Specifications of Dual-Purpose Kerosene

ACIDITY, TOTAL	mg KOH/gm	MAX	0.015	ASTM D 3242
ANILINE GRAVITY PRODUCT		MIN	4800	ASTM D611
OR				
NET SPECIFIC ENERGY	J/gm	MIN	42800	ASTM D 240
AROMATICS	VOL%	MAX	22	ASTM D 1319
COLOR SAYBOLT		MIN	+25	ASTM D156
CORROSION Cu STRIP				ASTM D130
2 HOURS, 100°C		MAX	NO. 1	
CORROSION SILVER STRIP				IP 227
4 HOURS, 50°C		MAX	NO. 1	
DENSITY, 15°C	kg/litre	MIN	0.775	
		MAX	0.83	
DISTILLATION				ASTM D 86
IBP	°C	MIN	145	
10% RECOVERED	°C	MAX	205	
95% RECOVERED	°C	MAX	275	
EP	°C	MAX	300	
FLASH POINT, ABEL	°C	MIN	40	IP 170
FREEZING POINT	°C	MAX	−47	ASTM D 2386
HYDROGEN CONTENT	Wt%	MIN	13.8	ASTM D 3701
MERCAPTAN SULFUR	Wt%	MAX	0.003	ASTM 3227
NAPHTHLENES	VOL%	MAX	3	ASTM D 1840
OLEFINS	VOL%	MAX	5	ASTM 1319
SMOKE POINT	mm	MIN	23	IP 27
SULFUR	Wt%	MAX	0.04	ASTM 1266
THERMAL STABILITY				ASTM D 3241
FILTER PRESSURE DIFFERENTIAL	mmHg	MAX	25	
TUBE DEPOSIT RATING, VISUAL		MAX	<3	
VISCOSITY KINEMATIC @ −20°C	Cst	MAX	8	ASTM D 445
WATER REACTION				ASTM 1094
INTERFACE RATING		MAX	1b	
SEPARATION RATING		MAX	2	

IBP = INITIAL BOILING POINT; EP = END POINT.

and converted into hydrogen sulfide and ammonia. Hydrodesulfurization reactions are exothermic and a cool hydrogen quench is added to the interbed areas to limit the temperature rise in the reactor. The reactor effluent is cooled in effluent/feed heat exchangers E-101 to E-104, next in

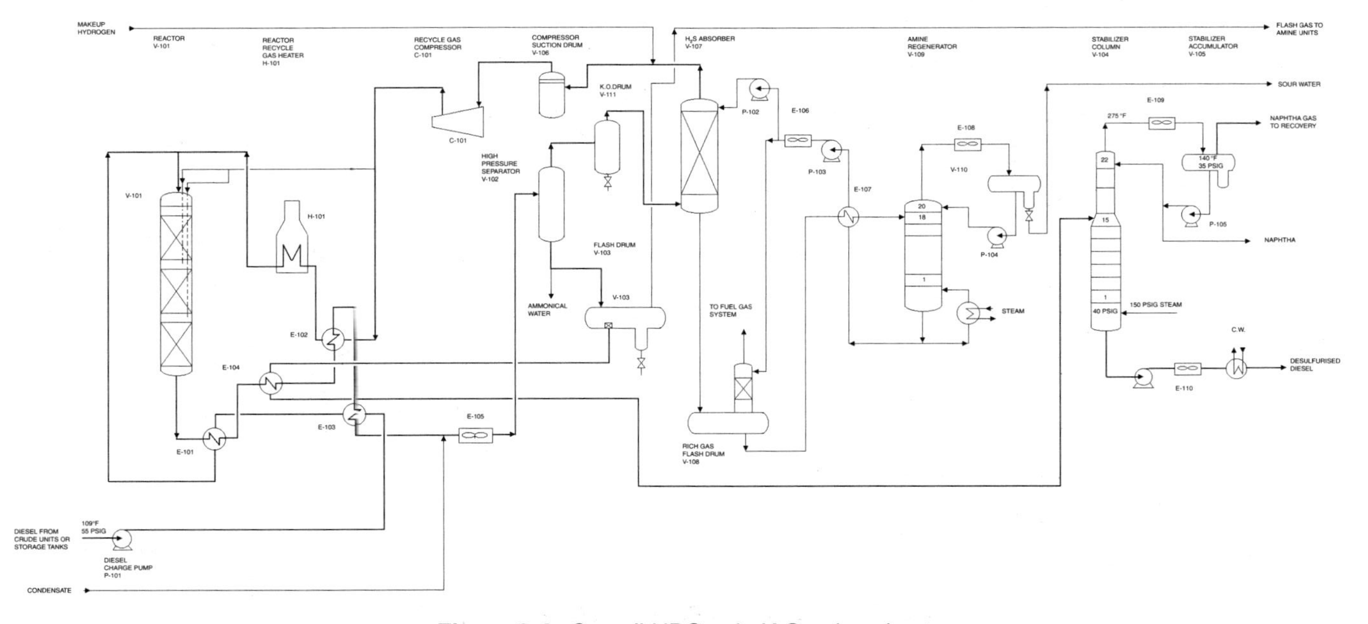

Figure 2-4. Gas oil HDS unit. K.O. = knockout.

air cooler E-105, and then is flashed into high-pressure separator drum V-102. Condensate is injected into the reactor effluent just before air cooler E-105 to dissolve and remove ammonium salts, formed as a result of nitrogen in the feed and the reaction of ammonia with H_2S, which can cause pipe blockages. The ammonium salt solution is removed from high-pressure separator drum V-102 and sent to the refinery wastewater system.

The hydrogen-rich gas from the HP separator, which is mainly hydrogen and some hydrogen sulfide, enters H_2S absorber V-107, where the H_2S is removed by a circulating diethanolamine (DEA) solution. The H_2S-free gas is mixed with makeup hydrogen coming from the hydrogen plant. The makeup and recycled hydrogen are compressed by centrifugal compressor C-101 to the reactor pressure and recycled to the HDS reactor. Part of the recycled gas is used as an interbed quench in the reactor, while the remaining hydrogen, after being heated in effluent/H_2 gas heat exchanger E-102 and fired heater H-101, joins the hydrocarbon feed before entering the reactor V-101.

The liquid hydrocarbon stream from separator V-102 is depressurized into flash drum V-103. The flash gas is sent for H_2S removal before going to refinery fuel system. The bottoms liquid stream from V-103 is preheated in the reactor effluent heat exchanger E-104 stream before flowing into the stabilizer column V-104.

Stabilizer column V-104 separates the gases, light ends, and naphtha formed as a result of HDS reactions in the reactor. The stabilizer overhead vapors are condensed in air-cooled exchanger E-109 into the overhead accumulator drum V-105. This liquid (unstabilized naphtha) is used as a reflux to the stabilizer column, and the excess is pumped out as naphtha product.

Heat is supplied to the stabilizer through medium-pressure steam. The stabilizer bottoms is the desulfurized diesel product. This stream is cooled in heat exchangers E-110 and E-111 before being sent to storage.

ULTRA-LOW SULFUR DIESELS

The production of diesel with ultra-low sulfur (<500 ppm) requires a high-severity operation. Small fluctuations in feedstock properties, unit operating parameters, and catalyst activity significantly affect the required operating conditions, catalyst deactivation, and cycle length.

Essential to deep HDS is good contacting efficiency between the catalyst and the liquid. Homogeneous gas and liquid distribution over the entire cross section of the top of the catalyst bed is essential.[1] Dense loading of catalyst is essential to ensure more homogenous catalyst loading, minimizing the risk of preferential flow through regions with lower catalyst loading density. The pressure drop at the start of the run is higher than that for sock loading but more stable during the cycle.

Hydrogen sulfide in the recycle gas inhibits the desulfurization activity of the catalyst. At 5 vol% H_2S, about 25–30% catalyst activity is lost. For deep HDS operations, H_2S scrubbing of the recycled gas is justified.[2]

Feedstock properties greatly affect the degree of desulfurization. Most of the sulfur in the middle distillates is present only as few types of alkyl-substituted benzothiophenes (BT) and dibenzothiophenes (DBT). These compounds differ greatly in their refractivity toward HDS. In the straight run gas oils, the sulfur concentration increases gradually over the boiling range, peaking at 662°F, followed by a decline. The Light Cycle Gas oil (LCO) containing feed shows a double peak structure consistent with sulfur existing predominently as benzothiophene and dibenzothiophenes.[3]

The effect of feed distillation on color is significant. The color bodies are concentrated in the tail end of the boiling range. A significant improvement in color can be obtained if the end point of the feedstock is reduced slightly. Also, the color of product increase with time, due to higher reactor temperatures.

Operating conditions of a gas oil HDS unit are shown in Table 2-11. The corresponding feed and product properties, unit yield, utility consumption, and diesel specifications are shown in Tables 2-12 to 2-15.

Table 2-11
Gas Oil HDS Operating Conditions

OPERATING PARAMETER	UNITS	
REACTOR INLET TEMPERATURE	°F	645
REACTOR ΔT	°F	55
REACTOR INLET PRESSURE	psig	2280
H_2 PARTIAL PRESSURE	psig	1728
REACTOR PRESSURE DROP	psi	34.5
LHSV*	hr^{-1}	1.42
H_2/OIL RATIO	scf/bbl	4575
RECYCLED GAS/FEED RATIO	scf/bbl	5065
QUENCH GAS/FEED RATIO	scf/bbl	971
H_2 CHEMICAL CONSUMPTION	scf/bbl	422.3
% DESULFURIZATION		98.9
% DENITRIFICATION		73.9

* NI-MO ON ALUMINA BASE CATALYST.

Table 2-12
Gas Oil HDS Feed and Product Properties

PROPERTY	UNITS	FEED	NAPHTHA	LIGHT DSL	HEAVY DSL
DENSITY	60°F	0.8967	0.7716	0.8597	0.8702
ANILINE POINT	°F	156		150	196
ASTM D-86, °F					
IBP	°F	450	98	444	615
10%	°F	575	172	520	660
30%	°F	645	234	566	690
50%	°F	685	284	594	725
90%	°F	795	386	652	805
95%	°F	810	410	666	815
EP	°F	825	456	692	835
SULFUR	ppmw	22900	3000	93	1158
NITROGEN	ppmw	800		196	450

IBP = INITIAL BOILING POINT; EP = END POINT.

Table 2-13
Gas Oil HDS Unit Yields

STREAM	WT FRACTION
FEED	
DIESEL FEED	1.0000
HYDROGEN	0.0071
TOTAL FEED	1.0071
PRODUCT	
OFF GAS	0.0030
ACID GASES	0.0236
CRACKED NAPHTHA	0.0109
LIGHT DIESEL	0.4372
HEAVY DIESEL	0.5324
TOTAL PRODUCT	1.0071

Table 2-14
Utility Consumption per Tons Feed

UTILITY	UNITS	CONSUMPTION
FUEL GAS	mmBtu	0.24
STEAM	mmBtu	0.11
POWER	kWhr	6.25
COOLING WATER	mig	0.35

Table 2-15
Typical Automotive Diesel Specifications

SPECIFICATION	UNITS			TEST METHOD
ACID NUMBER, STRONG	mg KOH/g	MAX	NIL	ASTM D 974
ACID NUMBER, TOTAL		MAX	0.1	
CARBON RESIDUE, RAMESBOTTOM ON 10% DISTILLATION	Wt%	MAX	0.2	ASTM D 482
CETANE INDEX		MIN	50	ASTM D 976
COLD FLOW PROPERTIES				
CLOUD POINT	°C	MAX	−5	ASTM D 2500
CFPP	°C	MAX	−15	IP 309
COLOR ASTM		MAX	1.5	ASTM D 1500
CORROSION COPPER STRIP @ 3 HOURS, 100°C		MAX	NO 1	ASTM D 130
DENSITY AT 15°C	kg/L	MIN	0.836	ASTM D 1298
		MAX	0.865	
DISTILLATION				
10%	°C	MAX	240	ASTM D 86
50%		MIN	240	
85%		MAX	350	
97%		MAX	370	
RECOVERY	VOL%	MIN	96	
FLASH POINT, PMcc	°C	MIN	66	ASTM D 93
SULFUR	Wt%	MAX	0.05	ASTM D 1552
VISCOSITY, KINEMATIC @ 37.8°C	Cst	MIN	2.5	ASTM D 445
		MAX	5	
WATER AND SEDIMENT	VOL%	MAX	0.01	ASTM D 2709

ATMOSPHERIC RESIDUUM DESULFURIZATION

The residuum produced from the crude unit has a high sulfur and metal (Ni, V, etc.) content. The atmospheric resids from most Middle Eastern crudes have a sulfur content of 4–5%. The purpose of the atmospheric residuum desulfurization (ARDS) unit (see Figure 2-5) is to reduce the sulfur content to less than 0.5 wt% sulfur to meet the quality criteria of products from downstream units. Also, a significant percentage of feed metals are removed during the process. The desulfurized atmospheric resid is used as a blend component for blending low-sulfur fuel oils or as feed for another processing unit; for example, a delayed coker unit.

THE PROCESS

In the crude unit, upstream of the ARDS unit, crude oil is treated in a two-stage desalting unit to reduce salt and sediment in the ARDS feed to a very low levels (>3 ppmw as sodium). This treatment is necessary to reduce catalyst bed fouling and catalyst deactivation in the ARDS reactors.

The cold resid feed is heated by heat exchange with ARDS distillate products to 450°F and the feed is next filtered in continuous backwashing feed filters. The filters remove all particles larger than 25 microns from the feed. The filtered resid flows to a surge drum with about 10 minutes hold up and is blanketed with fuel gas to prevent contamination of feed with air.

The feed is preheated first by exchange of heat with HP hot separator vapor then in feed heater H-101. Condensate-quality water is added to the feed at the heater inlet to increase the activity and stability of the catalyst. Water also prevents the deposit of solid ammonium salts in the effluent heat exchangers, downstream of the reactors. The heater outlet temperature is about 650°F. The temperature is kept below 700°F to prevent coking of heater tubes. Recycled gas plus makeup hydrogen are heated by exchange with the HP warm separator, HP hot separator, and in recycle gas heater H-102.

The hot recycled gas is mixed with heated resid upstream of guard reactor R-102. The guard reactor is the first reaction vessel and contains about 8% of the total catalyst in the unit. The function of the guard reactor is to remove sodium and other fouling material from the feed to minimize

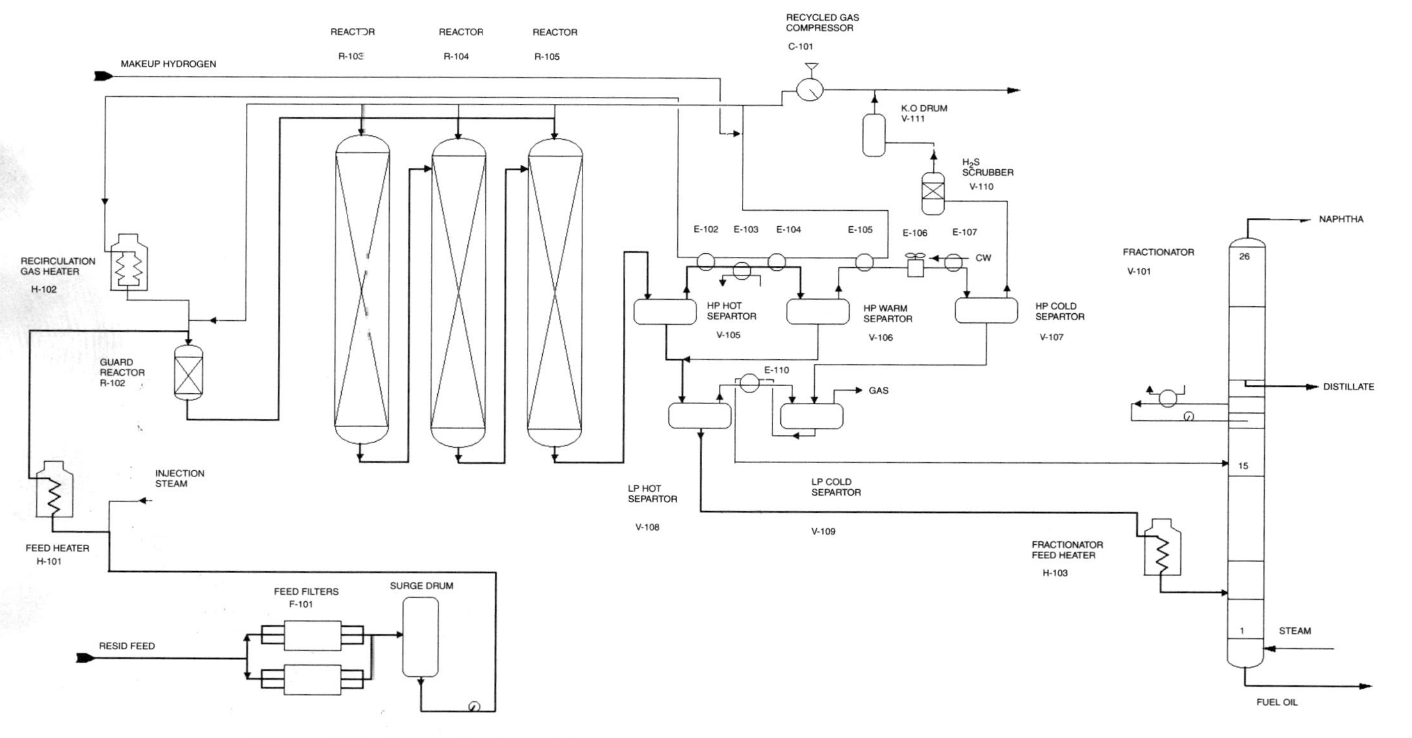

Figure 2-5. Atmospheric resid desulfuriser unit. LP = low pressure; K.O. = knockout.

plugging in the main reactors. The reactants leaving the guard chamber are quenched with cold recycled gas. The quenched mixture then flows into the first main reactor. The guard chamber catalyst is the same as in the main reactors. There are three main reactors in series. All reactors are the same size and contain only one bed of catalyst. The flow through the reactors is downward. Because desulfurization is an exothermic reaction, cold recycled gas is injected between reactors to control inlet temperatures to each succeeding reactor. Scale baskets and liquid distribution trays are provided in each reactor. In addition, layers of size-graded catalyst are installed in the top of each reactor. The upper 2 ft layer is of $\frac{1}{4}$-in. catalyst, the next 2 ft of $\frac{1}{8}$-in., and the main catalyst bed is of $\frac{1}{20}$-in. catalyst. The graded catalyst allows buildup in the top part of the bed without building up excessive pressure drop.

HIGH-PRESSURE SEPARATOR

The effluent from the last reactor flows to high-pressure hot separator V-105, which makes the necessary separation between the desulfurized liquid product and recycled gas. Vapor from HP hot separator is cooled to 500°F by heat exchange with recycled gas. The cooled, partially condensed HP hot separator vapors are fed to HP warm separator V-106. The function of the HP warm separator is to remove asphaltenes that may be entrained in the vapor leaving the HP hot separator. If the asphaltenes are not removed, an oil/water emulsion may form when water and oil condense in the high pressure loop. The HP warm separator operates at 500°F, which is well above the dew point of water, 330°F.

Liquid from the HP warm separator is combined with liquid from the HP hot separator and let down in pressure through a power recovery turbine. Vapor from the HP warm separator is cooled to 105°F by heat exchange with recycled gas, by air cooler E-106, and finally by water cooler E-107. The partially condensed HP warm separator vapor at 105°F feeds HP cold separator V-107. The final separation between the liquid and recycled gas is made in this vessel. The condensed water is also removed in this vessel. Water and oil are separated here, although both the streams are sent to the low-pressure (LP) cold separator. The separation of oil and water in the HP cold separator is necessary to avoid emulsification of an oil/water mixture across the pressure let-down valve in the line to the low-pressure cold separator.

LOW-PRESSURE SEPARATOR AND EXCHANGERS

Liquids from high-pressure hot and warm separators after let-down in pressure are fed to low-pressure hot separator V-108. The flashed mixture is separated in this drum. The liquid from the LP hot separator flows directly to fractionator feed heater H-103. Flashed vapor from the LP hot separator is partially condensed by heat exchange in E-110 and flow to LP cold separator V-109. The feeds to LP cold separator V-109 consist of separated oil and water streams from HP cold separator V-107. LP cold separator off-gas is sent to the hydrocarbon recovery system. Water from the LP cold separator is sent to sour water stripping unit. Hydrocarbon liquid from the LP cold separator is heated in low pressure hot separator (LPHS) vapor condenser E-110 and then flows to the fractionation section of the column.

HYDROGEN SULFIDE REMOVAL

The vapor from the high-pressure cold separator flows through the recycle gas filter, where the entrained oil and scale are removed to prevent foaming in the amine system. The vapor next flows to H_2S scrubber V-110, where it is in contact with a lean amine solution. The rich amine solution goes to amine regeneration unit.

RECYCLE GAS COMPRESSOR

The scrubbed recycled gas is sent to recycle gas compressor C-101, driven by a steam turbine. The compressed recycled gas is combined with makeup hydrogen and split into two streams. One stream is heated in the recycled gas preheater by heat exchange with HP warm separator and hot separator off-gases and finally in recycled gas heater H-102. The other stream of recycled gas from the compressor is utilized to quench the effluent from each reactor before entering the next reactor.

PROCESS WATER

Three uses are made of process water in the ARDS unit. Water is used to increase the activity of the catalyst and prevent the deposit of solids in the reactor effluent cooler. Water is also used to wash recycled gas before and after the amine wash. Water used to increase catalyst activity is added to the resid charge upstream of the charge heater and called *injection water.*

Deaerated condensate, stored under an inert gas blanket, is used for this purpose.

FRACTIONATION SECTION

The function of the fractionation section is to separate the desulfurized liquid stream from hot and cold low-pressure separators into the following products:

1. An overhead hydrogen-rich vapor to be processed in the gas handling plant.
2. A stabilized naphtha stream for processing in naphtha hydrotreater.
3. A distillate side cut product with an ASTM end point of 735°F.
4. A 680 plus bottom product.

The fractionator has 26 trays. The heated resid feed enters the column on tray 6. A small amount of the distillate from the low-pressure cold separator enters the column on tray 10. A stripping steam is introduced below the bottom tray to strip the material boiling below 680°F from the fractionator bottom.

CATALYST

The catalyst used in resid desulfurizers is cobalt, molybdenum, and nickel oxides on a high-purity alumina support. The catalyst for demetallization (HDM) and reduction of Conradson carbon have higher pore

Table 2-16
Properties of ARDS Catalyst

PROPERTY	HDM SERVICE	HDS SERVICE
COMPOSITION		
CoO, Wt%		3
MoO_3	12	14
NiO_3	2.5	0.2
BASE	Al_2O_3	Al_2O_3
SURFACE AREA, m^2/gram	115	200
PORE VOLUME, cm^3/gram	0.9	0.5
BULK DENSITY, gm/cc	0.51	0.67

volume that those used in the desulfurization (HDS) service. Properties of catalysts used in the resid service are shown in Table 2-16.

Catalyst is produced in $\frac{1}{4}$-, $\frac{1}{8}$-, $\frac{1}{20}$-in. diameter extrudates. Larger sizes are loaded in the top portion of the reactors to trap metal scales and other contaminants, while the lower portion of the reactor has a smaller-sized catalyst.

In the ARDS process, the high deactivation rate of the catalyst is due to the deposit of metals and carbon. A number of different catalysts are used, each playing a specific, complementary role. The catalyst in the guard reactor has the main job of retaining most of the metals contained in the feed. A demetallization catalyst with larger pore volume is used, which preferentially converts the resins and asphaltenes to which most metals are attached. In operation, these catalysts can retain metals of up to 50% of their body weight before they are completely deactivated.

The catalysts in the main reactors have hydrodesulfurization and hydrodenitrification as their main functions. At the end of run conditions, these may have adsorbed metals of up to 8–9% of their weight.

The operating conditions of an ARDS unit are shown in Table 2-17. The corresponding feed and product properties, unit yields, and utility consumption are shown in Tables 2-18 to 2-20.

ONSTREAM CATALYST REPLACEMENT

Resid desulfurizer units are expensive to build and operate because of severe operating conditions and high catalyst consumption compared to distillate desulfurization units. Thus, there is always an economic incentive to increase the onstream operating factors and run length, maximize throughput of an operating unit, and increase the conversion of resid to distillate and so minimize fuel oil production.

A major impediment in these objectives is the high metal content (mainly nickel and vanadium) of the feed, which rapidly deactivates the catalyst in the guard reactor, resulting in short run lengths, and limits the severity of operation of the unit. To appreciate the impact of high metal feeds, consider an ARDS unit processing 40 MBPSD of 680°F plus Middle Eastern atmospheric resid feed with the following characteristics:

API gravity = 13.5°
Sulfur = 4.2 wt%

Table 2-17
Operating Conditions for the Atmospheric Resid Desulfurizer Unit

OPERATING PARAMETER	UNITS	
REACTORS INLET TEMPERATURE	°F	681
REACTORS OUTLET TEMPERATURE	°F	716
CATALYST BED WABT	°F	700
FIRST REACTOR INLET PRESSURE	psig	1880
H_2 PARTIAL PRESSURE, INLET	psig	1475
HP HOT SEPARATOR PRESSURE	psig	1812
HP HOT SEPARATOR TEMPERATURE	°F	716
HP WARM SEPARATOR PRESSURE	psig	1785
HP WARM SEPARATOR TEMPERATURE	°F	500
HP COLD SEPARATOR PRESSURE	psig	1735
HP COLD SEPARATOR TEMPERATURE	°F	105
LP HOT SEPARATOR PRESSURE	psig	340
LP HOT SEPARATOR TEMPERATURE	°F	701
LP COLD SEPARATOR PRESSURE	psig	325
LP COLD SEPARATOR TEMPERATURE	°F	105
RECYCLED GAS RATE	scf/bbl	5060
H_2 CHEMICAL CONSUMPTION	scf/bbl	460
MAKEUP H_2	scf/bbl	771
PURGE RATIO, PURGE TO MAKEUP	Wt%	45
FRACTIONATOR		
FLASH ZONE TEMPERATURE	°F	694
OVERFLASH	VOL%	0.6
COLUMN TOP PRESSURE	psig	21.0
GAS OIL/RESID CUT POINT	°F	695

WABT = WEIGHTED AVERAGE BED TEMPERATURE.

Metals:

Ni = 58 ppmw
V = 18 ppmw

Assuming a run length of 12 months, a unit onstream factor of 0.98%, and a desulfurized resid metal content of 15 ppmw, it can be seen that, during the course of the run, feed will contain approximately 170 tons metals, out of which approximately 140 tons would be deposited on the catalyst. This may represent almost 30–35% of the fresh overall catalyst weight, assuming a liquid hourly space velocity (LHSV) of 0.29.

Table 2-18
ARDS Unit Feed and Product Properties

PROPERTY	UNITS	ATM RESID FEED	NAPHTHA	DIESEL	RESIDUUM
TBP CUT POINT	°F	680	C5-320	320–680	680+
END POINT	°F		345		
BROMINE NUMBER	MAX		2		
OCTANE, RON			62		
OCTANE, MON			60		
COLOR, MAX.	ASTM			1	
API GRAVITY		13.2		32.5	
S. GRAVITY		0.978			0.937
CON CARBON	% Wt	12			6.5
ASPHALTENE	% Wt	3.9			
NICKEL	ppmw	20.1			8
VANADIUM	ppmw	67			20
SODIUM	ppmw				
SULFUR	% Wt	4.2	0.01	0.05	0.5
NITROGEN	% Wt	0.245		0.02	0.13
VISCOSITY, 210	Cst	60			
VISCOSITY, 122	Cst	770			275
CETANE INDEX				45	
POUR POINT	°F			0	

Table 2-19
ARDS Unit Yields

FEED	
ATMOSPHERIC RESID	1.0000
HYDROGEN	0.0160
TOTAL INPUT	1.0160
PRODUCTS	
ACID GAS	0.0380
OFF-GASES	0.0200
NAPHTHA	0.0170
DIESEL	0.1860
FUEL OIL 680+	0.7550
TOTAL OUTPUT	1.0160

NOTE: ALL YIELDS ARE IN MASS FRACTION.

Table 2-20
ARDS Unit Utility Consumption per Ton Feed

UTILITY	UNITS	
FUEL	mmBtu	0.3000
ELECTRICITY	kWhr	17.0000
STEAM	mmBtu	0.2200
DISTILLED WATER	MIG*	0.0240
COOLING WATER	MIG*	0.6500

*MIG = THOUSAND IMPERIAL GALLONS.

To achieve a higher onstream factor and longer run lengths even while treating heavy resids with high metal content, many innovative reactor designs have been developed, such as the Shell "moving bed bunker technology" for processing vacuum resids and similar feeds. In these, the guard reactor is operated at high temperature and its catalyst deactivates at the fastest rate due to metal deposits. This is replaced either more frequently or continuously. Reactors downstream of guard reactor receive feed with much lower metal content and can therefore operate at much lower severity or increased conversion levels.

In practice, the on-line catalyst in the guard reactor is replaced using a variety of designs: two fixed-bed guard reactors in series (permutable reactors), continuous addition of fresh catalyst and withdrawal of spent catalyst in an ebullated type (OCR, or onstream catalyst replacement) guard reactor, or other, similar configurations. A reduction up to 50% in vanadium content is typical. A spherical catalyst with demetallization-promoting metals on an inert carrier with large pores (typically, nickel and molybdenum on macroporous alumina, pore volume $0.9\,cm^3/gm$) is used.

PERMUTABLE REACTOR SYSTEM

An permutable reactor system (IFP process) has two fixed-bed guard reactors in a swing arrangement and permutable operation (Figure 2-6). In the lead position, the deactivation rate is very high due to large amounts of metals, sodium, and sediment deposits on the catalyst. The catalyst deactivation rate in the reactor in second position is much smaller. The permutable reactors with high pore demetallization catalyst are designed

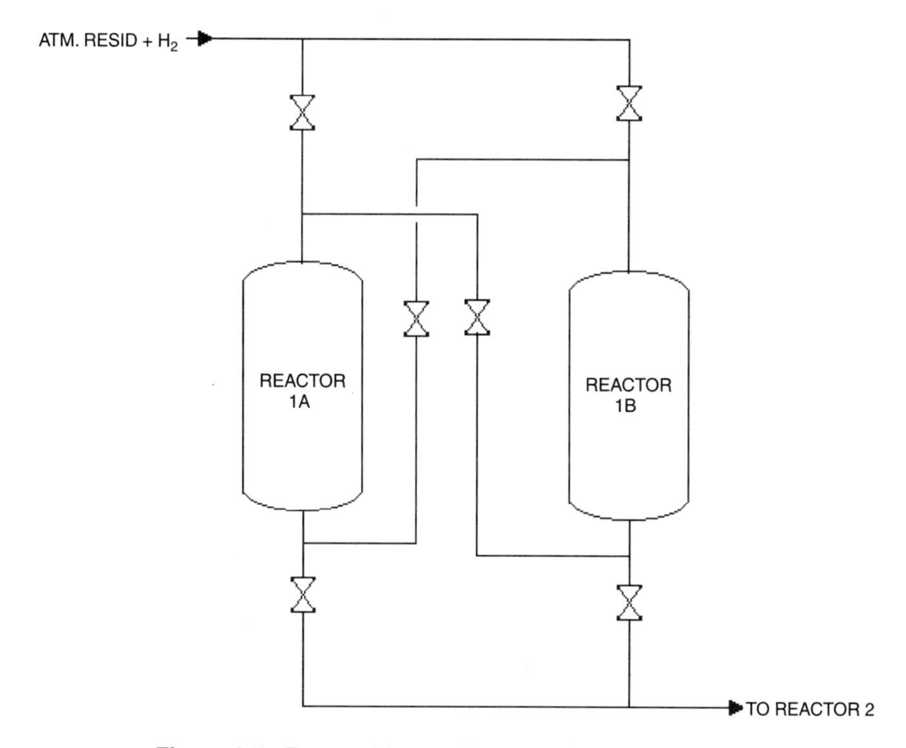

Figure 2-6. Permutable guard reactors for ARDS the unit.

for higher-temperature operation to counter the effect of metal deactivation and reach the end of run condition in shorter time.

Consider two reactors, 1A and 1B in the permutable arrangement, with reactor 1A in the first position and 1B in the second position. When the catalyst contained in the lead reactor (1A) is deactivated or the pressure drop exceeds the permissible limit, this reactor is bypassed without shutting down the unit, thereby avoiding a drop in unit throughput.

The deactivated catalyst in reactor (1A) is stripped and cooled, the catalyst is unloaded, and fresh catalyst is loaded. The catalyst is sulfided and activated and the reactor is heated before returning it to service.

The reactor (1A) with fresh catalyst is brought back on-line in the second or lag position in less than 15 days and this combination (1B and 1A) continues operation until the lead reactor (1B) reaches its EOR condition.

PERMUTATION CYCLE

The frequency of permutation is determined depending on the feed and unit design. If, for example, the overall run length of the ARDS unit is 64 weeks, the first permutation may be done after 20 weeks, when catalyst activity decrease or pressure drop increase in the lead reactor. For example,

- During weeks 1–20, both reactors (1A and 1B), with reactor 1A in the lead position, are in operation.
- During weeks 21 and 22, 1A is under regeneration and only reactor (1B) is in operation. The feed rate may be decreased to maintain performance.
- During weeks 23–32, reactor 1A is back in operation in second position. The reactor operation during this period is 1B and 1A. At the end of week 32, the catalyst in reactor 1B is completely deactivated.
- During weeks 33 and 34, reactor 1B is under regeneration and only reactor (1A) is in operation. The feed rate is decreased to maintain performance.
- During weeks 35–46, reactor 1B is back in service and put in the first position. The reactor operation is 1A and 1B. At the end of week 46, the catalyst in reactor 1A is completely deactivated.
- During weeks 47 and 48, the catalyst in reactor 1A is regenerated and the unit operation is based on only one guard reactor (1B).
- During weeks 49–64, reactor 1A is back in service, with reactor 1A again in the lead position instead of the lag or second position. The reactor operation is 1A and 1B.

At the end of week 64 (end of cycle), all the catalyst is fully deactivated and disposed of.

ONSTREAM CATALYST REPLACEMENT REACTOR

In OCR configuration (Chevron process technology), the guard reactor is an ebullated bed reactor. The feed is introduced at the bottom of the reactor and flows upward through the catalyst, slightly ($<3\%$) expanding the catalyst bed. Fresh catalyst is added at the top of the bed and spent catalyst is withdrawn at the bottom. Both catalyst addition and withdrawal are batch operations, usually done once a week, and this creates a plug flow of catalyst. Thus, the least active catalyst is in contact with most reactive feed at the bottom of the reactor. The upward flow of feed in reactor results in liquid as the continuous phase. Hydrogen is needed only for reaction and not for ebullating the bed. Pressure drop in the bed is low and constant. The catalyst used is spherical with high HDM activity, which allows for downstream fixed-bed reactors to operate at lower temperature.

A charge heater controls the feed temperature. The temperature inside the reactor is controlled by liquid quenching through a quench distributor located at different elevations in the reactor.

The advantages of a permutable or OCR reactor system for guard reactor are

1. Greater run length.
2. Higher space velocities or increased unit throughput.
3. Higher conversion of resid.
4. Increased stream factor.

NOTES

1. C. N. Satterfielf. "Trickle Bed Reactors." *American Institute of Chem. Engineering Journal* 21, no. 209 (1975).
2. Rautianen E. P. M and C. C. Johnson. "Commercial Experience with Ketjenfine 752." Akazo Catalyst Symposium, 1991.
3. X-Ma et al. "Hydrodesulfurisation Reactivities of Various Sulfur Compounds in Diesel Fuels." *Industrial Engineering and Chemistry Research* 33, no. 218 (1994).

CHAPTER THREE

Hydrocracking Processes

Distillate hydrocracking is a refining process for conversion of heavy gas oils and heavy diesels or similar boiling-range heavy distillates into light distillates (naphtha, kerosene, diesel, etc.) or base stocks for lubricating oil manufacture. The process consists of causing feed to react with hydrogen in the presence of a catalyst under specified operating conditions: temperature, pressure, and space velocity.

HYDROCRACKING REACTIONS

DESULFURIZATION

The feedstock is desulfurized by the hydrogenation of the sulfur containing compounds to form hydrocarbon and hydrogen sulfide. The H_2S is removed from the reactor effluent leaving only the hydrocarbon product. The heat of reaction for desulfurization is about 60 Btu/scf of hydrogen consumed:

$$\text{Thiophene } (R\text{—}C_4H_3S) + 4H_2 \xrightarrow{\text{CATALYST}} RCH_2CH_2CH_2CH_3 + H_2S$$

Thiophene   Paraffin   Hydrogen Sulphide

DENITRIFICATION

Nitrogen is removed from feedstock by the hydrogenation of nitrogen-containing compounds to form ammonia and hydrocarbons. Ammonia is later removed from the reactor effluent, leaving only the hydrocarbons in the product. The heat of reaction of the denitrification reactions is about

67–75 Btu/scf of hydrogen consumed, but the amount of nitrogen in the feed is generally very small, on the order of a few parts per million, and hence its contribution to overall heat of reaction is negligible:

$$R\text{-}CH_2CH_2NH_2 + H_2 \longrightarrow RCH\,CH_2CH_3 + NH_3$$

Amine — Paraffin — Ammonia

OLEFIN HYDROGENATION

The hydrogenation of olefins is one of the most rapid reaction taking place, and therefore almost all olefins are saturated. The heat of reaction is about 140 Btu/scf of hydrogen consumed. Olefin content is generally small for straight-run products, but for stocks derived from secondary/thermal processes such as coking, visbreaking, or resid hydrocracking (H-OIL* etc.), it can contribute a considerable amount of heat liberated in the hydrocracker reactor.

$$RCH_2CH = CH_2 + H_2 = RCH_2CH_2CH_3$$

Olefin — Paraffin

SATURATION OF AROMATICS

Some of the aromatics in the feed are saturated, forming naphthenes. Saturation of aromatics accounts for a significant proportion of both the hydrogen consumption and the total heat of reaction. The heat of reaction varies from 40 to 80 Btu/scf of hydrogen consumed, depending on the type of aromatics being saturated. In general, higher reactor pressure and lower temperature result in a greater degree of aromatic saturation:

H
H–C–H
(benzene ring) + $3H_2$ =
H
H–C–H
H_2 H_2
H_2 H_2
H_2

AROMATIC — NAPHTHENE

---

*H-OIL is a commercial processed for resid hydrocracking/resid desulfurisation, licensed by Hydrocarbon Research Inc. (USA).

HYDROCRACKING OF LARGE MOLECULES

Hydrocracking of large hydrocarbon molecules into smaller molecules occurs in nearly all processes carried out in the presence of excess hydrogen. These reactions liberate about 50 Btu/scf of hydrogen consumed. The heat released from the hydrocracking reactions contributes appreciably to the total heat liberated in the reactor. Cracking reactions involving heavy molecules contribute to lowering the specific gravity and forming light products, such as gas and light naphtha, in the hydrocracker products.

An example of a hydrocracking reaction is

$$RCH_2CH_2CH_2CH_3 + H_2 = RCH_3 + CH_3CH_2CH_3$$

The yield of light hydrocarbons is temperature dependent. Therefore, the amount of light end products produced increases significantly as the reactor temperature is increased to compensate for a decrease in catalyst activity toward the end of run conditions.

FEED SPECIFICATIONS

Hydrocracker feed is typically heavy diesel boiling above the saleable diesel range or vacuum gas oil stream originating from the crude and vacuum distillation unit, atmospheric resid desulfurizers, coker units, solvent deasphalting units, and the like. The hydrocracking catalyst is very sensitive to certain impurities, such as nitrogen and metals, and the feed must conform to the specifications laid down by the catalyst manufacturers to obtain a reasonable catalyst life.

FEED NITROGEN

Nitrogen in the feed neutralizes catalyst acidity. Higher nitrogen in the feed requires slightly more severe operating conditions, particularly the temperature, and causes more rapid catalyst deactivation.

FEED BOILING RANGE

A higher-than-designed feed distillation end point accelerates catalyst deactivation and requires higher reactor temperatures, thus decreasing catalyst life.

The feed properties have little direct effect on light product yield, but they affect the catalyst temperature required to achieve the desired conversion. The yield of light gases (C_{4-}) and naphtha boiling-range material is increased when the catalyst temperature is increased.

ASPHALTENES

In high cut point vacuum distillation, there is always a possibility that excessive high molecular weight, multiring aromatics (asphaltenes) can be found in vacuum gas oil distillates. In addition to causing excessive catalyst poisoning, asphaltenes may be chemically combined with the catalyst to deactivate the catalyst permanently.

METALS

Metals, particularly arsenic, and alkalies and alkaline earth deposit in the catalyst pores reduce catalyst activity. Common substances that can carry metallic catalyst contaminants include compounded lubricating oils or greases, welding fluxes, and gasketing.

Iron carried in with the feed is likely to be the most troublesome metallic catalyst contaminant. It may be chemically combined with heavy hydrocarbon molecule, or it may exist as suspended particulate matter. In either case, it not only deactivates the catalyst but also plugs the catalyst interstices such that excessive pressure drop develops. Normally, this plugging appears as a crust at the top of the first catalyst bed.

CHLORIDES

The feed may contain trace amounts of organic and inorganic chlorides, which combine with ammonia produced as a result of denitrification reactions to form very corrosive deposits in the reactor effluent exchanger and lines.

OXYGEN

Oxygenated compounds, if present in the feed, can increase deactivation of the catalyst. Also, oxygen can increase the fouling rate of the feed effluent heat exchangers.

CATALYST

Hydrocracking reactions can be divided into two groups: (1) desulfurization and denitrification—hydrogenation of polyaromatics and monoaromatics—are favored by the hydrogenating function of the catalyst (metals) and (2) hydrodealkylation, hydrodecyclization, hydrocracking, and hydroisomerization reactions are promoted by the acidic function of the catalyst (support). The support function is affected by the nitrogen content of the feed.

The catalyst employed in hydrocracking is generally of the type (Ni-Co-Fe), (Mo-W-U) on a silica/alumina support. The ratio of alumina to silica is used to control the degree of hydrocracking, hydrodealkylation, hydroisomerization, and hydrodecyclization. Cracking reactions increase with increasing silica content of the catalyst. Metals, in the form of sulfide, control the desulfurization, denitrification, and hydrogenation of olefins, aromatics, and the like.

The choice of catalyst system depends on the feedstock to be treated and the products required. Most of the time, the suitable system is obtained by the use of two or more catalysts with different acidic and hydrogenation functions. The reactor may also contain a small amount, up to 10%, of desulfurization and denitrification catalyst in the last bed of the reactor.

PROCESS CONFIGURATION

Hydrocracker units can be operated in the following possible modes: single-stage (once-through-mode) operation, single-stage operation with partial or total recycling, and two-stage operation. These operation modes are shown in Figures 3-1 and 3-2.

The choice of the process configuration is tied to the catalyst system. The main parameters to be considered are feedstock quality, the product slate and qualities required, and the investment and operating costs of the unit.

SINGLE-STAGE OPERATION

This operating mode has large effect on the product yield and quality. Single-stage operation produces about 0.3 bbl naphtha for every barrel of middle distillate. The single stage scheme is adapted for conversion of

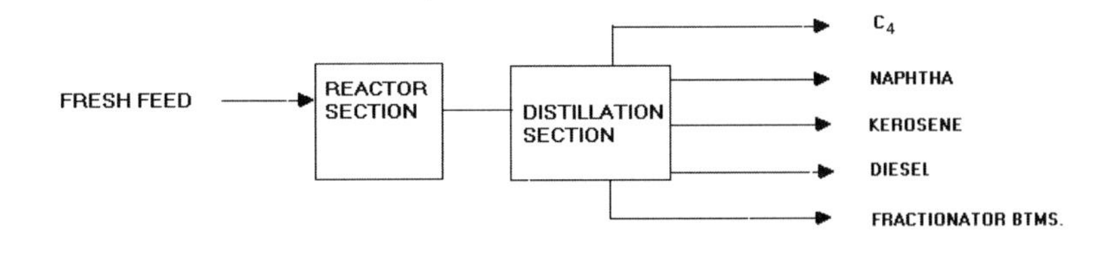

ONCE-THROUGH MODE

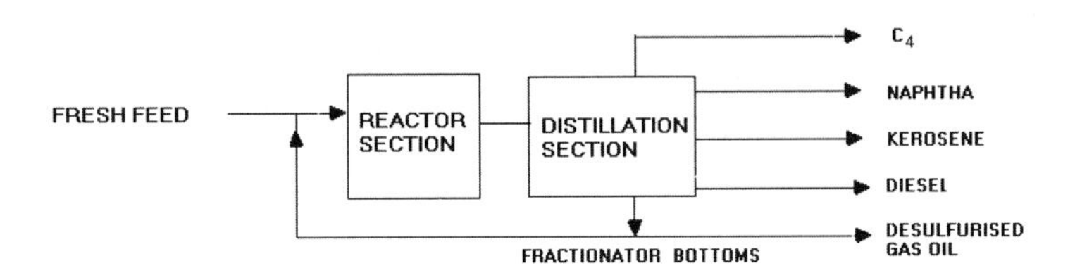

PARTIAL-RECYCLE MODE

Figure 3-1. Hydrocracker operation, once-through and partial-recycle modes.

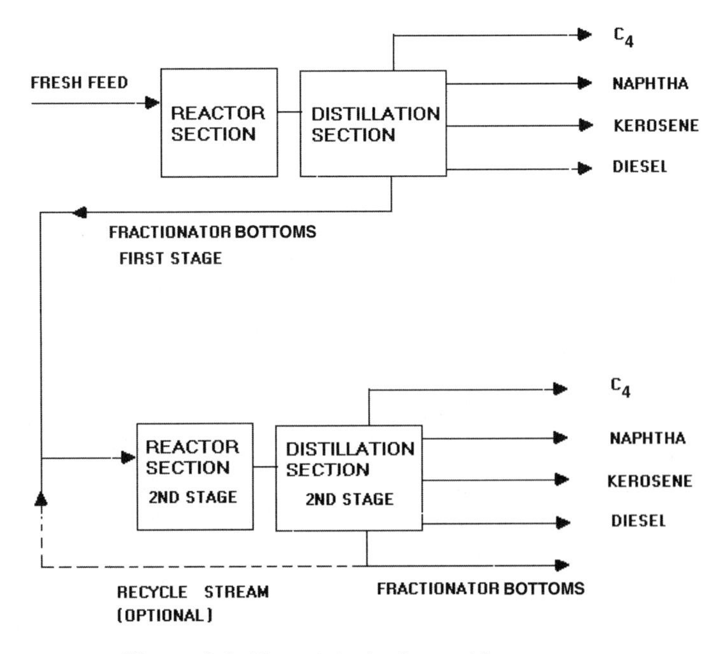

Figure 3-2. Two-stage hydrocracking process.

vacuum gas oils into middle distillate and allows for high selectivity. The conversion is typically around 50–60%. The unconverted material is low in sulfur, nitrogen, and other impurities and is used as either feed for fluid catalytic cracking units (FCCU) or a fuel oil blending component.

The single-stage process may be operated with partial or total recycling of the unconverted material. In total recycling, the yield of naphtha is approximately 0.45 bbl per barrel of middle distillate products. In these cases, the fresh feed capacity of the unit is reduced. Thus, increased conversion is achieved basically at the cost of unit's fresh feed capacity and a marginally increased utility cost. The partial recycling mode is preferable to total recycling to extinction, as the latter results in the buildup of highly refractory material in the feed to the unit, resulting in higher catalyst fouling rates.

TWO-STAGE OPERATION

In the two-stage scheme, the unconverted material from the first stage becomes feed to a second hydrocracker unit. In this case, the feed is already purified by the removal of sulfur, nitrogen, and other impurities; and the second-stage can convert a larger percentage of feed with better product quality.

A heavy gas oil feed contains some very high-boiling aromatic molecules. These are difficult to crack and, in feed recycling operation, tend to concentrate in the recycle itself. High concentration of these molecules increase the catalyst fouling rate. In a two-stage operation, the first stage is a once-through operation; hence, the aromatic molecules get no chance to concentrate, since there is no recycle. The first stage also reduces the concentration of these molecules in the feed to the second stage; therefore, the second stage also sees lower concentration of these high-boiling aromatic molecules.

The two-stage operation produces less light gases and consumes less hydrogen per barrel of feed. Generally, the best product qualities (lowest mercaptans, highest smoke point, and lowest pour point) are produced from the second stage of the two-stage process. The poorest qualities are from the first stage. The combined product from the two stages is similar to that from a single stage with recycling for the same feed quality.

The two-stage scheme allows more flexible adjustment of operating conditions, and the distribution between the naphtha and middle distillate is more flexible. Compared to partial and total recycling schemes, the two-stage scheme requires a higher investment but is overall more economical.

PROCESS FLOW SCHEME

The oil feed to the reactor section consists of two or more streams (see Figure 3-3). One stream is a vacuum gas oil (VGO) feed from the storage tank and the other stream may be the VGO direct from the vacuum distillation unit. Also, there may be an optional recycling stream consisting of unconverted material from fractionator bottom. The combined feed is filtered in filters F-01 to remove most of the particulate matter that could plug the catalyst beds and cause pressure drop problems in the reactor. After the oil has passed through surge drum V-02, it is pumped to the reactor system pressure by feed pump P-01.

Hydrogen-rich recycled gas from the recycling compressor is combined with oil feed upstream of effluent/feed exchangers E-01/02. The oil gas stream than flows through the tube side of exchanger 02A and 02B, where it is heated by exchange with hot reactor effluent. Downstream of the feed effluent exchangers, the mixture is further heated in parallel passes through reactor feed heater H-01. The reactor inlet temperature is controlled by the Temperature Recorder and Controller (TRC) by controlling the burner fuel flow to the furnace.

A portion of the oil feed is by passed around the feed effluent exchanger. This bypass reduces the exchanger duty while maintaining the duty of reactor feed heater H-01 at a level high enough for good control of reactor inlet temperature. For good control, a minimum of 50–75°F temperature rise across the heater is required.

Makeup hydrogen is heated on the tube side of exchanger E-01 by the reactor effluent. This makeup hydrogen then flows to the reactor.

Hydrocracker reactor V-01 is generally a bottle-type reactor. The makeup hydrogen after preheating in exchangers E-01 flows up through the reactor in the annular space between the reactor outside shell and an inside bottle. The hydrogen acts as a purge to prevent H_2S from accumulating in the annular space between the bottle and outside shell. It also insulates the reactor shell.

After the makeup hydrogen has passed upward through the reactor, it combines with the recycled gas and the heated oil feed from the feed heater in the top head of the reactor. The hot, vaporized reaction mixture then passes down the reactor. Cold quenching gas from the recycling compressor is injected to the reactor between the catalyst beds to limit the temperature rise produced by exothermic reactions.

The reactor is divided among a number of unequal catalyst beds. This is done to give approximately the same temperature rise in each catalyst

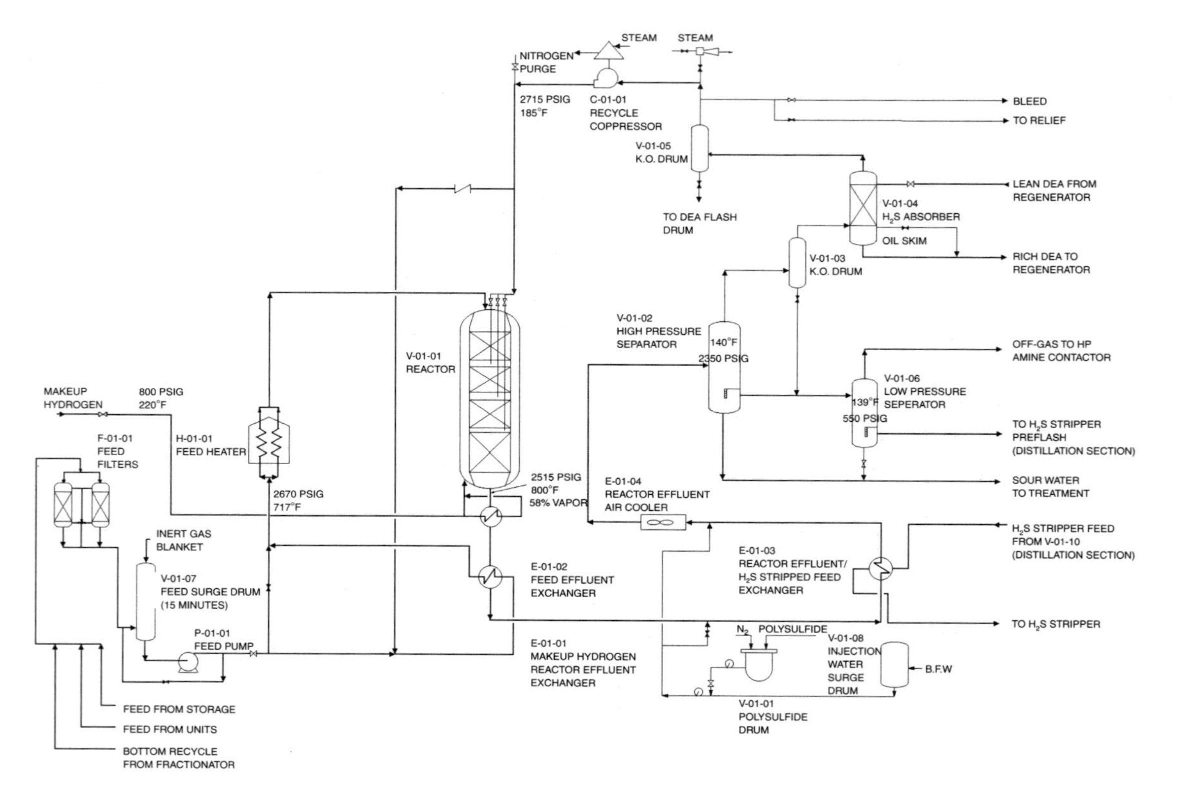

Figure 3-3. Distillate hydrocracker (reactor section). K.O. = knockout.

bed and limit the temperature rise to 50°F. Thus, the first and second beds may contain 10 and 15% of the total catalyst, while the third and fourth beds contain 30 and 45% of the total catalyst.

Reactor internals are provided between the catalyst beds to ensure thorough mixing of the reactants with quench and ensure good distribution of vapor and liquid flowing to each bed. Good distribution of reactants is of utmost importance to prevent hot spots and maximize catalyst life.

Directly under the reactor inlet nozzle is a feed distributor cone inside a screened inlet basket. These internal elements initiate feed stream distribution and catch debris entering the reactor. Below the inlet basket, the feed stream passes through a perforated plate and distributor tray for further distribution before entering the first catalyst bed.

Interbed internal equipment consists of the following:

- A catalyst support grid, which supports the catalyst in the first bed, covered with a wire screen.
- A quench ring, which disperses quenching gas into hot reactants from the bed above.
- A perforated plate for gross distribution of quenched reaction mixture.
- A distribution tray for final distribution of quenched reaction mixture before it enters the next catalyst bed.
- A catalyst drain pipe, which passes through interbed elements and connects each catalyst bed with the one below it.

To unload the catalyst charge, the catalyst from the bottom bed is drained through a catalyst drain nozzle, provided in the bottom head of the reactor. Each bed then drains into the next lower bed through the bed drain pipe, so that nearly all the catalyst charge can be removed with a minimum of effort.

Differential pressure indicators are provided to continuously measure pressure drop across top reactor beds and the entire reactor.

The reactor is provided with thermocouples located to allow observation of catalyst temperature both axially and circumferentially. Thermocouples are located at the top and bottom of each bed. The temperature measured at the same elevation but different circumferential positions in the bed indicate the location and extent of channeling through the beds.

EFFLUENT COOLING

The reactor effluent is at 800–850°F, start of run (SOR) to end of run (EOR), at reactor outlet. The reactor effluent is cooled in makeup hydrogen/reactor effluent exchanger E-01 and reactor effluent/H_2S stripper feed exchanger E-03. The reactor effluent is further cooled in reactor effluent/air cooler E-04 to about 140°F.

WATER AND POLYSULFIDE INJECTION

Condensate is injected into the reactor effluent just upstream of effluent air cooler. The function of the injection water is to remove ammonia and some H_2S from the effluent. The effluent temperature at the injection point is controlled to prevent total vaporization of the injected water and preclude deposition of solid ammonium bisulfide.

Trace amounts of cyanide ion in the reactor effluent contribute to corrosion in the effluent air cooler. A corrosion inhibitor such as sodium polysulfide is also injected to prevent cyanide corrosion.

HIGH-PRESSURE SEPARATOR

The high-pressure separator, V-02, temperature is controlled at approximately 140°F by a temperature controller, which adjusts the pitch of half the fans of the air cooler. The temperature of the separator is closely controlled to keep the downstream H_2S absorber temperature from fluctuating. The hydrogen purity is lower at higher temperatures. However, at lower temperature, poor oil/water separation occur in the separator drum.

LOW-TEMPERATURE SEPARATOR

The high-temperature separator liquid is depressurized through a HP separator level control valve to 550 psig and flashed again to low-pressure separator V-06. The low-pressure separator overhead vapors flow to high-pressure amine contractor V-04. The hydrocarbon stream leaving the separator is fed to H_2S stripper V-11 via a stripper preflash drum. The sour water drawn from the low-pressure separator is sent to sour water treating facilities.

RECYCLE GAS ABSORBER

DEA (diethanolamine) absorption is used to remove H_2S from the recycled gas in recycle gas absorber V-04. H_2S gas is absorbed by the DEA solution because of the chemical reaction of DEA with H_2S. Typical properties of DEA are shown in Table 8-19.

The amount of H_2S that can react depends on the operating conditions. The low temperature, high pressure, and high H_2S concentration in the H_2S absorber favor the reaction. In the DEA regeneration facilities, high temperature and low pressure are used to reverse the reaction and strip H_2S from the DEA solution.

About 90% of the H_2S formed by the desulfurization reactions is removed from the recycled gas in a high-pressure absorber by scrubbing the gas with aqueous diethanolamine solution. The absorber is a vertical vessel packed with stainless steel ballast rings. Recycled gas flows through a support plate and upward through the packing. A lean DEA solution from the DEA regenerator enters the top of the absorber through an inlet distributor and flows downward through the packing. Rich DEA from the bottom of the absorber is sent to the H_2S recovery unit.

RECYCLE GAS COMPRESSOR

Recycled gas is circulated by recycle gas compressor C-01, driven by a steam turbine. The largest portion of the recycled gas stream joins the oil feed stream upstream of feed effluent exchangers. A portion of the gas stream from the recycling compressor flows on temperature control to interbed quenching.

DISTILLATION SECTION

The purpose of distillation section (see Figure 3-4) is to remove H_2S and light ends from the first-stage reactor effluent and fractionate the remaining effluent into naphtha, kerosene, and diesel cuts. The bottom stream is either fed to the second stage of the hydrocracker, recycled to extinction with the fresh feed, or withdrawn as product.

Hydrocarbon liquid flows to H_2S stripper V-11 from stripper preflash drum V-10. The preflash drum removes some of the light ends and H_2S from the low-pressure separator oil before it is stripped and fractionated. The stripper column contains packed sections below the feed plate and two sieve trays above the feed inlet. Stripping is achieved with steam,

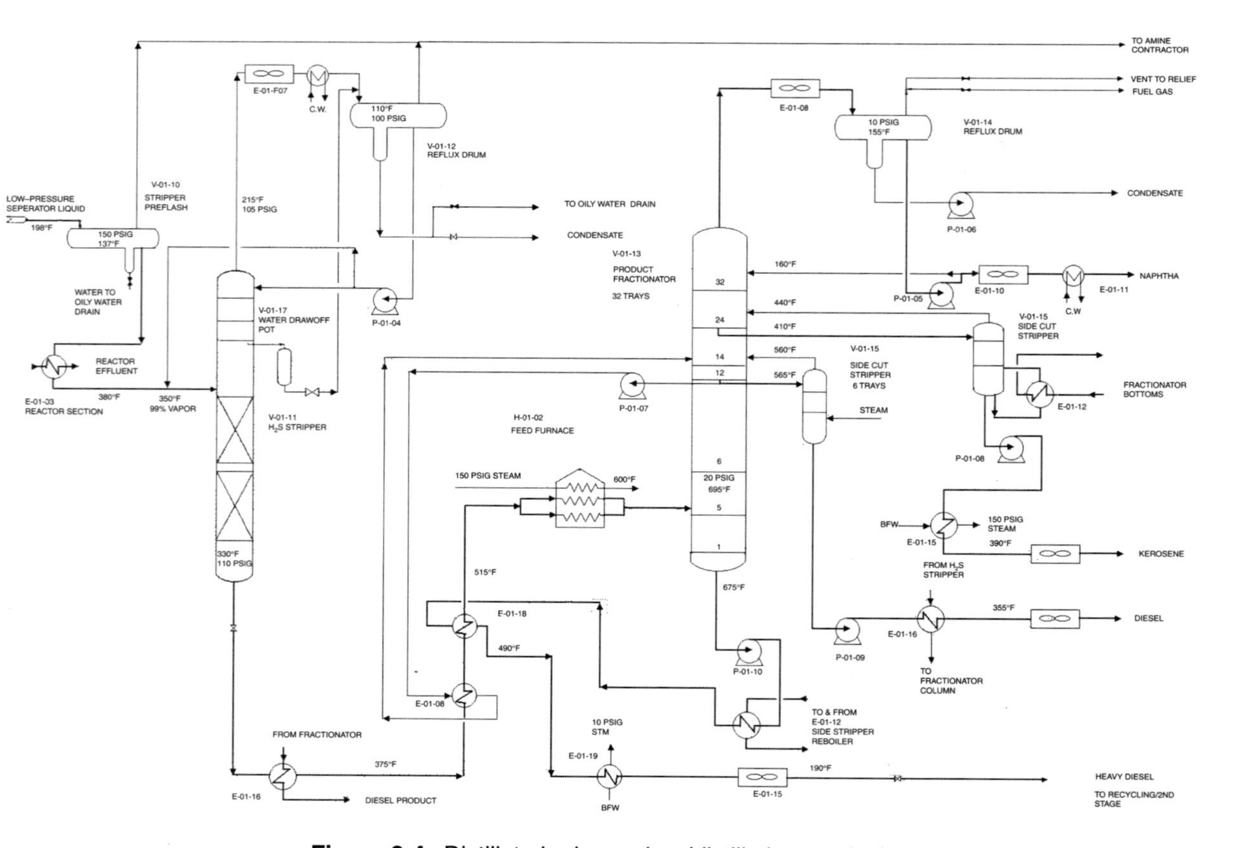

Figure 3-4. Distillate hydrocracker (distillation section).

which removes H_2S and light ends. The stripper overhead vapor is partially condensed in air cooler E-07 and a trim water cooler then flashed in reflux drum V-12. The sour gas from the reflux drum is sent to a low-pressure amine contactor. The condensed hydrocarbon liquid is refluxed back to the stripper. H_2S stripper bottoms are sent to product fractionator V-13 after heat exchange with diesel in E-16, circulating reflux in E-18, and fractionator bottoms in E-18. The fractionator feed is then brought to column temperature by heating in the feed heater H-02. After heating, the partially vaporized fractionator feed is introduced into the flash zone of product fractionator V-12. In the flash zone, the vapor and liquid separate. The vapor passes up through the rectifying section, containing approximately 27 trays.

Heat is removed from the fractionator column in the overhead condenser and in a circulating reflux heat removal system. Vapor leaving the top tray of the column is condensed in overhead condensers. The condensed overhead vapor is separated into hydrocarbon and water phases. Part of the hydrocarbon is recovered as overhead product, and the rest is sent back to the column as reflux to ensure good separation.

The portion of the column below the flash zone contains five trays. Superheated steam is injected below the bottom tray. As the steam passes up through the stripping section, it strips light components from the residual liquid from the flash zone.

OPERATING CONDITIONS

The operating conditions for single-stage hydrocracking are shown in Table 3-1. The yields and qualities for once-through operation, two-stage hydrocracking, and one-stage operation with partial recycling of unconverted material are shown in Tables 3-2 to 3-5. It must be stressed that the yields depend on the catalyst composition and process configuration employed, and these can vary significantly.

TEMPERATURE

The typical hydrocracker reactor operates between 775–825°F and 2600 psig reactor inlet pressure. The high temperatures are necessary for the catalyst to hydrocrack the feed. The high reactor pressure is

Table 3-1
Single-Stage Hydrocracker Operating Conditions

OPERATING PARAMETERS	UNITS	
CATALYST AVERAGE TEMPERATURE	°F	775
SPACE VELOCITY, LHSV	hr^{-1}	1.72
REACTOR INLET PRESSURE	psig	2600
REACTOR PRESSURE DROP	psi	50
HYDROGEN PARTIAL PRESSURE, INLET	psi	2000
HYDROGEN CHEMICAL CONSUMPTION	scf/bbl	1150
MAKEUP + RECYCLE AT REACTOR INLET, SOR	scf/bbl	5000
MAKEUP H_2 PURITY	VOL%	95
HP SEPARATOR TEMPERATURE	°F	140
HP SEPARATOR PRESSURE, SOR	psig	2415
BLEED RATE, SOR (100% H_2)	scf/bbl	200
RECYCLE COMPRESSOR SUCTION PRESSURE	psig	2390
RECYCLE COMPRESSOR DISCHARGE PRESSURE	psig	2715

necessary for catalyst life. Higher hydrogen partial pressure increases catalyst life. To keep the hydrogen partial pressure at a high level, high reactor pressure and high hydrogen content of the reactor feed are necessary. To accomplish this, an excess of hydrogen gas is recycled through the reactor. A makeup hydrogen stream provides hydrogen to replace the hydrogen consumed chemically in hydrocracking, olefin, aromatics saturation, and the hydrogen lost to atmosphere through purge or dissolved in oil. A cold hydrogen-rich gas is injected between the catalyst beds in the reactor to limit the temperature rise caused by exothermic hydrocracking reactions.

In the hydrocracking process, the feed rate, operating pressure, and recycle gas rate are normally held constant. The reactor temperature is the only remaining variable requiring close control to achieve the required liquid feed conversion.

As the catalyst activity declines with time on stream due to catalyst fouling, it becomes necessary to increase the reactor temperature to maintain the original liquid feed conversion rate. This rate of increase of reaction temperature with time is called the *fouling rate*. Additional temperature variation may be required to compensate for changes in reactor feed rate or feed properties, gas/oil ratio, hydrogen partial pressure, and the like (see Figure 3-5).

Table 3-2
Hydrocracker Feed and Product Qualities

PROPERTY	FEED	LIGHT NAPHTHA	HEAVY NAPHTHA	KEROSENE	DIESEL
SINGLE-STAGE OPERATION					
API	22.6	80	53	38	34
ANILINE POINT °F	174		106	124	176
ASTM	D-1160	D-86	D-86	D-86	D-86
IBP/5	590/-	110/-	215/-	350/-	440/-
10/30	715/790	115/125	225/245	375/410	545/590
50	840	135	260	445	610
70/90	885/940	150/170	270/290	475/515	640/685
95/FBP	960/1010	-/195	-/325	-/550	-/725
COMPOSITION, LV%					
PARAFFINS	16	76	30	20	26
NAPHTHENES	22	20	54	55	49
AROMATICS	62	4	16	25	25
SULFUR, ppmw	25000		<10	<10	10
MERCAPTANS, ppm			<5	<5	
NITROGEN, ppmw	970		1	1	5
SMOKE POINT mm				15	
FREEZE POINT °F				−50	
POUR POINT °F	75			−80	20
CLOUD POINT °F				−55	25
DIESEL INDEX				47	60
OCTANE CLEAR		76	61		
VISCOSITY, CST, 100°F	59			2	9

Table 3-2
Continued

PROPERTY	FEED	LIGHT NAPHTHA	HEAVY NAPHTHA	KEROSENE	DIESEL
SINGLE-STAGE WITH RECYCLE					
API	20.9	80	56	42.2	38
ANILINE POINT °F	175		124	148	193
ASTM	D-1160	D-86	D-86	D-86	D-86
IBP/5	740/760	110/-	195/-	330/-	485/-
10/30	775/835	115/125	215/230	350/390	560/585
50	875	135	250	420	600
70/90	915/975	150/170	265/290	460/520	620/655
95/FBP	975/1035	-/195	-/320	-/550	/705
COMPOSITION, LV%					
PARAFFINS		77	45	38	48
NAPHTHENES		21	47	45	43
AROMATICS		2	8	17	9
SULFUR, ppmw	31000	<5	10	6	8
MERCAPTANS, ppm					
NITROGEN, ppmw	1050		1	2	1.5
SMOKE POINT, mm				22	
CETANE NUMBER					68
FREEZE POINT °F				−50	
POUR POINT °F	90			−65	5
CLOUD POINT °F					
DIESEL INDEX					73.2
OCTANE CLEAR		80	54		

TWO-STAGE OPERATION					
API	81	56	43	40	37
ANILINE POINT °F		126	147	169	190
ASTM D-86					
IBP/5	110/-	210/-	345/365	375/400	460/520
10/30	115/125	225/240	375/410	415/475	545/590
50	135	255	446	540	610
70/90	150/170	270/295	475/515	595/655	635/675
95/FBP	-/195	305/325	530/550	675/700	695/720
COMPOSITION, LV%					
PARAFFINS	83	43	35	39	42
NAPHTHENES	15	49	50	46	42
AROMATICS	2	8	15	15	16
SULFUR, ppmw	<5	<5	<5	<5	6
MERCAPTANS, ppm	<5	<5	<5	<5	
NITROGEN, ppmw			<1	2	3
SMOKE POINT, mm			20		
FREEZE POINT °F			−54		
POUR POINT °F			−70	−15	10
CLOUD POINT °F			−65	−10	15
DIESEL INDEX			63	68	70
OCTANE CLEAR	75	54			
VISCOSITY, CST, 100°F			2	4	8

Table 3-3
Distillate Hydrocracker Yields

STREAM	ONCE THROUGH	PARTIAL RECYCLE	TWO-STAGE OVERALL
VGO FEED	1.0000	1.0000	1.0000
HYDROGEN	0.0225	0.0293	0.0345
TOTAL FEED	1.0225	1.0293	0.0345
GASES	0.0166	0.0257	0.0391
HP GAS	0.0070	0.0069	0.0127
ACID GAS	0.0216	0.0243	0.0221
CRACKED NAPHTHA	0.0969	0.1637	0.2132
KEROSENE	0.1372	0.1861	0.2636
DIESEL	0.2961	0.3955	0.4365
HEAVY DIESEL (ONCE-THROUGH MODE)	0.4471		
HEAVY DIESEL (PARTIAL-RECYCLE MODE)		0.2271	
BLEED FROM SECOND STAGE			0.0473
TOTAL PRODUCTS	1.0225	1.0293	1.0345

NOTE: ALL FIGURES ON W/W BASIS. W/W = WEIGHT/WEIGHT BASIS.

Table 3-4
Utility Consumption *(per Ton Feed)*

UTILITY	UNITS	SINGLE STAGE	PARTIAL RECYCLE MODE
FUEL GAS	mmBtu	0.6	0.76
POWER	kWhr	18	23
STEAM	mmBtu	0.12	0.15
DISTILLED WATER	MIG*	0.016	0.02
COOLING WATER	MIG*	0.33	0.42

*MIG = 1000 IMPERIAL GALLONS.

Table 3-5
Mild Hydrocracker Operating Conditions

OPERATING PARAMETERS	UNITS	
CATALYST AVERAGE TEMPERATURE	°F	775
SPACE VELOCITY, LHSV	hr^{-1}	1.4
REACTOR INLET PRESSURE	psig	1051
REACTOR PRESSURE DROP	psi	38
HYDROGEN PARTIAL PRESSURE, INLET	psi	749
HYDROGEN CHEMICAL CONSUMPTION	scf/bbl	358
MAKEUP + RECYCLE AT REACTOR INLET, SOR	scf/bbl	2766
MAKEUP H_2 PURITY	VOL%	92
HP SEPARATOR TEMPERATURE	°F	140
HP SEPARATOR PRESSURE, SOR	psig	850
BLEED RATE, SOR (100% H_2)	scf/bbl	10.5
RECYCLE COMPRESSOR SUCTION PRESSURE	psig	820
RECYCLE COMPRESSOR DISCHARGE PRESSURE	psig	1070
LEAN DEA TEMPERATURE	°F	150

CATALYST AVERAGE TEMPERATURE

The catalyst average temperature (CAT) is determined by the following equation:

$$\mathrm{CAT} = \left[A_1 \times \left[\frac{(T_I + T_O)}{2} \right] \text{top bed} + A_2 \times \left[\frac{(T_I + T_O)}{2} \right] \text{second bed} \right.$$

$$\left. + A_3 \times \left[\frac{(T_I + T_O)}{2} \right] \text{third bed} + A_4 \times \left[\frac{(T_I + T_O)}{2} \right] \text{fourth bed} + \cdots \right]$$

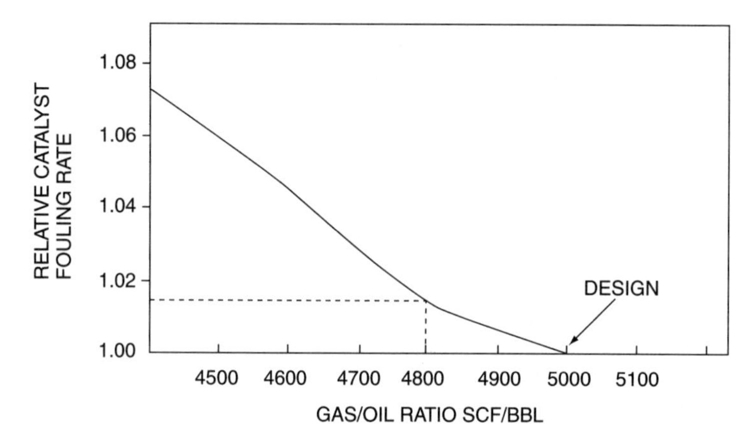

Figure 3-5. The effect of the gas/oil ratio on the catalyst fouling rate.

where T_1 is the bed inlet temperature; T_0 is the bed outlet temperature; and A_1, A_2, A_3, and A_4 are the volume fractions of the total reactor catalyst in the individual bed. A typical hydrocracker reactor temperature profile is shown in Figure 3-6.

CATALYST FOULING RATE

The design of a hydrocracker unit is based on a specified conversion rate of the feed and a specified catalyst life, usually 2–3 years between catalyst regeneration. During the course of the run, the activity of the catalyst declines due to coke and metal deposits, and to maintain the design conversion rate, the temperature of the catalyst has to be increased. The catalyst manufacturers specify a maximum temperature, called the *end of run* temperature, which signifies the EOR condition. When this temperature is reached, the catalyst must be regenerated or discarded.

The rate of increase in average reactor catalyst temperature (to maintain the design conversion rate) with time is called the *catalyst fouling rate*. It is an important parameter, used to make an estimate of time when the EOR conditions are likely to be reached. The refinery keeps a record of the reactor average temperature with time, starting from the day the feed is introduced into the reactor after the new catalyst has been loaded. A graph is drawn between the time on stream vs. the average reactor temperature (Figure 3-7). The data may show scatter, so a straight line is drawn through the data. From this curve, an estimate of the catalyst fouling rate and remaining life of the catalyst can be estimated.

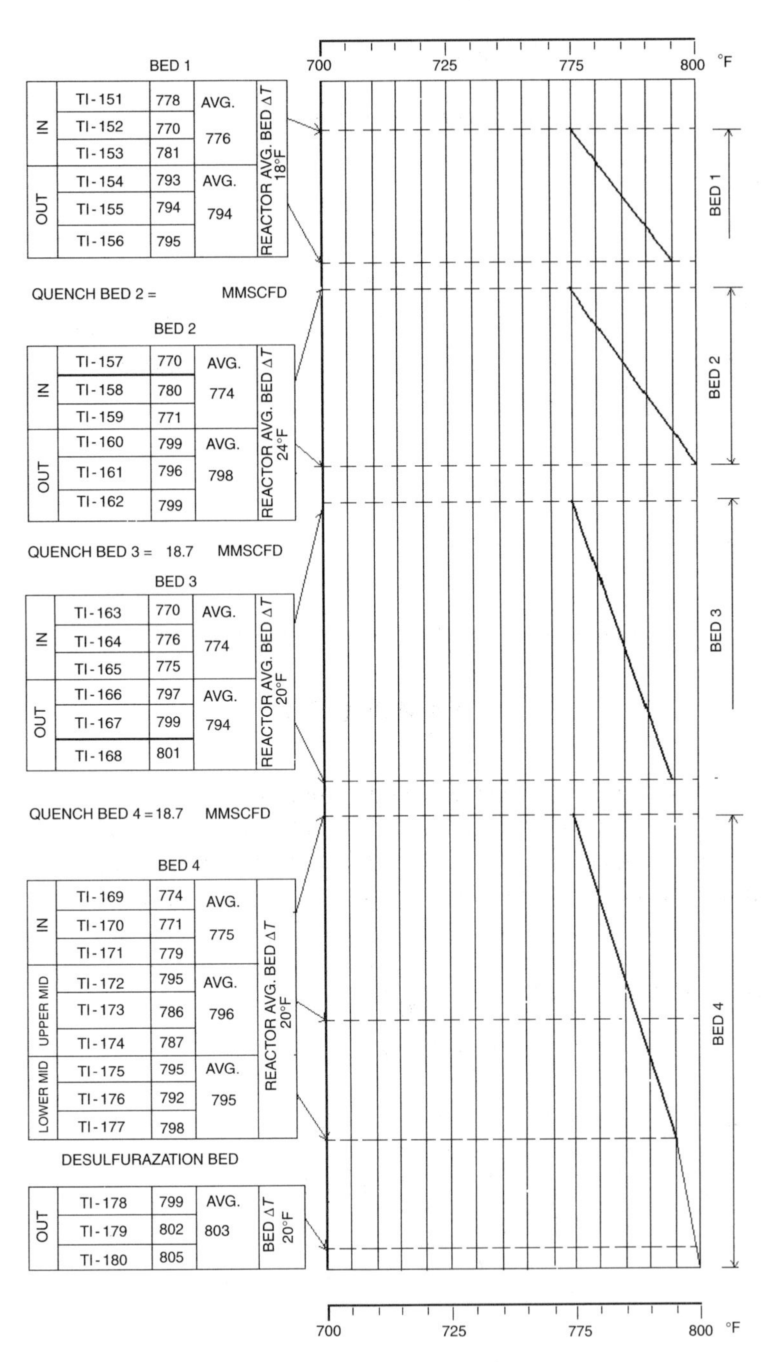

Figure 3-6. Hydrocracker reactor temperature profile.

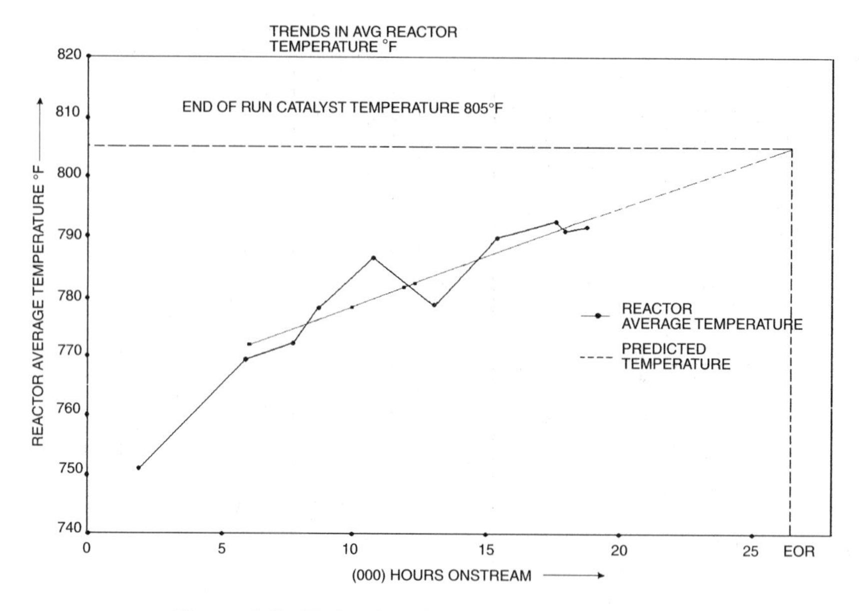

Figure 3-7. Estimating the remaining catalyst life.

EXAMPLE 3-1

The design start of run (SOR) temperature of a hydrocracker reactor is 775°F. After the unit is onstream for 12,000 hours, the CAT (weighted average bed temperature) is 800°F. Estimate the remaining life of the catalyst if the design EOR temperature of the catalyst is 805°F:

$$\text{Catalyst fouling rate} = \frac{(800 - 775)}{12{,}000}$$
$$= 0.00208°\text{F/hr}$$

$$\text{Remaining life} = \frac{(805 - 800)}{0.00208}$$
$$= 2404 \text{ hr or } 3.33 \text{ months}$$

HYDROGEN PARTIAL PRESSURE

The factors affecting hydrogen partial pressure in the hydrocracker reactor are

1. Total system pressure.
2. Makeup hydrogen purity.
3. Recycle gas rate.
4. HP gas bleed rate.
5. HP separator temperature.

Hydrogen partial pressure in the hydrocracker reactor is a basic operating parameter that provides the driving force for hydrocracking reactions. Also, the hydrogen partial pressure has an important effect on the catalyst fouling rate (Figure 3-8). An increase in the hydrogen partial pressure serve to suppress the catalyst fouling rate. In an operating unit, the hydrogen partial pressure is maximized to operate the unit at the lowest possible temperature. This increases the run length and minimize light ends production.

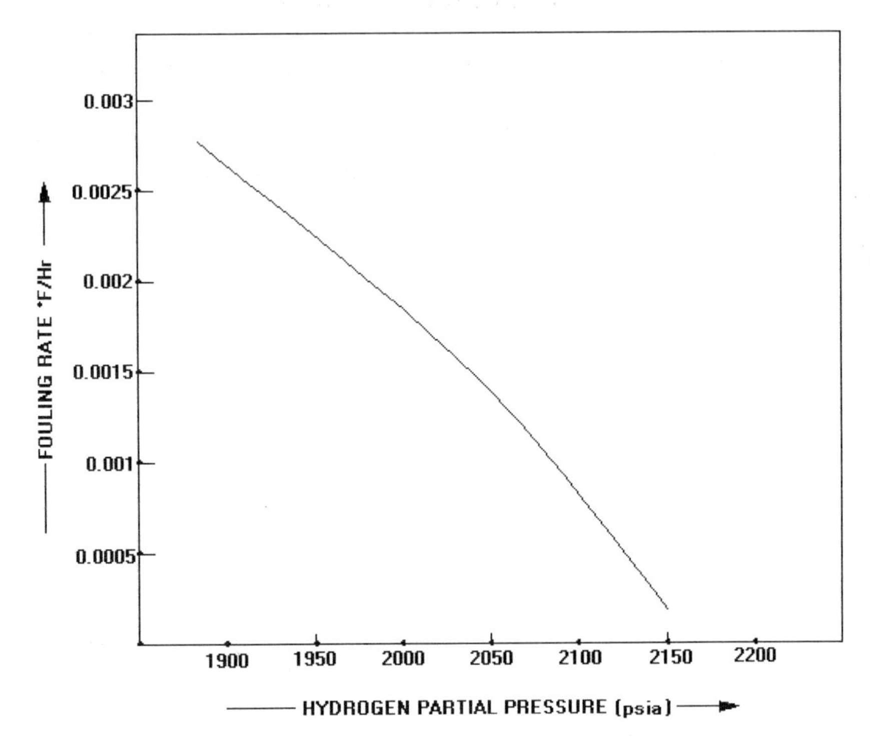

Figure 3-8. Effect of H_2 partial pressure on catalyst fouling rate (schematic).

FEED RATE

Increasing the feed rate requires an increase in average catalyst temperature to maintain the required feed rate conversion. The increased feed rate also causes an increase in the catalyst fouling rate and the hydrogen consumed chemically and dissolved in the high-pressure separator liquid. The effect of the feed rate on the catalyst fouling rate is shown in Figure 3-9.

FEED CHARACTERIZATION

A heavier feed, as characterized by the ASTM D1160 weighted boiling point, requires an increase in average catalyst temperature to maintain the desired level of feed conversion. An increase in the catalyst fouling rate also occurs. Further, a feed having a higher end point for a given weighted boiling point requires an increased temperature for desired conversion over that required by a lower-end point feed.

LIQUID FEED CONVERSION

The following equation shows how to determine liquid feed conversion:

$$\text{Conversion rate} = \frac{(\text{reactor liquid feed rate} - \text{fractionator bottom rate})}{\text{reactor liquid feed rate}}$$

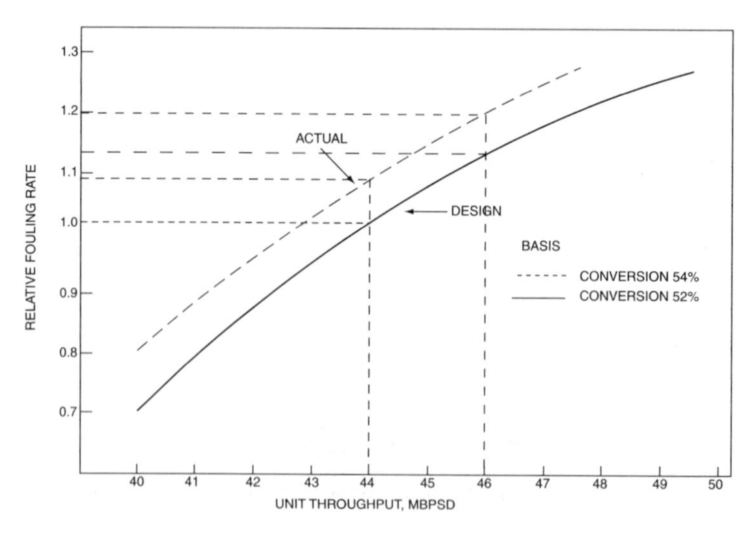

Figure 3-9. The effect of the feed rate on the catalyst fouling rate (schematic).

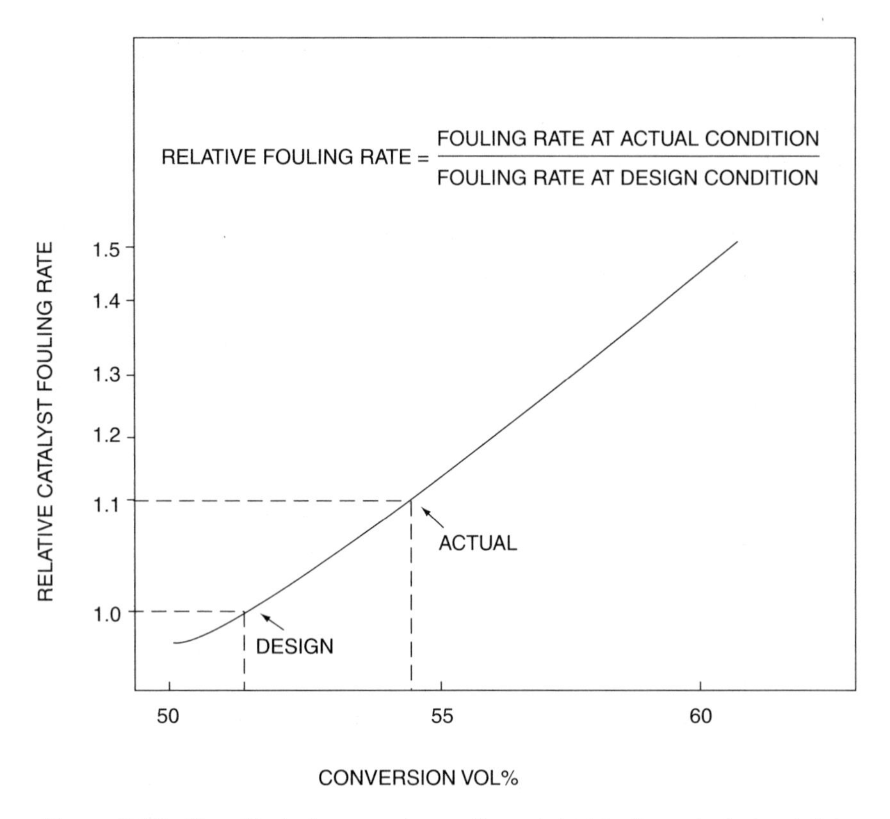

Figure 3-10. The effect of conversion on the catalyst fouling rate (schematic).

The liquid feed conversion depends strongly on temperature dependent and, as such, has a direct effect on the catalyst fouling rate (Figure 3-10).

GAS BLEED RATE

Bleeding minimizes the buildup of light hydrocarbons in the recycling gas, which lowers hydrogen partial pressure. Increasing the gas bleed rate lowers the light hydrocarbon concentration in the recycling gas and increases the hydrogen partial pressure. Decreasing the bleed rate allows light hydrocarbons in the recycle gas to build up to higher concentrations and thus lower hydrogen concentration. The bleed rate is typically kept at about 20% of the hydrogen chemical consumption.

CATALYST SULFIDING AND UNIT STARTUP

The catalyst is sulfided during startup. Before initiating the sulfiding, the distillation and the amine absorber are put into operation.

Sulfiding is the injection of a sulfur-containing chemical such as dimethyl sulfide into the hot circulating gas stream prior to the introduction of liquid feed. The chemical injection is performed with a chemical injection pump from a storage drum under an inert gas blanket. The reactor is first evacuated to 24–26 in. mercury to remove all air, tested for leaks, and purged with nitrogen a number of times.

The furnace is next purged and fired, and the reactor is heated to 450°F. When the reactor is ready for sulfiding, the reactor inlet temperature is increased to 500–525°F. When the catalyst temperature reaches 500–525° at the top bed, this temperature is held until the temperature is at least 500° at the inlet and about 450° at reactor outlet.

The feed heater fire is reduced. With recycle compressor operating, the reactor pressure is adjusted to about 300 psi. At 300 psig, the hydrogen is introduced slowly, through the makeup compressor and pressure increased to 700 psig at reactor inlet.

Sulfiding chemical or sour gas injection is then added into gas stream at the reactor inlet, at a rate equivalent to 0.5 mol% but no more than 1.0 mol% H_2S.

Addition of a sulfiding agent causes two temperature rises in the reactor, first due to reaction of hydrogen with the sulfiding chemical to form H_2S, which occurs at the reactor inlet, and second due to reaction of the sulfiding chemical with the catalyst, and this moves down through the catalyst bed. The catalyst temperature is closely monitored during sulfiding, and it is not allowed to exceed 600°F.

After the sulfiding temperature rise has passed through the reactor, the H_2S concentration of gases out of the reactor begin to rise rapidly. When 0.1% H_2S is detected in the effluent gas, the recycle gas bleed is stopped and sulfiding continued at a low injection rate until the concentration of H_2S in the circulating gas is 1–2 mol%. At the same time, the reactor inlet temperature is increased to 560° and the reactor outlet temperature to at least 535°F.

When no significant reaction is apparent in the reactor, the system pressure is increased by adding makeup hydrogen until a normal operating pressure is reached at the suction of the recycling compressor. The sulfiding medium is added batchwise to maintain the H_2S concentration of 1–2 mol% in the recycle gas.

When the reactor system has remained steady at the design pressure, at reactor inlet/outlet temperature of 560/530° and H_2S concentration of 1–2 mol%, for 2 hours, sulfiding is complete. The reactor is slowly cooled by reducing inlet temperature and adding quench between the catalyst beds until reactor temperature reaches about 425°F. No point in the reactor is allowed to cool below 425°F.

The startup temperature is next approached from the lower side by increasing the reactor inlet temperature to 450–475° and outlet temperature to 425–450°, the quench temperature controller is set at 450°F. The recycle gas rate is set at the design rate. Also the HP separator pressure is set at the design value. Condensate and polysulfide injection at the design rate is started.

The reactor is now ready for introduction of the feed. The feed pump starts pumping only 20% of the design feed rate to the reactor, shunting the rest of the feed back to the feed tank.

Heat is released due to adsorption of hydrocarbon when the feed passes over the catalyst for the first time. This shows up as a temperature wave that passes down through the catalyst. When the temperature wave due to hydrocarbon adsorption has passed through the reactor and a liquid level is established in the high pressure separator, the liquid is sent to a low-pressure separator and the distillation section.

As the reactor temperature and feed rate is increased in the steps that follow, amine circulation is established through the recycling gas absorber.

The reactor feed rate is increased by 10–15% of the design at a time, allowing the system to line out for at least 1 hour after each feed rate increase. The process is repeated until the feed is at 50% of the design rate.

Next, the temperature is increased in about 20°F increments or less at the inlet to each reactor bed. When increasing reactor temperature, the top bed should be adjusted first then each succeeding lower bed. The system is allowed to line out for at least 1 hour after each reactor temperature increase until all reactor temperatures are steady. Then, the reactor temperature is increased again until the desired conversion is reached.

When the desired operation (conversion and feed rate) has been reached and the unit is fully lined out, it is monitored so that recycle gas rate is steady and at the design rate. The H_2S absorber bypass is slowly closed, forcing all the recycling gas through the absorber. While closing this bypass, close attention is given to knockout drum level, absorber-packed bed ΔP, and absorber level to avoid compressor shutdown due to a sudden carryover of amine solution.

NORMAL SHUTDOWN

The following procedures are typical when a run is ended to replace spent catalyst or perform general maintenance. The same procedure is followed when a run is ended for catalyst regeneration, except that the reactor is not opened.

Care must be taken to avoid furnace and catalyst coking during shutdown and the formation of highly toxic nickel carbonyl when the reactor is cooled, the possibility of fires due to explosive hydrogen oxygen mixtures or exposure of pyrophoric material to air when the reactor is opened, and exposure of personnel to toxic or noxious conditions when the catalyst is drained or equipment is entered.

SHUTDOWN PROCEDURE

The following presents a general procedure to be followed for a normal shutdown.

1. Gradually reduce the liquid feed and adjust the catalyst temperature downward, so that feed conversion remains at the desired level. When a very low feed rate is reached, stop all liquid feed to the reactor but continue to circulate recycle gas rate at the normal rate. As the liquid feed rate is cut, the reactor feed and effluent rates will be out of balance for short periods. The feed must be gradually reduced to prevent temperatures greater than the maximum design temperature occurring in the effluent/feed heat exchangers.
2. To strip off as much of adsorbed hydrocarbon as possible from the catalyst, raise the reactor temperature to the normal operating temperature after the feed is stopped. Circulate hot hydrogen until no more liquid hydrocarbon appears in the high-pressure separator. If the shutdown is for only brief maintenance, which does not require stopping the recycle compressor, continue circulation at the normal operating temperature and pressure. After the maintenance work is complete and unit is ready for feed, lower the reactor temperature and introduce the feed following the normal startup procedure.
3. If the shutdown is for catalyst regeneration, catalyst replacement, or maintenance that requires stopping the recycle compressor or depressurizing the reactor system, continue circulation at the normal operating pressure for 2 hours. Reduce the furnace firing

rate and start gradually reducing the reactor temperature to 400°F. Any cooldown design restrictions for the reactor must be adhered to, to avoid thermal shock.

4. Add quench gas as required to evenly cool the reactor.
5. To maintain heater duty high enough for good control and speed up cooling after heater fires are put out, bypass reactor feed gas around the feed/effluent exchanger, as required.
6. While cooling, remove the bulk of hydrocarbon oil from the high-pressure separator by raising the water level and pressure to the low-pressure separator. Also, increase the pressure of any low-pressure liquid remaining in the H_2S absorber to the amine section. Purge and block the liquid hydrocarbon line from the high-pressure separator. Block in the H_2S absorber and drain the amine lines. Drain all vessels and the low points of all lines to remove hydrocarbon inventory.
7. If the shutdown is for catalyst regeneration, hold the reactor at 400°F.
8. If the shutdown is for catalyst replacement or maintenance that requires opening the reactor, determine the CO content of the recycle gas. If the CO content is below 30 ppm CO, continue the cooling procedure; if more than 30 ppm, proceed as follows:

   a. If more than 30 ppm CO has been detected in the recycle gas and the reactor is to be opened, the CO must be purged before cooling can continue to eliminate possible formation of metal carbonyls. Stop recycling gas circulation while the catalyst temperature is still above 400°F. Do not cool any portion of the bed below 400°F. Before depressurizing the system, check that all these valves are closed:

      - Suction and discharge valves on charge pump and spares.
      - Block lines on liquid feed lines.
      - Block valves on the high-pressure separator, liquid product and water lines.
      - Recycle and makeup compressor suction and discharge lines.
      - Chemical injection and water lines.
      - Makeup hydrogen line.
      - H_2S absorber.

   b. Depressure the system. Do not evacuate below atmospheric pressure, as there is the danger of explosion if air is drawn into the system. For effective purging of the reactor system, the nitrogen gas is injected at the recycle compressor discharge and follows the normal flow path through the catalyst beds.

c. Pressurize the system with dry nitrogen to 150 psig. Purge the compressor with nitrogen.
d. Depressurize the system to 5 psig and purge with nitrogen for 5 minutes.
e. Repressurize the system to 150 psig and purge all blocked lines as mentioned previously.
f. Pressurize the system with nitrogen from nitrogen header and recycle compressor.

9. Relight the furnace fires and maintain the reactor at a temperature of 400°F for at least 2 hours. Then analyze the recycling gas for CO content. If the CO content is still above 30 ppm, repeat the preceding procedure.
10. If the CO content is below 30 ppm, the reactor can be cooled below 400°F following the cooling rates restrictions for the hydrocracker vessel metallurgy. Stop the furnace fires and continue cooling by circulating nitrogen until the temperature has fallen to 120°F.
11. Stop nitrogen circulation and ensure that the system is blocked in preparation to depressurizing. Depressurize and purge the system again with nitrogen. Maintain the system under slight nitrogen pressure, so that no oxygen is admitted.

CATALYST REGENERATION

The activity of the catalyst declines with time on stream. To recover the lost activity, the catalyst is regenerated in place at infrequent intervals.

In the regeneration process, the coke and sulfur deposits on the catalyst are burned off in a controlled manner with a dilute stream of oxygen. The burning or oxidation consists of three burns at successively higher temperatures. This is followed by a controlled reduction of some of the compounds formed during the oxidation step, in a dilute hydrogen stream. The catalyst is next stabilized by sulfiding. During the combustion phase, sulfur oxides are liberated and some metal sulfates are formed as the sulfided catalyst is converted to the oxide form. During the reduction phase, the residual metal sulfates are reduced to the metal sulfides with liberation of sulfur dioxide.

During both the combustion and the reduction phases, a dilute caustic quenching solution is mixed with reactor effluent to cool the gases through the H_2O-SO_3 dew point and to neutralize the SO_2/SO_3 present. Cooling the gases in this manner prevents corrosion of heat exchangers and other downstream equipment. Also, sulfur oxides must be neutralized to prevent damage to the catalyst.

During regeneration, maximum use is made of equipment used in the normal operation. The recycle compressor is used to recirculate the inert gas. Compressed diluted makeup air is mixed with recycled inert gas flowing to the reactor inlet to provide oxygen to sustain a burn wave in the top catalyst beds. The reactor inert gas is preheated in the feed effluent exchanger then combined with diluted makeup air and heated in the feed furnace before flowing to the reactor inlet.

The reactor/effluent gases are cooled in the feed effluent exchanger and mixed with recirculating dilute caustic quench solution before being cooled in the effluent/air cooler. The cooled mixture then flows to the high-pressure separator to separate out the vapor and liquid phases. A portion of the inert gas from the separator is bled to the atmosphere and the reminder recirculated. The dilute caustic quench solution then flows to the low-pressure separator. This allows the use of a lean DEA pump for quencher solution circulation. Some of the dilute caustic quench solution is bled to limit the concentration of dissolved solids. Fresh caustic and process water are added to maintain the quench solution at a proper pH.

The regeneration is carried out at the pressure that allows the maximum flow rate as limited by the maximum power of the recycle compressor or the discharge temperature of the makeup compressor (1150 psig). The makeup air is diluted with the recycling inert gas to avoid forming a combustible mixture in the lubricated makeup compressor. With hydrogen present the maximum O_2 concentration is 4 mol% maximum at a compressor outlet temperature of 265°F.

After the regeneration burn has started and the recycling gas stream is free of hydrogen, the O_2 concentration through the makeup compressor is raised to 7%.

The concentration of O_2 in the first burn is kept at 0.5%. The initial burn is started at an inlet temperature of 650°F or less. This produces a temperature rise of 100°F. The burn is concluded when the last bed temperature starts to drop off and oxygen appears in the recycle gas.

After the first burn is completed, the reactor inlet temperature is increased to 750°F and air is introduced to obtain 0.5 vol% oxygen at the reactor inlet. The oxygen may not be totally consumed as it passes through the reactor during this burn and the oxygen content of recycling gas at the reactor outlet may remain at fairly constant level, even though oxygen is being consumed in the reactor. This burn normally produces only a small temperature rise and sometimes no temperature rise is apparent.

For the third or final burn, the temperature at reactor inlet is raised to 850°F ± 25°F and the oxygen content of the inlet gas is slowly increased to 2% by volume. This condition is maintained at least for 6 hours or until the oxygen consumption is essentially nil, as indicated by a drop in inlet oxygen content of less than 1% for 2 hours with makeup air stopped. No point in the reactor should be allowed to exceed 875°F. The reactor temperature should never be allowed to exceed 900°F. Throughout the regeneration, a circulating dilute caustic solution is injected into reactor effluent gases.

REDUCTION PHASE

After the hydrocarbon, coke, and sulfur have been burned from the catalyst, a reduction step is necessary before introducing the sulfiding and feed. During oxidation, a portion of sulfur on the catalyst is converted into sulfates. These sulfates may be reduced when exposed to high-pressure hydrogen, and the reduction can occur at temperatures below those required for sulfiding. A large heat release accompanies these reactions, which can result in loss of catalyst activity or structural damage to the catalyst and reactor.

The oxidized catalyst is reduced in a very dilute hydrogen atmosphere so that the heat of reduction is released gradually under controlled conditions.

The reactor is pressurized with nitrogen while heating to 650°F. Hydrogen is then introduced to obtain a concentration of 1–2 mol% at the reactor inlet. This causes a controlled reduction reaction to move through the reactor, as indicated by a temperature rise of 25–40°F per percent hydrogen.

If, after 1 or 2 hours, the reduction is proceeding smoothly, slowly increase concentration of hydrogen to 2–3 mol% at the reactor inlet. If the temperature rise exceeds 100°F or the reactor temperature at the reduction wave exceeds 750°F, stop increasing the hydrogen content. Do not exceed 3% hydrogen at the reactor inlet or allow the temperature to exceed 750°F at any point in the reactor during this procedure.

The reduction reaction is controlled by limiting the quantity of hydrogen available. The catalyst temperature is held below 750°F during the reduction step to avoid damage to the catalyst.

When hydrogen breaks through the bottom of the reactor, reduce or stop hydrogen addition. Hold the reactor temperature at 650° and 2–3% hydrogen at the reactor inlet for at least 8 hours and until hydrogen consumption has dropped to a low rate. When the wave has passed through the reactor and no more hydrogen is consumed, the reactor is cooled to about 400°F.

Some sulfur oxides are generated by the reduction reactions and consequently a dilute quench is circulated as during the oxidation phase to neutralize the quenched effluent gases. However, since the quantity of sulfur oxides is relatively small compared to that liberated during the carbon/sulfur burn, an ammonia quench may be used instead of a caustic one. This arrangement uses equipment used during normal operation rather than the quench circulation, piping, and mixing header. It is important to note that a catalytic reformer hydrogen is unsuitable for reduction and the first phase of subsequent sulfiding. Accordingly, the procedure requires manufactured or electrolytic hydrogen.

MILD HYDROCRACKING

Mild hydrocracking, as the name suggests, operates at much lower pressure and much milder other operating conditions than the normal hydrocracking process. The objective of the process is basically to desulfurize the VGO to make it suitable for FCCU feed. Other impurities, like nitrogen, are also removed and about 30% of the feed is converted into saleable diesel. Compared to normal hydrocracker units, mild hydrocrackers require much less initial investment. The operating conditions and yield from a mild hydrocracker unit are shown in Tables 3-5 and 3-6, respectively.

RESIDUUM HYDROCRACKING

Resid hydrocracking is designed to convert straight-run residual stocks and atmospheric or vacuum resids into distillates by reacting them with hydrogen in the presence of a catalyst. The process operates under severe operating conditions of high temperature and pressure, comparable to

Table 3-6
Mild Hydrocracker Yields

VGO FEED	1.0000
HYDROGEN	0.0061
TOTAL FEED	1.0061
GASES	0.0191
CRACKED NAPHTHA	0.0134
KEROSENE	0.0110
DIESEL	0.3117
HEAVY DIESEL	0.6509
TOTAL PRODUCTS	1.0061

NOTES: ALL FIGURES ON A W/W BASIS.
FEED SG = 0.9169, 2.7% SULFUR
PRODUCT.
DIESEL: 0.07% S, 43 DI.
HEAVY DIESEL: 0.12% S.

Table 3-7
Resid Hydrocracking Catalyst Characteristics

PROPERTY	UNITS	
CATALYST COMPOSITION		NI-MO ON ALUMINA BASE
SHAPE		EXTRUDATES 0.963 mm × 3.93 mm
BULK DENSITY	lb/ft^3	45
CATALYST DENSITY, WITH PORES	lb/ft^3	105
CATALYST DENSITY, EXCLUDING PORES	lb/ft^3	186

distillate hydrocracker units. A part of the resid is converted into distillates. Also, the resid is partially desulfurized and demetallized. The distillates produced are separated in a distillation column into naphtha, kerosene, and light diesel. The heavy products from the reaction section and fractionation tower are separated in a vacuum distillation section into heavy diesel, heavy vacuum gas oil, and vacuum resid (see Tables 3-7 and 3-8). The major reactions that occur are shown in Figure 3-11.

RESID HYDROCRACKER REACTOR

The resid hydrocracking reactions are conducted in an ebullated or fluidized bed reactor to overcome the problems associated with a fixed

Table 3-8
Resid Hydrocracking Spent Catalyst Composition

COMPONENT	UNITS	WT%
NICKEL	Wt%	2.1
MOLYBDENUM	Wt%	3.73
COBALT	Wt%	0.006
CARBON	Wt%	12.18
SULFUR	Wt%	11.9
VANADIUM	Wt%	7.1

bed (see Figure 3-12). A liquid phase passes upward through a bed of catalyst at a velocity sufficient to maintain the catalyst particles in continuous random motion. This liquid velocity is achieved by circulating a liquid recycling stream by means of an ebullating pump external to the reactor.

An ebullating bed system offers the following advantages over the conventional fixed-bed system:

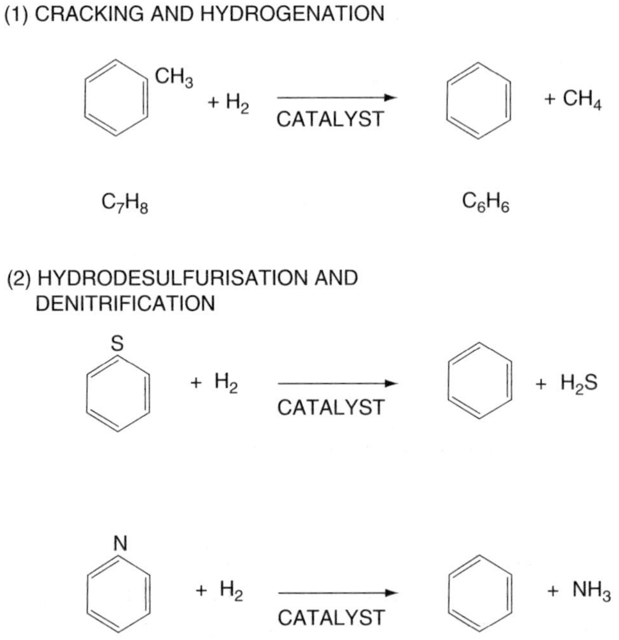

(1) CRACKING AND HYDROGENATION

C_7H_8 + H_2 —CATALYST→ C_6H_6 + CH_4

(2) HYDRODESULFURISATION AND DENITRIFICATION

S + H_2 —CATALYST→ + H_2S

N + H_2 —CATALYST→ + NH_3

Figure 3-11. Resid hydrocracking reactions.

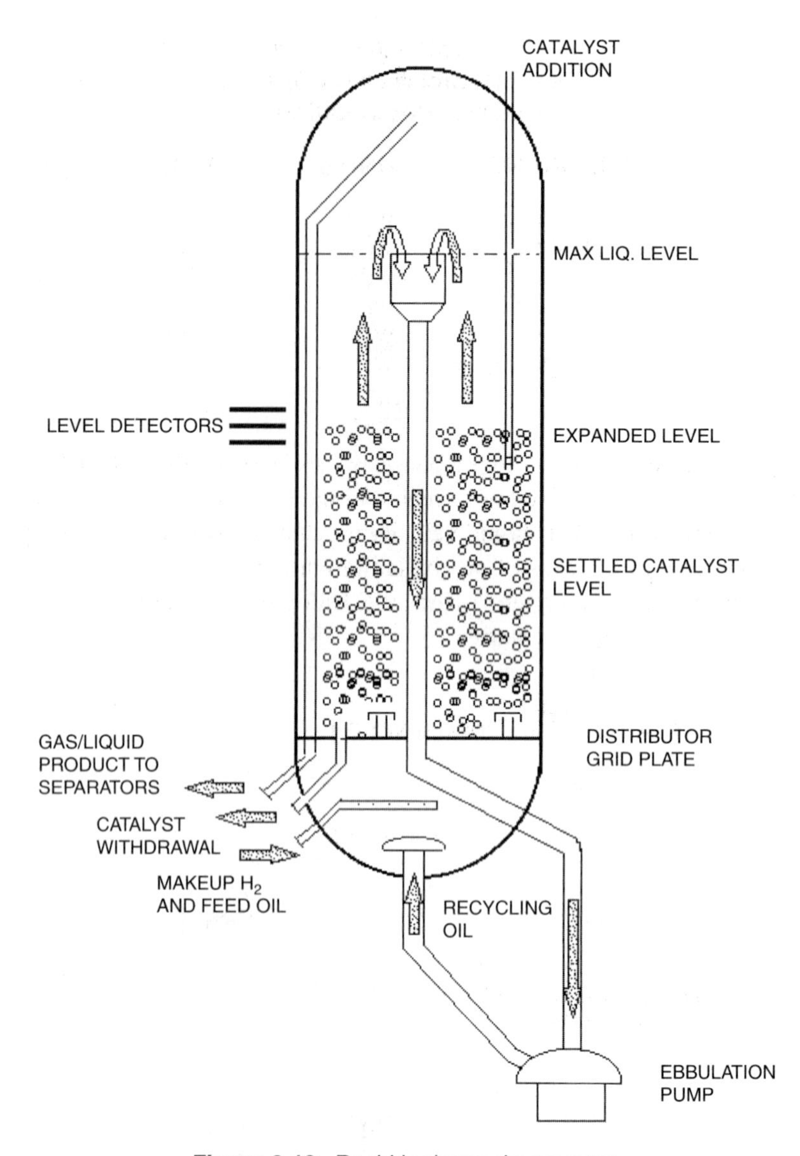

Figure 3-12. Resid hydrocracker reactor.

1. *Isothermal reactor conditions.* The mixed conditions of this reactor provide excellent temperature control of the highly exothermic reactions without the need for any quench system. Undesirable temperature sensitive reactions are controlled.

2. *Constant pressure drop.* Since the catalyst is in a state of constant random motion, there is no tendency for pressure drop to build up as a result of foreign material accumulation.
3. *Catalyst addition and withdrawal.* The catalyst can be added or withdrawn from an ebullating bed on either a continuously or intermittent basis. This feature permits operation at an equilibrium activity level, thereby avoiding change in yield and product quality with time encountered in fixed-bed reactors due to aging of the catalyst.

Within the reactor, the feed enters the lower head of the reactor through a sparger to provide adequate distribution of the reactant stream. The ebullating stream is distributed through the lower portion of the reactor by an individual sparger. These spargers effect a primary distribution of the feed stream across the reactor cross-sectional area. The feed then passes through a specially designed distributor plate, which further ensures uniform distribution as the vapor and liquid flow upward through the catalyst bed.

The oil and hydrogen dissolved in liquid phase under relatively high hydrogen partial pressure react with each other when brought in intimate contact with active catalyst above the distribution plate. The primary reactions taking place are hydrocracking, hydrogenation, hydrodesulfurization, and denitrification. In addition, the organometallic compounds in the feed are broken down under high temperature and high hydrogen partial pressure and are, in part, adsorbed on the catalyst, the remainder passing through the catalyst bed, ultimately ending up in fuel oil. The metal buildup on the catalyst would result in complete deactivation of the catalyst. Therefore, the activity level of the catalyst is maintained by addition of fresh catalyst and withdrawal of spent catalyst in a programmed manner.

The temperature and the catalyst activity level control the conversion level. The other variables such as hydrogen partial pressure, circulating gas rate, reactor space velocity, and ebullating rate are unchanged in an operating unit.

The reactor average temperature is varied by the amount of preheating performed on the total oil and gas streams that pass through separate fired heaters. Normally, the oil heater outlet is maintained at a moderate temperature to minimize skin cracking of the oil, and the adjustment of the reactor temperature is done primarily by increasing the preheating of the hydrogen feed gas to the maximum temperature possible within the heater design limitation.

The ebullating oil flow is controlled by varying the speed of the ebullating pump. Within the reactor, the ebullating liquid is drawn into a conical collecting pan, located several feet above the catalyst bed interface to ensure catalyst-free liquid circulation down the internal stand pipe and into external ebullating pump. The ebullating liquid is distributed through the reactor bottom by its individual sparger.

The height of the fluidized bed of the catalyst in the reactor is related to the gas flow rate, the liquid flow rate, and the physical properties of the fluid, which in turn are affected by the operating temperature and pressure, conversion, size, density, and shape of the catalyst particles. All these parameters are maintained within the constraints imposed by these correlations.

RESID HYDROCRACKER UNITS

The feed to resid hydrocracker is crude unit vacuum bottoms. Fresh feed is mixed with a heavy vacuum gas oil diluent. The combined feed enters surge drum V-33, is pumped to feed heater H-01 then mixed with high-pressure recycled hydrogen gas, is preheated in H-02 in mixers, then fed to the reactor V-01 at about 750°F and 2450 psig (see Figure 3-13).

Ebullation pump P-02 maintains circulation to keep catalyst particles in suspension. The catalyst bed level is monitored by a radioactive source. Fresh catalyst is added to the reactor and spent catalyst is withdrawn from the reactor periodically to maintain the catalyst level.

The reactor effluent flows as a vapor/liquid mixture into vapor/liquid separator V-02. The vapor from separator drum is a hydrogen-rich stream containing equilibrium quantities of hydrocarbon reaction products. The liquid is composed essentially of heavier hydrocarbon products from resid hydrocracking reactions and unconverted feed. The hot flash liquid contains an appreciable amount of dissolved hydrogen and light ends. The flash vapor leaving V-02 is cooled with light distillate in exchanger E-02. The cooled vapor effluent is then fed to primary distillate knockout drum V-03 for removal of condensed liquid. Vapor from V-03 is further cooled to about 140°F in air cooler E-03 and then enters distillate separator drum V-04. On leaving separator V-04, the hydrogen-rich gas is split into two streams. The larger stream is compressed to approximately 2700 psig in a recycle gas compressor and returned to the reactor V-01 as recycled gas. The smaller purge gas stream is withdrawn as purge to maintain the purity of the recycled gas.

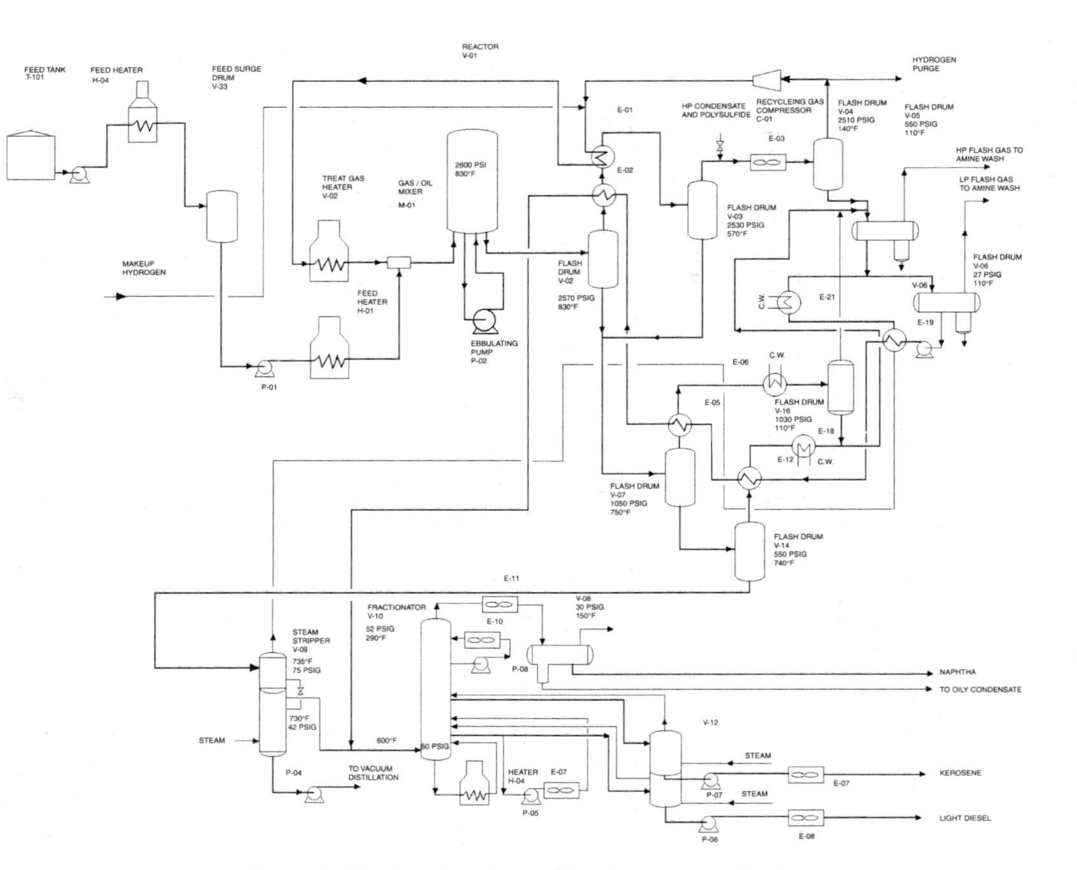

Figure 3-13. Resid hydrocracking (reactor section).

Condensed liquid from V-03 joins with reactor effluent liquid from V-02, then the combined stream is flashed to remove hydrogen in three successive stages.

The first stage let down occurs at 1050 psig in V-07. The resulting hydrogen-rich vapor is cooled in exchanger E-05 then cooled with cooling water to approximately 110°F in E-06. This cooled stream is flashed again in flash gas drum V-16. The vapor from V-16 is let down to a pressure of about 550 psig and joins the high-pressure flash gas from V-05. The liquid from V-16 is further let down to a pressure of 550 psig in high-pressure flash drum V-05.

Ammonical condensate is injected upstream of E-03 and E-06 to dissolve ammonium bisulfide deposits. The ammonical water from V-04 and V-05 are returned to the ammonical water treatment plant to strip away H_2S and NH_3.

The second stage of let down of the combined liquid stream to about 550 psig occurs in high-pressure, high-temperature flash drum V-14. Vapor from this flash drum is cooled with light distillate in exchanger E-12 and cooling water in E-18. It is then fed to flash drum V-05, in addition to high-pressure condensate from V-04 and V-16. High-pressure flash gas from V-05 and gas let down from V-16 are sent to an amine wash.

The final let down of the combined reactor liquid stream occurs at 75 psig in steam stripper V-09. The resulting vapor is cooled through light distillate exchanger E-19 to 300°F and then through a water cooler to the ambient temperature before it is fed to low-pressure flash drum V-06 at about 30 psig, along with condensate from V-05. The low-pressure flash gas from V-06 is sent to the amine wash unit.

The light distillate from V-06 is pumped by P-03 through exchangers E-19 and E-12, E-05, and E-02, preheated to approximately 600°F, and fed to the fractionator V-10.

The hot flashed reactor liquid from V-14 flows into stripper V-09, where it is steam stripped to remove middle distillates. The stripped overhead vapors are fed to fractionator tower V-10 along with preheated light distillate from V-06.

The fractionator tower feed consists of light and middle distillates from V-06 and stripped gases from V-09. The fractionator produces two sidestreams and overhead vapor and liquid streams. A light diesel cut is fed to diesel side stripper V-13 for removal of light ends by steam stripping. The raw light diesel stream is pumped by P-06 and cooled in air cooler E-07/8 before sending it to storage.

A kerosene cut is withdrawn to stripper V-12 for removal of light components by steam stripping. The raw kerosene stream is sent by P-07, cooled in air cooler E-07, and sent to storage.

A pumparound sidestream is circulated near the top of the fractionator to provide internal reflux to upper section with P-08 and air cooler E-10.

A lower pumparound is taken at light diesel drawoff by pump P-05, passing through air cooler E-07 to the fractionator to provide internal reflux in the lower section of the tower.

The fractionator overhead vapor is cooled and partially condensed in air cooler E-11 before going to naphtha accumulator drum V-08. Vapor from this drum is sent to an amine wash. The unstabilized liquid naphtha from V-08 is sent to storage by pump P-07.

The separated water withdrawn from V-08 is sent to the oily water condensate system.

PRODUCT FRACTIONATION

The fractionator bottom product is pumped by P-16 to a vacuum feed surge drum, then the charge is heated to a temperature of about 780°F before it enters the vacuum tower (see Figure 3-14). The flash zone is at 700°F and 26 inch vacuum maintained by steam ejectors.

The feed is fractionated into heavy diesel, heavy vacuum gas oil, and vacuum bottoms.

The heavy diesel is withdrawn as side cut, cooled in air cooler E-52, and sent to storage. The heavy vacuum gas oil (HVGO) side cut is cooled by raising 150 psig steam then in air cooler E-54. A part of HVGO is used as diluent in the resid hydrocracker feed, and the rest is used in fuel oil blending.

The vacuum bottoms from the unit are pumped by P-53 to off-site storage tanks and used in fuel oil blending.

The typical operating conditions of resid hydrocracker units are shown in Table 3-9, and yields and product qualities are shown in Tables 3-10 to 3-12.

The liquid hourly space velocity is calculated based on the flow volume of the liquid feed at the reactor operating conditions divided by the volume of the catalyst up to the fluidized bed height. Only the fresh feed flow is considered, since the net recycle flow is zero. Also, the volume of the makeup and recycled hydrogen is neglected.

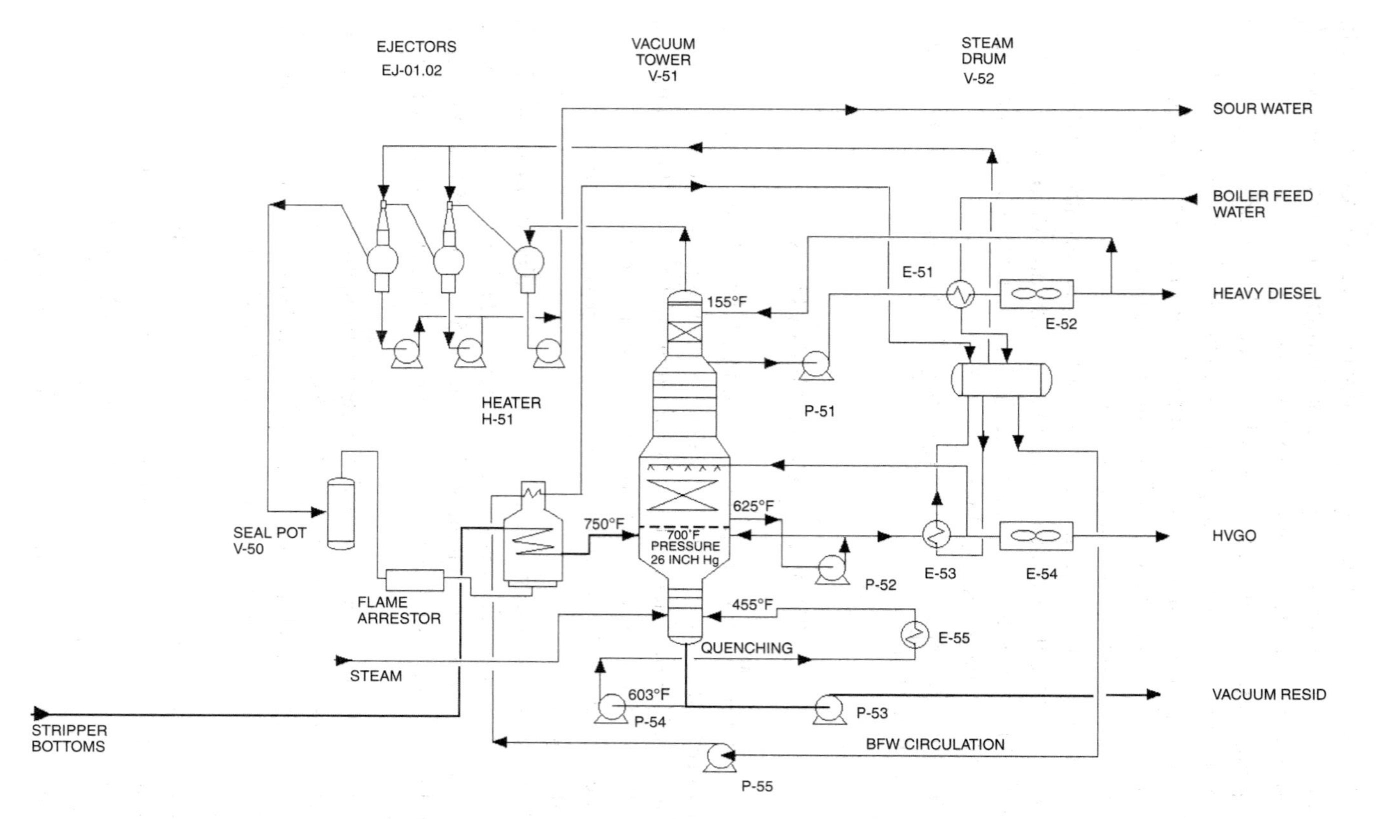

Figure 3-14. Residuum hydrocracking (distillation section). BFW = boiler feedwater.

Table 3-9
Resid Hydrocracker Operating Conditions

OPERATING VARIABLE	UNIT	
REACTOR AVERAGE TEMP	°F	827
REACTOR SYSTEM PRESSURE	psig	2400
HYDROGEN PARTIAL PRESSURE	psig	1895
MAKEUP HYDROGEN	scf/bbl	1400
RECYCLING GAS FLOW	scf/bbl	4300
BLEED GAS FLOW	scf/bbl	700
HYDROGEN CONSUMPTION	scf/bbl	660
LIQUID HOURLY SPACE VELOCITY	(hr^{-1})	0.906
HP FLASH GAS PRESSURE	psig	500
LP FLASH GAS PRESSURE	psig	43
FEED/H_2 MIX TEMPERATURE	°F	748
PLENUM CHAMBER TEMPERATURE	°F	811
CATALYST ADDITION RATE	lb/bbl	0.106

EXAMPLE 3-2

Calculate the LHSV for a resid hydrocracker reactor with the following operating conditions:

$$\text{Resid feed rate} = 24{,}150\,\text{bpsd}$$
$$\text{Diluent in feed} = 3000\,\text{bpsd}$$
$$\text{Total reactor feed} = 27{,}150\,\text{bpsd}, 60^\circ\text{F}$$
$$\text{Reactor cross section} = 133\,\text{ft}^2$$
$$\text{Reactor temperature} = 850^\circ\text{F}$$
$$\text{Fluidized bed height} = 47\,\text{ft}$$
$$\text{Specific gravity of oil at } 60^\circ\text{F} = 0.900$$
$$\text{Specific gravity at } 850^\circ\text{F} = 0.553$$
$$\text{Makeup } H_2 \text{ flow} = 33.7\,\text{mmscf}$$
$$\text{Recycled } H_2 \text{ flow} = 69.5\,\text{mmscf}$$

Neglecting the volume of the makeup and recycled hydrogen gas, the superficial liquid linear velocity in the reactor is due to total temperature

Table 3-10
Resid Hydrocracker Feed and Product Qualities

PROPERTY	UNITS	FEED	NAPHTHA	KEROSENE	LIGHT DIESEL	HEAVY DIESEL	VGO	VACUUM BOTTOM
API GRAVITY		7.5	64.5	40.9	30.9	24.8	18.9	6.6
DENSITY		1.018	0.7219	0.8208	0.8713	0.9065	0.9408	1.0245
ASPHALTENES	Wt%	6.85						7.3
CON CARBON	Wt%	17.2						20.8
BROMINE NUMBER			30	22	20			
DISTILLATION, ASTM	°F							
IBP		725	102	318	365	440	615	800
5%		870	144	360	468	540		880
10%		930	158	372	520	575	740	925
20%		975	184	380	548	600	760	—
50%			244	422	592	710	815	
70%			278	452	622	765	850	
90%			310	490	688	785	905	
95%			322	504	720	800	945	
EP			344	522	744	820	970	
METALS	ppmw							
Ni		32						30
Na		19						11
V		89						70
NITROGEN	ppmw	2575	155	729	1295	1875	2000	4120
SULFUR	Wt%	4.82	0.19	0.43	0.97	1.6	2	3.59
VISCOSITY	Cst, 210°F	775						363

Table 3-11
Resid Hydrocracker Yields

STREAM	YIELD WT FRACTION
FEED	0.8840
DILUENT	0.1160
HYDROGEN	0.0360
TOTAL	1.0360
PRODUCTS	
GASES	0.0626
H_2S	0.0200
NAPHTHA	0.0600
KEROSENE	0.0913
LIGHT DIESEL	0.0655
HEAVY DIESEL	0.0797
HEAVY GAS OIL	0.1833
RESIDUE	0.4647
LOSSES	0.0089
TOTAL	1.0360

corrected volume of fresh oil plus that of recycled oil. Since the net recycle flow is zero, this can be neglected.

$$\begin{aligned}\text{Flow rate in the reactor} &= 27{,}150 \times 0.90 \times 5.6146/(24 \times 0.553) \\ &= 10{,}337\ \text{ft}^3/\text{hr} \\ \text{Reactor velocity} &= 10{,}337/(133 \times 3600) \\ &= 0.0215\ \text{ft/sec}\end{aligned}$$

$$\begin{aligned}\text{Catalyst volume up to fluidization level} &= 47 \times 133 \\ &= 6251\ \text{ft}^3 \\ \text{Liquid hourly space velocity} &= 6251/10{,}337 \\ &= 0.6047\end{aligned}$$

From the data in Table 3-10, we see that virtually all products from the unit require some further treatment before blending. Naphtha, kerosene, and diesels have high bromine numbers and high nitrogen content, requiring further hydrofinishing treatment to make them suitable for blending. Similarly, the heavy diesel cut requires further hydrotreating to reduce its nitrogen content and make it suitable for FCCU or hydrocracker feed.

Table 3-12
Utilities Consumption *(per Ton Feed)*

UTILITY	UNITS	CONSUMPTION
ELECTRICITY	kWh	5.70
FUEL	mmBtu	1.06
STEAM	mmBtu	0.20
COOLED WATER	MIG*	1.56
CATALYST	LBS	0.71

*MIG = 1000 IMPERIAL GALLONS.

CHAPTER FOUR

Gasoline Manufacturing Processes

CATALYTIC REFORMING

Catalytic reforming of heavy naphtha is a key process in the production of gasoline. The major components of petroleum naphthas are paraffins, naphthenes, and aromatic hydrocarbons. The relative amount of these hydrocarbons depend on the origin of the crude oil. The aromatic content of the reforming feed is usually below 20% of the total hydrocarbons whereas the paraffins and naphthenes vary between 10 and 70% depending on the origin of the crude oil.

The aim of catalytic reforming is to transform, as much as possible, hydrocarbons with low octane to hydrocarbons with high octane. The chemical reactions that lead to these changes are guided by a catalyst under well-defined operating conditions.

From the octane view point, the best hydrocarbon fuels for an internal combustion engine are isoparaffinic and aromatic hydrocarbons. For example, the aromatic hydrocarbons from C_7 to C_{10} have research octane numbers (RON) of 118 to 171, whereas the corresponding cyclohexanes have octane numbers of 43 to 104. A similar comparison can be made between isoparaffins and normal paraffins.

The chemical reactions of catalytic reforming are grouped according to the respective hydrocarbon type.

REFORMING REACTIONS

Dehydrogenation

The naphthenic hydrocarbons are dehydrogenated to form aromatics (see Figure 4-1). The reaction is extremely fast, and the yields obtained

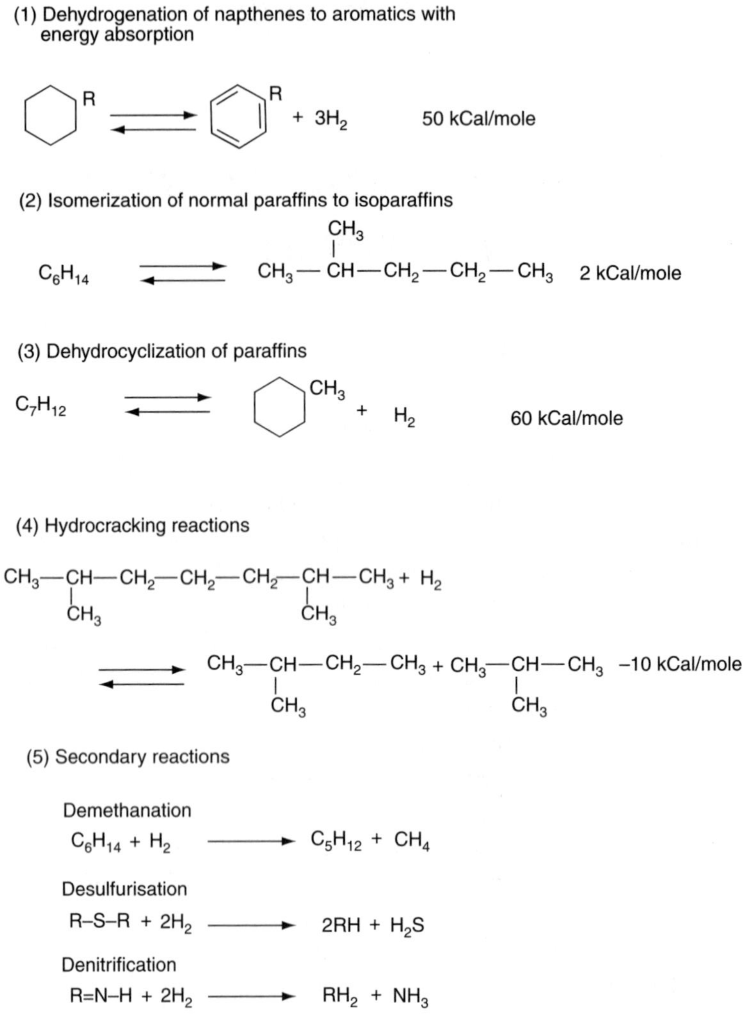

(1) Dehydrogenation of napthenes to aromatics with energy absorption

$$\text{C}_6\text{H}_{11}\text{R (cyclohexane-R)} \rightleftharpoons \text{C}_6\text{H}_5\text{R (benzene-R)} + 3H_2 \qquad 50 \text{ kCal/mole}$$

(2) Isomerization of normal paraffins to isoparaffins

$$C_6H_{14} \rightleftharpoons CH_3-CH(CH_3)-CH_2-CH_2-CH_3 \qquad 2 \text{ kCal/mole}$$

(3) Dehydrocyclization of paraffins

$$C_7H_{12} \rightleftharpoons \text{C}_6\text{H}_{11}\text{CH}_3 \text{ (methylcyclohexane)} + H_2 \qquad 60 \text{ kCal/mole}$$

(4) Hydrocracking reactions

$$CH_3-CH(CH_3)-CH_2-CH_2-CH_2-CH(CH_3)-CH_3 + H_2$$

$$\rightleftharpoons CH_3-CH(CH_3)-CH_2-CH_3 + CH_3-CH(CH_3)-CH_3 \qquad -10 \text{ kCal/mole}$$

(5) Secondary reactions

Demethanation

$$C_6H_{14} + H_2 \longrightarrow C_5H_{12} + CH_4$$

Desulfurisation

$$R\text{–}S\text{–}R + 2H_2 \longrightarrow 2RH + H_2S$$

Denitrification

$$R{=}N\text{–}H + 2H_2 \longrightarrow RH_2 + NH_3$$

Figure 4-1. Reforming reactions.

are almost those predicted by thermodynamics. Also, the reaction is endothermic, ΔH 50 kCal/mole. Dehydrogenation reactions are very important, because they increase the octane number and the reactions produce hydrogen. The only disadvantage is their endothermicity. Due to the large heat absorption, the feed has to be reheated several times, requiring a number of furnaces and reactors.

Isomerisation

Isomerisation of paraffins is also a fast reaction. The reaction is almost thermoneutral, ΔH 2 kCal/mole. This reaction has a negligible effect on the final octane number.

Dehydrocyclization

The dehydrocyclization of paraffins is the key reaction for producing high-octane gasoline. It is highly endothermic, ΔH 60 kCal/mole. The yield of this reaction is limited by kinetics. The reaction rate is much slower than the naphthene dehydrogenation. Its contribution to increasing the octane number is extremely important because the change of a paraffin mixture to corresponding aromatics lead to increase in octane from 60 to 80. A lower rate of this reaction leads to more severe operating conditions and increase in coke formation.

Hydrocracking

Hydrocracking reactions (see Figure 4-1) are important in the reforming reactors. Unlike other reforming reactions, hydrocracking is exothermic with a heat release of 10 kCal/mole. Compared to hydrodecyclization, the reaction rate is small at low temperature and conversion rates. However, the rate increases with higher temperatures and increases in aromatic content when it becomes an important competitor of hydrodecyclization reactions. The reaction products appear in the reformate and in the gases. The presence of light components C_4 and C_5 gives important volatility properties to reformate. Also, hydrocracking decreases the liquid yield and increases the aromatic content due to a concentration effect.

FEED QUALITY

Typically, the feed to a cat reformer unit for gasoline production is a heavy straight-run naphtha with an initial boiling point (IBP) of 194°F and final boiling point (FBP) of 284°F. Benzene is an undesirable component in the gasoline because of environmental pollution concerns. It is therefore important to minimize or exclude any benzene precursors in the cat reformer feed by keeping the feed IBP higher than 180°F. The cat reformer feed is hydrotreated in a naphtha hydrotreater unit to remove any sulfur, nitrogen, and other impurities which can poison the reforming catalyst.

CATALYST

The catalyst used for cat reforming consists of a high-purity alumina base impregnated with platinum and metallic activators. The platinum content is approximately 0.35% by weight. The catalyst is generally in spheres of 2 mm diameter. In semi-regenerative-type units, the life cycle is about 1 year, after which catalyst is regenerated by burning off carbon.

SEMI-REGENERATIVE REFORMING UNIT

A simplified process flow diagram of a semi-regenerative reforming unit is shown in Figure 4-2. The unit consists of three reactors, containing a reforming catalyst, fired heaters, a hydrogen recycling system with a gas drier and product debutanizer.

Hydrotreated naphtha charge from the naphtha unifiner is combined with a hydrogen-rich recycling stream and passes through a low-pressure drop feed effluent heat exchanger, heater, then the first reactor in the series at about 900°F. As the reactions in the reactors are endothermic due to dehydrogenation of naphthenes to aromatics and dehydrocyclization of paraffins to aromatic carbons, the outlet temperature of the reactor effluent drops, and this stream is reheated and enters the second reactor. Similarly, the effluent of the second reactor is again heated before it enters the third reactor.

The reactor effluent from the last reactor is cooled in the feed reactor effluent heat exchanger, followed by air and water coolers. It is then separated into a liquid product and a hydrogen-rich gas. A part of the separated gas is recycled. The rest of hydrogen-rich gas is bled off to maintain system pressure.

The liquid product from the separator drum is stabilized in the debutanizer column, where lighter components such as C_1, C_2, C_3, and C_4 gases are removed as overhead product and used as refinery fuel. The condensed liquid from debutanizer overhead drum is returned to the column as reflux. The debutanizer bottom product, after heat exchange with debutanizer feed, is sent to reformate storage tanks.

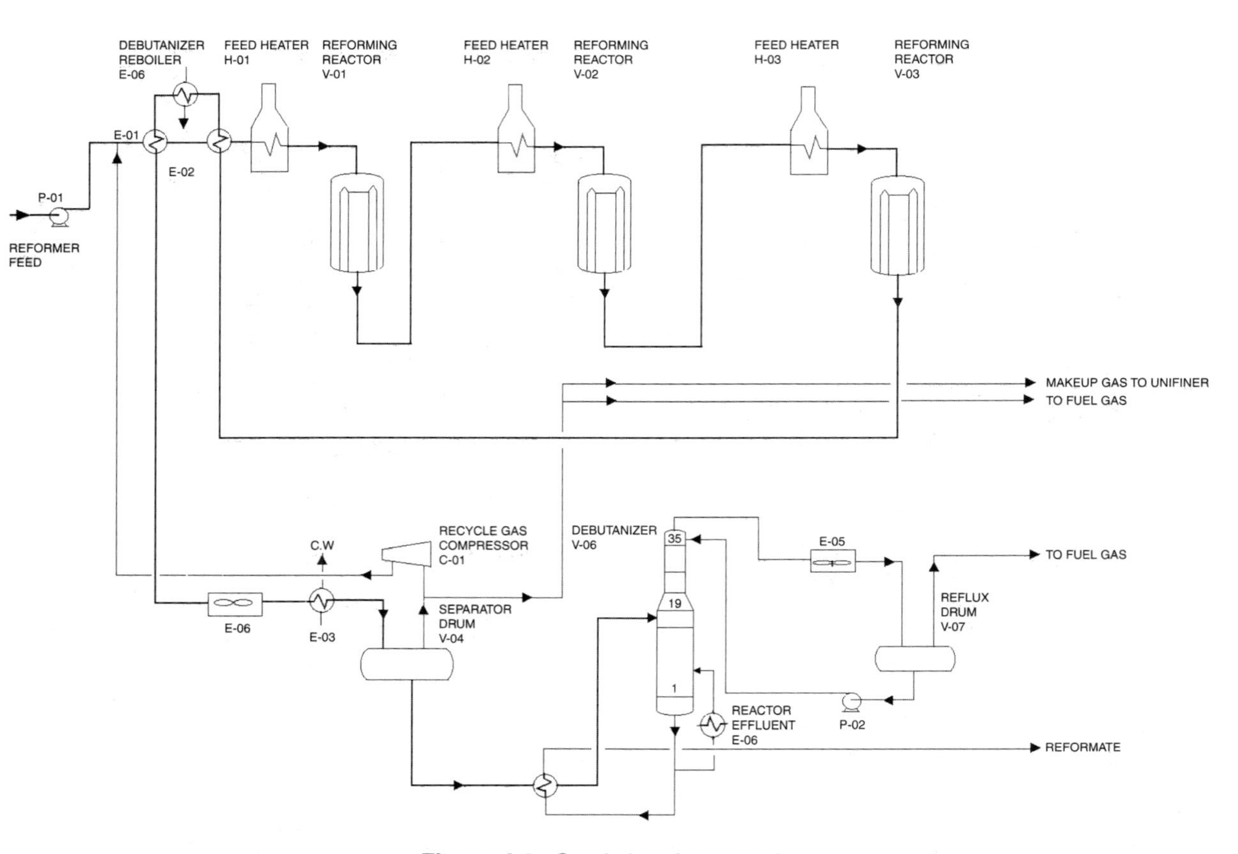

Figure 4-2. Catalytic reformer unit.

CONTINUOUS REGENERATION

Catalyst reforming technology is available in which fresh catalyst is added to the first reactor and moves through all the reactors. A part of the catalyst, about 5%, is continuously withdrawn and regenerated batchwise in a fixed-bed regenerator in the same manner as in semi-regenerative procedures. The regenerated catalyst is returned to the first reactor.

The advantage of the scheme is that the reforming operation can be conducted at higher temperatures and lower pressures, resulting in greater reformate yield, even at high severities. The activity of the catalyst remains constant throughout the run. No downtime for periodic regeneration of the catalyst is required as for semi-regenerative reformer units.

OPERATING CONDITIONS AND YIELDS

The typical operating conditions of a semi-regenerative cat reformer unit are shown in Table 4-1. The yields from the unit and properties of the reformate are shown in Tables 4-2 to 4-5.

FLUID CATALYTIC CRACKING

Fluid catalytic cracking (FCC) is an effective refinery process for conversion of heavy gas oils into gasoline blend components. Cracking

Table 4-1
Catalytic Reformer Operating Conditions

OPERATING VARIABLE	UNITS	
REACTOR WEIGHTED AVERAGE		
INLET TEMPERATURE, SOR	°F	934
INLET TEMPERATURE, EOR	°F	1013
SEPARATOR PRESSURE	psig	185.00
SEPARATOR TEMPERATURE	°F	130.00
RECYCLE RATIO		
MOLES H_2/MOLES FEED		4.5
SPACE VELOCITY (WHSV)		2.75

WHSV = LIQUID HOURLY SPACE VELOCITY ON WEIGHT BASIS.

Table 4-2
Reformer Feed and Product Properties

PROPERTY	UNITS	
FEED (HYDROTREATED HEAVY NAPHTHA, TBP 194–284°F)		
DISTILLATION		
IBP	°F	194
10%		203
30%		221
50%		239
70%		257
90%		275
EP		284
SULFUR	ppmw	0.5
NITROGEN	ppmw	0.5
AS	ppb	<1
PB	ppb	<1
CU	ppb	<1
WATER	ppm	<10
PONA		
P	VOL%	69
N		20
A		11
SPECIFIC GRAVITY		0.734
REFORMATE		
SPECIFIC GRAVITY		0.778
RON		96
ASTM DISTILLATION	°F	
IBP		140
30%		194
50%		230
70%		257
FBP		311
RVP	psi	3.4

IBP = INITIAL BOILING POINT; EP = END POINT; FBP = FINAL BOILING POINT; RVP = REID VAPOR PRESSURE.

is achieved at high temperatures in contact with powdered catalyst without the use of hydrogen. After separation of the catalyst, the hydrocarbons are separated into the desired products by fractionation. The main products of the FCC process are gasoline, distillate fuel oil, and olefinic C_3/C_4 liquefied petroleum gas (LPG). By-product coke, which is deposited on the catalyst during the reaction, is burned off in the regenerator. The heat

Table 4-3
Reformer Yields

COMPONENT	96 RON W/W
FEED	
HEAVY NAPHTHA	1.0000
TOTAL FEED	1.0000
PRODUCTS	
H_2	0.0193
C_1	0.0085
C_2	0.0138
C_3	0.0269
IC_4	0.0180
NC_4	0.0228
IC_5	0.0276
NC_5	0.0184
C_{6+}	0.8447
TOTAL PRODUCT	1.0000

Table 4-4
Cat Reformer Product Yields, W/W
(*Continuous Catalyst Regeneration Unit*)

	100 RON	102 RON
FEED		
HEAVY NAPHTHA	1.0000	1.0000
TOTAL FEED	1.0000	1.0000
PRODUCTS		
H_2	0.0310	0.0320
C_1	0.0120	0.0140
C_2	0.0200	0.0230
C_3	0.0290	0.0330
IC_4	0.0170	0.0190
NC_4	0.0230	0.0260
C_{5+}	0.8680	0.8530
TOTAL	1.0000	1.0000

liberated during the combustion of coke supplies the heat required to vaporize the feedstock and heat of reaction.

FCC gasoline historically has been the principal blend component for gasoline formulations. Distillate cut is a diesel boiling range material

Table 4-5
Cat Refomer Utility Consumption *(per Ton Feed)*

UTILITY	UNITS	SEMI-REGENERATIVE REFORMER	CONTINUOUS REFORMER
FUEL GAS	mmBtu	2.32	2.3200
STEAM	mmBtu	0.47	0.4700
POWER	kWhr	5.3	17.0000
COOLING WATER	MIG	0.7	1.2800

used as a diesel blend component after hydrotreating. Heavier cuts such as light cycle oil, heavy cycle oil, and clarified oils are used as fuel oil blend components and are excellent cutter stocks for vacuum resids. The olefinic LPG produced in the FCC process can be used in downstream alkylation and polymerization processes to yield more gasoline.

The feed to FCC unit (FCCU) is heavy diesel and vacuum gas oil from the crude distillation unit. The properties of a typical FCCU feed are shown in Table 4-12. Feed contaminants such as sulfur, nitrogen, and trace metals, such as nickel and vanadium, affect the yield and qualities of the product and also the catalyst consumption of the unit.

Higher sulfur in the feed is reflected in the products. While all cracked products contain some sulfur, sulfur distribution varies widely in the cracked product and cannot be controlled by operating parameters or catalyst design.

Nitrogen compounds temporarily deactivate the catalyst, resulting in lower conversion levels. This effect is reversible and can be controlled by operating at the highest possible reactor temperature.

However, the nickel and vanadium cause the most problems. Their effects on the cracking catalyst are quite different. Nickel increases the coke and gas yield, while vanadium is deactivator of zeolite catalyst, causing higher makeup catalyst addition rates.

CATALYST

FCC catalyst is a fine powder made up primarily of silica and alumina and containing acid sites that enable the catalyst to crack heavy hydrocarbons to gasoline and lighter products without formation of excessive amount of coke. Catalyst particles are of 50–60 microns in diameter. Earlier FCC catalysts were natural occurring clays, which suffered from low cracking activity and poor stability. Later synthetic silica/alumina catalysts containing 25% alumina were developed, which were more active and stable.

These were replaced with present day zeolite catalysts with greatly increased activity, stability, and improved selectivity. It has been realized for some time that better FCCU yield could be achieved with shorter contact time between the feedstock and the catalyst by using zeolite catalysts. With the earlier catalysts of low activity and long residence time, some of the gasoline formed cracked further in the catalyst bed to LPG resulting in lower gasoline yield. Development of zeolite catalyst made possible short contact time cracking in a riser, yielding in higher gasoline and lower coke.

The present day FCCU catalyst consists of basically three components: zeolites, active matrix, and a binder. It is possible to alter the process yield by adjusting the zeolite/matrix ratio. Optimizing the zeolite/active matrix ratio is done to achieve the best overall yield for a given unit, product slate, and feedstock. The use of zeolites in the FCCU has contributed most significantly to this advance. The coke yield decreases dramatically as the zeolite content of the catalyst increases.

The overall activity of the catalyst increases for a given zeolite content. The effect is more pronounced with an easier-to-crack paraffinic feed. With a paraffinic feed, a significant increase in conversion is observed with increasing zeolite content. Aromatic feed however is less susceptible to higher zeolite activity.

The large pore structure of the active matrix portion of the catalyst provides for easy access of large, heavy oil molecules, thereby providing their effective conversion. The matrix surface area is also a factor in strippability: The ability to strip hydrocarbons from the catalyst surface increases as pore diameter increases. With aromatic feeds, coke yield increases dramatically with increasing matrix contribution. The paraffinic feed shows a similar but less dramatic response. LPG yields are not significantly affected by matrix contribution, but dry gas yields for both paraffinic and aromatic feeds are directly related to matrix activity. Gasoline selectivity decreases with increasing conversion and increasing matrix activity.

Both FCCU design and catalyst improvements have combined to yield higher conversion/lower coke ratios over the years.

OPERATING CONDITIONS

The operating conditions for processing vacuum gas oil feed to maximize the production of light cat gasoline are shown in Tables 4-6 and 4-7.

Table 4-6
FCCU Reactor Operating Conditions

VARIABLE	UNITS	
FEED TEMPERATURE	°F	446
CATALYST/OIL RATIO		5.4
CATALYST CIRCULATION RATE*	tons/min	21.7
CATALYST MAKEUP RATE*	tons/day	2.5
RISER OUTLET TEMPERATURE	°F	991
DISPERSION STEAM	Wt% F. FEED	0.9
STRIPPING STEAM	tons/ton CATALYST	0.0213
REACTOR PRESSURE	psig	30
REGENERATOR PRESSURE	psig	33
REGENERATOR TEMPERATURE	°F	1341
FLUE GAS TEMPERATURE	°F	1355

* FOR UNIT FEED RATE OF 40 mbpsd.

Table 4-7
Regenerator Cyclone Operating Conditions

VARIABLE	UNIT	
GAS MOL WT		29.4
GAS VISCOSITY	C.P.	0.042
GAS DENSITY	lb/ft^3	0.071
REGENERATOR SUPERFICIAL VELOCITY	ft/sec	2.92
FIRST-STAGE CYCLONE INLET VELOCITY	ft/sec	0.16
SECOND-STAGE CYCLONE INLET VELOCITY	ft/sec	0.16

CP = Centipoise (viscosity units).

PRODUCT YIELDS

The yield of products in FCC depends on the feedstock quality, type of catalyst, and operating conditions. FCC units are usually operated to maximize the yield of gasoline; however, the process is versatile and can be operated to maximize middle distillate or LPG, both of which are at the expense of gasoline. The yields are shown in Table 4-8. Table 4-9 shows the composition of FCCU gases. Table 4-10 shows yields for a low-severity operation to maximize middle distillates. Table 4-11 shows the utility consumption for a FCCU unit. A summary of FCCU product qualities is presented in Table 4-12.

Table 4-8
FCCU Yields (Gasoline Operation)

STREAM	V/V	W/W
VGO	1.0000	1.0000
TOTAL FEED	1.0000	1.0000
PRODUCTS		
H_2S		0.0015
HYDROGEN		0.0006
METHANE		0.0165
ETHYLENE		0.0127
ETHANE		0.0139
PROPYLENE	0.0750	0.0434
PROPANE	0.0306	0.0173
BUTYLENES	0.0912	0.0617
ISOBUTANES	0.0617	0.0386
n-BUTANES	0.0208	0.0135
TOTAL GASES	0.2793	0.2197
LIGHT CAT NAPHTHA	0.4353	0.3419
HEAVY CAT NAPHTHA	0.1638	0.1561
DISTILLATE	0.1360	0.1410
LIGHT CYCLE GAS OIL	0.0196	0.0210
CLARIFIED OIL	0.0552	0.0656
COKE		0.0547
TOTAL		1.0000

Light Cat Naphtha

The light cat naphtha cut from FCCU has a high mercaptan sulfur content, about 120 ppm: therefore, further treatment of this cut is necessary to reduce the mercaption content to below 5 ppm by a Merox or equivalent process, before it can be blended into gasoline formulations. The light cat naphtha or FCCU light gasoline has a RON of 91–92 and MON (motor octane numbers) of approximately 80–81. Where gasoline specifications are not very stringent, light cat naphtha forms the major blend component of gasoline blends. However, if MON and other constraining specifications exist, other blend components, such as cat reformate, alkylate, or MTBE (methyl tertiary butyl ether), may be required in the blend to compensate for the low MON of light cat naphtha.

Table 4-9
FCCU Gases Composition

COMPONENT	DRY GAS	LPG
H_2O	0.60	
N_2	8.10	
CO	0.47	
CO_2	1.40	
H_2S	1.48	0.14
H_2	11.88	
C_1	38.62	
$C_2=$	16.97	
C_2	16.92	0.38
$C_3=$	2.02	29.29
C_3	0.60	11.24
IC_4		19.47
NC_4		6.55
$C_4=$		32.04
C_{4+}	0.94	
C_{5+}		0.89
TOTAL	100.00	100.00
MOL WT	21.6	51.20
NET HEATING VALUE, BTU/SCF	1008	2625.00

Heavy Cat Naphtha

Most heavy cat naphtha or gasoline is disposed of as a gasoline blend component at present. However, because of its high end point (375°F) and higher density (API 33.4 compared to 68.5 for light cat naphtha), future gasoline specs will find it increasingly difficult to include this material in gasoline blends and can pose serious disposal problems.

Distillate

The distillate cut, boiling in the diesel range has a very low cetane index of 28, compared to a typical value of 53 for straight-run diesels. Also, the untreated distillate cannot be blended into diesel blends because of the storage stability problems they may create. Blends with distillate from the FCCU must be hydrotreated before they are blended into finished diesels.

Table 4-10
FCCU Yields to Maximize Middle Distillate Operation

	W/W
STREAM	
VGO	1.0000
TOTAL FEED	1.0000
PRODUCTS	
H_2S	0.0014
HYDROGEN	0.0026
METHANE	0.0072
ETHYLENE	0.0030
ETHANE	0.0033
PROPYLENE	0.0220
PROPANE	0.0090
BUTYLENES	0.0346
ISOBUTANES	0.0199
n-BUTANES	0.0058
TOTAL GASES	0.1088
LIGHT CAT NAPHTHA	0.2765
HEAVY CAT NAPHTHA	0.1265
DISTILLATE	0.3566
LIGHT CYCLE GAS OIL	0.0533
CLARIFIED OIL	0.0344
COKE	0.0439
TOTAL	1.0000

Light-Cycle Gas Oil/Clarified Oil

These cuts are excellent cutter stocks for fuel oil blending because of its high aromatic content.

Table 4-11
Utility Consumption *(per Ton Feed)*

UTILITY	UNITS	CONSUMPTION
FUEL	mmBtu	0
STEAM	mmBtu	−0.093
POWER	kWhr	5.2
COOLING WATER	MIG*	2.92

*MIG = 1000 IMPERIAL GALLONS.

Table 4-12
FCCU Feed and Product Properties

PROPERTY	UNITS	FEED	LCN	HCN	DISTILLATE	LCO	CLARIFIED OIL
API		25.6	68.5	33.4	20	15.3	5
BROMINE NUMBER			66.5	16			
CETANE INDEX					27.9		
CON CARBON	%Wt	0.29					4.5
DISTILLATION	°F						
IBP			105	265	400	625	
10%			130	285	440	640	675
50%			180	330	525	655	815
90%			150	375	600	675	920
EP					640	725	
SPECIFIC GRAVITY		0.901	0.708	0.858	0.934	0.964	1.037
MERCAPTANS	ppmw		120	300			
SULFUR	%Wt	0.4	0.03	0.12	0.55	0.95	1.34
METALS	ppmw						0.1
NICKEL + VANADIUM	wppm	0.9					
NITROGEN	ppmw	880	13	45			
PONA ANALYSIS	% VOL						
PARAFFINS			37	19			
OLEFINS			35	10			
NAPHTHENES			13	12			
AROMATICS			15	59			
POUR POINT	°F				0	32	63
RON			91.6	92			
MON			80.4	80.5			
RVP	Psi		8.4	0.4			
VISCOSITY 122	Cst				2.9	9.5	110

RESID PROCESSING IN THE FCCU

It is possible to process resid feed in the FCCU in place of vacuum gas oil. The resid feed is no less crackable than vacuum gas oil. Paraffinic resid crack easily yields higher-valued products, while aromatic feeds is difficult to crack. Increasing the aromatic resid level in the feed increases coke and dry gas yields and decreases overall conversion and gasoline yield. However, the metals (nickel and vanadium) and Conradson (Con) carbon of the resid from most crudes make processing of resid in the FCCU unattractive.

With pure VGO feed, the coke yield is about 5% of the feed. If now resid is added to this feed, displacing some VGO, and the severity of FCCU operation maintained, the coke yield will increase, depending on the Con carbon content of the resid. About 40% of the Con carbon is converted to coke. Metals in the resid increase this quantity to almost 50%. As the coke yield increases, regenerator temperatures increase. At a 6% coke yield, the regenerator temperature increases to 1360°F and some form of catalyst cooling is required. At about 8% coke yield, it may be difficult to find an outlet for the steam generated from catalyst cooling, and this limits the quality of resid processed in the unit. Increasing, Ni and V in the feed due to resid injection increase the catalyst deactivation rate and hence the catalyst addition rate. Also, the conversion rate falls. Thus, resid processing in FCCU is economical with resid from crudes with very low metal and Con carbon contents, such as Bombay High, Brent, Murban, etc.

THE REFINERY FCCU UNIT

The VGO feed from storage or the vacuum distillation unit flows to feed surge drum V-107 and is heated to about 445°F (see Figure 4-3). The preheated feed is mixed with a small amount of dispersion steam and enters FCCU reactor V-101, where a stream of hot catalyst coming from regenerator V-102, at 1360°F, contacts the feed. The catalyst/oil ratio is about 5.4. The catalyst, oil, and steam mixture flows up through the reactor tube, where cracking reactions occur during its short residence at a temperature of approximately 990°F. At the end of the reactor, the cracked oil/catalyst mixture is separated by cyclones. The reactor effluent passes through one or more cyclones to separate any entrained catalyst particles; next any entrained hydrocarbons are stripped from the catalyst by medium pressure (MP) steam. The effluent flows out of the reactor to

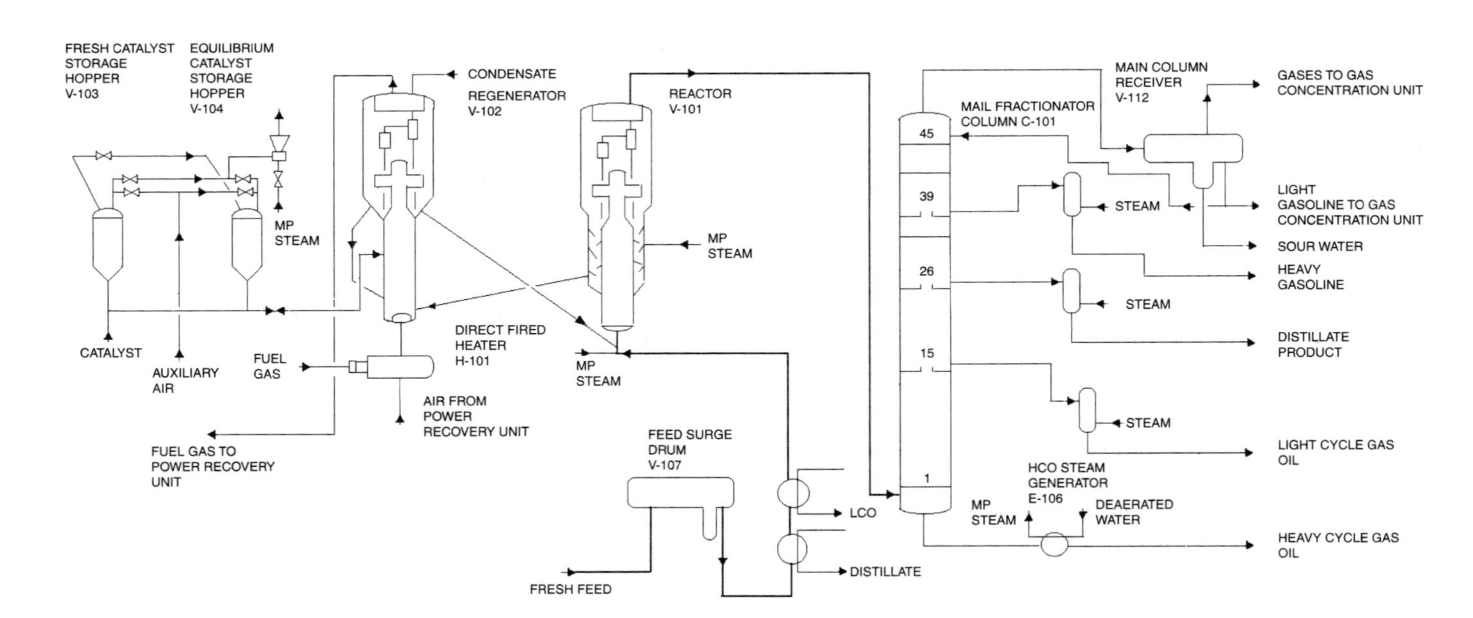

Figure 4-3. Fluid catalytic cracking unit. LCO = light cycle oil.

distillation column C-101, where the cracked effluent is separated into different products. About 5% of the feed is converted to carbon, which is deposited on the catalyst.

After steam stripping, the hot, separated catalyst is transported to FCCU regenerator V-102 by a stream of air. In the regenerator, the carbon deposited on the catalyst is burned off in a stream of hot air. Air for combustion is preheated to 435°F. Due to combustion of carbon, the catalyst temperature rises to about 1340°F, and the carbon on the catalyst is burned off. Flue gases pass through a number of cyclones to reduce particulate emission, produced from the attrition of the catalyst particles. The hot, regenerated catalyst flows back to FCCU reactor V-101 to continue the cycle. The hot flue gases, at 1340°F, generated by combustion of coke on the catalyst, are sent to a power recovery turbine. The power generated is used in an air blower to supply air to the regenerator.

Makeup catalyst is added to the regenerator to compensate for the loss of catalyst due to particle attrition and emitted to atmosphere.

The FCCU reactor effluent is sent to fractionation column C-101, very similar to a crude distillation column, with side strippers for the side cuts. The column has about 45 trays. The feed is introduced at the bottom of the column. The main cuts are;

- Vapors from the overhead reflux drum, which are sent to the gas concentration unit. The liquid, the light gasoline, is also sent to the gas concentration unit. Part of this stream provides the reflux to the fractionating column.
- Heavy gasoline.
- Distillate.
- Light cycle oil.
- Heavy cycle oil.

The broad cuts of heavy gasoline, distillate, and light cycle oil are drawn from the main column to their respective steam strippers, with six plates in each, and stripped light ends are returned to the column. The stripped side cuts are withdrawn as product.

THE FCCU GAS CONCENTRATION UNIT

The FCCU main fractionator column overhead receiver V-112 operates at about 20 psig and 104°F (see Figure 4-4). The vapors from this

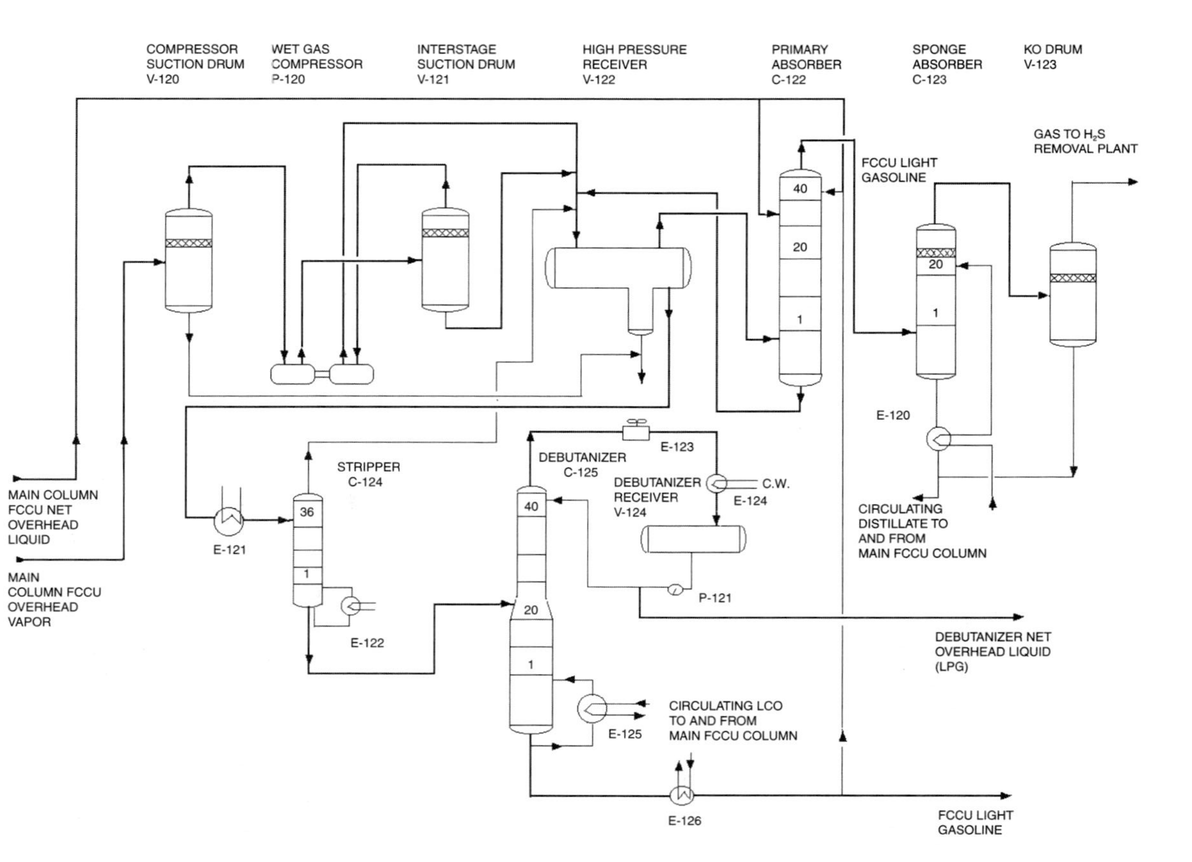

Figure 4-4. FCCU gas concentration unit. KO = knockout; C.W. = cooling water; LCO = light cycle oil.

drum are compressed in wet gas compressor P-120 to 205 psig. A part of the vapor condenses by compression and is collected in high-pressure receiver V-122.

The vapor from the high pressure receiver is fed to the bottom of primary absorber C-122, a column with 40 plates and operating at about 200 psig. The liquid separated in overhead drum V-112 of main FCCU column is used as the absorbent medium. The rich effluent containing the absorbed gas flows to high pressure receiver V-122.

The unabsorbed gases from C-122 flow to a second absorber C-123, the sponge absorber, a column with 20 plates and operating at 197 psig. The bottom stream from the sponge absorber goes back as reflux to the main FCCU fractionator column.

The liquid from high-pressure receiver V-122 goes to stripper C-124, with 35 plates, to strip off light ends. The stripper bottoms next go to debutanizer column C-125 with 40 plates, operating at 160 psig. Here, any LPG contained in the feed is separated as an overhead product, and the light gasoline stream is recovered as the bottom stream from the debutanizer column.

ALKYLATION

Alkylation is an important refining process for the production of alkylate, a high-octane gasoline blending component. Alkylate product is a mixture of branched hydrocarbons of gasoline boiling range. Alkylate has a motor octane (MON) of 90–95 and a research octane (RON) of 93–98. Because of its high octane and low vapor pressure, alkylate is considered an excellent blending component for gasoline.

Alkylates are produced by the reaction of isobutylene with isobutane in the presence of a catalyst (sulfuric acid) at low temperatures, as per the reactions shown in Figure 4-5. Propylene present in the feed in small concentration also reacts with isobutane to form isoheptane.

Under reaction conditions unfavorable to alkylate formation, propylene may also polymerize to form the undesirable product polypropylene. Amylenes can undergo a similar reaction to form alkylate, but since amylenes have a high octane number to start with, their conversion to alkylate is not as advantageous, as in the case of butylenes. Another side reaction that can negatively affect the formation of alkylate is ester formation by the reaction of olefin with sulfuric acid.

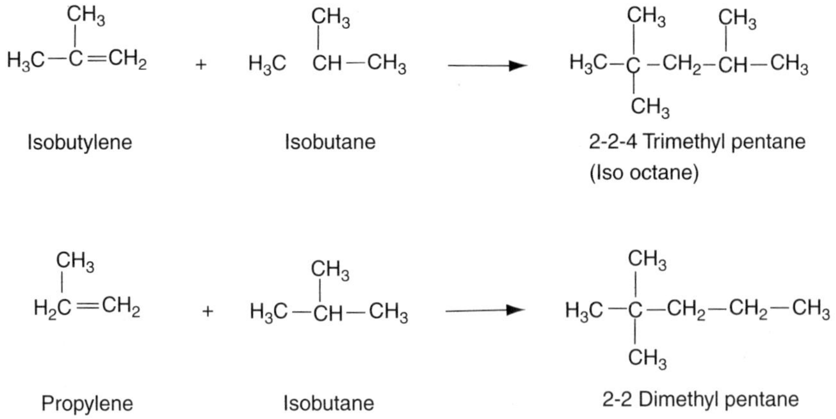

$$H_3C-C(CH_3)=CH_2 + H_3C-CH(CH_3)-CH_3 \longrightarrow H_3C-C(CH_3)_2-CH_2-CH(CH_3)-CH_3$$

Isobutylene + Isobutane → 2-2-4 Trimethyl pentane (Iso octane)

$$H_2C(CH_3)=CH_2 + H_3C-CH(CH_3)-CH_3 \longrightarrow H_3C-C(CH_3)_2-CH_2-CH_2-CH_3$$

Propylene + Isobutane → 2-2 Dimethyl pentane

Figure 4-5. Alkylation reactions.

PROCESS VARIABLES

The process variables that influence the quality of alkylate product (octane, distillation, and density) and acid consumption rate are the olefin type, isobutene concentration, temperature, mixing, space velocity, and acid strength and composition. These are described next.

Olefin Type

The type of olefin in the alkylation feed, especially the ratio of butylene to propylene affects the product quality and acid consumption rate. In propylene alkylation, the octane is about 5 numbers lower and acid consumption about twice those of butylene alkylation. The heat of the reaction, isobutane consumption, and alkylate yield vary with olefin type.

Isobutane Concentration

In the alkylation reaction, a molecule of olefin reacts with a molecule of isobutane to form an alkylate molecule. This reaction occurs in the presence of a sulfuric acid catalyst. A side reaction that competes with the alkylation reaction is polymerization. In this reaction, two or more olefin molecules react with each other to form a polymer. Polymerization causes lower product octane and increased acid consumption.

Both alkylation and polymerization occur in the acid phase. Because olefins are extremely soluble in the acid phase and isobutene is only slightly soluble, a large excess of isobutene must be maintained within the reaction zone to ensure that enough isobutene is absorbed by the acid to react with the olefins.

Temperature

Minimizing the reaction temperature reduces polymerization rate relative to the alkylation rate, resulting in a higher octane number and lower acid consumption. The optimum temperature for sulfuric acid alkylation is 45–50°F. Temperatures below 40°F are avoided to prevent freezing the sulfuric acid. Also, very low temperatures retard settling rates in the acid settler and can result in acid carryover.

The temperature in the contactor depends on the olefin feed rate, which in turn determines the heat of the reaction generated. Removal of heat from the reactor is a function of compressor suction pressure. Lower suction pressure produce more vaporization of the contactor effluent stream within the tubes of the contactor, which in turn produces more cooling and a lower contactor temperature.

Mixing

Vigorous mixing of hydrocarbon and acid is very beneficial for the reaction. Increased mixing produces a finer dispersion of hydrocarbon droplets in the acid continuous phase emulsion, increasing the surface area for mass transfer of isobutane into the acid catalyst. This improves the product quality and reduces acid consumption.

Space Velocity

Because alkylation occurs almost instantaneously, the residence time of reactants is not a limiting parameter. The space velocity (*SV*) in this case may be defined as follows:

$$SV = \frac{\text{olefin in contactor (bbl/hr)}}{\text{acid in contactor (bbl)}}$$

The term is simply a measure of the concentration of olefins in the acid phase of the reactor. As the olefin space velocity increases, octane tends to decrease and acid consumption tends to increase.

Acid Strength and Composition

The minimum acid strength required to operate the system is 85–87 wt%, although this varies somewhat depending on the olefin type and spent acid composition. At acid strengths lower than this, polymerization becomes so predominant that the acid strength cannot be maintained and the plant is said to be in an *acid runaway* condition. To provide a sufficient safety margin, an acid strength of 89–90% H_2SO_4 is used. However, the composition of acid diluents, as well as acid strength, is important. Water lowers the acid catalytic activity three to five times faster than hydrocarbon diluents. Some water is necessary to ionize the acid. The optimum water content is approximately 0.5–1% by weight. Impurities present in the olefin feed stream either react with or are absorbed in the acid catalyst, causing a decrease in strength and a need for increased acid makeup.

ALKYLATION PROCESS

Feed Sources

The feed to alkylate unit is usually cracked LPG from the FCCU unit. The FCCU's LPG is fractionated into a C_3/C_4 splitter to remove propane and lighter components (see Figure 4-6). The bottom stream from the splitter is charged to MTBE unit, if the refinery has one, which along with its primary product produces a raffinate stream that contains isobutane and butylene in a ratio suitable for alkylate manufacture. The composition of the raffinate stream is roughly as follows:

COMPONENT	VOL%
ISOBUTANE	47.3
NORMAL BUTANE	11.4
ISOPENTANE	1.4
BUTYLENE	38.9
AMYLENES	0.8
BUTADIENE	0.2

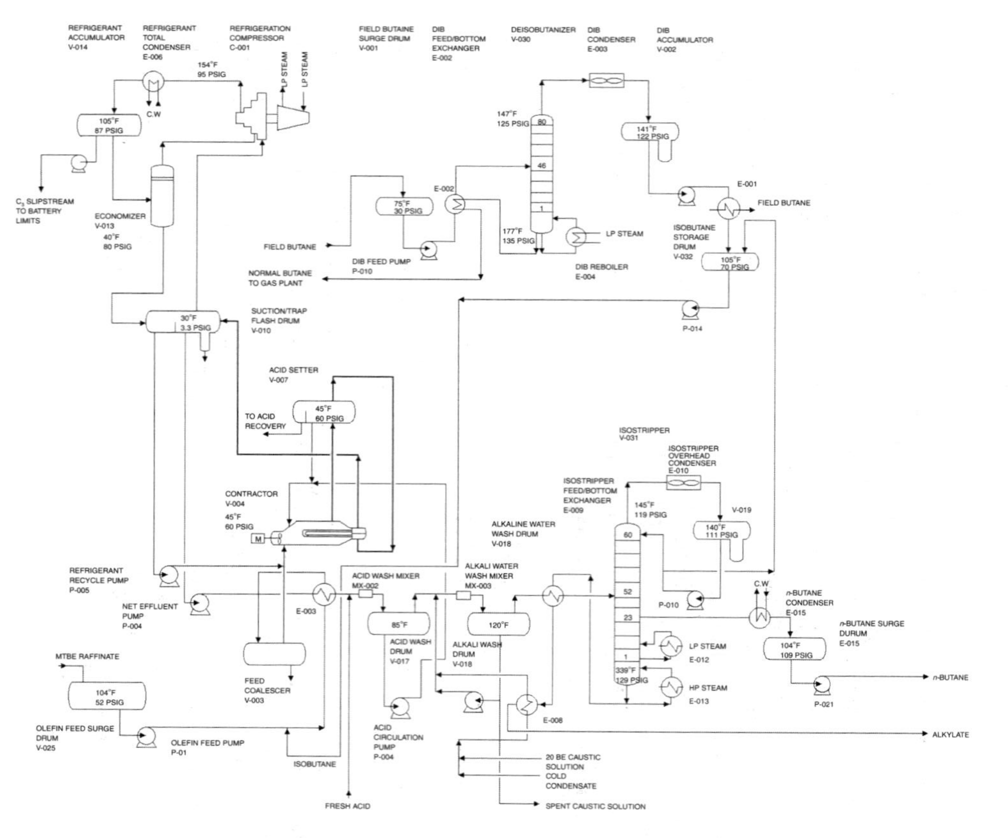

Figure 4-6. Alkylation process flow diagram.

If the refinery has no MTBE unit, the isobutane content of the FCCU gases may be insufficient to react to all the isobutylene in the feed and isobutane from external sources may be required.

Reaction Zone

The olefin feed is combined with the recycled isobutane stream before being charged to the reaction zone. The combined feed is at 104°F and saturated with dissolved water. The combined feed stream is cooled to 57°F by exchanging heat in E-003 with the contactor net effluent stream, which is then fed to alkylation reactor.

The alkylation reactor or contactors are horizontal pressure vessels containing a mixing impeller, an inner circulation tube, and a tube bundle to remove heat generated by the alkylation reaction.

Feed is injected into the suction side of the impeller inside the circulation tube. The impeller rapidly disperses the hydrocarbon feed in the acid catalyst to form an emulsion. The emulsion is circulated at high rates within the reactor. A portion of the emulsion is withdrawn from the reactor at the discharge side of the impeller and flows to acid settler V-007, where the acid and the hydrocarbon phases are allowed to separate. Being heavier, the acid settles to the bottom of the vessel and is returned to the suction side of the reactor impeller. Thus, the impeller acts as emulsion pump between the reactors and settlers.

The sulfuric acid present within the reaction zone serves as a catalyst to alkylation reactions. A certain amount of acid is consumed as a result of side reactions and feed contaminants. To maintain the desired spent acid strength, a small amount of fresh acid (98.5 wt% H_2SO_4) is continuously charged to the reactor and an equivalent amount of 90% H_2SO_4 spent acid is withdrawn from the aftersettler.

The fresh acid is drawn from acid wash drum V-017 in the effluent-treating section. The spent acid is sent to an acid aftersettler drum to recover hydrocarbons which are sent back to acid settlers. The spent acid goes down to an acid blow-down drum for degassing before being sent out of the battery limit for storage or disposal.

The acid-free hydrocarbon phase flows from the top of the acid settler through a back-pressure control valve into the tube side of the reactor tube bundle. The back-pressure control valve is set at about 64 psig to maintain the contents of the settler in liquid phase. As the hydrocarbon stream passes through the control valve, its pressure is reduced to

4.3 psig, flashing a portion of the stream's lighter components and cooling this stream to about 30°F. When the two-phase stream passes through the tube bundle, additional vapor is generated as a result of heat generated by alkylation reactions.

Refrigeration Section

After leaving the tube bundles, the contractor effluent stream flows into the suction trap side of the suction trap/flash drum V-010, where the liquid and vapor portions are separated. The suction trap/flash drum is a two-compartment vessel with a common vapor space. The net reactor effluent is accumulated on the suction trap side of a separation baffle and pumped to the effluent treating section. Cold refrigerant condensate from the refrigeration system is accumulated on the other side of the baffle. This refrigerant recycling stream consists mostly of isobutane and is returned to the reactor by refrigerant recycle pump P-005. The vapor portions of both streams are combined and flow to the suction side of refrigeration compressor C-001. The compressor increases the pressure of the refrigerant vapor to 95 psig. At this pressure, the compressor discharge stream is condensed by the refrigerant total condenser with cooling water at the ambient temperature.

The refrigerant condensate is collected in refrigerant accumulator drum V-014. A part of the refrigerant condensate is treated in a caustic wash followed by a coalescer to remove any traces of acidic components before being charged to an external C_3/C_4 splitter column.

This slipstream to the external C_3/C_4 splitter purges propane from the alkylation unit and prevents propane concentration from increasing within the unit. Most of isobutane present in the propane slipstream is recycled to the alkylation unit by way of C_3/C_4 splitter bottoms.

The remaining refrigerant condensate from the refrigerant accumulator drum flows to economizer drum V-013. As the condensate enters the economizer drum, it is flashed through a control valve to a pressure of 45 psig. At this pressure, a portion of the stream is vaporized. The economizer vapor flows to an intermediate stage of the compressor, while the cooled economizer liquid flows to the flash drum side of the suction trap/flash drum V-010, where further vaporization and cooling occurs.

Effluent Treatment

The liquid phase from the contactor tube bundle is collected on the suction trap side of suction trap/flash drum V-010. This stream contains traces of acid and neutral esters formed by the reaction of sulfuric acid with olefins. The esters are corrosive and must be removed to prevent corrosion in the downstream equipment. These esters are removed in the effluent treating section by washing the stream with fresh sulfuric acid followed by washing with a dilute alkaline water stream.

The net effluent from the suction trap is at 35°F. This stream is pumped through feed/effluent exchanger E-003 to heat it to 85°F. The net effluent stream is next contacted with fresh acid in acid mixer MX-002. The circulating acid is mixed with net effluent just ahead of the acid wash mixer. The hydrocarbon and acid phases are separated with the aid of electrostatic precipitator EP-01 in acid wash drum V-017. Some of the acid from this drum flows to the reaction section, while the remainder is recirculated. Fresh acid is pumped from storage at a continuous rate to maintain the acid level in this drum at 25%.

The treated hydrocarbon effluent flows from the top of acid wash drum V-017 to the alkaline water wash drum V-018 to neutralize any remaining esters as well as any entrained acid carried over from the acid wash. The alkaline water wash is operated at 120°F to thermally decompose the remaining esters.

Heat is added to the alkaline washwater by heat exchange with the isostripper bottoms stream. The circulating alkaline water is heated to 160°F before mixing with the hydrocarbon stream. The hydrocarbon/alkaline water mixture flows through alkaline water wash mixer MX-003 and then into the alkaline water wash drum, where the hydrocarbon and aqueous phases are separated by gravity settling.

Isobutane Stripper

The treated net contractor effluent is heated by heat exchange with isostripper bottom stream in exchanger E-009 to 165°F and fed to the isobutane stripper, where isbutane is stripped off from the alkylate product.

The overhead isobutane vapor is condensed in air cooler E-010 and collected in accumulator drum V-019. A portion of this liquid is refluxed back to the column, while the bulk of isobutane is recycled to the reaction section via isobutane storage drum V-032.

Normal butane is removed from the column as a vapor side draw from 22nd plate and sent outside the unit battery limit.

The isostripper bottom liquid is the alkylate product. The hot alkylate product is cooled by heat exchange with isostripper feed in E-009, circulating alkaline water heater in E-008, and finally in alkylate cooler E-014 before being sent to storage.

The operating conditions of an alkylation unit are shown in Table 4-13. Table 4-14 shows the effect of olefin type on alkylation. The process yields, utility consumption, and feed and product properties are shown in Tables 4-15 to 4-17.

ISOMERIZATION OF C_5/C_6 NORMAL PARAFFINS

GENERAL

Most gasoline formulations require inclusion of some light naphtha to meet the front-end distillation and octane specs. However, C_5/C_6 normal paraffins in this boiling range have low octane, which make them very

Table 4-13
Alkylation Unit Operating Conditions

OPERATING PARAMETERS	UNITS	
CONTACTOR/SETTLER		
Contactor Feed Temperature	°F	39
Contactor/Settler Temperature	°F	45
Settler Pressure	psig	60
Mixer speed	rpm	600
% Acid in the Contactors	LV%	50–60
Acid (H_2SO_4) Strength	Wt%	93–96
REFRIGERATION SECTION		
Compressor Suction Temperature	°F	30
Compressor Suction Pressure	psig	1.3
Compressor Discharge Temperature	°F	148
Compressor Discharge Pressure	psig	95
Economizer Vapor Temperature	°F	79
Economizer Vapor Pressure	psig	43
DEISOBUTANIZER		
Feed Temperature	°F	141
Feed Pressure	psig	119
Column Top Temperature	°F	147
Column Top Pressure	psig	125
Column Bottom Temperature	°F	177
Column Bottom Pressure	psig	135
Number of Plates		60

Table 4-14
Effect of Olefin Type on Alkylate Yield and Quality

PROCESS PARAMETER	UNITS	PROPYLENE	BUTYLENE	AMYLENE
TRUE ALKYLATE YIELD	bbl/bbl OLEFIN	1.77	1.72	1.58
REACTION ISOBUTANE	bbl/bbl OLEFIN	1.30	1.13	1.00
HEAT OF REACTION	Btu/lb OLEFIN	840	615	500
ALKYLATE RON		90	93	93
ALKYLATE MON		89	93	93
ACID CONSUMPTION	lb/bbl ALKYLATE	38	21	29

Table 4-15
Alkylation Process Yields

STREAM	YIELD WT%
FEED	
RAFFINATE EX MTBE PLANT	1.0000
PRODUCTS	
C_3 PURGE STREAM	0.0069
C_4 TO BOILERS	0.0596
ALKYLATE PRODUCT	0.9290
LOSS	0.0045
TOTAL	1.0000

Table 4-16
Alkylation Unit Utility Consumption
(per Ton Feed)

UTILITY	UNITS	
POWER	kWhr	64.085
STEAM	mmBtu	2.0402
COOLING WATER	MIG	12.065

difficult to include in the gasoline formulation. Branched chain C_5 and C_6 hydrocarbons have higher octane, making them more suitable for inclusion in gasoline (Table 4-18).

Table 4-17
Properties of Feed and Product

PROPERTY	UNITS	
FEED (MTBE RAFFINATE)		
COMPOSITION	Wt%	
PROPANE		0.06
I-BUTANE		45.53
N-BUTANE		11.36
I-PENTANE		1.52
N-PENTANE		0.03
$C_{4=}$		40.51
$C_{5=}$		0.76
1,3 BD		0.23
TOTAL		100.00
FEED TOTAL SULFUR	ppmw	20
TOTAL OXYGENATES	ppmw	800
ALKYLATE PRODUCT		
RON		96
MON		94
ASTM DISTILLATION		
IBP	°F	102
10%		160
30%		212
50%		223
70%		230
90%		258
EP		401

Table 4-18
Properties of Normal (*N*-) and Isoparaffins (*I*-)

HYDROCARBONS	BP, °F	RVP, psia	RON	MON
N-C_5	96.8	15	62	62
I-C_5	82.4	20	92	89.6
N-C_6	156.2	5	25	26
2-METHYL PENTANE	140	7	73	73
3-METHYL PENTANE	145.4	6	75	73
2–2 DIMETHYL BUTANE	122	10	92	93
2–3 DIMETHYL BUTANE	136.4	7	102	94

NOTE:
BP = BOILING POINT.
RVP = REID VAPORIZATION PRESSURE.

The isomerization process is designed for continuous catalytic isomerization of pentanes, hexanes, and their mixtures. The process is conducted in an atmosphere of hydrogen over a fixed bed of catalyst and at operating conditions that promote isomerization and minimize hydrocracking.

Ideally, the isomerization catalyst should convert the feed paraffins to the high-octane-number branched molecules—C_5 to isopentanes, C_6 to 2-3 dimethyl butane, and the like—however, the isomerization reaction is equilibrium limited with low temperatures favoring the formation of branched isomers. Under commercial conditions, the ratio of isoparaffin to normal paraffin in the reactor effluent is 3:1 and 9:1, respectively, for pentane and hexanes.

Dehydrocyclization is partially promoted and the reactor effluent contains less naphthenes than in the feed. The C_6 isomer equilibrium distribution is split approximately 45/55 between the higher-octane dimethyl butanes and lower-octane methyl pentanes. For typical C_5/C_6 feeds, the equilibrium limits the product research octane to approximately 83–85 clear on a one-pass basis.

Feed to the isomerization unit is a hydrotreated C_5/C_6 cut, free from sulfur, nitrogen, and water. The bulk of the straight-run C_7 cut is normally included in the cat reformer feed, due to its high content of aromatic precursors and because some of it cracks to C_3 and C_4 under isomerization unit operating conditions. The isomerization unit catalyst is otherwise unaffected by a high concentration of C_7 in the feed. Benzene, if present in the feed, is hydrogenated to cyclohexane, which is then isomerized to an equilibrium mixture of methyl cyclopentane, cyclohexane, and partially converted to isoparaffin. This represents an octane loss but liquid volume increase.

Feed to the isomerization unit needs to be hydrotreated to remove sulfur. Sulfur reduces the isomerization rate and therefore the octane number. Its effect, however, is temporary, and the catalyst resumes its normal activity once the feed sulfur concentration falls. Water is the only potential contaminant that can poison the catalyst and shorten its life.

The isomerization process normally has two reactors in a series flow configuration, each containing an equal volume of catalyst. Valves and piping are provided, which permit reversal of the process positions of the two reactors and isolation of either for catalyst replacement. Over time, the catalyst becomes deactivated by water. When the catalyst in the lead reactor is spent, the reactor is taken off-line for reloading. During the short period when the reactor is out of service, the second reactor is used to maintain continuous operation until the catalyst reloading is completed, allowing an almost 100% onstream factor.

Both isomerization and benzene hydrogenation are exothermic reactions, and the temperature increases across the reactor. Equilibrium requires that the reactor temperature be as low as the activity of the catalyst permits. Therefore, the effluent from the first reactor is cooled by heat exchange with cold incoming feed before entering the second reactor. Thus, the two-reactor system permits the imposition of an inverse temperature gradient. Most of the isomerization is accomplished at a high rate in the first reactor and under more favorable equilibrium conditions in the second reactor.

CATALYST

The catalyst used in the isomerization process have a zeolites base or are platinum-impregnated chlorinated alumina, although the latter are preferred. Due to their chlorinated nature, these are very sensitive to feed impurities, particularly water, elemental oxygen, sulfur, and nitrogen. The reactor operating temperature is 300–340°F. The reactor is operated at approximately 450 psig.

Organic chloride promoter (CCl_4) is continuously added with the feed (measured in parts per million) and converted to hydrogen chloride in the reactor. It is not necessary to provide separate equipment for recovery and reuse of hydrogen chloride. It is permitted to leave the unit by way of the stabilizer gas. The quantity of stabilizer gas is small due to minimum hydrocracking of the feed. The stabilizer gas is scrubbed for removal of hydrogen chloride before entering the refinery fuel system.

Single-pass isomerization of C_5/C_6 straight-run feed gives a product of 83–85 research octane clear. This can be increased to 92–93 by recycling unconverted C_5/C_6 back to the reactor.

The coking propensity of isomerization catalyst is low, thus there is no need for hydrogen recycling. The only hydrogen present is that required for saturation of aromatics and a small excess for adjustment to changes in feed composition.

HYDROCARBON CONTAMINANTS

The tendency of the catalyst to coke or sludge is minimal. The process therefore offers great flexibility with respect to the amount of hydrocarbons other than C_5/C_6. Sharp fractionation is not required to prevent C_6

cyclics and C_7 from entering the isomerization reactor. The effect of some of these hydrocarbons follows.

Olefins

The isomerization catalyst can tolerate up to 2% of C_5/C_6 olefins. Therefore, the feed from an FCCU or thermal cracker cannot be handled in an isomerization unit. Large quantities of olefins, if present in the feed, physically coat the catalyst following polymerization.

Cyclic Compounds

Cyclic compounds, if present in the feed, are adsorbed on the catalyst, reducing the active sites available for paraffin isomerization. Therefore, if the feed contains significant amounts of cyclic compounds, such as benzene, the catalyst inventory in the reactor has to be increased. Unsaturated cyclic hydrocarbons consume considerable quantities of hydrogen, resulting in exothermic reactions, which is undesirable from an isomerization equilibrium view point. Benzene is rapidly hydrogenated and converted to cyclohexanes. Cyclohexanes and other C_6 naphthenes are partially converted to C_6 paraffins.

C_7 Hydrocarbons

C_7 hydrocarbons crack readily to C_3 and C_4, and those that do not hydrocrack are isomerized to a mixture having a lower octane number than C_5 or C_6. C_7 naphthenes have an effect similar to C_6 naphthenes.

ISOMERIZATION PROCESS

Light naphtha feed is charged by charge pump P-101 to one of the two drier vessels, D-101 and 102, filled with molecular sieves, and designed to remove water to protect the catalyst (see Figure 4-7). The makeup hydrogen is compressed by makeup gas compressor C-101 to 500 psig. The gas then flows to gas driers D103 and D104, similar to those for liquid feedstock before it is combined with fresh feed. The feed is mixed with makeup hydrogen, heated through heat exchange with reactor effluent in E-101 and E-102 and a steam heater E-103, heated with medium pressure steam, and sent to the reactors. In normal operation, two reactors in series are employed.

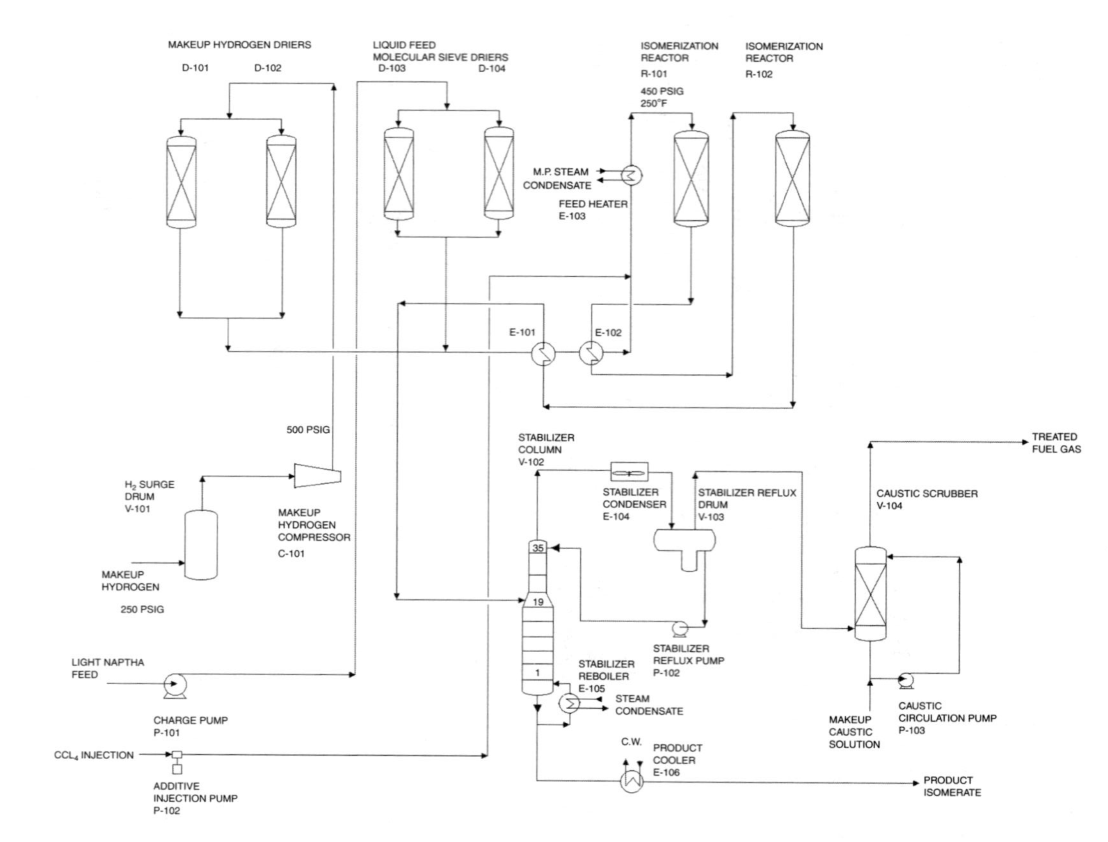

Figure 4-7. C_5/C_6 paraffins isomerization process. M.P. = medium pressure; C.W. = cooling water.

Table 4-19
C_5/C_6 Isomerization Operating Conditions

VARIABLE	UNITS	
REACTOR PRESSURE	psig	450
REACTOR INLET TEMP	°F	300–340
H_2/HC MOLE RATIO		0.05
LHSV	hr^{-1}	2

The reactor effluent is heat exchanged with fresh feed in E-101 and taken directly to product stabiliser column V-102. The column overhead vapor product flows to the off-gas caustic wash column. In this column, the off gas is washed with dilute caustic, circulated by a pump at the bottom of the column, to remove traces of chlorides before the gas flows to the fuel system. The stabilized, isomerized liquid product flows from the bottom of the column and is transferred to the gasoline blending system. Alternatively, the stabilizer bottom can be separated into normal and isoparaffin components to recycle low-octane normal paraffin. Product octanes in the range of 88–92 RON can be obtained by this method. Recycle of normal paraffins and, if possible, methyl pentanes is required to increase octane gain.

The efficiency of separation by distillation is limited, however, because normal C5 boils between i-C_5 (isopentane) and C_6 isomers, separation by a molecular sieve is more effective. The molecular sieve selectively adsorbs normal paraffins, due to their smaller pore diameter, while excluding the larger-branched molecules.

The operating conditions of an isomerization unit are shown in Table 4-19. The process yields, utility consumption, and feed and product properties are shown in Tables 4-20 to 4-22.

METHYL TERTIARY BUTYL ETHER

Methyl tertiary butyl ether (MTBE) is produced by the reaction of isobutene with methanol (see Figure 4-8). Its main use is as a gasoline blending component, due to its high octane level (RON = 115–135, MON = 98–100). Any hydrocarbon stream containing isobutene can be used for the production of MTBE. In refineries with a cat cracker unit, C_4 cut from cat cracking is the principal source of isobutene for

Table 4-20
Isomerization Unit Yields

STREAM	YIELD WT%
FEED	
LIGHT NAPHTHA FEED	1.0000
HYDROGEN	0.0040
TOTAL FEED	1.0040
PRODUCT	
ISOMERATE	0.9940
GASES	0.0100
TOTAL PRODUCT	1.0040

Table 4-21
Isomerization Unit Utility Consumption
(per Ton Feed)

UTILITY	UNITS	CONSUMPTION
ELECTRICITY	kWhr	7.22
MP STEAM	mmBtu	0.536
COOLING WATER	MIG	6.15

MTBE production. In petrochemical plants, C_4 cut from steam cracking after butadiene extraction can be used for MTBE manufacture.

CHEMICAL REACTIONS

Methanol reacts with isobutene to form MTBE. A number of secondary reactions can also occur, depending on the operating conditions and feed impurities (see Figure 4-9).

The etherification reaction is conducted in the presence of a catalyst that is cationic resin and strongly acidic. These resins are produced by sulfonation of a copolymer of polysryrene and divinyl benzene. These catalysts are very sensitive to impurities, which could destroy their acidic function, and high temperatures, which could remove the sulfonic bond. The activity of the catalyst allows operation below 195°F.

To obtain optimum yields, the reactor temperature is kept as low as possible to minimize side reactions and maximizing MTBE yield.

Table 4-22
Isomerisation Unit Feed and Product Properties
Feed (Light Natural Naphtha)

FEED COMPOSITION	WT FRACTION
ISOPENTANE	0.2431
N-PENTANE	0.3616
2,2 DIMETHYL BUTANE	0.0122
2,3 DIMETHYL BUTANE	0.0225
2,METHYL PENTANE	0.0970
3,METHYL PENTANE	0.0644
N-HEXANE	0.0225
METHYL CYCLOPENTANE	0.0592
CYCLOHEXANE	0.0695
BENZENE	0.0246
C7 AND HEAVIER	0.0234
TOTAL FEED	1.0000
FEED S.G	0.6502
FEED RVP, PSI	11.30

PRODUCT PROPERTIES
C4+ ISOMERATE

PROPERTY	UNITS	
RON		84.2
MON		81.6
SPECIFIC GRAVITY		0.6390
RVP	PSI	14.5

C5+ ISOMERATE

PROPERTY	UNITS	
RON		84
MON		81.3
SPECIFIC GRAVITY		0.641
RVP	psi	13.5
ASTM D-86 DISTILLATION		
IBP	°F	91.8
10%	°F	94.1
20%	°F	97.4
50%	°F	104.4
70%	°F	127.1
90%	°F	146.4
95%	°F	157.1
FBP	°F	184.4

TYPICAL MAXIMUM FEED CONTA-MINANT LEVEL (*FOR CHLORINATED ALUMINA CATALYSTS*)

CONTAMINANT	UNITS	
SULFUR	ppmw	1
NITROGEN	ppmw	1
WATER	ppmw	0.5
OXYGENATES	ppmw	0.5

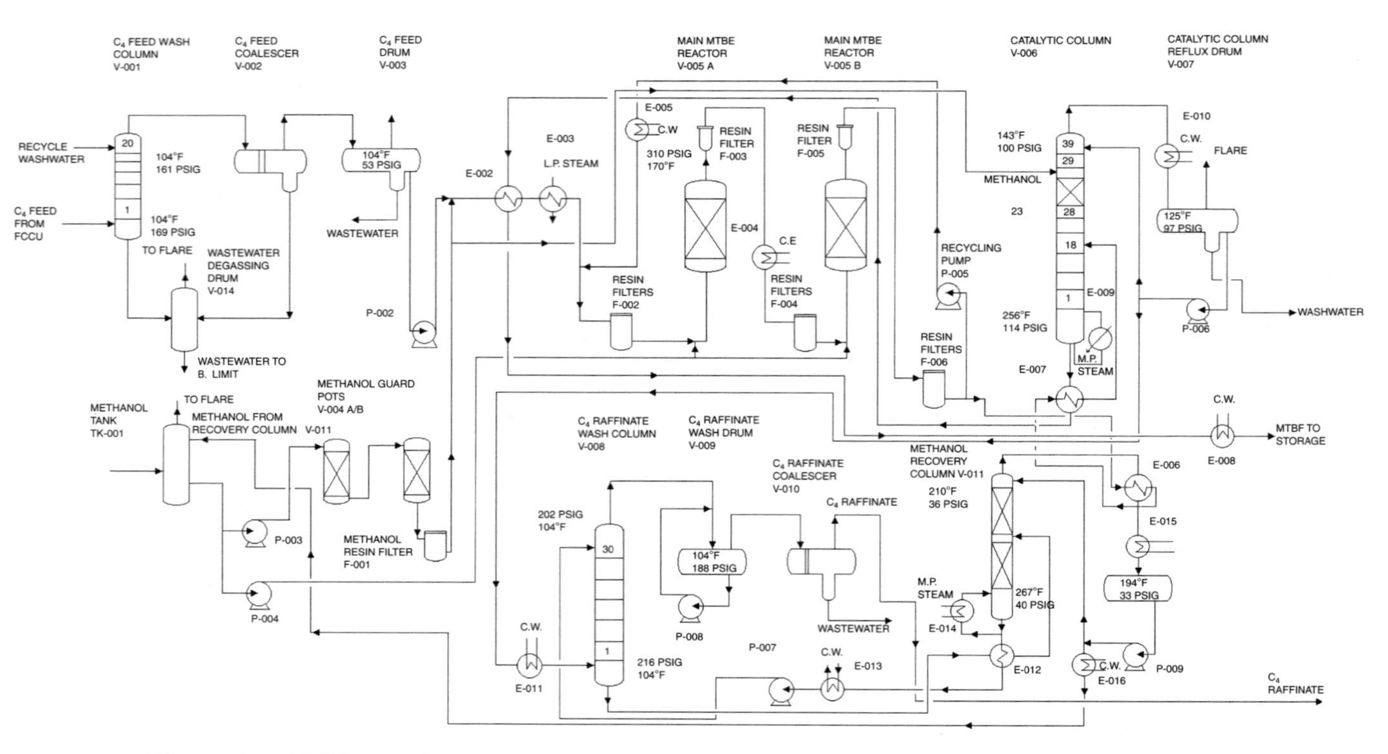

Figure 4-8. MTBE manufacture. L.P. = low pressure; M.P. = medium pressure; C.W. = cooling water.

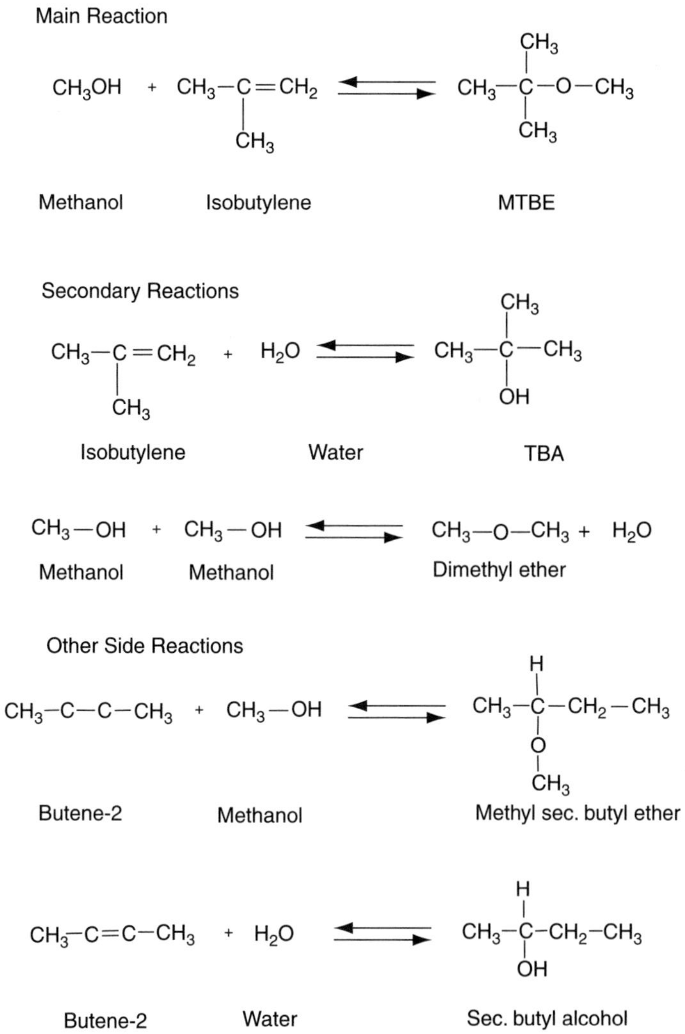

Main Reaction

$$CH_3OH + CH_3{-}\underset{\displaystyle CH_3}{\underset{|}{C}}{=}CH_2 \rightleftharpoons CH_3{-}\overset{\displaystyle CH_3}{\overset{|}{\underset{\displaystyle CH_3}{\underset{|}{C}}}}{-}O{-}CH_3$$

Methanol Isobutylene MTBE

Secondary Reactions

$$CH_3{-}\underset{\displaystyle CH_3}{\underset{|}{C}}{=}CH_2 + H_2O \rightleftharpoons CH_3{-}\overset{\displaystyle CH_3}{\overset{|}{\underset{\displaystyle OH}{\underset{|}{C}}}}{-}CH_3$$

Isobutylene Water TBA

$$CH_3{-}OH + CH_3{-}OH \rightleftharpoons CH_3{-}O{-}CH_3 + H_2O$$

Methanol Methanol Dimethyl ether

Other Side Reactions

$$CH_3{-}C{-}C{-}CH_3 + CH_3{-}OH \rightleftharpoons CH_3{-}\overset{\displaystyle H}{\overset{|}{\underset{\displaystyle \underset{\displaystyle CH_3}{\underset{|}{O}}}{\underset{|}{C}}}}{-}CH_2{-}CH_3$$

Butene-2 Methanol Methyl sec. butyl ether

$$CH_3{-}C{=}C{-}CH_3 + H_2O \rightleftharpoons CH_3{-}\overset{\displaystyle H}{\overset{|}{\underset{\displaystyle OH}{\underset{|}{C}}}}{-}CH_2{-}CH_3$$

Butene-2 Water Sec. butyl alcohol

Figure 4-9. MBTE reactions.

Pressure has no effect on the reaction, but the pressure is chosen to keep the system in a liquid phase. The pressure of the catalyst column is adjusted to keep the temperature at a minimum level and optimum conversion rate to MTBE.

Conversion increases with higher methanol at a constant temperature. However, higher methanol results in increased processing costs to recover excess methanol, and an economic optimum has to be determined between increased conversion and increased production cost.

PROCESS DESCRIPTION

The MTBE process consists of the following sections: feed wash, main reaction, catalyst distillation, C_4 raffinate wash, and methanol washing and recovery.

The Feed Wash Section

C_4 feed washing is done in 20-tray sieve column V-001 to remove actonitrile in addition to other impurities; here, C_4 feed is washed with cold condensate. The tower is operated full of liquid with the water/ hydrocarbon interface maintained above the top tray. Water effluent from V-001 is sent to V-014 for degassing. C_4 from the column top flows to C_4 coalescer drum V-002 and C_4 surge drum V-003.

The Reaction Section

The main reactors are vertical vessels, each having expandable bed of an ion-exchange-resin-type catalyst. Methanol makeup, mixed with methanol recycle, is pumped from TK-001 by pump P-003, through methanol guard pots V-004 to remove basic and cationic compounds.

The C_4 feed and a part of methanol respectively pumped through P-002 and P-003 under flow and flow ratio control are mixed with recycle before feeding the main reactors V-005 A/B. The two reactors are installed in series. The feed is heated with LP steam in feed preheaters E-002 and E-003. Interreactor cooler E-004 is used during EOR conditions of column catalyst to increase conversion.

The catalyst is fluidized by recycling a part of reactor effluent through P-005. The temperature of the reactor is controlled by cooling the recycled stream with cooling water in exchanger E-005, which removes the heat of reaction and heat added during preheat in E-003. Filters F-002/4/6 are installed at both reactor inlet and outlets to collect fine particles from the resins. The conversion in the reactors is up to 80%.

Cartridge filters with an acrylic wool fiber filter element are installed at the inlet of reactors to trap circulating resin particles of 75+ micron size. Similarly, strainers are placed at reactor outlets to trap circulating fine particles originating from the catalyst bed.

The Catalytic Distillation Section

Isobutylene final conversion is achieved in a catalytic column V-006, where the reaction and distillation are simultaneous. This column includes a fractionating tower yielding MTBE product at the bottom and unconverted C_4 raffinate/methanol azeotrope at top. Additional conversion is made possible by enhanced contact between reactants in a number of packed beds. In addition to packed beds, the column is equipped with approximately 40 trays for distillation.

The main reactor effluent is preheated in methanol recovery column condenser E-006, then in feed/bottom exchanger E-007 before entering column V-006. The methanol injection at the top of first resin bed is controlled by the methanol analyzer on the bottom of the column. Also, methanol is injected at other points to improve conversion. Bottom MTBE product is sent to storage after cooling via feed/bottom heat exchange in E-007, feed preheater E-002, and trim cooler E-008. The overhead of the column is condensed in cooling water condenser E-010. One part of the reflux drum V-007 liquid is sent to the column through P-006, and the other part to the methanol recovery section. C_4 distillate is sent to the fuel gas header.

The C_4 Raffinate Wash

The C_4 catalyst column distillate from distillate reflux drum V-007, after cooling in trim cooler E-011, is sent to wash column V-008 to remove the methanol. Washing is done with a countercurrent water stream pumped from the methanol recovery column through P-007. The C_4 overhead raffinate is freed from methanol by a second water wash, where water is recirculated by pump P-008, and freed from water in coalescer V-010 before being sent to the battery limit. Spent water is sent from V-009 to feed washing column V-001.

The Methanol Fractionation Section

V-008 bottom preheated in feed/bottom exchanger E-012 is sent to methanol recovery column V-011 under interface-level control. Bottom water product is recirculated to a second washing after cooling in feed/bottom exchanger E-012 and trim cooler E-013.

The overhead column vapors are condensed in E-006 with the catalytic column feed and in trim cooler E-015. One part of the reflux drum liquid is sent to the column through P-009, and the other part is sent to methanol tank V-051. The column is reboiled with medium pressure steam in thermo-syphon reboiler E-014.

PROCESS VARIABLES

Reactor Outlet Temperature

The optimum temperature for MTBE synthesis has been found to be approximately 175°F. Lower temperatures retard conversion to MTBE but minimize side reactions.

Table 4-23
MTBE Operating Conditions

OPERATING PARAMETER	UNIT	
REACTOR		
REACTOR TEMPERATURE	°F	175
REACTOR PRESSURE	psig	310
SPACE VELOCITY, LHSV	hr^{-1}	4.9
CATALYTIC COLUMN		
COLUMN TOP TEMPERATURE	°F	145
COLUMN TOP PRESSURE	psig	100
REFLUX DRUM TEMP.	°F	125
REFLUX DRUM PRESSURE	psig	97
COLUMN BOTTOM TEMPERATURE	°F	276
COLUMN BOTTOM PRESSURE	psig	114
NUMBER OF PLATES		40
C_4 FEED WASH COLUMN		
COLUMN TOP PRESSURE	psig	169
COLUMN TEMPERATURE	°F	104
NUMBER OF PLATES		20
RAFFINATE WASH COLUMN		
COLUMN TOP TEMPERATURE	°F	104
COLUMN TOP PRESSURE	psig	202
COLUMN BOTTOM TEMPERATURE	°F	104
COLUMN BOTTOM PRESSURE	psig	216
NUMBER OF PLATES		30
METHANOL RECOVERY COLUMN		
COLUMN TOP PRESSURE	psig	36
COLUMN TOP TEMPERATURE	°F	210
COLUMN BOTTOM TEMPERATURE	°F	287
COLUMN BOTTOM PRESSURE	psig	40

Reactor Pressure

The reactor pressure is chosen to keep the reactants in liquid state. The optimum pressure is approximately 310 psig. A lower pressure induces vaporization of the C_4 feed, a poor reaction rate, and heat transfer in the reactor.

Recycle Flow

The recycle flow is required to keep the bed in the expanded state. The expanded level of the bed is approximately 130% of the static level of the catalyst in the reactor. The superficial liquid velocity depends on the temperature and composition of the liquid and catalyst particle size, shape, and density.

The operating conditions of an MTBE unit are shown in Table 4-23. The process yields, utility consumption, and feed and product properties are shown in Tables 4-24 to 4-26.

Table 4-24
MTBE Process Yields

FEED	YIELD % WT
$C_4/C_4 =$ MIX FEED	1.0000
METHANOL	0.0731
TOTAL FEED	1.0731
PRODUCT	
MTBE PRODUCT	0.2338
RAFFINATE	0.8393
TOTAL PRODUCT	1.0731

Table 4-25
MTBE Utility Consumption *(per Ton C_4 Feed)*

UTILITY	UNITS	CONSUMPTION
POWER	kWhr	5.978
STEAM	mmBtu	0.87
COOLING WATER	MIG*	0.509

*MIG = 1000 IMPERIAL GALLONS.

Table 4-26
MTBE Feedstock and Product Properties

	UNITS	C_4 FEED	METHANOL	MTBE	C_4 RAFFINATE
SPECIFIC GRAVITY		0.584	0.792	0.74	0.551
MOL WT		57.4	32	88	57.3
VAPOR PRESSURE	100°F, psia	62		9.6	
TOTAL SULFUR	ppmw	17			20
C_4 FEED COMPOSITION	MOL%				
PROPYLENE		0.010			
PROPANE		0.070			
BUTYLENES		47.730			
BUTANES		50.400			
PENTANES		1.590			
DIOLEFINS, ETC.		0.200			
MTBE COMPOSITION	Wt%				
MTBE WT%				98.9	
METHANOL				0.2	
DIMER				0.5	
TBA				0.3	
MTBE RON				114	

CHAPTER FIVE

Hydrogen Production and Recovery

Hydrogen is required in refineries for a large number of hydrotreating and hydrocracking processes, to remove sulfur, nitrogen, and other impurities from hydrotreater feed and to hydrocrack the heavier gas oils to distillates. A limited quantity of hydrogen is produced in the catalytic reforming of naphthas, but generally the quantity is insufficient to meet the requirements of hydrocracker and hydrotreating units. As hydrogen production is capital intensive, it is always economical to recover hydrogen from low-purity hydrogen streams emanating from hydrotreating and hydrocracking units and minimize production from hydrogen units. In the absence of hydrogen recovery, these streams end up in fuel gas or sent to flare.

Most refinery hydrogen is produced by the steam reforming of natural gas. The conventional hydrogen production in the refineries involve the following steps: natural gas desulfurization, steam reforming, high- and low-temperature shift conversion and trace CO and CO_2 removal by methanation.

NATURAL GAS DESULFURIZATION

Compressed natural gas at 110°F and 480 psig, after passing through knockout (K.O.) drum V-101, flows to the bottom of sulfinol absorber C-101 (see Figure 5-1). Here, the gas is treated with sulfinol solvent flowing down from the top to remove most of the H_2S and CO_2 contained in the feed. The treated gas, which contain less than 1 ppmw H_2S, passes through knockout drum V-102 and is washed with water in C-102 to remove any traces of the sulfinol solvent. The sulfinol solvent is regenerated in sulfinol regeneration column C-103 with a steam reboiler. The rich sulfinol solvent flows down the column and the acid gas leaves the column at the top.

Table 5-1
Primary Reformer Operating Conditions

OPERATING CONDITIONS	UNITS	
GAS INLET TEMPERATURE	°F	700
GAS OUTLET TEMPERATURE	°F	1575
GAS INLET PRESSURE	psig	360
GAS OUTLET PRESSURE	psig	310
ESTIMATED HEAT TRANSFER	Btu/hr ft^2	23317
THEORETIC H_2 SPACE VELOCITY	hr^{-1}	2627
TYPICAL FEED AND PRODUCT COMPOSITION	**INLET % MOL**	**OUTLET % MOL**
CH_4	79.10	2.6
C_2H_6	1.49	
C_3H_8	0.33	
C_4H_{10}	0.18	
C_5H_{12}	0.10	
CO	0.00	7.882
CO_2	0.00	12.324
H_2	5.90	73.703
N_2	12.90	3.49
AR	0.00	0
TOTAL	100.00	100.00
STEAM/GAS MOLAR RATIO	5.6	1.19

CATALYST
HIGH GEOMETRIC SURFACE RINGS
SIZE: DIAMETER 0.59–0.65 IN.
COMPOSITION:
Ni = 12–14 WT%
Al_2O_3 = 80–86 Wt%
BULK DENSITY, lb/ft^3 = 70
SURFACE AREA, m^2/gm = 1.5–5.0
PORE VOLUME, cm^3/gm = 0.10–0.20

The desulfurized feed gas is next heated in the convection section of primary reformer H-101 to 700°F, mixed with a small quantity of hydrogen (about 5%), and flows to sulfur absorbers R-101 and R-102, containing mainly zinc oxide with a small quantity of desulfurization catalyst. For the steam reforming process, it is essential to remove all sulfur in the feedstock before it enters the primary reformer, to prevent poisoning the catalyst. If only H_2S is present, there is no need to employ the hydrogenation step. In this case, H_2S is adsorbed on ZnO, which can pick up to 20% of its weight when operated at 750°F.

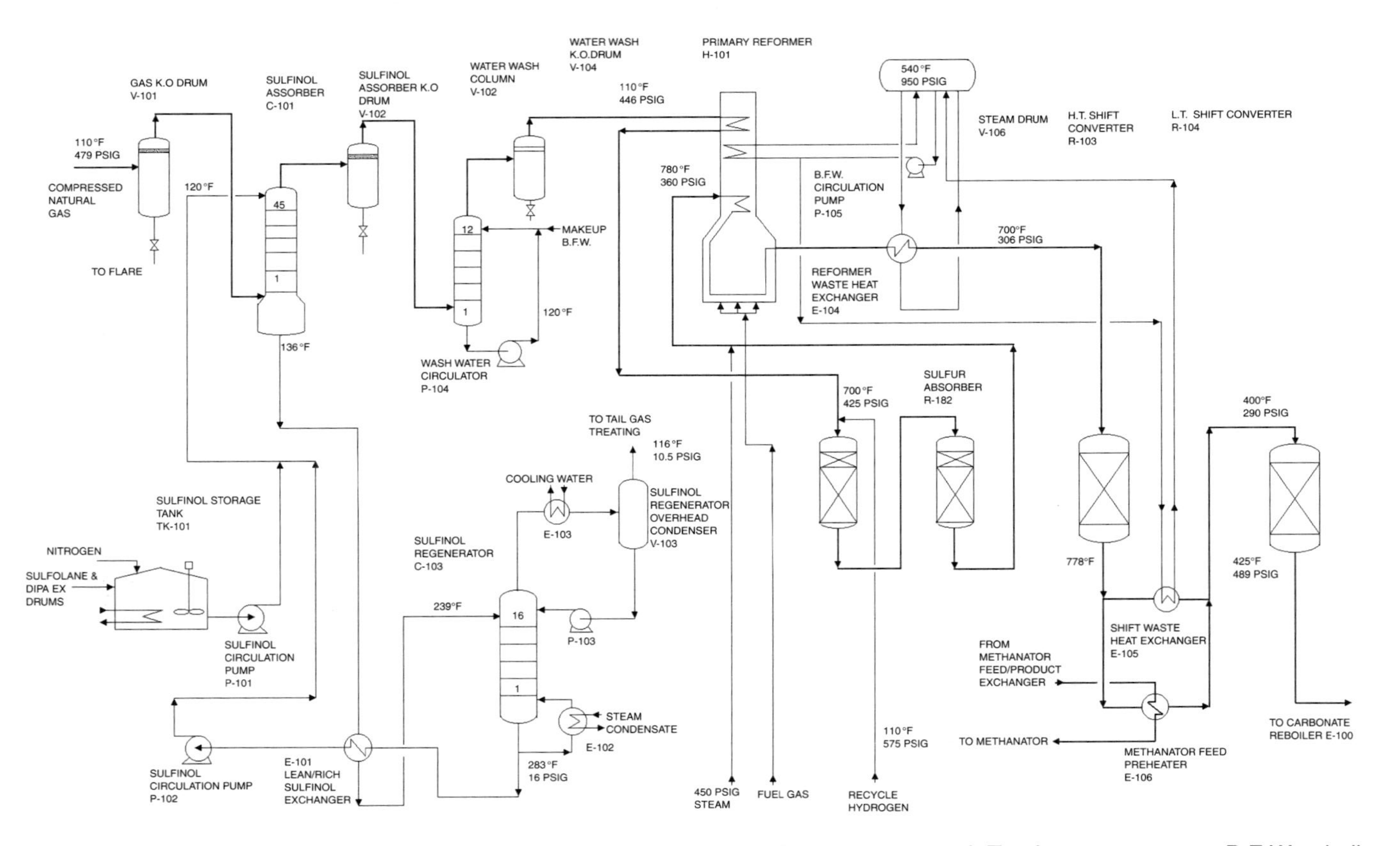

Figure 5-1. Hydrogen plant (reforming and shift conversion). H.T. = high temperature; L.T. = low temperature; B.F.W. = boiler feedwater.

Table 5-2
High-Temperature Shift Converter

OPERATING PARAMETER	UNITS	
FEED INLET TEMPERATURE	°F	700
EFFLUENT FROM HT CONVERTER	°F	770
FEED INLET PRESSURE	psig	306.0
REACTOR OUTLET PRESSURE	psig	304.0
DRY GAS INLET SPACE VELOCITY	hr^{-1}	1638
REACTOR FEED AND EFFLUENT COMPOSITION	**INLET MOL%**	**OUTLET MOL%**
CH_4	3.70	3.40
CO	8.88	2.00
CO_2	11.42	15.90
H_2	72.76	75.10
N_2	3.23	3.60
AR	0.00	0.00
DRY TOTAL	100.00	100.00
H_2O	0.921	0.8
TYPICAL CATALYST COMPOSITION	**WT%**	
Fe_2O_3	89–91	
CuO	2	
Cr_2O_3	8–10	

If mercaptans and some other easily hydrogenated sulfur compounds, such as disulfides, are present, then by adding 2–5% hydrogen to the natural gas, ZnO alone both hydrogenates and adsorbs sulfur compounds. If nonreactive sulfur compounds like thiophene are present, a Co-Mo or Ni-Mo hydrogenation catalyst plus 5% recycle hydrogen is used to ensure complete conversion to H_2S, which is fully absorbed on ZnO. Any trace of chloride, which is another catalyst poison, is also removed in this step.

STEAM REFORMING

In the desulfurization stage, the sulfur in the feed is reduced to less than 0.5 ppm. The purified gas is mixed with superheated steam and further heated to about 900°F before entering the catalyst-filled tubes of primary reformer H-101. A steam/carbon molar ratio of 3–6 is used. The steam/carbon ratio is an important process variable. A lower steam/carbon

Table 5-3
Low Temperature Shift Converter

OPERATING PARAMETER	UNITS	
FEED INLET TEMPERATURE	°F	400
EFFLUENT FROM HT CONVERTER	°F	425
FEED INLET PRESSURE	psig	292.0
REACTOR OUTLET PRESSURE	psig	290.0
DRY GAS INLET SPACE VELOCITY	hr^{-1}	2487
REACTOR FEED AND EFFLUENT COMPOSITION	**REACTOR IN MOL%**	**REACTOR OUT MOL%**
CH_4	3.47	3.40
CO	2.00	0.00
CO_2	17.02	15.50
H_2	74.48	77.90
N_2	3.03	3.30
AR	0.00	0.00
DRY TOTAL	100.00	100.10
H_2O	0.8	0.767

CATALYST FORM	PALLETS
Dimensions	5.2 mm × 3.5 mm
Composition	wt%
CuO	33.0
ZnO	33.0
Al_2O_3	33.0

ratio requires higher operating temperatures of the reformer and higher fuel consumption per unit hydrogen. The reforming furnace is fired with natural gas. The hot flue gases leave the radiant section of the reforming furnace at 1740–1830°F. The reformed natural gas exits the reformer tubes at 1560°F and 270 psig. The waste heat in the flue gases leaving the radiant section of the reformer are used to preheat reactants and superheat steam in a waste heat reboiler. The flue gas from the primary reformer is discharged to the atmosphere at approximately 300°F. The operating conditions of primary reformer are listed in Table 5-1.

CARBON MONOXIDE CONVERSION

The hot gas leaving the reforming tubes of H-101 is cooled to 650–700°F in waste heat boiler E-104 and enters a high-temperature shift converter R-103. In this reactor, about 75% of the CO is converted into CO_2 over an iron and chromium oxide catalyst. The gas from the high-temperature

Table 5-4
Methanator Reactor

OPERATING VARIABLE	UNITS	
FEED INLET TEMPERATURE	°F	600
EFFLUENT FROM METHANATOR REACTOR	°F	660
REACTOR INLET PRESSURE	psig	274.0
REACTOR OUTLET PRESSURE	psig	273.9
DRY GAS INLET SPACE VELOCITY	hr^{-1}	5927
REACTOR FEED AND EFFLUENT COMPOSITION	**REACTOR IN MOL%**	**REACTOR OUT MOL%**
CH_4	4.17	4.50
CO	0.18	0.00
CO_2	0.10	0.00
H_2	91.90	91.82
N_2	3.65	3.69
AR	0.00	0.00
DRY TOTAL	100.00	100.00
H_2O	0.011	0.015
CO		<0.45 ppm
CO_2		<9.55 ppm

CATALYST FORM	SPHERES
Dimensions	3/16–5/16 in. diameter
Composition,	wt%
Ni	34.0
Al_2O_3	50–55
CaO	5–7
Life	4–5 Years

shift converter is cooled to 430°F by raising steam in E-105 and preheating the methanator feed in E-106 and boiler feed water before it enters the low-temperature CO converter R-104 at approximately 400°F. In this reactor, most of the remaining carbon monoxide is converted to CO_2 over a copper-base catalyst. The operating conditions of high and low temperature shift converters are listed in Tables 5-2 and 5-3.

CARBON DIOXIDE REMOVAL

CO_2 from the process gas from the shift reactors is removed by a Benfield CO_2 removal system (see Figure 5-2). This process has two conversion

Table 5-5
Hydrogen Plant Overall Yield

STREAM	YIELD WEIGHT FRACTION
FEED	
NATURAL GAS	1.0000
TOTAL FEED	1.0000
PRODUCT	
HYDROGEN	0.5640
LOSS	0.4360
TOTAL PRODUCT	1.0000

Table 5-6
Utility Consumption *(per Ton Feed)*

UTILITY	UNITS	CONSUMPTION
POWER	kWhr	450.80
FUEL	mmBtu	35.90
STEAM	mmBtu	1.70
COOLING WATER	MIG*	16.10

*MIG = 1000 IMPERIAL GALLONS.

steps: an absorption step at elevated pressure, in which CO_2 is absorbed in an aqueous solution of potassium carbonate, and a regeneration step at near atmospheric pressure, in which CO_2 is stripped from the carbonate solution to make it suitable for reuse in the absorption step. The basic reaction involved is as follows:

$$K_2CO_3 + H_2O + CO_2 = 2KHCO_3$$

On absorption of CO_2, potassium carbonate in solution is converted into potassium bicarbonate, as shown by the equation. The reaction is reversed in the regenerator by steam stripping the solution and lowering the pressure.

After leaving the low-temperature CO converter, the gas is cooled to approximately 335°F by reboiling in Benefield solution regenerator E-109. The gas leaves this reboiler at approximately 265°F and enters a knockout drum, where the condensed water is removed. The gas next enters CO_2 absorber C-104, where CO_2 is reduced to 0.2% by contacting

Table 5-7
Feed and Product Properties, Natural Gas Feed

COMPONENT	MOL%
COMPOSITION	
CARBON DIOXIDE	2.20
NITROGEN	0.65
METHANE	89.10
ETHANE	6.31
PROPANE	0.34
ISOBUTANE	0.30
N-BUTANE	0.60
ISOPENTANE	0.20
N-PENTANE	0.20
HYDROGEN SULFIDE, ppmv	400
MERCAPTANS, ppmv	600
TOTAL	100.00
TEMPERATURE, °F	110
PRESSURE, psig	480

Table 5-8
Product Hydrogen

COMPONENT	
HYDROGEN, MOL%	97
$CO + CO_2$, ppmv MAX.	20
CO, ppmv MAX.	10
METHANE & NITROGEN	REMAINDER

with Benfield solution flowing from the top of the column. CO_2 is removed from the gas stream by a solution of potassium carbonate, promoted by DEA and containing a vanadate corrosion inhibitor (approximately 25.7% potassium carbonate and 4.76% diethanol amine).

Part of the potassium carbonate is converted to bicarbonate on each pass. Since the cooler solution has a lower carbon dioxide vapor pressure, the top part of the CO_2 absorber is provided with a cooled stream of solution. The remainder of the solution enters the absorber lower down with no cooling.

The hot, rich solution flows to the top of CO_2 stripper C-105 (Benfield solution regenerator), where it is regenerated by contact with steam rising

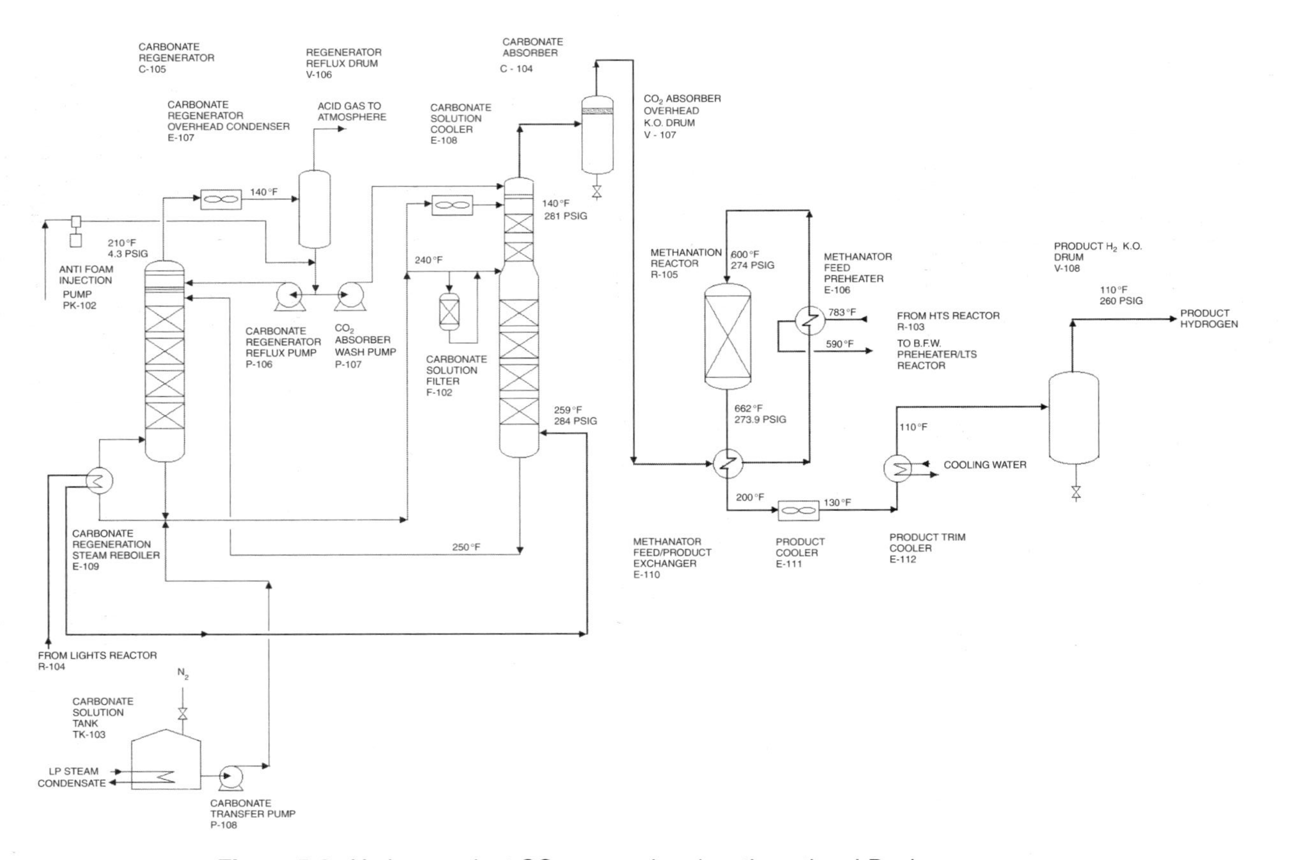

Figure 5-2. Hydrogen plant CO_2 removal and methanation. LP = low pressure.

from solution reboilers attached to the base of the column. A large proportion of the reboiler heat is provided from the gas stream, and the remainder is obtained from low-pressure steam (65 psia) condensed in the steam reboiler.

CO_2 and steam leaving the top of the column is cooled sufficiently to maintain the system water balance and vented to atmosphere. In locations where visible plume is not permitted, it is cooled 20–30°F above ambient temperature before venting it to atmosphere. Surplus condensate in this situation is degassed and sent to an off-site area for treatment.

The solution from the base of the CO_2 stripper is pumped to a CO_2 absorber. A part of the solution is cooled and sent to the top of the column, while the remainder enters the bottom half of the column with no cooling. Both columns are packed with metal pall rings packing.

METHANATION

The gas that leaves the carbonate absorber at about 160°F is heated to 600°F by heat exchange with the process gas in two exchangers, E-110 and E-106, before entering methanation reactor R-105. Here, any remaining oxides of carbon are catalytically converted to methane over a nickel-alumina catalyst. The exit gas, the product hydrogen, is used to partially preheat the feed before being cooled to 20–30°F above ambient temperature. A small quantity of water is condensed and removed before the gas is sent to the battery limit for use. Operating conditions of methanation reactor are shown in Table 5-4.

Figure 5-3 shows the chemical reactions involved in H_2 manufacture. Tables 5-5 to 5-8 show process yield, utility consumption and feed/product qualities.

PRESSURE SWING ADSORPTION ROUTE

The pressure swing adsorption (PSA) route is simpler than the conventional route, in that the low-temperature CO conversion, CO_2 removal by liquid scrubbing, and methanation to catalytically remove the remaining oxides of carbon are replaced by a molecular sieve system. This system works by adsorbing CO_2, CO, CH_4, N_2, and H_2O at normal operating pressure while allowing hydrogen to pass through. The molecular sieve is regenerated by lowering the pressure and using some of the product to sweep out the desorbed impurities. Due to this pressure cycling, it is commonly referred to as a *pressure swing adsorption system.*

A FEED GAS PURIFICATION

$$RSH + H_2 = H_2S + RH$$

$$C_2H_2 + 2H_2 = C_2H_6$$

$$H_2S + ZnO = ZnS + H_2O$$

B STEAM REFORMING

$$CH_4 + H_2O = CO + 3H_2$$

$$CO + H_2O = CO_2 + H_2$$

C HIGH TEMPERATURE & LOW TEMPERATURE SHIFT REACTIONS

$$CO + H_2O = CO_2 + H_2$$

D CO_2 ABSORPTION & REGENERATION

(with Benfield solution)

$$K_2CO_3 + CO_2 + H_2O \rightleftharpoons 2KHCO_3$$

(with monomethanolamine)

$$(CH_2CH_2OH)NH_2 + CO_2 + H_2O = (CH_2CH_2OH)NH_3^+ + HCO_3^-$$

E METHANATION

$$CO + 3H_2 = CH_4 + H_2O \quad \Delta H = -93000 \text{ BTU/LBMOLE}$$

$$CO_2 + 4H_2 = CH_4 + 2H_2O \quad \Delta H = -76800 \text{ BTU/LBMOLE}$$

Figure 5-3. Chemical reactions involved in H_2 manufacture.

Steps involved in PSA route are as follows;

DESULFURIZATION

The natural gas is heated to 750°F and desulfurized in exactly the same way as in the conventional route. The only variation being that recycle gas used must be pure hydrogen.

REFORMING

The desulfurized natural gas is mixed with superheated steam to give a molar ratio of 3.0 moles of steam for each mole of carbon in the natural gas. The mixture is preheated to 895°F before being distributed to the catalyst-filled tubes of the reforming furnace. The reformed gas leaves the tube at 1562°F and approximately 315 psia and enters the waste heat boiler, where it is cooled to 660–680°F before it enters the high-temperature CO converter.

The bulk of the fuel gas for the reformer is obtained from the purge gas from the PSA unit.

REMOVAL OF IMPURITIES

When gas leaves the high-temperature CO converter, it contains about 4 mol% carbon monoxide and 5 mol% of CO_2 plus methane. As the gas is cooled down to ambient temperature, most of the water vapor is removed by condensation. The gas then enters one of the adsorption vessels, where all the carbon compounds, residual water vapor, any nitrogen, and a small amount of hydrogen are adsorbed.

Most of the hydrogen passes through, leaving as a very pure gas. After some time, the molecular sieve adsorber becomes saturated, and the feed is switched to another vessel, containing a freshly regenerated molecular sieve.

The saturated vessel is depressurized very slowly to a low pressure of approximately 3–5 psig. The gas is then swept out using the smallest possible quantity of hydrogen product. The vessel is then repressurized by hydrogen, and it is ready to be swung on-line for its next period as adsorber. Commercial systems have a minimum of three or four vessels to give a smooth operation. With this number, about 74% of hydrogen in the raw gas can be recovered.

The purged gas flow is intermittent and of varying composition over the cycle. A surge vessel is required to ensure good mixing and even outflow. The purge gas is used as fuel in the reforming furnace.

PARTIAL OXIDATION PROCESS

Partial oxidation is a noncatalytic process[1] for the manufacture of hydrogen from heavy feedstocks, such as vacuum resides and asphaltic pitch. The heavy feed is partially burned using oxygen in a reactor. Due to high temperature, the remaining part of the feed is cracked. The composition of the hot synthesis gas leaving the reactor is mainly CO and H_2, with smaller quantities of CO_2, Ar, N_2, CH_4, and H_2S along with some soot and ash. All the sulfur in the feed is converted to H_2S. Argon in the synthesis gas originates from air from which oxygen is separated. Ash comes from the metals in the residual feed. Tables 5-9 to 5-11 list the parameters of this process.

Table 5-9
Typical Operating Conditions,
Synthesis Gas Reactor

PARAMETER	UNITS	
TEMPERATURE	°F	2000–2800
PRESSURE	psig	1200–2000

Table 5-10
Patrial Oxidation Process, Overall Yields

PROPERTY	
FEED ORIGIN	VACUUM RESID FROM WAFRA CRUDE
API	4.3
SPECIFIC GRAVITY	1.0420
SULFUR, WT%	6.2
VISCOSITY, 210°F CST	4000
STREAM FEED	**YIELD WT FRACTION**
VACUUM RESID	1.0000
OXYGEN	1.1140
BOILER FEEDWATER	1.7154
TOTAL	3.8294
PRODUCT	
HYDROGEN	0.2105
CH_4	0.0504
A	0.0061
N_2	0.0031
CO_2 + WASTE	3.4956
H_2S	0.0637
TOTAL	3.8294

Table 5-11
Utility Consumption *(per Ton Feed)*

UTILITY	UNITS	CONSUMPTION	
		1*	2*
POWER	kWhr	102.4	0
FUEL	mmBtu	4.2	9.3
STEAM	mmBtu	−3.7	0
COOLING WATER	MIG	12.2	49.7

1* = OXYGEN *IMPORT AT 1400 psig.*
2* = IN-*PLANT GENERATION OF OXYGEN. AIR SEPARATION UNIT ON STEAM DRIVE.*

The synthesis gas is cooled by either direct quenching or raising steam in waste heat boilers. The cooled synthesis gas from the partial oxidation reactor is sent to a single-stage shift converter,[2] which converts most of the carbon monoxide to carbon dioxide over a Co-Mo catalyst, by reaction of the carbon monoxide with steam. Using a single-stage shift prevent relying on low-temperature shift catalysts, which are sensitive to water, sulfur, and chlorides. Acid gas removal and separation are carried out next. The H_2S recovered is sent to a Claus unit for production of elemental sulfur. The final purification step is methanation, which reduces carbon oxides to less than 10 ppm.

SYNTHESIS GAS GENERATION

The feed is heated to 500°F by high-pressure steam and pumped to the synthesis gas reactor at a pressure of 500−2000+ psig, depending on the required product hydrogen pressure (see Figure 5-4). Major improvement results from operating at a pressure of 2000 + psig, eliminating product gas compression. Oxygen from an air separation plant with a purity of 99.5% or more is preheated, mixed with process steam, and fed through a special burner to a refractory-lined combustion chamber in pressure vessel V-101. In the reactor, the heavy fuel oil is partially burned in an atmosphere of pure oxygen. Steam acts as moderator. CO_2 or N_2 can also be used as moderator. Due to the high temperature (2000–2800°F) in the reactor, the remaining part of the feed is cracked. The gas leaving the reactor consists of CO, H_2, CO_2, CH_4, Ar, H_2S, and some carbon in the form of soot. The hot synthesis gases

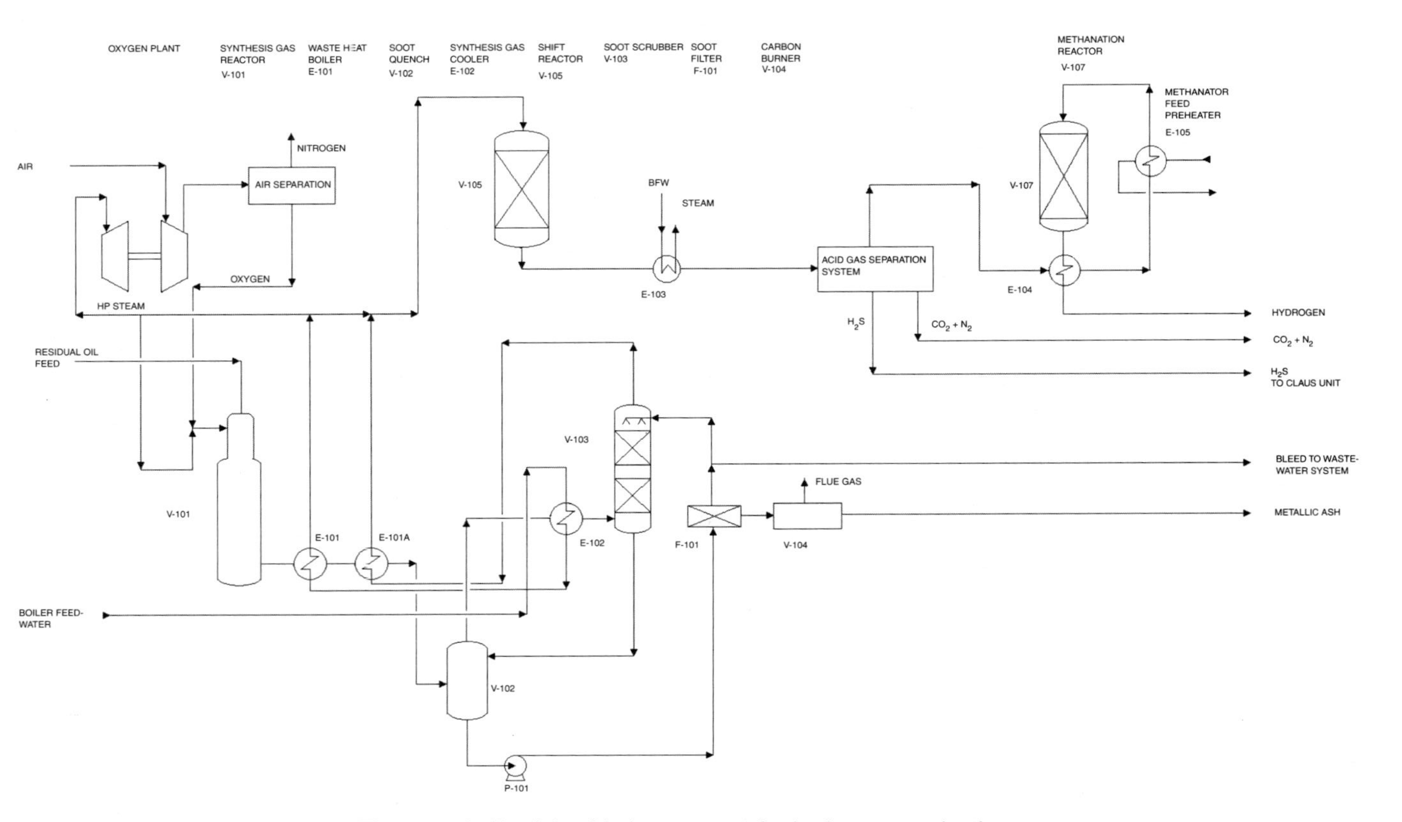

Figure 5-4. Partial oxidation process for hydrogen production.

are cooled in waste heat boiler E-101A and 101B, by incoming boiler feedwater, and a large quantity of high-pressure steam is generated in this step. Direct quenching can also be used if gas is required at higher pressures due to elimination of hydrogen compression and increased effectiveness of steam generation in the quenching.

The synthesis gas contains approximately 0.5–1 wt% of the feed as carbon particles or soot. Synthesis gas is next quenched with water coming from soot scrubber V-103, in soot quencher V-102. This removes the bulk of solids, such as carbon particles and ash (consisting mostly of metals in the form of oxides) entrained in the reactor effluent. The synthesis gas still contains entrained carbon particles and enters a soot scrubber, where the synthesis gas is washed with water to remove the remaining entrained soot. Synthesis gas leaving scrubber contains less than 1 ppmv soot.

The water/soot slurry is filtered in F-101 to remove carbon particles and ash, and the clear water is recycled to the soot scrubber. The filter cake is incinerated to burn off carbon, leaving only ash (mainly metallic oxides), which is disposed off-site.

In some process configurations,[3] the soot/water slurry is contacted with extraction naphtha and decanted into a soot/naphtha stream and gray water stream. The gray water is flashed off and returned to the soot scrubber. A small bleed stream of gray water is removed from the system to control dissolved solids and chloride ions. The soot/naphtha slurry is contacted with a portion of reactor feed, and the naphtha is stripped from the resulting stream for recycling. The feed/carbon mixture is sent back to the reactor feed tank. In this manner, the soot is entirely consumed.

SHIFT CONVERSION

The synthesis gas consists mainly of carbon monoxide, hydrogen, carbon dioxide, H_2S, and water vapor, with minor amounts of impurities such as methane, argon, and nitrogen. The synthesis gas is passed over a catalytic bed in shift converter V-105, where carbon monoxide reacts with steam over a sulfur-resistant catalyst to produce carbon dioxide. The reaction is exothermic:

$$CO + H_2O = CO_2 + H_2$$

ACID GAS REMOVAL

Carbon dioxide and H_2S are next removed from the synthesis gas by absorption in a variety of solvents such as MEA solution, cooled methanol (Rectisol process),[4] or potassium carbonate solution (Benfield process).[5] The process removes essentially all H_2S and CO_2 from the synthesis gas and delivers a mixture of H_2S and CO_2 to the Claus unit, recovering about 95–97% of the sulfur contained in the feedstock as saleable sulfur.

METHANATION

The absorber effluent is heated by heat exchange with the methanation reactor effluent and enters methanation reactor V-107, where the remaining small quantities of carbon oxides in the feed are converted to methane by passing through a methanation catalyst.

Gas leaving the methanator is product hydrogen of 98+% plus purity. The hydrogen purity depends on the oxygen purity, generator pressure, and conversion level in the shift reactors.

The principle advantages of the partial oxidation process over that of steam reforming are these:

- Good feedstock flexibility. The process can operate on feedstocks from natural gas to residual fuel oil, whereas the steam reforming process is limited to hydrocarbons no heavier than naphtha.
- No catalyst required in the synthesis gas generation, which also increases feedstock availability by tolerating impurities in the feedstock.
- Fewer process heat requirements.
- Hydrogen generation at much higher pressure than in a steam-reforming unit.

The principle disadvantage is that process requires a supply of high-purity oxygen, necessitating capital investment in the oxygen plant. Also, large quantities of high-pressure steam are generated, for which there may be no use, unless steam is used to drive the air separation unit compressor. Therefore, the cost of hydrogen production by the partial oxidation process does not compare favorably with that of gas based steam-reforming units.

The partial oxidation process is suitable for refineries and petrochemical plants that must process heavy, high-sulfur, high-metal crude oils and, at the same time, meet the environmental pollution standards. In this case, the partial oxidation process can convert its heavy resids into clean synthesis gas for use in furnaces and also convert a part of this synthesis gas into hydrogen. An alternative for these plants or refineries would be investment in the stack gas-cleanup units to control SO_2 emissions from burning high-sulfur resids. Also, the partial oxidation process may be of greater interest for ammonia and urea manufacture where by-products nitrogen (from the oxygen plant) and CO_2 can be used.

The typical operating conditions, feed properties, yields, and utility consumption for a partial oxidation unit are shown in Tables 5-9 to 5-11. Stoichiometry and product yields vary with the rate of water or CO_2 fed to reactor.

HYDROGEN RECOVERY

The objective of hydrogen recovery unit (see Figure 5-5) is to recover hydrogen from hydrogen-rich off gases released from several units in a refinery; for example, the hydrocracker, various desulfurizing units (naphtha, kerosene, diesel, and fuel oil), or the cat reformer.

Hydrogen-rich gas recovered in a hydrogen recovery unit (HRU) is gathered into a hydrogen header together with hydrogen gas manufactured by the hydrogen production unit to supply makeup hydrogen to several hydrogen-consuming units. See Tables 5-12 to 5-15 for the unit's operating parameters and yields.

The HRU consists of four sections: feed gas treatment, feed gas compression, the pressure swing adsorption system, and tail gas compression.

FEED GAS TREATMENT

The feed gas coming from various desulfurization and hydrocracking units (purge gases, stripper off gases, etc.) usually contain a high volume percentage of H_2S (2.9 vol% max, which must be brought down to less than 50 ppmv to prepare the feed for the PSA unit. The feed gas is first fed to feed gas knockout drum V-102, at 240 psig, where any condensed liquid is separated out.

The hydrogen sulfide is next removed by scrubbing with ADIP solution in Adip absorber column V-101. The ADIP solution is a 2-molar solution of DIPA (di-isopropyl amine) in water. The ADIP solution enters the top of the column at 114°F and 290 psig pressure. The following reactions take place.

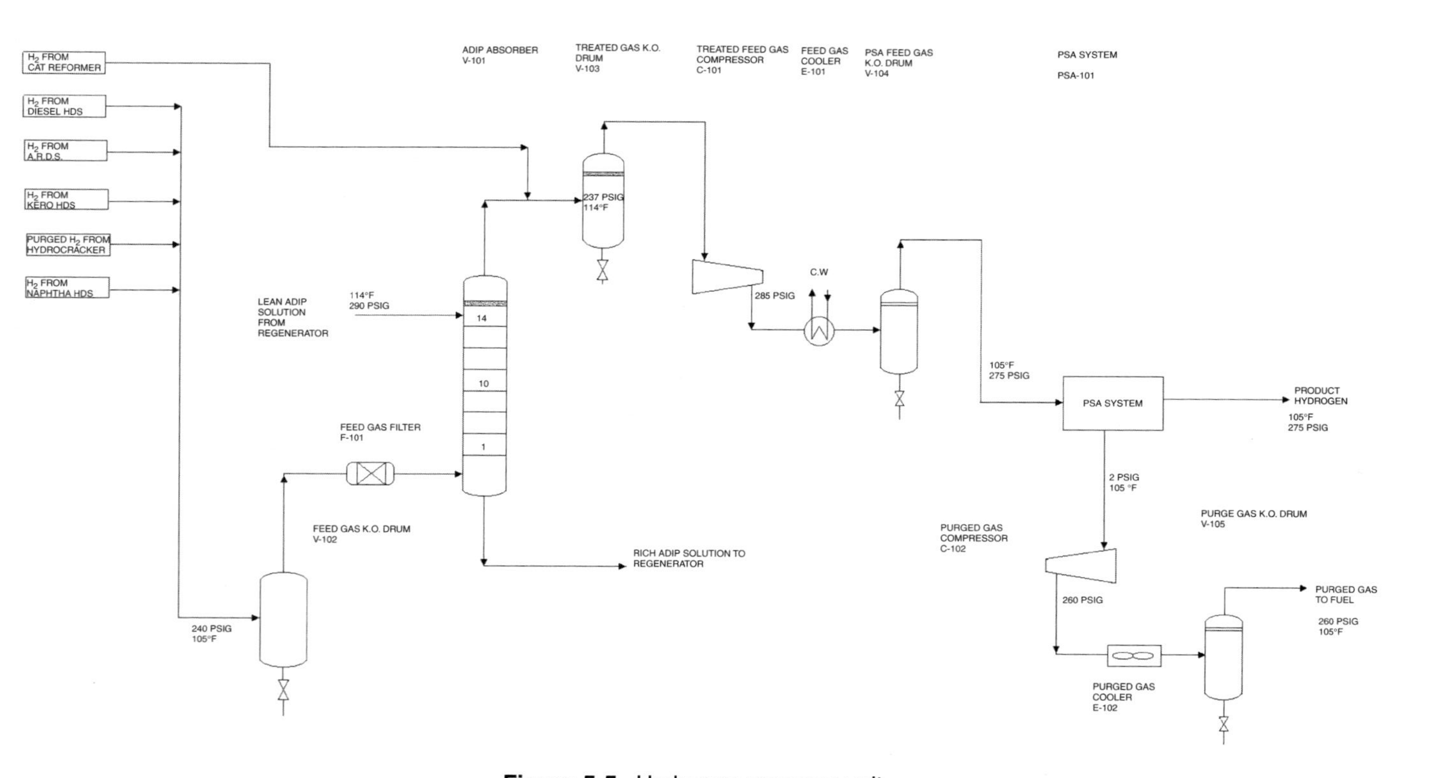

Figure 5-5. Hydrogen recovery unit.

Table 5-12
Operating Conditions

OPERATING PARAMETER	UNITS	
ADIP ABSORBER		
FEED GAS INLET H_2S CONTENT	VOL%	2.9
TREATED GAS H_2S CONTENT	ppmv	10
FEED GAS TEMPERATURE	°F	105
FEED GAS INLET PRESSURE	psig	240
ADIP SOLUTION INLET TEMPERATURE	°F	114
ADIP SOLUTION INLET PRESSURE	psig	290
PSA SYSTEM		
ADSORPTION CYCLE		
PRESSURE	psig	275
TEMPERATURE	°F	105
DESORPTION CYCLE		
PRESSURE	psig	2
TEMPERATURE	°F	105
TAIL GAS COMPRESSION		
GAS INLET PRESSURE	psig	2
GAS OUTLET PRESSURE	psig	260

With H_2S, a hydrosulfide is formed:

$$H_2S + R_2NH = HS^- + R_2N^+ + H_2$$

With CO_2, a carbamate is formed:

$$CO_2 + 2\ R_2NH = R_2NCOO^- + R_2N^+ + H_2$$

The absorber has 14 trays. The treated gas passes out of the column to a knockout drum at 114°F and 237 psig, where any entrained adip solution is separated from the gas. Hydrocarbon liquid collected in the feed knockout drum is discharged to the flare by draining.

The rich ADIP solution leaves the bottom of the absorber under level control. Because of the exothermic nature of the reaction, the rich adip solution is at 16°F higher temperature than the lean ADIP solution temperature.

To prevent the condensation of hydrocarbons in the absorber, the temperature of the lean solution fed to the absorber is at least 10°F higher

Table 5-13
Feedstock Properties *at the Inlet of the PSA System*

	UNITS	
COMPOSITION		
H_2	vol%	78.46
C_1	vol%	11.56
C_2	vol%	4.30
C_3	vol%	3.00
IC_4	vol%	0.73
NC_4	vol%	0.86
C_5	vol%	0.24
C_6	vol%	0.46
C_7	vol%	0.33
C_8	vol%	0.05
C_9+	vol%	0.01
TOTAL		100.00
IMPURITIES		
H_2O CONTENT		SATURATED
H_2S	ppm	50
MERCAPTANS & COS	ppm	2
HCL	ppm	3
NH_3	ppm	100
CO	ppm (MAX)	50
CO_2	ppm (MAX)	50
BENZENE	ppm	1000
TOLUENE	ppm	1000
XYLENE	ppm	500
PRODUCT HYDROGEN		
HYDROGEN	vol%	99
$CO + CO_2$	ppmv	20
CH_4	ppmv	1

than that of the feed gas temperature. The rich ADIP solution is sent to a regenerator column, where H_2S and other absorbed impurities are stripped off. The regenerated apip solution is returned to the absorber.

THE PSA SYSTEM

The feed gas, after leaving the knockout drum of the ADIP absorber is compressed from 225 psig to 285 psig by feed gas compressor. After

Table 5-14
Overall Hydrogen Recovery Unit Yields

STREAM	YIELD WT%
FEED	
HYDROGEN-RICH GAS	1.0000
PRODUCTS	
HYDROGEN	0.1510
RICH PURGE GAS TO FUEL	0.8420
LOSS (H_2 TO FUEL)	0.0070
TOTAL	1.0000

Table 5-15
Unit Utility Consumption *(per Ton Feed)*

UTILITY	UNITS	CONSUMPTION
POWER	kWhr	1.08
STEAM	mmBtu	2.25
COOLING WATER	MIG	13.24

removing the heat of compression by the feed gas aftercooler and knocking out condensate in the feed gas cooler knockout drum, the feed gas is routed to the PSA system at 105°F and at 275 psig.

The pressure swing adsorption system works on the principle of physical adsorption, in which highly volatile compounds with low polarity (as represented by hydrogen) are practically unadsorbable compared to light and heavy hydrocarbon molecules, which are adsorbed on molecular sieves. Thus, most impurities contained in the hydrogen-containing feed stream are selectively adsorbed and a high-purity hydrogen product obtained.

The pressure swing adsorption works at two pressure levels: The adsorption of impurities is carried out at high pressure (275 psig, 105°F); and the desorption or regeneration of the molecular sieves is carried out at low pressure (2.8–5.6 psig).

Typically two batteries, each containing six adsorbers working as an integral unit, are used. Out of six adsorbers, two (in parallel) are continuously on adsorption, while remaining adsorbers perform different regeneration steps.

After the adsorption step, the adsorber is regenerated in the following manner:

- The adsorber is depressurized to a low pressure level, usually 2 psig. The gas is then swept out by smallest quantity of hydrogen product.
- The vessel is then repressurized with hydrogen and ready to be swung on-line for its next period as adsorber.

All absorption, desorption, and depressurization cycles are controlled and optimized by a microprocessor. The low-pressure tail gas is collected in a surge drum and boosted to 260 psig by a compressor.

The hydrogen recovery is about 83% of the hydrogen contained in the feed.

TAIL GAS COMPRESSION

The HRU tail gas is routed to a tail gas compressor suction drum and compressed from 2 psig to 260 psig. In the first compressor stage, the gas is compressed to 60 psig and cooled to 105°F by the compressor intercooler to knock out condensate in the tail gas to intercooler knockout drum. In the second stage, the tail gas is compressed to 260 psig, cooled to 105°F to knock out condensate, and sent to the refinery fuel gas system or the hydrogen plant.

NOTES

1. C. P. Marion and W. L. Slater. "Manufacture of Tonnage Hydrogen by Partial Combustion, the Texaco Process." Sixth World Petroleum Congress, Frankfurt, Germany 1963.
2. W. Auer. "A New Catalyst for the Co Shift Conversion of Sulfur-Containing Gases." 68th National Meeting of American Institute of Chemical Engineers, Houston, Texas, February 1971.
3. J. M. Brady and L. Nelson. "Heavy Residue Gasification Schemes." Conference on New Opportunities for Fuel Oil in Power Generation, Institute of Petroleum, London, February 19, 1990.
4. Process licensors are Linde A. G. and Lurgi GmbH.
5. H. E. Benson and R. E. Parrish. "Hi Pure Process Removes CO/H_2S," *Hydrocarbon Processing* 53, no. 4 (April 1974), p. 81.

CHAPTER SIX

Residuum Processing

DELAYED COKING

Delayed coking is a thermal process in which the vacuum residue from crude distillation is heated in a furnace then confined in a reaction zone or coke drum under proper operating conditions of temperature and pressure until the unvaporized portion of the furnace effluent is converted to vapor and coke.

Delayed coking is an endothermic reaction, with the furnace supplying the necessary heat for the coking reactions. The reactions in the delayed coking are complex. In the initial phase, the feed is partially vaporized and cracked as it passes through the furnace. In the next step, cracking of the vapor occurs as it passes through the drum. In the final step, successive cracking and polymerization of the liquid confined in the drum takes place at high temperatures, until the liquid is converted into vapor and coke.

The coke produced in the delayed coker is almost pure carbon containing some of the impurities of the feed, such as sulfur and metals.

THE DELAYED COKING PROCESS

The reduced crude or vacuum resid enters the coke fractionator bottom surge zone (see Figure 6-1). The feed is mixed with recycle condensed in the bottom section of the fractionator and pumped by heater charge pump P-04 through coke heater H-01, where the charge is rapidly heated to the desired temperature for coke formation in the coke drums. Steam is injected in each heater coil to maintain the required minimum velocity and residence time and suppress the formation of coke in the heater coils.

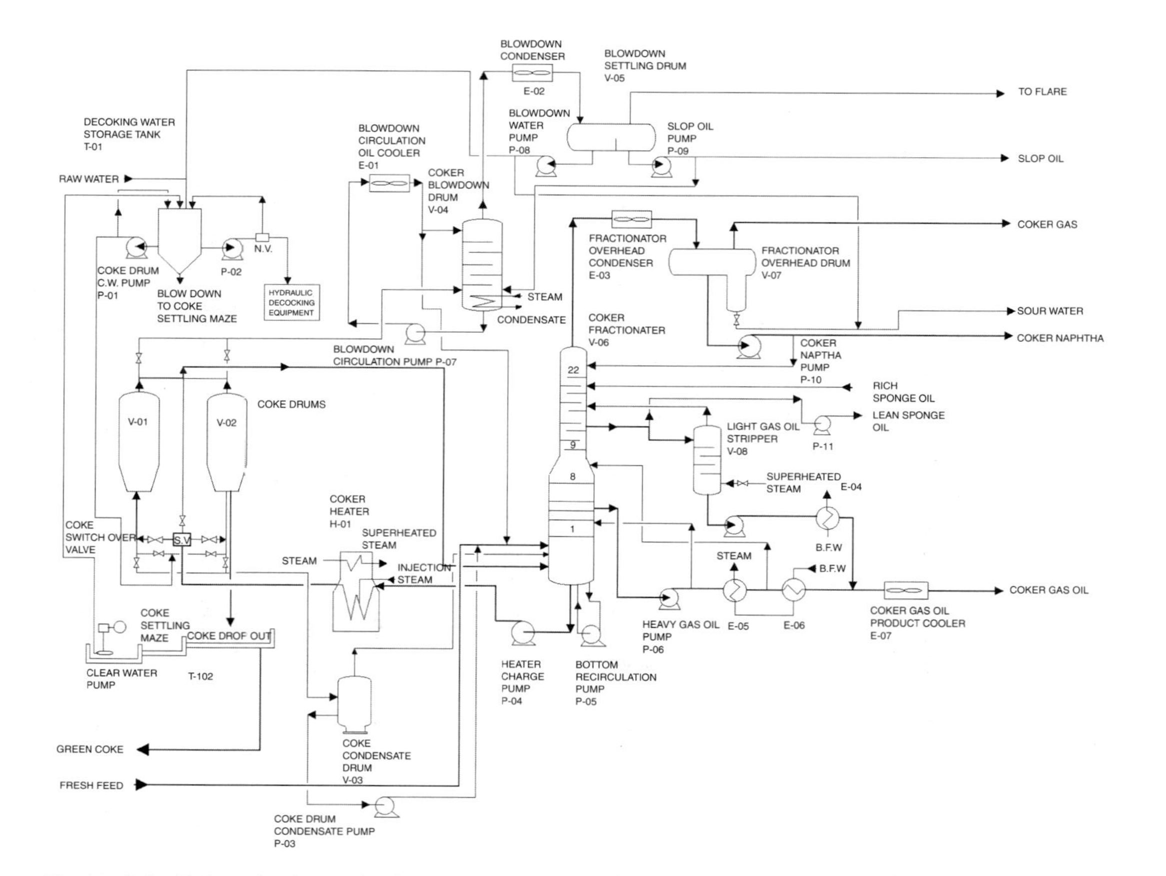

Figure 6-1. Delayed coker unit. C.W. = cooling water; SV = coker switch value; NV = by-pass value.

The vapor/liquid mixture leaving the furnace enters coke drums V-01 or V-02, where the trapped liquid is converted into coke and light hydrocarbon vapors. The vapors make their way through the drum and exit it. A minimum of two drums are required for operation. One drum receives the furnace effluent, which it converts into coke and gas, while the other drum is being decoked.

The coke overhead vapors flow to the coker fractionator and enter below the shed section. The coke drum effluent vapors are quenched with fresh feed then washed with hot gas oil pumpback in the wash section trays. These operations clean and cool the effluent product vapors and condense a recycled stream at the same time. This recycled stream, together with fresh feed, is pumped from the coker fractionator to the coker furnace. The washed vapor passes through the rectifying section of the tower. A circulating heavy gas oil pumparound stream withdrawn from the pumparound pan is used to remove heat from the tower, condensing a major portion of the heavy gas oil and cooling the ascending vapors. The hot pumparound stream of heavy gas oil withdrawn from the fractionator is used for steam generation. The heavy gas/oil product is partially cooled via exchange with steam generation then with an air cooler to storage temperature. The light gas/oil product is steam stripped for removal of light ends, partially cooled via steam generation and air cooled to storage temperature.

Lean sponge oil is withdrawn from the fractionator and cooled by heat exchange with rich sponge oil, then air cooled, before flowing to the top of the sponge absorber. Rich sponge oil is returned to the fractionator after preheat by exchange with lean sponge oil.

The overhead vapors are partially condensed in fractionator overhead condenser before flowing to the fractionator overhead drum. The vapor is separated from the liquid in this drum. The vapor flows under pressure control to the suction of the gas compressor in the gas recovery section. The top of the fractionator is refluxed with part of the condensed hydrocarbon liquid collected in the overhead drum. The balance of this liquid is sent with the vapors to the gas recovery plant. Sour water is withdrawn from the overhead drum and pumped to off-site treating facilities.

VAPOR RECOVERY SECTION

The vapors from the fractionator overhead drum flow to the compressor suction drum and then to two-stage coker gas compressor C-151 (see Figure 6-2). The first stage discharge flows through the compressor

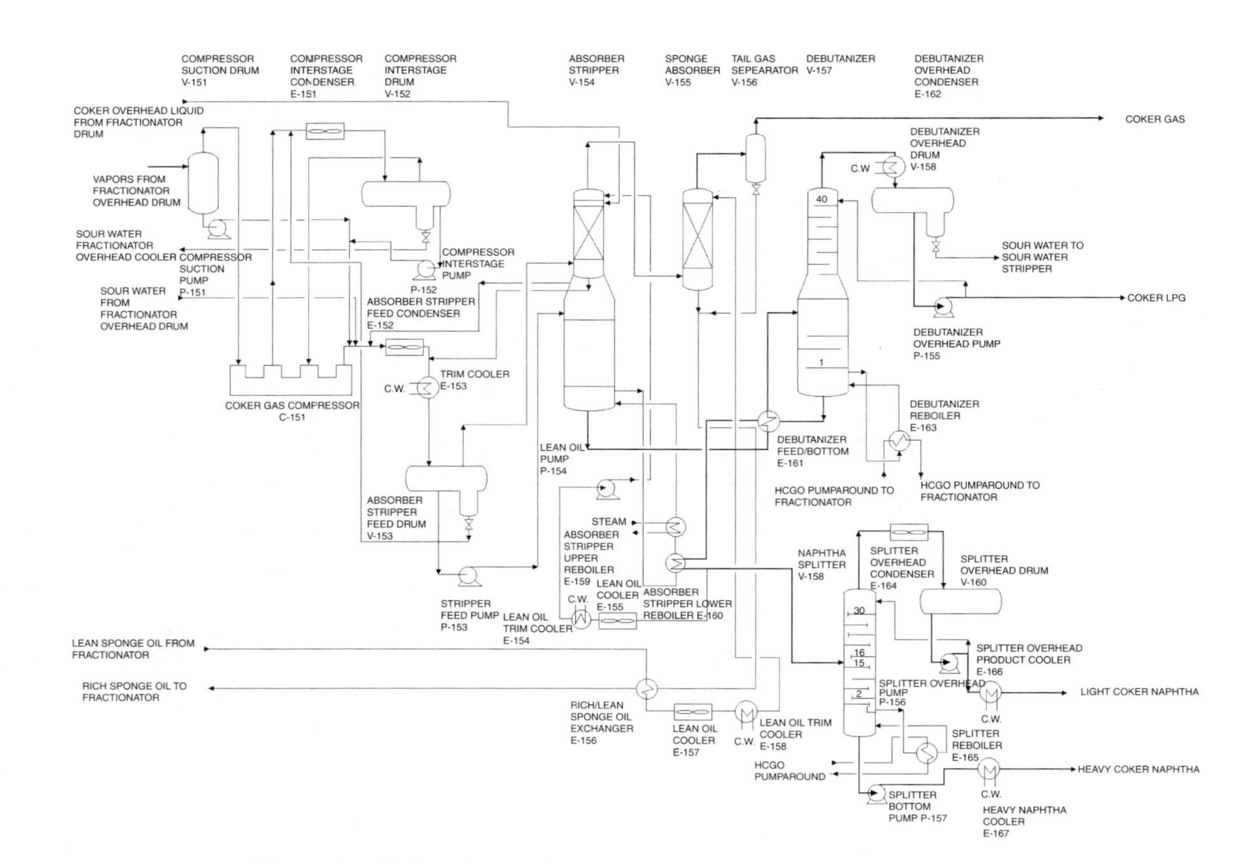

Figure 6-2. Delayed coker (vapor recovery section). C.W. = cooling water.

interstage condenser to the compressor interstage drum. Sour water from absorber stripper feed drum V-153 is injected into the compressor first-stage discharge to prevent cyanide-induced corrosion in the downstream equipment. Vapors from the compressor interstage drum enter the second stage of the compressor.

Liquid hydrocarbon from the interstage drum is pumped to absorber stripper feed condenser E-152, and sour water flows to the coker fractionator overhead cooler. Vapor from the second stage of the compressor is combined with the following: overhead vapor from the stripper, sour water from fractionator overhead drum, and hydrocarbon liquid from compressor interstage drum.

The combined flow is cooled in absorber stripper feed trim condenser E-152. The cooled vapor/liquid mixture from trim cooler E-153 flows to absorber stripper feed drum V-153. The hydrocarbon liquid from the feed drum is pumped by stripper feed pump to the absorber stripper. Sour water from the absorber stripper feed drum flows to the compressor inter stage condenser.

In the stripper, light hydrocarbons are stripped out. The liquid from the bottom of the absorber stripper flows through debutanizer feed bottom exchanger E-161 to debutanizer V-157. Stripper overhead vapors return to absorber stripper feed condenser E-152. The stripper is reboiled with a steam reboiler. Sour water withdrawn from the stripper enters a water separator (not shown in Figure 6-2) then to the battery limit.

Vapor from the absorber stripper feed drum flows to the bottom of the absorber. This vapor is contacted in countercurrent flow with the lean oil. This lean oil consists of unstabilized gasoline from the fractionator overhead drum and a stream of cooled naphtha from the bottom of the debutanizer V-157. Rich oil from the bottom of the absorber flows to absorber stripper feed trim condenser E-153.

The vapor from the top of the absorber flows to the bottom of sponge absorber V-155, where it is contacted in counercurrent flow with cool lean sponge oil. The sponge absorber minimizes the loss of naphtha to the fuel gas. The tail gas from the top of the sponge absorber flows to tail gas separator V-156, where entrained liquids are removed before flowing to battery limits. Rich sponge oil from the bottom of the sponge absorber is sent to the coke fractionator.

Debutanizer V-157 splits the liquid from the stripper into two streams, a C_3/C_4 overhead and a naphtha bottom stream. The debutanizer is reboiled using heavy coker gas oil (HCGO) pumparound from the fractionator.

Naphtha from the bottom of the debutanizer is cooled as it passes through the debutanizer feed/bottom exchanger and absorber stripper reboiler. The net naphtha product flows under level control to splitter V-159. A stream of cooled naphtha is recycled by the lean oil pump to the absorber for use as lean oil. The overhead vapor from debutanizer is condensed in debutanizer overhead condenser E-162 and enters debutanizer overhead drum V-158. Any uncondensable vapor from the overhead drum is vented to the compressor suction drum. Hydrocarbon liquid from the debutanizer overhead drum flows to the debutanizer overhead pump, which handles both the debutanizer reflux and the C_3/C_4 product stream.

Naphtha splitter V-159 separates the net naphtha product from the debutanizer into a light coker naphtha and heavy coker naphtha product. The naphtha splitter is reboiled with HCGO pumparound. The naphtha splitter overhead vapor is condensed in the splitter overhead condenser and collected in the splitter overhead drum. The splitter overhead pump pumps both the splitter reflux and the light coker naphtha stream. The heavy coker naphtha from the bottom of the naphtha splitter is pumped through the splitter bottom product cooler to the battery limits by the splitter bottom product pump.

The light hydrocarbon vapor from the blowdown settling drum are sent to the fuel gas recovery system.

STEAM GENERATION

The net heat removed from the fractionator by the heavy gas oil pumparound stream is used to reboil towers in the vapor recovery section. Additional heat may be removed from the pumparounds by steam generation and preheating boiler feed water.

DECOKING

The decoking operation consists of the following steps:

1. *Steaming*. The full coke drum is steamed to remove any residual oil liquid. This mixture of steam and hydrocarbon is sent first to the fractionator and second to coker blowdown system, where the hydrocarbons are recovered.
2. *Cooling*. The coke drum is filled with water, allowing it to cool below 200°F. The steam generated during cooling is condensed in the blowdown system and may be recovered outside battery limits.

3. *Draining*. The cooling water is drained from the drum and recovered for reuse.
4. *Unheading*. The top and bottom heads are removed in preparation for coke removal.
5. *Decoking*. Hydraulic decoking is the most common cutting method. High-pressure water jets are used to cut coke from the coke drum. Water is separated from the coke fines and reused.
6. *Heading and testing*. After the heads are replaced, the drum is tightened, purged, and pressure tested.
7. *Heating*. Steam and vapors from hot coke drum are used to heat the cold coke drum. The condensed water is sent to the blow down drum. Condensed hydrocarbons are sent to either the coke fractionator or the blowdown drum.
8. *Coking*. The heated coke drum is placed on stream and the cycle is repeated for the other drum.

OPERATING CONDITIONS

Operating conditions for a delayed coker unit processing a vacuum resid from a light Arabian/Kuwait mix are shown in Table 6-1. The basic variables contributing to yield and product quality are temperature, pressure, and recycle ratio. The effect of temperature, pressure, and recycle ratio and their correlation with coker yields have been discussed in detail by Castiglioni[1] and Deblane and Elliot.[2]

Temperature

The temperature is used to control the volatile combustible material (VCM) of the coke product. Coke is generally produced with a VCM

Table 6-1
Delayed Coker Operating Conditions

OPERATING PARAMETER	UNIT	
HEATER COIL OUTLET TEMPERATURE	°F	927
DRUM OUTLET TEMPERATURE	°F	802
DRUM PRESSURE	psig	24.5
FLASH ZONE TEMPERATURE	°F	749
COMBINED FEED RATIO (CFR)		1.24
TOWER TOP TEMPERATURE	°F	204
ACCUMULATOR PRESSURE	psig	10.7

ranging between 6 and 8%. This produces harder and, if the impurities level is acceptable, more desirable anode-grade coke. At constant drum pressure and recycle ratio, the coke yield decreases as drum temperature is increased. If the coking temperature is too low, the coking reactions do not proceed far enough and produce pitch or soft coke. When the temperature is too high, the coke formed is generally too hard and difficult to remove from the coke drum with hydraulic decoking equipment.

As the furnace supplies all the necessary heat to sustain endothermic coking reactions, higher temperature increases the potential of coking furnace tubes and transfer lines. Therefore, in actual practice, the furnace outlet temperature and corresponding coke drum vapor temperature must be maintained within a narrow limit.

Pressure

Pressure increases at constant temperature cause more heavy hydrocarbons to be retained in the coke drum. This increases the coke yield and slightly increases the gas yield. Therefore, delayed cokers are designed with the lowest possible operating pressure to minimize coke and increase distillate yields. Lower pressure results in increased expense for the vapor handling system. The range of pressure used is between 15 and 35 psig.

Recycle Ratio

The recycle ratio has the same general effect as pressure on product distribution. As recycling is increased, the coke and gas yields increase while the liquid yield decreases. Recycling is used primarily to control the end point of the coker gas oil. In general, the refinery operates at as low a recycle ratio as product quality and unit operation permit.

COKER RUN LENGTH

The run length between decoking while processing heavier, high Conradson carbon feed stocks is an important variable. Run length is affected by feedstock quality and operating conditions. Although it is not possible to predict a run length, values generally encountered are between 9 and 12 months for a maximum radiant heat flux of 10,000 to 12,000 Btu/(hr)(ft^2) and feed cold velocities in the tube of around 6 ft/sec. Multiple injections of steam are used to adjust coil residence time and velocity.

COKER YIELDS AND QUALITY

The yield and quality from a typical coker unit processing Middle Eastern crude are shown in Tables 6-2 to 6-7.

COKE PROPERTIES AND END USES

The properties of the coke produced and its end uses can vary widely, depending on the properties of the feedstock to the delayed coker unit and

Table 6-2
Delayed Coker Utility Consumption *(per Ton Feed)*

UTILITY	UNIT	CONSUMPTION
FUEL	mmBtu	0.8571
STEAM	mmBtu	−0.07
POWER	kWhr	7.35
COOLING WATER	MIG*	1.63

*MIG = 1000 IMPERIAL GALLONS.

Table 6-3
Delayed Coker Yield, Vacuum Resids

COMPONENT	FEED 1 WT%	FEED 2 WT%
FEED		
UNDESULFURIZED VACUUM RESID	1.0000	
DESULFURIZED VACUUM RESID		1.0000
TOTAL FEED	1.0000	1.0000
PRODUCTS		
ACID GAS	0.0141	0.0050
COKER OFF GAS	0.0809	0.0830
COKER LIGHT NAPHTHA	0.0430	0.0200
COKER HEAVY NAPHTHA	0.0850	0.0695
COKER KEROSENE		0.1750
COKER DIESEL	0.2220	0.2550
COKER GAS OIL	0.2360	0.1690
COKE	0.3190	0.2135
LOSS		0.0100
TOTAL PRODUCT	1.0000	1.0000

FEED 1: FROM LIGHT ARABIAN/KUWAIT MIX.
FEED 2: DESULFURIZED VACUUM RESID FEED FROM KUWAIT CRUDE.

Table 6-4
Delayed Coker Gas Yield, Undesulfurized Vacuum Resid Feed

COMPONENT	WT% FEED
H_2	0.03
H_2S	1.41
C_1	1.94
$C_2 =$	0.27
C_2	1.85
$C_3 =$	0.62
C_3	1.58
IC_4	0.24
$C_4 =$	0.73
NC_4	0.83
TOTAL (INCLUDING ACID GAS)	9.50

Table 6-5
Feed and Product Qualities, Undesulfurized Feed

PROPERTY	UNITS	FEED	C5-175°F	175–355°F	350°F+	COKE
GRAVITY	°API	7.3	74	56.6	26.8	
SULFUR	Wt%	4.7	0.4	0.9	2.9	6.2
NITROGEN	ppmw		100	200	1500	
PONA	vol%		50/46/3/1	30/40/20/10		
BROMINE NUMBER			100	70	25	
RON			81			
VISCOSITY	122, cst				4	
VANADIUM	ppmw				0.2	229
NICKEL	ppmw				0.3	50
CON. CARBON	Wt%	21.2			0.1	
VOLATILES	Wt%					10

processing conditions employed. Coke product from coker units is known as *green coke*. Green coke is composed mainly of carbon but also contains 10–15% volatile hydrocarbons, together with other impurities, such as sulfur, vanadium, nickel, or nitrogen. If green coke has sufficiently low levels of impurities, such as sulfur and metals, it may be suitable for calcining. The higher-quality green coke is often said to be of "anode quality." Calcining involves heating the green coke to drive off volatile

Table 6-6
Feed and Product Qualities, Desulfurised Feed

PROPERTY	UNITS	FEED	LIGHT NAPHTHA	HEAVY NAPHTHA	KEROSENE	DIESEL	GAS OIL	COKE
GRAVITY	°API	14	82.5	63.6	44.2	29.8	19.2	
REAL DENSITY	gm/cc							2.105
ASTM D86								
	10%, F	1015	118	192	331	548	717	
	50%		126	213	390	609	831	
	90%		147	255	476	697	932	
SULFUR	Wt%	1.51	0.1	0.1	0.27	0.79	1.2	2.67
NITROGEN	ppmw		100			2300		
PONA	vol%							
BROMINE NUMBER								
FREEZE POINT	°F				−54.6			
POUR POINT	°F					38.1		
RON			81.8	73.7				
VISCOSITY	122, cst					25.6		
VANADIUM	ppmw							150
NICKEL	ppmw							80
SODIUM	ppmw							90
CON. CARBON	Wt%	12.3						
VOLATILES	Wt%							9.5

Table 6-7
Typical Coke Specifications

TYPE	UNITS	SPONGE COKE	NEEDLE COKE
USE		ANODES	ELECTRODES
GREEN COKE			
SULFUR	Wt%	<3	<1.5
METALS			
V	ppmw	<350	
Ni	ppmw	<300	
Si	ppmw	<150	
Fe	ppmw	<270	
VOLATILE MATTER	Wt%	<12	<6
CALCINED COKE			
MOISTURE	Wt%	<0.5	<0.5
VOLATILE MATTER	Wt%	<0.5	
ASH	Wt%	0.5	0.5
SULFUR	Wt%	<3.0	<1.50
METALS			
V	ppmw	<350	
Ni	ppmw	<300	
Si	ppmw	<150	
Fe	ppmw	<270	
DENSITY	gm/cc		
200 MESH, R.D.		2.04–2.08	>2.12
VIBRATED BULK DENSITY		>0.80	
CTE*	$1/°C \times 10^{-7}$	<40	<4

*CTE = COEFFICIENT OF THERMAL EXPANSION.

components and improve its electrical conductivity. Calcined coke produced is used in aluminum smelting, titanium dioxide production, and to increase the carbon content in iron and steel production. Petroleum coke not suitable for calcining is used as fuel in various applications and said to be of "fuel grade." Its high sulfur content frequently limits its scope for use in power generation. A major consumer is the cement industry, where impurities present in fuel are absorbed in the cement product and not released to atmosphere.

Petroleum coke can be broadly classified into two categories, sponge coke and needle coke, depending on its physical properties, such as its texture, density, porosity, electric resistivity, and coefficient of thermal conductivity. Typical properties of sponge and needle coke are shown in Table 6-7.

Sponge Coke

Sponge coke is average-quality petroleum coke produced from mainly nonaromatic feedstocks. The aluminum industry is the largest single user for this coke, where it is used for anode making. Roughly 0.4–0.5 ton anode is consumed per ton of aluminum produced. Anodes are manufactured by blending petroleum coke aggregates with coal tar pitch. The characteristics of calcined coke follow:

- *Sulfur.* Anode-grade calcined petroleum coke varies in sulfur content between 0.5 and 3%. Although sulfur in calcined petroleum coke improves anode performance because it inhibits the negative side reactions of air and carbon dioxide, the allowable sulfur level in most cases is determined by environmental regulations in areas where the smelting plants are located.
- *Metals.* Metals are contaminants with a negative impact on the purity of aluminum and other products produced. Vanadium promotes air reactivity and inhibits conductivity of the aluminum produced. Calcium and sodium have a significant negative impact on CO_2 and air reactivity.
- *Density.* The apparent and bulk density are variables that determine the density of the resulting anode constructed from the material. They are also important parameters for the amount of pitch (coal tar) used as a coke binder during anode construction. Higher density values are desirable.
- *Air and CO_2 reactivity.* At an anode operating temperature of approximately 1800°F, the carbon reacts and is consumed by the hot air around the anode and CO_2 being generated as a part of the process. Values of 10% and 0.1% per minute of air and CO_2 reactivity are typical.
- *Particle size.* In general, calcined coke needs to have 30–35% of the particles larger than ¼ in. to produce a suitable aggregate required for the production of anodes.

Needle Coke

Needle coke is used in the manufacture of large diameter (24–28 in.) graphite rods for ultra high-power furnace electrodes. The material used must be of high density, low resistivity, high strength, and very low coefficient of thermal expansion (CTE).

This material is produced from aromatic feedstocks to the coker unit. Also, the feed must be very low in sulfur, metals, and asphaltenes. Refinery-derived feedstocks such as thermal tars, decant or slurry oil from catalytic cracking, and extracts from the lube solvent extraction unit are most suitable.

The operating conditions for needle coke require higher drum temperatures, which decrease the CTE of the coke produced. This, however, increases furnace fouling rate and coke cutting time, as the coke produced is harder. A 15°F increase in drum temperature is known to decrease the CTE by 30% but triples the coke cutting time.[3]

VISBREAKING

Visbreaking is a mild thermal cracking process.[4] The function of a visbreaking unit is to produce lower viscosity and low-pour resid for blending to fuel oil. In this cracking process, cracked gas, gasoline/naphtha, gas oil, and thermal tar are produced. The gas oil is blended back into the thermal tar to yield fuel oil. Thermal cracking reduces the viscosity and pour point of the resid and hence the cutter stock requirement for blending this resid to fuel oil. Thus, the overall production of fuel oil is reduced. A second consideration is the removal of some feed sulfur. Although visbreaking is an inefficient process in this respect, sulfur removal does occur to some extent.

THE VISBREAKING PROCESS

The feed to the unit is vacuum resid at 670°F coming from the vacuum distillation unit (see Figure 6-3). The charge stream is pumped by a charge pump P-101 to visbreaker heater H-101A and 101B, where it is heated to 920°F. The visbreaker heater is a single-pass, controlled-gradiant, box-type heater with a preheater and two soaking coils. The temperature profile of a visbreaker heater is shown in Figure 6-4. Water injection connections are provided at the inlet of the soaking coils, although during normal operation, no water is injected. Normally, one heater is in operation while the other one is on standby.

During the course of the run, coke forms in the tube. Therefore, permanent steam/air decoking facilities are provided. The tube metal temperature is assumed to reach a maximum of 1400°F during decoking. The coil is designed for approximately 133% of the clean tube pressure drop.

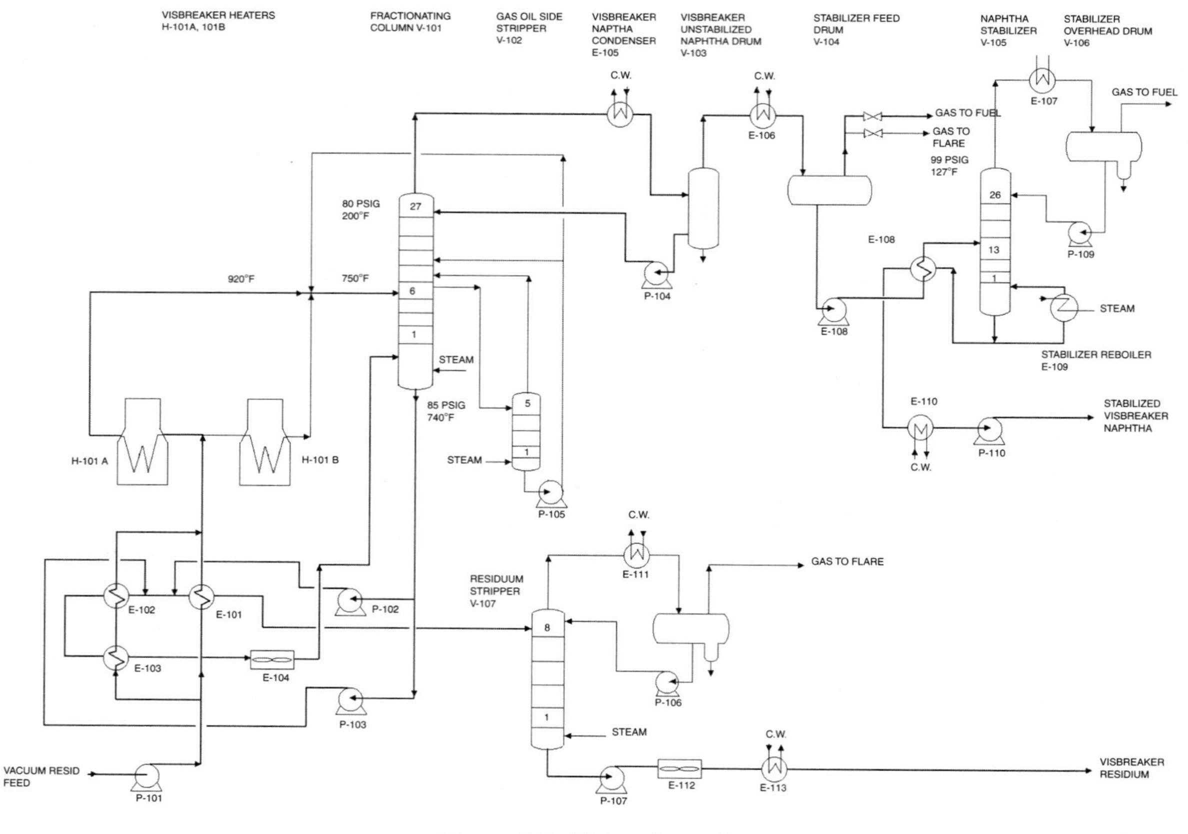

Figure 6-3. Visbreaker unit.

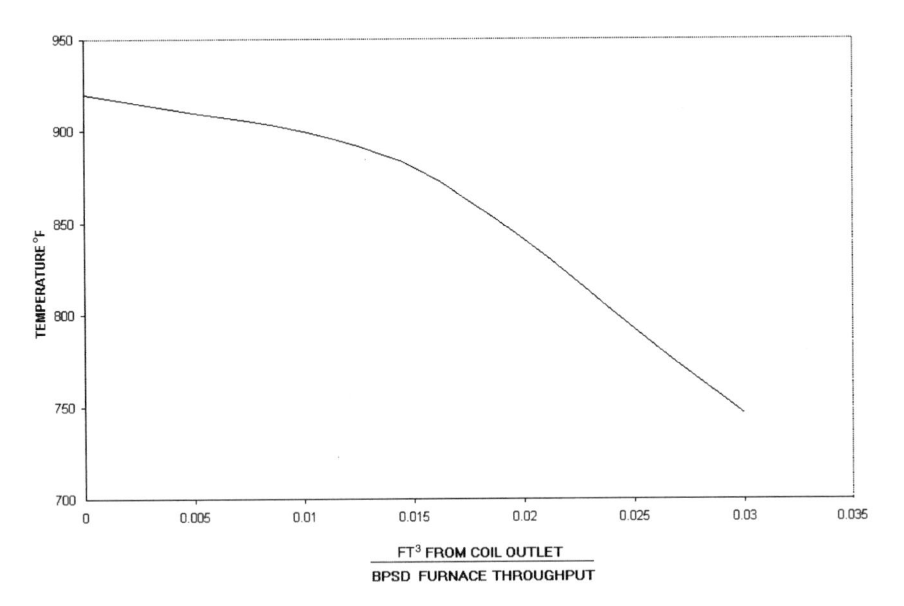

Figure 6-4. Visbreaker temperature gradient.

The effluent from the heater is quenched to 750°F, with cracked gas oil coming from gas oil side-stripper. The quenched visbreaker effluent flows to main fractionator column V-101, with about 27 plates. The vapor is flashed off and separated into gas, gasoline, cracked naphtha, and a gas oil cut. The overhead vapor condensed in condenser E-105 accumulates in unstabilized naphtha drum V-103, which provides reflux to the main fractionator column. The uncondensed vapor is further cooled in E-103 and accumulates in stabilizer feed drum V-104. The unstabilized naphtha from V-104 flows to a stabilizer or debutanizer column V-105, with 26 plates, which removes any C_4 gases as the top stream and a stabilized naphtha stream as the bottom product.

A gas oil sidestream is withdrawn from the main fractionator column and steam stripped in V-102, the gas oil stripper. This stream is pumped by P-105 and quenches the visbreaker heater effluent at the heater outlet.

To minimize coke formation in the bottom sump section, the fractionator column is maintained at 650°F by removing a slipstream of thermal tar and cooling it by heat exchange with incoming fresh feed and an air cooler before returning it to the column.

The bulk of visbroken thermal tar is transferred by P-102 to residuum stripper column V-107, where it is steam stripped with medium-pressure steam to remove light ends and gas. The visbroken tar is withdrawn from the stripper column and cooled in exchangers E-112 and E-113 before being sent to storage.

The operating conditions of a visbreaker unit are shown in Table 6-8. The yield from a visbreaker unit, the unit's utility consumption, and the properties of feed and product are shown in Tables 6-9 to 6-12.

The important process variables are charge stock quality, cracking temperature, and residence time of oil in the coil. Feed to visbreaker units is either reduced crude or vacuum resids. Resids from paraffinic crudes are the most suitable because of the larger reduction in pour point achieved. The coil temperature profile and residence time are very closely controlled (Figure 6-4) to monitor the severity of operation, usually measured by the amount of gasoline produced. Higher than optimum temperatures can produce unstable tar and more frequent coking of visbreaker coil or shorter run lengths.

Table 6-8
Visbreaker Operating Conditions, *Basis 7634 BPSD Vacuum Resid from Darius (Iranian) Crude*

OPERATING PARAMETER	UNITS	
PREHEATER SECTION (CONVECTION + ONE RADIANT COIL)		
FEED INLET TEMPERATURE	°F	670
OUTLET TEMPERATURE	°F	800
DUTY	mmBtu/hr	10.73
HEAT FLUX	Btu/hr °F	10000
FIRST SOAKING SECTION		
INLET TEMPERATURE	°F	800
OUTLET TEMPERATURE	°F	883
DUTY	mmBtu/hr	9.00
COIL VOLUME	ft^3	93
SECOND SOAKING SECTION		
INLET TEMPERATURE	°F	883
OUTLET TEMPERATURE	°F	920
DUTY	mmBtu/hr	12.9
COIL VOLUME	ft^3	93
STEAM SUPERHEATER SECTION (*CONVECTION COIL*)		
INLET PRESSURE	psia	170
INLET TEMPERATURE	°F	370
OUTLET PRESSURE	psia	160
OUTLET TEMP	°F	600

Table 6-9
Visbreaker Unit Yields

COMPONENT	VACUUM RESID FEED, W/W	ATMOSPHERIC RESID, W/W
FEED		
ATM RESID FEED		1.0000
VACUUM RESID FEED	1.0000	
TOTAL FEED	1.0000	1.0000
PRODUCTS		
H_2	0.0001	
H_2S	0.0053	
C_1	0.0047	
$C_2 =$	0.0006	
C_2	0.0047	
$C_3 =$	0.0026	
C_3	0.0051	
IC_4	0.0015	
$C_4 =$	0.0063	
NC_4	0.0069	
IC_5	0.0061	
$C_5 =$	0.0061	
NC_5	0.0023	
GASES, TOTAL		0.0250
LIGHT GASOLINE	0.0198	
NAPHTHA	0.0437	0.0750
GAS OIL	0.1455	0.2400
700°F + TAR	0.7387	0.6600
TOTAL	1.0000	1.0000

Table 6-10
Visbreaker Utility Consumption *per Ton Vacuum Resid Feed*

UTILITY	UNITS	CONSUMPTION
FUEL	mmBtu	1.8010
STEAM	mmBtu	−0.6520
POWER	kWhr	6.1840
DISTILLED WATER	MIG*	0.0257
COOLING WATER	MIG*	1.3011

*MIG = 1000 IMPERIAL GALLONS.

Table 6-11
Feed and Product Properties from Visbreaker, *Vacuum Resid Feed from an Iranian Crude*

PROPERTY	UNITS	FEED	LIGHT CRACKED GAS	CRACKED NAPHTHA	CRACKED GAS OIL	VISBREAKER TAR
GRAVITY	°API	4.8	73.9	52.3	34	4.4
SPECIFIC GRAVITY		1.0382	0.6889	0.7699	0.8550	1.0412
TBP RANGE	°F		C_5–180	180–330	330–670	
SULFUR	Wt%	6.28	0.34	0.57	1.35	6.32
VISCOSITY	Cst, 122°F	450000			2	300000
	Cst, 210°F	2300				1200
	Cst, 300°F	130				72
	Cst, 500°F	7.2				
MOL WT		690	77	106.7	193	480
UOP "K"		11.6				
POUR POINT	°F	135			30	60

Table 6-12
Feed and Product Properties from Visbreaker, *Atmospheric Resid Feed*

PROPERTY	UNITS	FEED	VISBREAKER TAR
GRAVITY	°API	17.7	21.5
SPECIFIC GRAVITY		0.9484	0.9248
VISCOSITY	Cst, 122°F	175	
	Cst, 210°F	22	10
POUR POINT	°F	50	40
DRY SEDIMENT	Wt%		0.15

HEATER DECOKING

Decoking is done by first introducing 450 psig steam, shutting off the burners and charge pump, then allowing several minutes to elapse while the heater tubes are blown to the column. The 450 psig steam is shut off and the heater blocked with blinds.

The decoking pipes are next connected to both the inlet and outlet of the heater and 150 psig steam is introduced at the inlet. One or more burners are lit in the preheat section to bring the steam temperature to 800°F. Water is introduced in the effluent line to limit the temperature to 250°F. After an hour has elapsed, to allow any volatile matter in the coke to be stripped off, air (105 psia) is very gradually introduced to burn off carbon. The progress of the burn off can be observed through heater peepholes as a relatively narrow band of red hot tubing. This band gradually progresses down the heater coil as the coke is burned off. The air rate is adjusted as necessary to maintain the dark red appearance of the tube.

The piping is arranged to permit reversing the direction of flow of air and steam, should a localized burning section becomes too hot (>1300°F). When the decoking step is complete, the tubes are cooled with steam, followed by air cooling, and finally water is very gradually introduced then increased in rate to wash out any residual fly ash. Afterward, the water is drained from the heater.

REGENERATION TIME

Regeneration or decoking time of visbreaker heater coil can be calculated by simple calculations.

EXAMPLE 6-1

The particulars of an actual visbreaker coil follow. We are to estimate the time required for decoking this coil. Assume that coking takes place in tubes operating above 800°F. Neglect heat losses and hydrogen content of coke.

$$\text{Visbreaker coil nominal diameter} = 4\,\text{in.}$$

$$\text{Visbreaker coil pipe schedule} = 80$$

$$\text{Visbreaker coil internal diameter} = 3.826\,\text{in.}$$

$$\text{Total length of visbreaker pipe} = 5080\,\text{ft}$$

$$\text{Length of pipe operating above } 800^\circ\text{F} = 2300\,\text{ft}$$

SOLUTION

$$\text{Volume of coke deposited assuming } {}^{3}/_{16}\text{-in. coke layer (ft}^3) = \frac{3.14 \times 3.826 \times 2300 \times 3}{(12 \times 12 \times 16)}$$

$$= 35.97$$

$$\text{Density of coke (assume)} = 75\,\text{lb/ft}^3$$

$$\text{Weight of coke deposited} = 35.97 \times 75$$

$$= 2700\,\text{lb}$$

Heat of Combustion

Assume coke deposits in the coil consist of 92% carbon and 8% sulfur.

$$\text{Heat from carbon} = 2700 \times 0.92 \times 14{,}100$$

$$= 35.02\,\text{million Btu}$$

$$\text{Heat from sulfur} = 2700 \times 0.08 \times 4000$$

$$= 0.864\,\text{million Btu}$$

$$\text{Total hat} = 35.884\,\text{million Btu}$$

Time of Regeneration

Assume superheated steam at 800°F is used for decoking and the available steam supply is 2500 lb/hr. In the operation, steam leaving the combustion zone is at 1250°F. The specific heat of steam is assumed at 0.51. The time of coil decoking can be calculated as follows:

$$\text{Time required} = \frac{35.884 \times 1{,}000{,}000}{(450 \times 0.51 \times 2500)}$$
$$= 62.54\,\text{hr}$$

Heater Duty

$$\text{For steam superheating from 340 to 800°F} = 2500\ (1432 - 1194)$$
$$= 535{,}000\,\text{Btu/hr}$$

Quench Water Requirements

The amount of water (90°F) required to quench 2500 lbs/hr steam from 1250 to 250°F is estimated as follows:

$$\text{Quench water} = \frac{2500 \times (1250 - 250) \times 0.5}{(1164 - 58)}$$
$$= 1130\,\text{lb/hr}$$
$$= 2.25\,\text{GPM}$$

SOLVENT DEASPHALTING

Historically, solvent deasphalting of vacuum residues has been used in the manufacture of lubricating oil to separate out the heavy fraction of crude oil beyond the range of economical commercial distillation, using propane as solvent. The feed to the deasphalting unit is usually a vacuum resid with a 950°F TBP cut point. Over time, this process has come to be used to prepare catalytic cracking feeds, hydrocracking feeds, hydrodesulfurizer feeds, and asphalts.

Studies have shown that high yield of oil can be obtained, while limiting asphaltenes and metals, by using the proper heavier solvent. Thus, extraction rates from 65 to 85% of deasphalted oil (DAO) have been obtained. Whereas vacuum residue is a very difficult feed stock for catalytic processes, DAO can be easily processed, like other heavy distillates. The asphalt produced can be blended with straight-run asphalts or blended back with fuel oil.

Modern solvent deasphalting units usually use a blend of light hydrocarbon solvents (C_5-C_6 paraffinic cut) to allow maximum operating flexibility. The solubility of oil in solvent at a fixed temperature increases as the concentration of heavier components in the solvent increase.

Selectivity is the ability of the solvent to separate paraffinic and sometimes resinous oils from the asphalt or vacuum resid feed. As the metals, sulfur, and nitrogen are generally concentrated in the larger molecules, the metal, sulfur, and nitrogen content of deasphalted oil is considerably reduced, as shown in Figure 6-5. Maximum selectivity subject to economic constraints is always the processing objective. Selec-

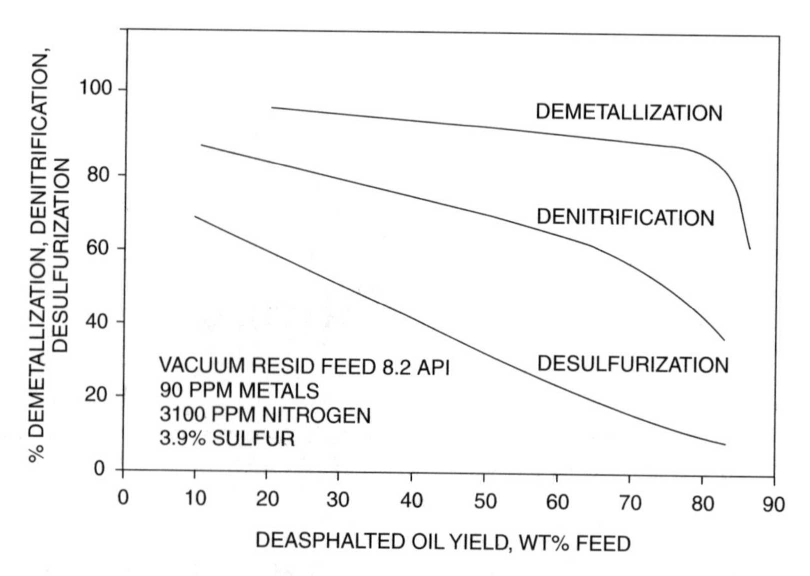

Figure 6-5. Percent of demetallization and desulfurization as a function of a desasphalted oil yield.

tivity can be improved by increasing the solvent/oil ratio at a constant DAO yield. Since considerable energy is required to recover the solvent, there is always an optimum solvent/oil ratio for each operation. The yield of DAO and asphalt from vacuum resid feed is shown in Table 6-13. Table 6-14 shows the deasphalting unit's utility consumption.

The solubility of oil in the solvent decreases with increasing temperature and this variable provides the major method of day-to-day operational control of the process.

Table 6-13
Solvent Deasphalting Yields and Product Qualities

PROPERTY	UNITS	FEED 1	FEED 2
FEED			
GRAVITY	API	8.2	13.8
SULFUR	Wt%	3.9	0.6
NITROGEN	wppm	3100	2100
NICKEL	wppm	19	2
VANADIUM	wppm	61	11
CON CARBON	Wt%	19	13
VISCOSITY	Cst, 210°F	500	375
DEASPHALTED OIL			
YIELD LV%		83	75
GRAVITY	API	14.4	19.5
SULFUR	Wt%	3.55	0.47
NITROGEN	wppm	2000	1200
NICKEL	wppm	3	0.3
VANADIUM	wppm	11	1.3
CON CARBON	Wt%	8.4	4.7
VISCOSITY	Cst, 210°F	92	79
ASPHALT			
YIELD	LV%	17	25
SPECIFIC GRAVITY		1.125	1.0839
SULFUR	Wt%	5.6	0.94
SOFTENING POINT	R & B, °F	302	200
VISCOSITY	Cst, 400°F		100

NOTE:
FEED 1, START RUN VACUUM RESID FROM MIDDLE EASTERN CRUDE.
FEED 2, VACUUM RESID EX ARDS UNIT, MIDDLE EASTERN CRUDE.

Table 6-14
Solvent Deasphalting Unit Utility Consumption

UTILITY	UNITS	CONSUMPTION
FUEL	mmBtu	0.533
LP STEAM	mmBtu	0.632
ELECTRIC POWER	kWhr	18.1

THE DEASPHALTING PROCESS

Extraction Section

Fresh feed (vacuum residue) is pumped into the solvent deasphalting unit and combined with a small quantity of predilution solvent to reduce its viscosity (see Figure 6-6). The combined vacuum residue and predilution solvent at the desired extraction temperature flow into the middle of rotating disk contactor (RDC) V-101. Solvent streams from HP and LP solvent receivers V-106 and V-107 are combined, and a portion of the combined stream is used as predilution solvent. The major portion of the solvent is regulated to the desired temperature by solvent heater E-101A and cooler E-101B then flows into the bottom section of the RDC. The desired temperature gradient across the RDC is maintained by steam coils located at the top of the tower. The highest temperature is at the top of the RDC, and the lowest temperature is at solvent inlet. In the top section of the RDC, relatively insoluble, heavier material separates from the solution of DAO. The material flowing back down the RDC provides internal reflux and improves the separation between DAO and asphalt.

DAO RECOVERY SECTION

DAO plus the bulk of the solvent (DAO mix) leaves the top of the RDC tower and flows to the DAO recovery system. The pressure at the top of the RDC is controlled by a back-pressure control valve. The DAO mix flows to fired heater H-101, to vaporize a portion of the solvent. The DAO mix then flows into HP flash tower V-104, where the solvent vapor is taken overhead. The DAO mix is held under level control in the bottom of this tower. The liquid mix then flows to pressure vapor heat exchanger (PVHE) E-102, where it is heated by condensing solvent from high-pressure flash. The solvent, totally condensed, flows to HP solvent

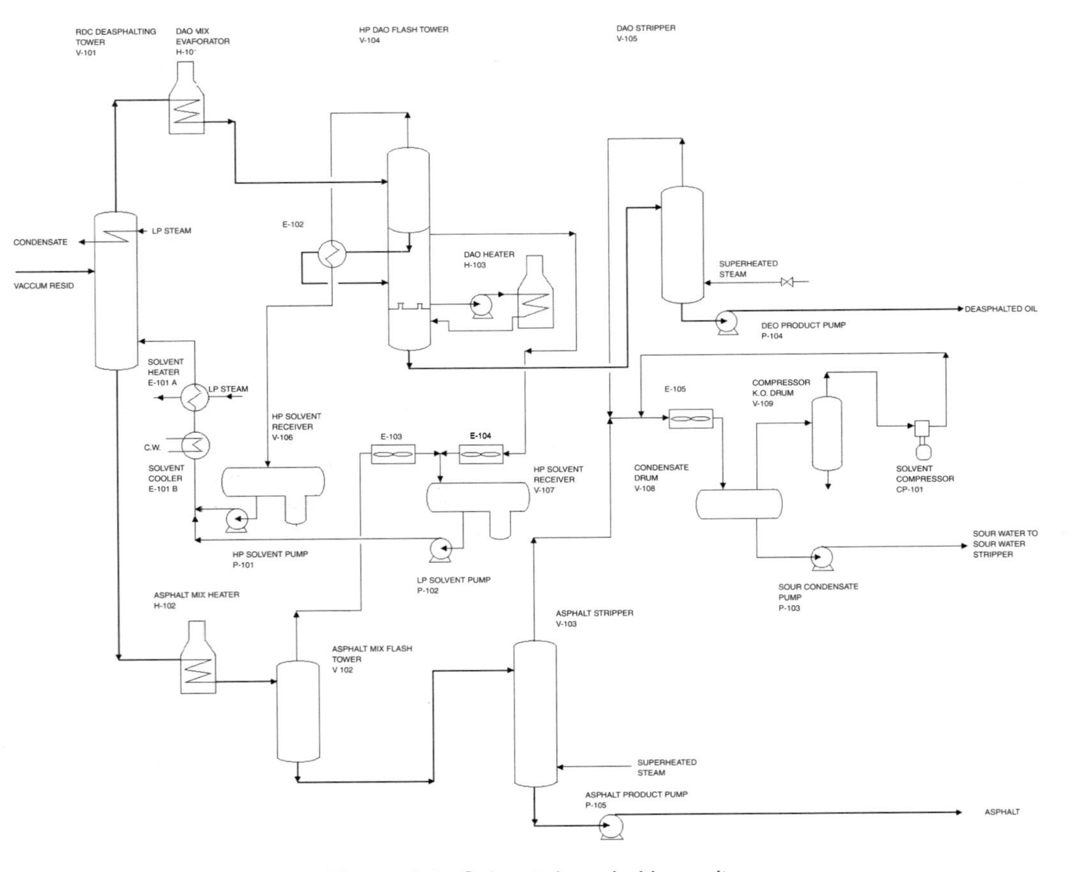

Figure 6-6. Solvent deasphalting unit.

receiver V-106. The DAO mix flows from PVHE to an LP flash tower, where again solvent vapors are taken overhead. The DAO mix flows down the tower, where it is further heated by rising solvent vapors from the reboiler. The remaining DAO mix is circulated through fired heater reboiler H-103 and returns to the bottom of the LP flash tower. The solvent vapor rise through the tower and the liquid mix is held under level control before flowing into DAO stripper V-105.

DAO RECOVERY

The mix enters the top tray of the stripper, and the remaining solvent is stripped overhead with superheated steam, which enters below the bottom tray. The DAO product is pumped from the stripper bottom by P-104 to the battery limits.

ASPHALT RECOVERY

The solvent/asphalt mix from the RDC tower flows at a controlled rate to asphalt mix heater H-102. The hot two-phase asphalt mix from the heater is flashed in asphalt mix flash tower V-102. The solvent vapors are taken overhead. The remaining asphalt mix flows to asphalt stripper V-103 and enters the tower on the top tray. Superheated steam is used to strip the remaining solvent from the asphalt. The wet solvent vapor overhead combines with the overhead vapors from the DAO stripper. The asphalt product is pumped from the stripper bottom by P-105 to the battery limits.

SOLVENT SYSTEM

The solvent evaporated in the LP flash tower is condensed and flows to the LP solvent receiver. Solvent from the asphalt flash tower, which operates at the same pressure, is condensed in a separate condenser and flows to LP solvent receiver. The reason for segregating the two solvent streams is the potential for the accidental fouling by asphalt carried over from the flash tower.

Vapor overhead from the two strippers is cooled, and most of the water is condensed and recovered in stripper condensate drum V-108. The water collected is considered sour and pumped from the unit to the sour water stripper. The noncondensed solvent vapor from this drum flows to solvent compressor suction drum V-109, where any entrained liquid is knocked out. The vapor is then compressed by solvent compressor CP-101 and joins the vapors from the asphalt flash drum, upstream of asphalt solvent condenser. Makeup solvent is pumped to the LP solvent receiver from the off-site solvent tank as required.

BITUMEN BLOWING

Industrial grades of bitumen with industrial uses, such as road paving, waterproofing, insulation, are manufactured from heavy crudes, either by vacuum reduction alone or air blowing of vacuum resids. However, the vacuum resids from all crudes are not suitable for making bitumen.

Vacuum distillation of some crudes under specified conditions may yield resid that meets the specifications of a bitumen for certain paving grades, but for low penetration and higher softening point grades, air blowing the vacuum resid under specified operating conditions is required. Pilot plant tests are generally necessary to establish whether a given crude can yield good-quality bitumen and determine the optimum operating conditions for a given bitumen grade. Subjecting the bitumen to high temperatures during processing can affect the ductility of the product and care is taken not to subject asphalt to more than approximately 750°F.

In the air blowing operation, the aromatic and polar compounds in the feed are condensed to form higher-molecular-weight chemical species. The process increases the asphaltene level, while the level of aromatics and polar aromatics are reduced. The change is also indicated by a significant increase in the average molecular weight of the blown asphalt. The saturates in the feed, consisting mainly of cyclosaturates, are not affected by the air blowing step.

When certain additives, such as ferric chloride, are added to the air blowing step, the cyclosaturates are dehydrogenated. There is increase in the level of aromatics, which is a major contributor to higher penetration values. These aromatics, in turn, are converted to asphaltenes.

It has been shown that the weight percentage of material with molecular weight 490 is related to the softening point of the asphalt, while the weight percent of material at molecular weight 2160 is related to penetration. The relationship between the softening point and lower-molecular-weight material (490) contained in asphalt and also between high molecular weight (2160) and penetration has been presented by Dark.[5] A balance of material at these two molecular weights is required to obtain air blown asphalts with desired final characteristics. Composition of asphalts, however, is also important. The ratio of saturates to asphaltenes determines the weathering characteristics of asphalts. Good-quality asphalts generally have a saturates/asphaltene ratio of between 2 and 3.

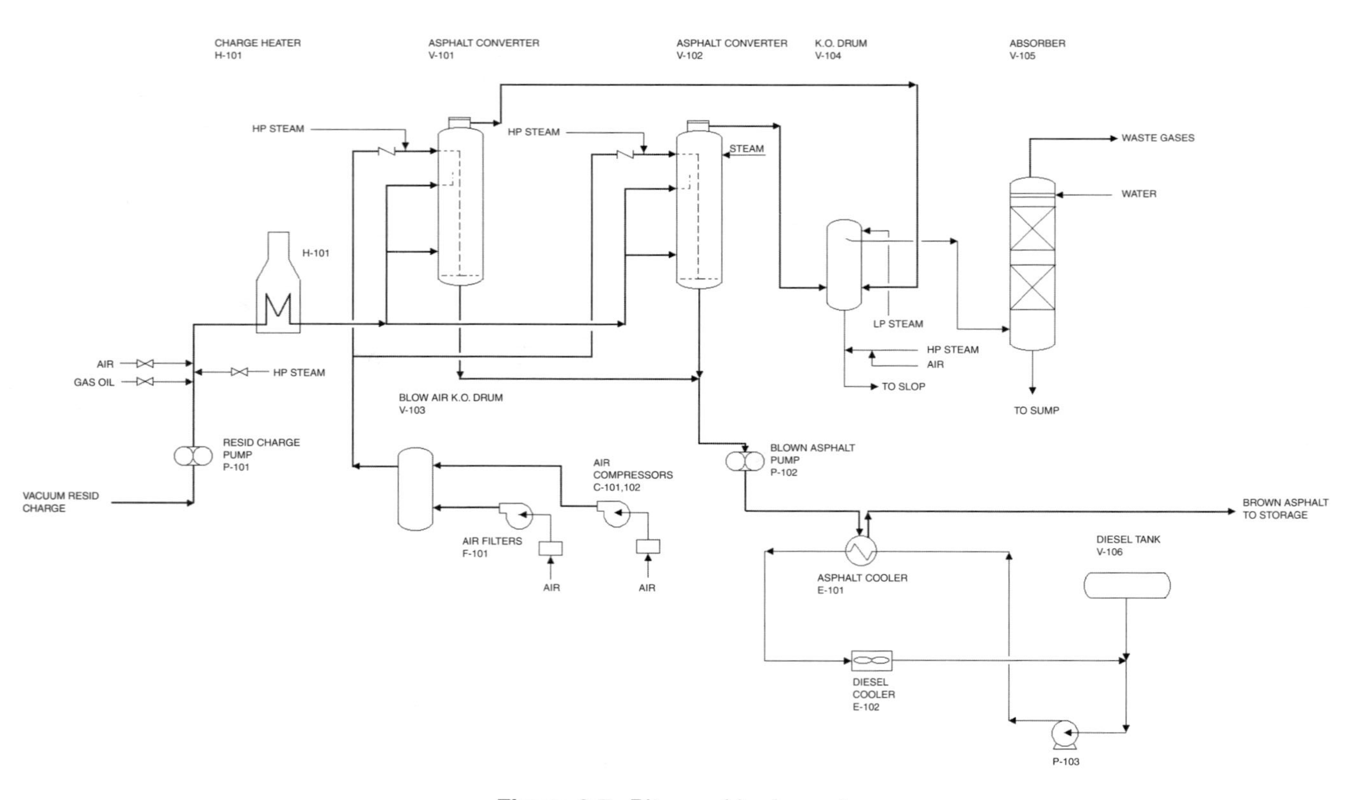

Figure 6-7. Bitumen blowing unit.

THE BITUMEN BLOWING PROCESS

Vacuum resid from the vacuum distillation unit is charged to asphalt converters V-101 and V-102 through steam-jacketed gear pump P-101 and charge heater H-101 (see Figure 6-7). The temperature of the feed to asphalt converters is controlled at 390–410°F. Compressed air at 100 psig mixed with HP steam is bubbled at a controlled rate through the vacuum resid using an internal sparger, causing exothermic reactions in the asphalt converter. The temperature of the bed is controlled by the air/steam injection rate in the range of 500–520°F. Approximately 2% of the charge is carried overhead as fumes and vapor. The waste gases pass through knockout drum V-104, where any condensed liquid separates out and is sent to slop. The vapors from V-104 are scrubbed with water in absorber V-105 and sent to refinery flare.

Finished blown asphalt from the asphalt converter is pumped through E-101, where its temperature is brought down to approximately 250°F by heat exchange with a closed-circuit diesel stream. The diesel stream, in turn, is cooled in an air cooled exchanger. The asphalt stream at 250°F is sent to storage through steam-jacketed pump P-102.

OPERATING CONDITIONS

The typical operating conditions for a bitumen blowing unit processing a 24 API Middle Eastern crude are listed in Table 6-15. The properties of the feed and product from the bitumen blowing unit are shown in Table 6-16. The overall yields from the unit and utility consumption are listed in Tables 6-17 and 6-18. The product specifications for various end uses and test methods used are shown in Tables 6-19 and 6-20.

Table 6-15
Bitumen Blowing Unit Operating Conditions

OPERATING VARIABLE	UNITS	
VACUUM RESID (FEED) TEMPERATURE	°F	390–410
BED TEMPERATURE	°F	500–520
CONVERTER PRESSURE	psig	95
AIR BLOWING RATE*	scft/bbl	166

*FOR 60/70 PEN. ASPHALT.

Table 6-16
Properties of Feed and Products

PROPERTY	FEED	PRODUCT
PENETRATION	220	60–70
SPECIFIC GRAVITY	1.016	1.028
KINEMATIC VISCOSITY		
210, Cst	800–1100	6400
250, Cst		1400
300, Cst		335
400, Cst		35
520, Cst		5

Table 6-17
Bitumen Blowing Unit Yields

STREAM	WT FRACTION
FEED	
VACUUM RESID	1.0000
TOTAL FEED	1.0000
PRODUCTS	
BITUMEN, 60/70 PEN.	0.9800
LOSSES	0.0200
TOTAL PRODUCT	1.0000

Table 6-18
Bitumen Blowing Unit Utility Consumption *(per Ton Feed)*

		CONSUMPTION	
UTILITY	UNITS	60/70 PEN.	40/50 PEN.
ELECTRICITY	kWhr	7	10
FUEL	mmBtu	0.04	0.06
STEAM	mmBtu	0.09	0.09
AIR	scft	1500	2300
DISTILLED WATER	MIG*	0.11	0.11
COOLING WATER	MIG*	1.3	1.3

*MIG = 1000 IMPERIAL GALLONS.

Table 6-19
Properties of Penetration Grade Bitumens

PROPERTY	TEST METHOD	GRADE										
			15 PEN.	25 PEN.	35 PEN.	40 PEN.	50 PEN.	70 PEN.	100 PEN.	200 PEN.	300 PEN.	450 PEN.
PENETRATION AT 25°C	IP 49	10–20	20–30	28–42	30–50	40–60	60–80	80–120	170–230	255–345	385–515	
SOFTENING POINT, MIN.	IP 58		63	57	52	58	47	44	41	33	30	25
MAX.		76	69	64	68	58	54	51	42	39	34	
LOSS ON HEATING FOR 5 HOURS AT 63°C	IP 45											
(A) LOSS BY MASS, MAX		0.1	0.2	0.2	0.2	0.2	0.2	0.5	0.5	1.0	1.0	
(B) DROP IN PENETRATION, %, MAX		20	20	20	20	20	20	20	20	25	25	
SOLUBILITY IN TRICHLORO ETHYLENE	IP 47											
% BY MASS, MIN.		99.5	99.5	99.5	99.5	99.5	99.5	99.5	99.5	99.5	99.5	
PERMITTIVITY AT 25°C AND 1592 Hz, MIN.				2.630	2.65	2.65	2.65					

Table 6-20
Properties of Cutback Bitumen

PROPERTY	TEST METHOD	GRADE		
		50 SEC.	100 SEC.	200 SEC.
VISCOSITY (STV)* AT 40°C, 10 mm CUP	IP 72	40–60	80–120	160–240
DISTILLATION				
(A) DISTILLATE AT 225°C (vol%, MAX.)	IP 27	1	1	1
DISTILLATE AT 360°C (vol%, MAX.)		8–14	6–12	4–10
(B) PENETRATION AT 25°C OF RESIDUE	ASTM D5-73	100–350	100–350	100–350
FROM DISTILLATION TO 360°C	IP 49			
SOLUBILITY IN TRICHLORO ETHYLENE, % MASS, MIN.	IP 47	99.5	99.5	99.5

*STANDARD TAR VISCOMETER

NOTES

1. B. P. Castiglioni. "How to Predict Coke Yield." *Hydrocarbon Processing* (September 1983), p. 77.
2. R. Deblane and J. D. Elliot "Delayed Coking; Latest Trends." *Hydrocarbon Processing* (May 1982), p. 99.
3. H. R. Jansen. "Conoco Coking/Calcining Process." Kellogg Symposium on Heavy Oil Upgrading, Nice, France, September 1982.
4. A. Rhoe and C. DeBlignieres, *Hydrocarbon Processing* (January 1979); H. Martin, "Visbreaking, a Flexible Process", *Oil and Gas Journal* (April 13, 1981); F. Stolfa, *Hydrocarbon Processing* (May 1982); R. Hournac, Kuhn, *Visbreaking; More Feed for FCCU Hydrocarbon Processing* (December 1979).
5. W. A. Dark. *"Asphalt Tests are Correlated." Hydrocarbon Processing* (September 1983), p. 104.

CHAPTER SEVEN

Treating Processes

Thioalcohols or thiols, more commonoly known as *mercaptans*, are a family of organic sulfur compounds present in a wide variety of untreated petroleum distillates, such as LPG, naphtha, kerosene, and gas oils. Specific mercaptans found in petroleum distillates originate in the crude or may form during subsequent crude refining. The concentration of these mercaptans in the crude distillate depends on the origin and sulfur distribution in the crude.

Mercaptans are undesirable in petroleum products. In the lower boiling range, they are moderately acidic and characterized by an extremely offensive odor. These properties diminish as mercaptan molecular weight increases. Thiophenol, which is an aryl mercaptan and more acidic than alkyl mercaptan, is found principally in cracked hydrocarbons. Thiophenol is undesirable in finished gasoline because it produces an unstable gasoline by promoting hydroperoxidation of olefins to gum. In summary, mercaptans are undesirable in finished petroleum products as they adversely affect the product's odor, stability, and quality, apart from being corrosive to the refining and handling equipment.

GENERAL PRINCIPLES

In the early days of refining industry, mercaptan removal was done with the classic "Doctor solution." The Doctor treatment[1] consists of contacting the oil with a little sulfur and alkaline sodium plumbite solution, as follows:

$$2\ RSH + Na_2PbO_2 = (RS)_2Pb + 2\ NaOH$$

$$(RS)_2Pb + S = R_2S_2 + PbS$$

Lead sulfide is reconverted to plumbite by heating the alkaline solution to 150–175°F and blowing with air, which converts sulfide to plumbite.

Lead sulfide itself was also used as sweetening agent.[2] The overall reaction is same as that given by the Doctor solution. The process consists in contacting sour distillate stream with sodium sulfide and lead sulfide suspended in caustic solution and air.

These treating processes are associated with high process losses in the form of leaded sludge. Also, water used in washing operations contain lead sulfide, which makes them no longer acceptable for any use and presents a disposal problem due to environmental concerns.

The UOP Merox process is a catalytic chemical treatment for petroleum distillates to remove mercaptants or convert them to disulfides. The process is based on the ability of catalysts composed of iron group metal chelates to promote the oxidation of mercaptans to disulfides using air as the source of oxygen. The overall reaction is as follows:

$$4RSH + O_2 = 2RSSR + 2H_2O$$

In this equation, R represents a hydrocarbon radical, which may be aliphatic, aromatic, or cyclic, saturated or unsaturated. The reaction is carried out in an alkaline medium, in the presence of a Merox catalyst at 90–120°F.

FCCU LIGHT GASOLINE

Light gasoline from the FCCU may contain 130 ppm or more mercaptan sulfur, which must be reduced to 5 ppm before this stock can be considered suitable for blending in gasoline grades (see Figure 7-1).

FCCU light gasoline is brought directly into the Merox reactor from the bottom of the FCCU debutanizer column on flow control reset by debutanizer level control. Tables 7-1 and 7-2 show the unit's feed and product properties.

A 2–5° Be (1–3 wt% NaOH) caustic from the caustic storage tank is sprayed into gasoline through an atomizing nozzle. The hydrocarbon/caustic mixture next enters air mixer MX-101, where a metered amount of air is injected from air compressor C-101 continuously into the gasoline by diffusion through a sintered steel cylinder. The effluent from the air mixer then flows into Merox reactor V-102. The feed is distributed

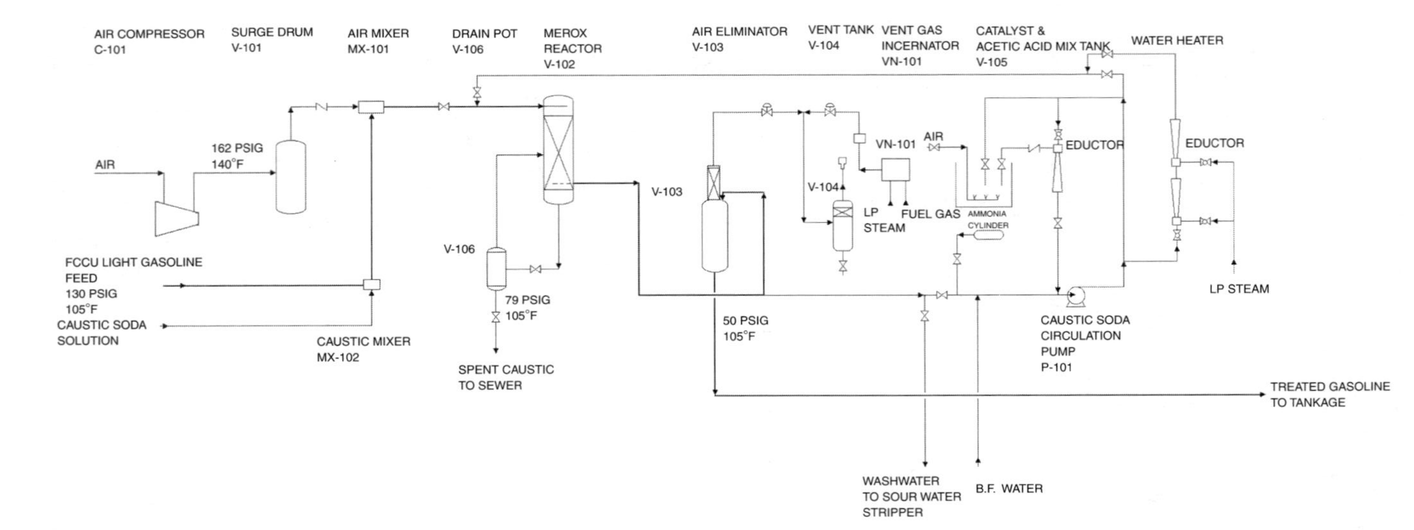

Figure 7-1. Light gasoline Merox treating.

Table 7-1
Feed and Product Properties, *FCCU Light Gasoline Merox*

PROPERTY	UNITS	FEED	PRODUCT
API GRAVITY		69	69
DISTILLATION ASTM			
IBP	°F	105	105
FBP	°F	280	280
H_2S CONTENT	WT. ppm, MAX	5	0
MERCAPTAN SULFUR	WT. ppm, MAX	130	5
TOTAL SULFUR	Wt%	0.04	0.04
CORROSION, Cu STRIP @ 50°C			MAX NO. 1
PEROXIDE NUMBER*			<0.3
EXISTENT GUM	mg/100 ml, max		<2
POTENTIAL GUM	mg/100 ml, max		<6
ENTRAINED Na OH	max, ppm		1

* UOP TEST METHOD 33.

Table 7-2
Feed and Product Properties, *FCCU Heavy Gasoline Merox*

FEED	UNITS	FEED	PRODUCT
API GRAVITY		33.7	33.7
DISTILLATION ASTM			
IBP	°F	265	105
50%	°F	325	325
90%	°F	380	380
H_2S CONTENT	WT. ppm, MAX	5	0
MERCAPTAN SULFUR	WT. ppm, MAX	450	5
TOTAL SULFUR	Wt%	0.187	0.187
CORROSION, Cu STRIP @ 50°C			Max No. 1
PEROXIDE NUMBER*			<0.3
EXISTENT GUM	mg/100 ml, max		<2
POTENTIAL GUM	mg/100 ml, max		<6
ENTRAINED Na OH	ppm, max		1

* UOP TEST METHOD 33.

uniformly across the reactor by a distributor assembly. The reactor is a packed column of activated charcoal over which Merox catalyst has been deposited.

The sweetening reaction takes place in the reactor as the gasoline feed, caustic, and air flow downward through the catalyst bed. The reactor is operated at approximately 130 psig and 105°F to keep all air dissolved in the gasoline. Any undissolved air can be manually vented to atmosphere from the top of the reactor.

The treated gasoline exits from the reactor through a fine screen located near the bottom of the reactor. The gasoline from the reactor next passes through air eliminator drum V-103, where dissolved air in the gasoline is allowed to disengage at a reduced pressure. Air/hydrocarbon vapor is burned off by mixing with fuel gas. Gasoline is sent to storage after injection of an inhibitor.

The small amount of caustic solution injected into the feed as well as water formed during the reaction are coalesced by the reactor charcoal bed and drop by gravity to the bottom of the reactor. They pass through the drain screen assembly and are sent to disposal under level control.

Impregnation of granular charcoal with Merox catalyst is done in auxiliary equipment. This includes a circulating pump for circulating ammonia water (which is the catalyst carrier), a small drum V-105 for mixing catalyst, an eductor to draw catalyst into ammonia water, and necessary piping to circulate ammonia water. For subsequent catalyst reactivation, a continuous water heater is provided to heat water for washing the catalyst bed of foreign materials adsorbed by charcoal.

JET FUEL (ATK) SWEETENING

The feed to the unit is, preferably, taken direct from the crude distillation unit with no intermediate storage (see Figure 7-2). Storage of raw kerosene results in unnecessary preaging of naphthenic acids and ingress of oxygen before prewashing. Foaming and color loss problems are enhanced. Raw kerosene feed is heated to 130°F by low-pressure steam in kerosene feed heater H-101. Charge pump P-101 discharges the heated feed to caustic prewash vessel V-102.

In this vessel, the feed is prewashed with a 1.5–2% solution of caustic soda to neutralize both the H_2S and naphthenic acids present in the feed. Also, to achieve sufficient contact between the caustic soda and the kerosene, the caustic is recycled by recirculating the caustic exiting the prewash vessel and mixing it with the main incoming kerosene stream via charge pump P-101.

The effluent from prewash vessel V-102 should contain no more than 0.005 mg KOH/g as the sum of acidity and sum of equivalent sodium

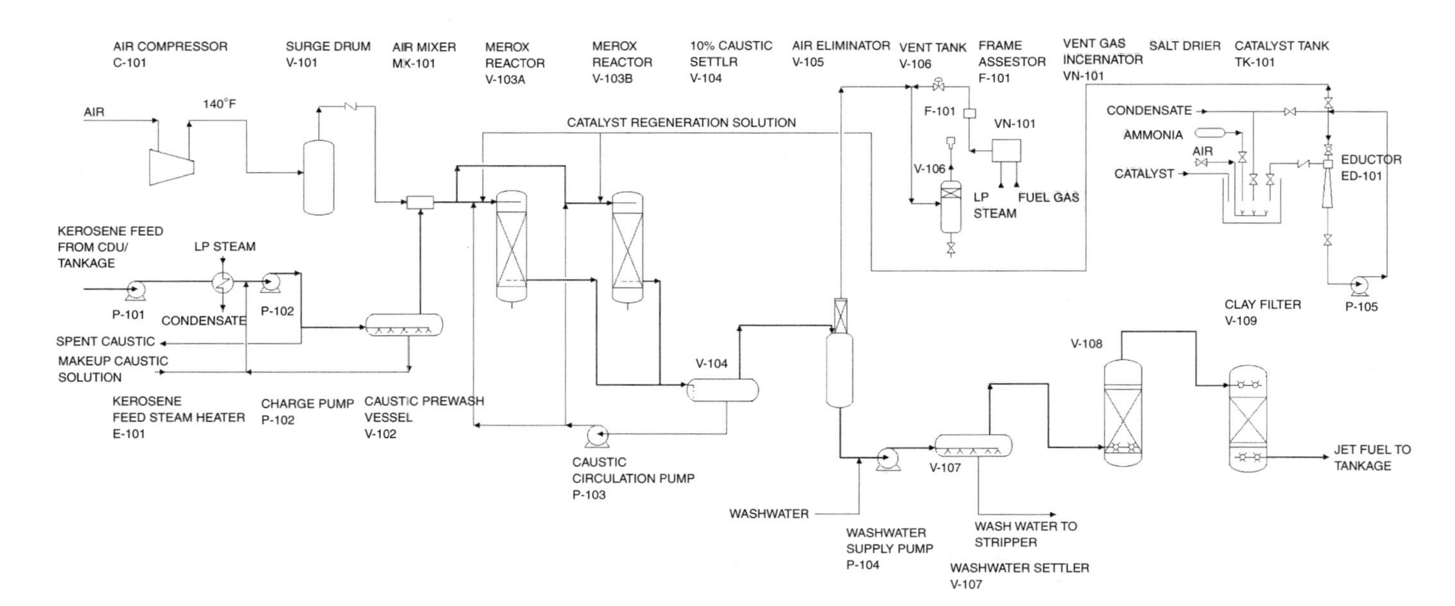

Figure 7-2. Aviation turbine kerosene (ATK) Merox treatment.

naphthenate present. If naphthenes still exist in the final product, thermal stability and WSIM test values can be adversely affected.

Air is mixed with prewashed kerosene feedstock before entering Merox reactor V-103. The plant air (70 psig) passes through a filter to remove any scale or dirt that would otherwise block the distributor nozzle in the reactor. From the filter, the air passes through a regulator to reduce the pressure to about 60 psig. The air flow, controlled and metered, is next mixed with prewashed kerosene feed through mixer MX-101, which diffuses air into kerosene stream.

The reactor is a packed column of activated charcoal over which Merox catalyst has been deposited. The reactor bed, kept alkaline by periodic circulation of caustic without interruption of the unit feed, also washes off accumulated contaminants. The sweetening reaction takes place in the reactor as the feed, caustic, and air flow downward through the catalyst bed. The reactor is operated at sufficient pressure to keep all air dissolved in the gasoline. Any undissolved air can be manually vented to atmosphere from the top of the reactor.

The treated kerosene exits the reactor through a fine screen located near the bottom of the reactor. The process control consists of checking product mercaptan level, minimizing air injection, and maintaining catalyst alkalinity.

Merox reactor effluent next flows into caustic soda settler V-104, in which any entrained caustic is settled and drained out. The operating pressure of the settler is in the range of 35–40 psig. Tables 7-3 and 7-4 show the jet fuel Merox operating conditions and feed and product properties.

The effluent from the caustic settler next passes through air eliminator drum V-105, where dissolved air in the kerosene is allowed to disengage at a reduced pressure. The vapors from the air eliminator are burned off after mixing with fuel gas.

The air-free product is mixed with washwater and pumped to washwater vessel V-107 to wash out any trace of caustic soda. The quantity of washwater used is approximately 10% of the feed to the vessel. The water used is of boiler feed water quality, free from contaminants. The caustic water that settles down in V-107 is sent to refinery foul water stripper.

The water-washed kerosene is next, it flows to salt filter V-108, where the kerosene is passed through a course bed (3–6 mesh) rock salt for final water elimination.

The final step in the finishing of jet fuel is clay filtration, wherein dried jet fuel is passed through a bed of activated clay in V-109 containing

Table 7-3
JET FUEL Merox Operating Conditions

OPERATING PARAMETER	UNITS	RANGE
CAUSTIC PREWASH VESSEL		
TEMPERATURE	°F	125–130
PRESSURE	psig	40–43
CONCENTRATION OF CAUSTIC WASH SOLUTION	VOL/VOL	1.5–2.0%
CAUSTIC CIRCULATION RATIO		
CAUSTIC SODA/KEROSENE	VOL/VOL	5–15%
AIR MIXING SECTION		
AIR/KEROSENE FEED RATIO		
ft^3 PER min AIR/1000 bbl KEROSENE	VOL/VOL	1.1–1.3
REACTOR		
TEMPERATURE	°F	120–125
INLET PRESSURE	psig	40–42
PRESSURE DROP	psig	2–7
CAUSTIC CONCENTRATION		
FOR REACTION	Wt%	10
LHSV	hr^{-1}	2.5
CAUSTIC SODA SETTLER		
PRESSURE	psig	25–30
TEMPERATURE	°F	120–125
WASHWATER SETTLER		
TEMPERATURE	°F	105
PRESSURE	psig	25–30
SAND FILTER		
TEMPERATURE	°F	105
PRESSURE INLET	psig	60
PRESSURE OUTLET	psig	58
CLAY TREATER		
(ATTAPULGUS CLAY, 30–60 MESH PARTICLE SIZE)		
BULK DENSITY	kg/m^3	685
SURFACE AREA	m^2/gm	95–103
OPERATING TEMPERATURE	°F	110
OPERATING PRESSURE	psi	60–70
SPACE VELOCITY, LHSV	hr^{-1}	1.74
SALT DRIER		
(CRYSTALLINE ROCK SALT)		
PARTICLE SIZE (U.S. SIEVE)		
3 MESH		PASS THROUGH
6 MESH		RETAINED
BULK DENSITY	kg/m^3	1050–1250
PARTICLE DENSITY	gm/cm^3	2.16
SPACE VELOCITY, LHSV	hr^{-1}	1.00

Table 7-4
Jet Fuel Merox Feed and Product Properties

	UNITS	KEROSENE FEED	PREWASHED KEROSENE	PRODUCT AFTER MEROX TREATING	SPECIFICATIONS
DISTILLATION, ASTM					
IBP		320	320	320	
10%	°F	352	352	352	
20%	°F	356	356	356	
50%	°F	378	378	378	
90%	°F	432	432	432	
FBP	°F	468	468	468	
RESIDUE	vol%	1.0	1.0	1.0	
DENSITY	kg/m^3	0.790	0.790	0.790	
FLASH POINT	°C	48	48	48	
FREEZE POINT	°C	−65	−65	−65	
SMOKE POINT	mm	28	28	28	
VISCOSITY −20°C	cst	4.2	4.2	4.2	
WATER SEPARATION					
INDEX MODIFIED WSIM				75	MIN 85
COLOR	SAYBOLT	+30	+26	+24	MIN 21
SULFUR TOTAL	Wt%	0.17	0.17	0.17	
SULFUR MERCAPTANS	ppm	130	130	5–20	20 MAX.
ASTM TOTAL ACIDITY	mg KOH/g	0.01	0.0015	0.0007	0.015 MAX

IBP = INITIAL BOILING POINT; FBP = FINAL BOILING POINT.

30–60 mesh activated clay. The object of clay filtration is to remove by adsorption nonionic (oil soluble) surfactants and other impurities. Removal of surfactants helps meet WSIM specifications for removal of oil fields chemicals, corrosion inhibitors finding their way in kerosene feed, and incomplete removal of naphthenate in Merox feed pretreatment. Thermal stability of the product is also improved by removal of impurities and organometallic copper compounds and other colored molecules formed as a result of oxidation of naphthenates due to excess air or temperature.

The Merox-treated product is superior to other chemically treated products, because it is free from copper, lead, and elemental sulfur. Merox-treated jet fuel is superior to hydrotreated product with respect to the product's lubricity and stability.

NOTES

1. C. D. Lowry, Jr. *Plumbite Sweetning of Gasoline*. Universal Oil Product Co., Booklet 242, May 1940.
2. W. L. Nelson. *Petroleum Refinery Engineering*, 4th ed. New York: McGraw-Hill, 1964.

CHAPTER EIGHT

Sulfur Recovery and Pollution Control Processes

Most crude oil contain varying amounts of sulfur. Hydrotreating various distillates from these crudes generate hydrogen sulfide, which is converted to elemental sulfur to minimize atmospheric pollution. In the absence of sulfur recovery, the only option would be to burn this gas in refinery furnaces, releasing huge amounts of sulfur dioxide into the atmosphere, an option no longer acceptable due to environmental concerns.

Sulfur recovery processes recover hydrogen sulfide from various refinery gaseous streams by scrubbing with an amine solution and convert it to elemental sulfur by controlled oxidation. Stack or flue gas desulfurization removes sulfur dioxide, generated as a result of burning high-sulfur fuel oil as refinery fuel, from refinery stacks. The process consists of scrubbing stack gases with aqueous ammonia to remove sulfur dioxide, which is reduced to elemental sulfur by reaction with hydrogen sulfide.

SULFUR RECOVERY FROM ACID GAS

The sulfur recovery unit recovers elemental sulfur from the acid gas containing H_2S by the Claus process. The chemical reaction involved is

$$H_2S + \frac{1}{2}O_2 = S + H_2O$$

This reaction is, in fact, a summation of the two following reactions:

$$H_2S + \frac{3}{2}O_2 = SO_2 + H_2O$$

$$2H_2S + SO_2 = 3S + 2H_2O$$

The reaction of SO_2 and H_2S also occur in the thermal stage at 2500°F by burning one third of H_2S to SO_2 to obtain H_2S/SO_2 stoichiometric ratio of 2 to 1. At high temperature, conversion up to 70% into sulfur may be reached for the Claus reaction. In the presence of inerts (mainly CO_2) in the feed, lower conversion values are obtained.

To increase sulfur recovery, outlet gases from thermal stage are cooled down:

- To condense and remove most of the sulfur, thus shifting the equilibrium of Claus reaction to the right.
- To operate at a lower temperature, in the most favorable thermodynamic range. Under these conditions, a catalyst is needed to ensure an optimum reaction rate.

The exhaust gases from the thermal stage are sent to a sulfur condenser, then to a catalytic converter to increase sulfur recovery. Most Claus units are designed with two or three catalytic converters with activated alumina catalyst to increase sulfur recovery.

The inlet temperature to the catalytic converter is an important variable. The temperature must be high enough to prevent sulfur condensation on the catalyst, which would decrease catalytic activity. But, if the temperature is too high, the exothermicity of the Claus reaction increases the temperature along the catalyst bed, thus limiting conversion.

THE CLAUS SULFUR RECOVERY PROCESS

Acid gas from upstream treating units is mixed with a controlled amount of air and burned in combustion reaction furnace H-01 (see Figure 8-1). The gaseous product emerges from the furnace at 2500°F and passes through waste heat boiler B-01, producing HP steam at 900 psig. After leaving the boiler, the sulfur-laden gas enters first condenser E-01, where sulfur is condensed and collected in a sulfur sump.

The gas leaving the first condenser is reheated by a burning fuel gas to the reaction temperature and passes through a catalyst converter with activated alumina balls, where additional sulfur is formed. Sulfur thus formed is removed again in the second condenser and collected in the sump. The gas is again reheated and passed through a second catalyst converter for further conversion to elemental sulfur, which is removed in the third condenser. The total sulfur recovered is about 90–95% of the

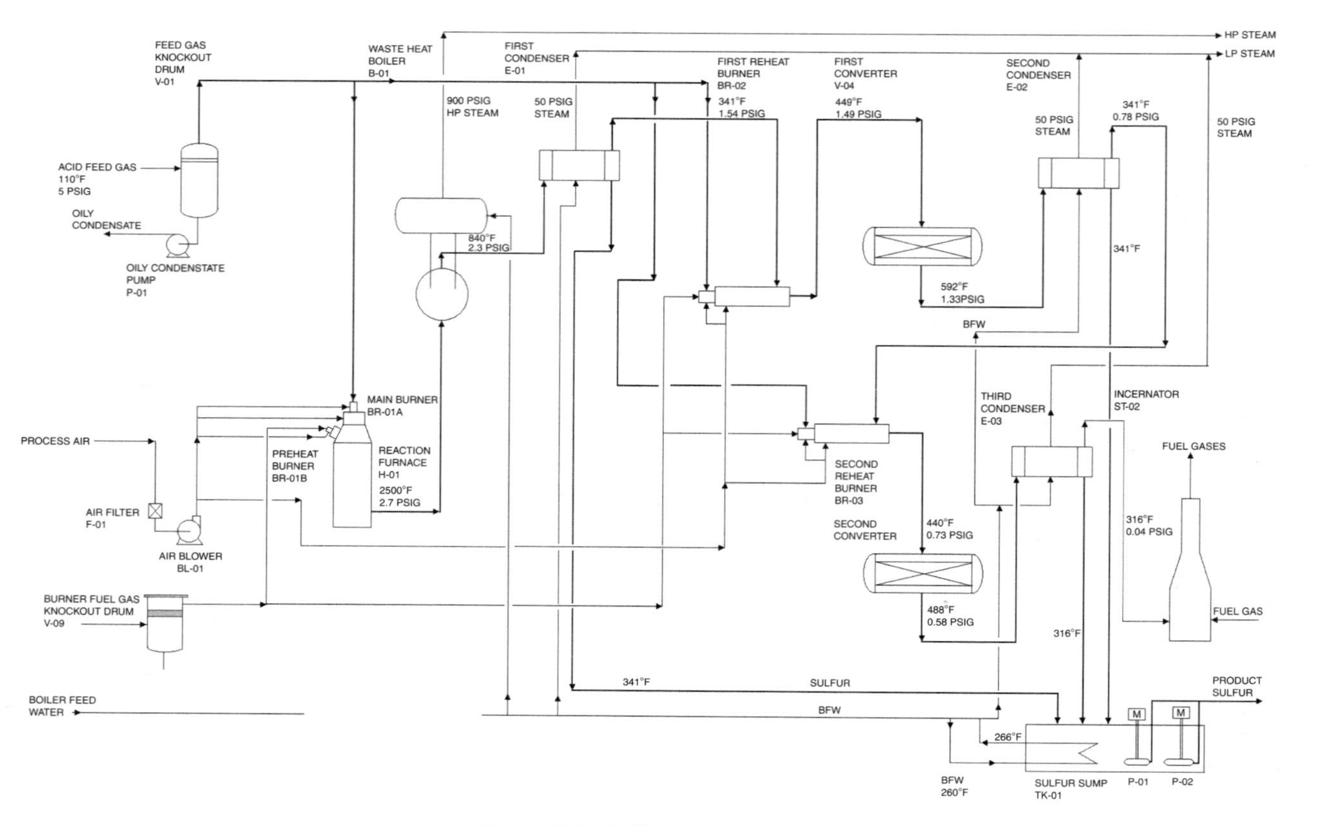

Figure 8-1. Sulfur recovery unit.

sulfur in the feed to the unit. After leaving the final condenser, the gas is burned in waste gas incinerator ST-02.

The sulfur from the waste heat boiler and all condensers is drained into sulfur sump TK-01, where it is kept in molten condition. From the sulfur sump, sulfur is transferred to liquid sulfur storage tanks. From the tanks, sulfur may be shipped in molten condition or pumped to a sulfur flaking unit to produce solid sulfur for shipments.

Operating conditions and feed composition of a Claus sulfur unit are shown in Table 8-1. The unit yield and utility consumption are shown in Tables 8-2 and 8-3.

CLAUS TAIL GAS TREATMENT

Sulfur recovery in a conventional Claus sulfur unit depends on the number of catalytic stages in the unit. An average Claus plant with three catalytic stages may recover approximately 95–97% of the sulfur contained in the acid gas fed to the unit, letting off as much as 3–5% of sulfur in the air in the form of stack gases after incineration. With increasing concern for environmental air quality, this practice is becoming unacceptable. Tail gas treatment removes most of the sulfur contained in the tail gas of a sulfur recovery unit in the form of elemental sulfur. The treated gas after incineration contain less than 1500 ppm SO_2.

THE CHEMISTRY OF THE SYSTEM

The Claus tail gases are contacted with solvent polyethylene glycol (PEG) containing a dissolved catalyst for accelerating Claus reaction:

$$2H_2S + SO_2 = 3S + H_2O$$

The reaction is exothermic, the heat of reaction being 35,000 kCal/kg mole SO_2 reacted. The catalyst used is a 22% solution of benzoic acid, sodium, and potassium hydroxides in water. A typical composition of the catalyst solution is shown in Table 8-4.

The reaction occurs in the liquid phase between dissolved H_2S and SO_2 in the PEG by way of a catalyst complex. It is carried out at approximately 250–300°F, thus ensuring continuous elimination of sulfur and water of reaction (sulfur's melting point is 246°F).

Table 8-1
Sulfur Recovery Unit Operating Conditions

	UNITS	
ACID GAS FEED TEMPERATURE	°F	110
ACID GAS FEED PRESSURE	psia	19.7
THERMAL STAGE		
FURNACE GASES OUTLET TEMPERATURE	°F	2520
FURNACE GASES OUTLET PRESSURE	psia	17.41
WASTE HEAT BOILER		
EFFLUENT GASES OUTLET TEMPERATURE	°F	840
EFFLUENT GASES OUTLET PRESSURE	psia	16.98
FIRST SULFUR CONDENSER		
INLET TEMP	°F	840
OUTLET TEMP	°F	341
INLET PRESSURE	psia	16.98
OUTLET PRESSURE	psia	16.24
FIRST REHEATER		
INLET TEMPERATURE	°F	341
OUTLET TEMPERATURE	°F	449
INLET PRESSURE	psia	16.98
OUTLET PRESSURE	psia	16.24
FIRST CONVERTER		
INLET TEMPERATURE	°F	449
OUTLET TEMPERATURE	°F	592
INLET PRESSURE	psia	16.19
OUTLET PRESSURE	psia	16.03
SECOND SULFUR CONDENSER		
INLET TEMP	°F	592
OUTLET TEMP	°F	341
INLET PRESSURE	psia	16.03
OUTLET PRESSURE	psia	15.48
SECOND REHEAT BURNER		
INLET TEMP	°F	341
OUTLET TEMP	°F	449
INLET PRESSURE	psia	15.48
OUTLET PRESSURE	psia	15.43
SECOND CONVERTER		
INLET TEMPERATURE	°F	449
OUTLET TEMPERATURE	°F	488
INLET PRESSURE	psia	15.43
OUTLET PRESSURE	psia	15.28
THIRD SULFUR CONDENSER		
INLET TEMP	°F	488
OUTLET TEMP	°F	316
INLET PRESSURE	psia	15.28
OUTLET PRESSURE	psia	14.8

ACID GAS FEED COMPOSITION:

H_2S 86.03 mol%

CH_4 0.51 mol%

CO_2 5.92 mol%

H2O 6.99 mol%

C_2H_6 0.28 mol%

C_3+ 0.27 mol%

Table 8-2
Sulfur Recovery Unit Yields

STREAM	YIELD WT%
FEED	
H_2S GAS	1.0000
TOTAL FEED	1.0000
PRODUCT	
SULFUR	0.8478
LOSS	0.1522
TOTAL	1.0000

Table 8-3
Sulfur Recovery Unit Utility Consumption *(per Ton Feed)*

UTILITY	UNIT	CONSUMPTION
FUEL	mmBtu	2.8
POWER	kWhr	65
STEAM	mmBtu	−7.5
DISTILLED WATER	MIG*	0.02
COOLING WATER	MIG*	0.2

*MIG = 1000 IMPERIAL GALLONS.

Table 8-4
Catalyst Composition

COMPONENT	WT%
PURE BENZOIC ACID	15.20
PURE POTASSIUM HYDROXIDE	1.00
PURE SODIUM HYDROXIDE	5.80
DEMINERALIZED WATER	78.00
TOTAL	100.00

NOTE: THE CATALYST IS A 22% SOLUTION OF BENZOIC ACID, SODIUM HYDROXIDE, AND POTASSIUM HYDROXIDE WITH THIS COMPOSITION. INDUSTRIAL-GRADE CHEMICALS CAN BE USED.

The level of catalytic activity of the system is characterized by a soluble sodium and potassium (Na^+ and K^+) content and free acidity (H^+) of the solvent. With the gas at steady conditions (flow and temperature) and a predetermined H_2S/SO_2 ratio, these levels can be maintained by catalyst makeup.

Sodium and potassium salts (K^+ and Na^+ content) in the PEG are maintained in the range of 20–50 millimole/kg. Free acidity of PEG (H^+ content) is maintained at 30–70 millimole/kg.

The benzoic acid consumption corresponds to its losses in off gases by vapor pressure. Sodium and potassium hydroxides are finally degraded to sodium and potassium sulfates, which are insoluble and deposited on the packing. The amount of sodium sulfate deposited on the catalyst can be estimated from the cumulative amount of pure KOH and NaOH injected into the solvent since the last washing. The washwater required is approximately 3.3 times the amount of sodium sulfate to be washed off the packing.

For a given packing volume and catalyst composition, the operating variables are

1. Total acid gas flow.
2. ($H_2S + SO_2$ content) in the acid gas flow.
3. The ratio H_2S/SO_2.
4. Solvent flow rate.

The temperature is held constant at the design value for optimum operation and to minimize PEG loss.

Any increase in acid gas flow decreases conversion. The optimum performance of the unit is reached by minimizing the SO_2 content in the off gas. This is achieved by a molar ratio H_2S/SO_2 equal to or greater than 2.3 in the feed gas by controlling operation of the upstream Claus unit operation. Any excess SO_2 decreases conversion, increases potassium and sodium consumption, and affects the color of the product sulfur. Long-duration SO_2 excess can damage the solvent, which may require charcoal treatment to restore its quality and required conversion rates. Flow of the solvent has an effect on the interfacial area available for mass transfer and conversion: Increasing the solvent flow increases interfacial area and conversion.

THE TAIL GAS TREATMENT PROCESS

The acid gas is introduced in reactor R-101 near the bottom (see Figure 8-2). The reactor, where gas/liquid contact is effected, is a packed column designed for a low pressure drop while ensuring the interfacial area

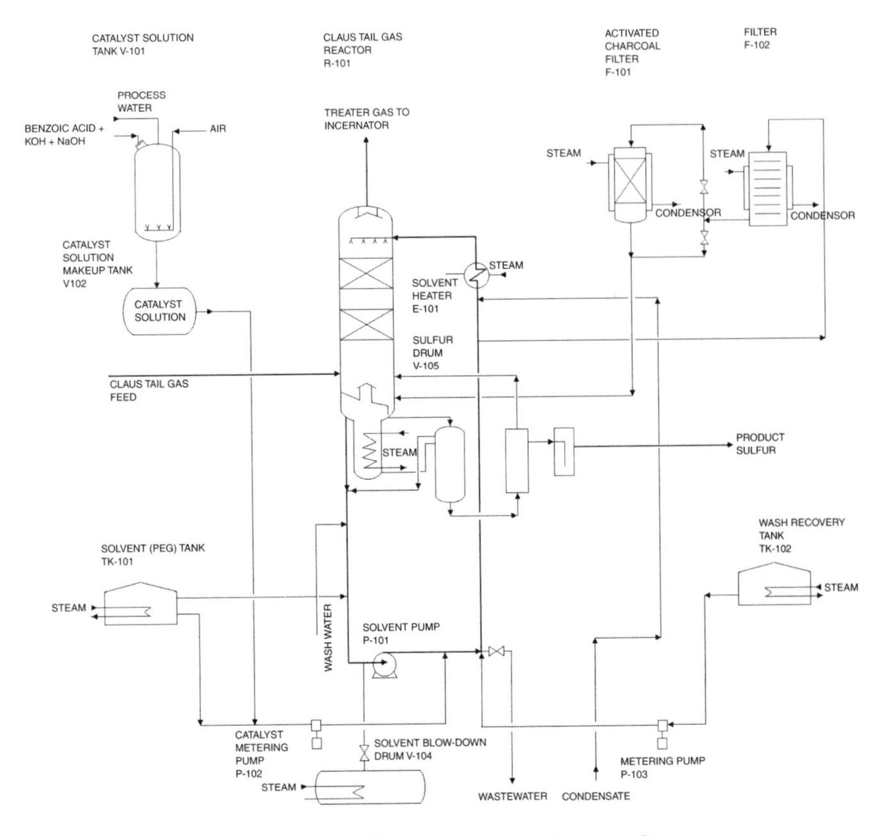

Figure 8-2. Claus tail gas treatment unit.

necessary for the desired conversion. The packing volume is divided into a number of beds separated by liquid redistributors and packing support plates to ensure maximum packing efficiency all along the reactor. At the top of the tower, a deflector plate is provided to minimize solvent entrainment in the off gases.

Acid gas flows countercurrently to the solvent to ensure the concentration gradient between the bottom and top of the reactor. An incorporated system of decantation at the bottom of the tower separates the liquid sulfur produced from the solvent. The sulfur is next sent to a sulfur pit by way of a seal leg. To keep sulfur in a liquid state, sulfur lines and valves are steam-jacketed.

The solvent at the bottom is recycled back to the top of the reactor by pump P-101. This pump is designed to operate on pure liquids or crystalline sulfur slurries. Steam-heated mechanical sealing with seal liquid taken at pump discharge is used. This pump is also used for filling and emptying the unit with solvent.

A sidestream is drawn from the solvent loop to filter a small part of the solvent through filter F-102 and the activated charcoal bed in F-101.

Although the Claus reaction is exothermic, the precise heat balance of the unit depends mainly on the heat losses in the reactor system. The reactor temperature is controlled by either additional condensate injection or steam-heating in exchanger E-101. The catalyst consumption is compensated by continuous injection of catalyst solution from V-102 through metering pump P-102.

Heat exchanger E-101 on the solvent circulation line heats the packed reactor and its PEG content to the reaction temperature in a period of 48 hr during start-up. It is placed at a level above the reactor bottom solvent inlet to prevent solid sulfur deposits between E-101 and the reactor. This exchanger is also used to compensate for heat losses during normal operation. It is used during shutdown to cool the unit and during washing to heat the washwater.

Solvent let-down tank T-101 is provided for PEG storage during filling and emptying and as a reserve for makeup solvent. Washing recovery tank T-102 is provided to recover the PEG/water mixture during the unit washing. The mixture is stored in the tank by solvent pump P-101 and reinjected intermittently in the system by metering pump P-103 to compensate for solvent losses and deposits on the packing.

Underground solvent blow-down drum V-104 is used for PEG recovery during purges of pumps, lines filters, and the like. Steam coils are used to heat the solvent in the tanks to reduce viscosity for pumping.

Operating conditions and feed composition of a Claus sulfur unit are shown in Table 8-5. The catalyst composition, unit yield, and utility consumption are shown in Tables 8-6 to 8-11.

FLUE GAS DESULFURIZATION

Flue gas desulfurization aims to reduce sulfur oxides emissions from the stack gases of the refineries and power plants to 150–1500 ppm range, for pollution control. The process is particularly useful for refineries and power plants burning high-sulfur heavy fuel oil.

Table 8-5
Sulfur Plant Tail Gas Treatment
Operating Conditions

OPERATING VARIABLE	UNITS	
REACTOR TEMPERATURE		
INLET GAS	°F	285–300
OUTLET GAS	°F	248–252
REACTOR INLET PRESSURE	psig	0.5–2.0
REACTOR OUTLET PRESSURE	psig	0.1–1.6
SOLVENT CIRCULATION RATE*	GPM	700

*BASIS ACID GAS FEED RATE OF 180 MSCF/HR.

Table 8-6
Catalyst Component Properties

PROPERTY	BENZOIC ACID	SODIUM HYDROXIDE	POTASSIUM HYDROXIDE
CHEMICAL FORMULA	C_6H_5COOH	NaOH	KOH
MOLECULAR WEIGHT	122	40	56
MELTING POINT, °F	252	604	752
SOLID SPECIFIC GRAVITY	1.266	2.13	2.04

Table 8-7
Solvent

SPECIFICATIONS	UNITS	
MOLECULAR WEIGHT		380–420
FREEZING POINT	°F	39–46
SPECIFIC GRAVITY		1.125
REFRACTIVE INDEX		1.466
VISCOSITY	122°F, cst	27.85
	212°F, cst	6.5–8.2
	275°F, cst	3.94
SULFUR SOLUBILITY		
	68°F, gm/liter	2–10
	248°F, gm/liter	27
	275°F, gm/liter	35

NOTE: THE SOLVENT USED IS POLYETHYLENE GLYCOL WITH AN AVERAGE MOLECULAR WEIGHT OF 400.

Table 8-8
Claus Tail Gas Treatment Unit Yields

STREAM	WT%
FEED	
ACID GAS FEED	1.0000
CONDENSATE	0.0481
TOTAL FEED	1.0481
PRODUCT	
SULFUR	0.0463
TREATED GAS	1.0019
TOTAL PRODUCT	1.0481

Table 8-9
Utility Consumption, *per Tons Sulfur Produced*

UTILITY	UNITS	CONSUMPTION
ELECTRICITY	kWhr	463
CONDENSATE	TONS	2.23

Table 8-10
Claus Tail Gas Treatment Unit Feed and Treated Gas Composition

COMPONENT	MOL WEIGHT	ACID GAS FEED, WT%	TREATED GAS, WT%	SULFUR, WT%
H_2S	34	0.0370	0.0053	
SO_2	64	0.0298	0.0000	
CO_2	44	0.1171	0.1168	
C_3	44	0.0089	0.0089	
N_2	28	0.6019	0.6005	
H_2O	18	0.2035	0.2678	
S_{6-8}	246	0.0019	0.0008	
LIQUID SULFUR	32			1.0000
TOTAL		1.0000	1.0000	1.0000

The process can be divided into four steps:

1. Ammonia scrubbing of flue gas.
2. Sulfitic brine treatment: recovery of gaseous SO_2 and NH_3 from aqueous sulfite/sulfate solution.

Table 8-11
Product Sulfur Properties

PROPERTY	
CARBONILE MATTER (PEG CONTENT)	<1000 ppm
ASH CONTENT	<200 ppm
SULFUR PURITY	>99.7%
COLOR	BRIGHT YELLOW

3. Reduction of a part of the SO_2 produced to H_2S, a step avoided if an outside source of H_2S is available.
4. Conversion of SO_2 into sulfur by reacting it with H_2S.

AMMONIA SCRUBBING

In the scrubbing step, absorption of SO_2 by aqueous ammonia solution takes place according to the following chemical reaction;

$$NH_3 + H_2O + SO_2 \rightleftharpoons (NH_4)HSO_3$$
$$2NH_3 + H_2O + SO_2 \rightleftharpoons (NH_4)_2SO_3$$

Owning to the presence of SO_3 in the flue gas or oxidation of a part of SO_2, sulfate may also be formed.

$$2NH_3 + H_2O + SO_3 = (NH_4)_2SO_4$$

Ammonia scrubbing of flue gas results in a treated flue gas leaving the stack with less than 150 ppm SO_2. Also an aqueous ammonium sulfate/sulfite solution is generated.

SULFITIC BRINE TREATMENT

Sulfitic brine is partly decomposed back into NH_3, SO_2, and H_2O in an evaporator according to the preceding equations.

The bottom liquid containing the bulk of sulfates is fed to a sulfate reduction reactor, where sulfates, thiosulfates, and remaining sulfites are reduced to SO_2, NH_3, and H_2O by reaction with molten sulfur:

$$(NH_4)_2SO_4 \rightleftharpoons NH_4HSO_4 + NH_3$$

$$2NH_4HSO_4 + S = 3SO_2 + 2NH_3 + 2H_2O$$

$$(NH_4)_2S_2O_3 = SO_2 + S + 2NH_3 + H_2O$$

This reaction takes place in a liquid bath at a temperature of approximately 660°F, using a molten salt bath. The reaction is endothermic; so heat must be supplied from external sources. The gases in the molten bath contain a certain amount of SO_3. This gas is reduced to SO_2 by reaction with fuel gas containing CO plus H_2 contained in the fuel gas. The reaction is carried out over a solid catalyst.

If refinery H_2S stream is available, the sulfur dioxide produced from sulfate reduction and sulfite decomposition is reacted with H_2S in liquid phase to produce elemental sulfur. The ratio of H_2S to SO_2 must be greater than 2.3 for this reaction to proceed to completion:

$$2H_2S + SO_2 = 2H_2O + 3S$$

High-purity liquid sulfur is produced as final product. If H_2S stream is unavailable (as in power plants), a part of SO_2 can be reduced to H_2S using a reducing gas ($CO + H_2$) in a catalytic converter.

PROCESS DESCRIPTION

The stack gases of the refinery are scrubbed with an aqueous solution of ammonia in column C-101 (see Figure 8-3). The ammonia solution is introduced into the column at the top, while flue gases enter the column near the bottom. The rich effluent containing dissolved sulfur oxides, called *sulfitic brine*, enters tank TK-101, where any dust contained in the stack gases is separated out by settling.

The sulfitic brine is pumped by P-101 to heat exchanger E-101, heated by medium-pressure steam, and flashed in drum V-101, where partial decomposition of sulfites take place at 300°F and 30 psia pressure. The gases produced as a result of decomposition of ammonium sulfites, SO_2

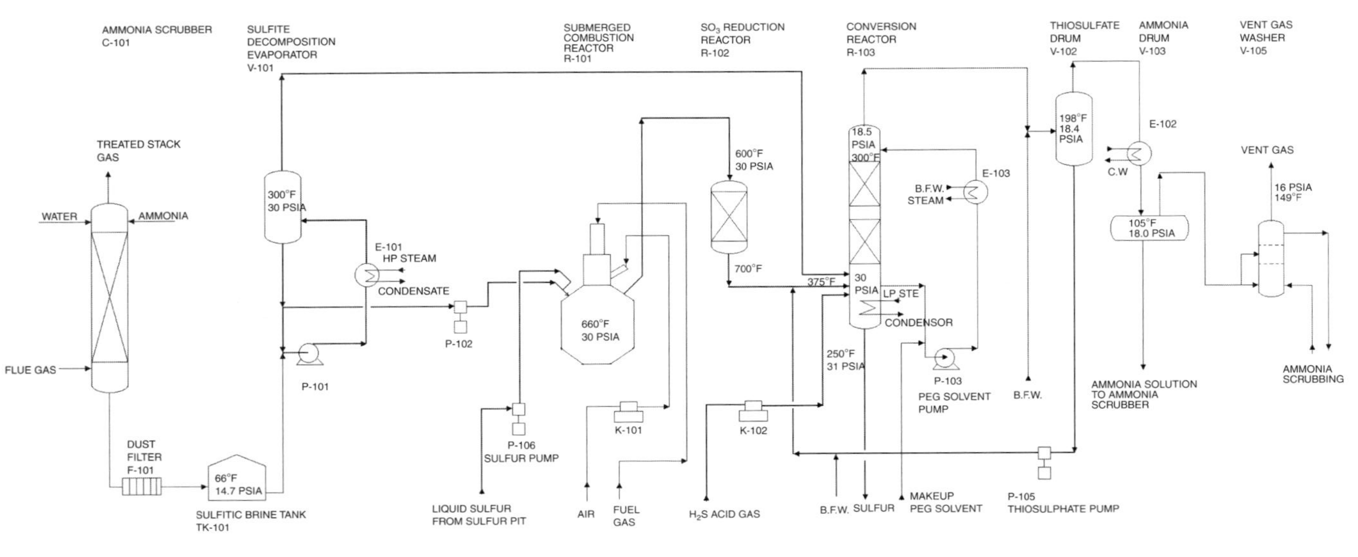

Figure 8-3. Flue gas desulfurization. B.F.W. = boiler feedwater.

and NH_3, are sent to R-103, the conversion reactor. The concentrated brine is sent by P-102 to R-101.

The concentrated brine, which consists of all ammonium sulfates, thiosulfates, and unconverted sulfites, is fed by metering pump P-102 to a sulfate decomposition reactor R-101, which is a bath of molten ammonium sulfates, where sulfate reduction takes place. Also, a stream of molten sulfur is pumped by metering pump P-106 to R-101. The temperature of the molten bath is maintained at 660°F. The reactions in the sulfates reduction reactor are endothermic. The heat is supplied to the system by submerged flame combustion of fuel gas and air. The hot gases are sent from a special burner directly into the molten bath via a graphite tube. This combustion takes place under slightly reducing conditions, implying that air/fuel gas inlet ratio is maintained slightly lower than stoichiometric (90%). The reduction of sulfates start at 570–590°F. The operating temperature is not allowed to exceed 750°F, when ammonia starts to react with medium. The operating temperature of the bath can be adjusted according to the thiosulfate/sulfate ratio in the bath. The molten salt in the bath is very corrosive. The reactor is therefore lined with acid-resisting refractory.

The gases above the molten bath (80% NH_4HSO_4 and 20% $(NH_4)_2SO_4$) contain a certain amount of SO_3 due to vapor pressure. The SO_3 is reduced to SO_2 by the following exothermic reactions. The CO and H_2 originate from unburned fuel gas. The reactions are carried over a solid state (Co-Mo on alumina) catalyst:

$$SO_3 + CO = SO_2 + CO_2$$

$$SO_3 + H_2 = SO_2 + H_2O$$

The off gases from reactor R-102 at about 700°F are quenched to 375°F by injection of boiler feedwater.

The SO_2-rich gases coming from V-101, R-101, and R-102 are mixed with H_2S from an outside source, such as a refinery amine plant, through compressor K-102 and enter conversion reactor R-103 near the bottom. The H_2S/SO_2 feed ratio must be maintained higher than 2, to protect the chemical system.

The reactor is an absorber column with 13 perforated trays. The solvent polyethylene glycol (PEG) is circulated by pump P-103 through cooler E-103 to remove the heat of reaction, feed enthalpy, and to keep the tower temperature at 300°F. Heat is removed by thermosyphon recirculation of boiler feedwater through E-103 and raising LP steam.

The PEG solvent flows down the column. The Claus reaction takes place in a liquid solvent PEG (polyethylene glycol, mol. wt. 400). Ammonia in the gas stream acts as a catalyst. The molten sulfur produced is collected by decantation at the bottom of the reactor. Makeup solvent is added to the cooling loop to compensate for the solvent losses.

As the Claus reaction in R-103 does not reach 100% completion, gases coming out from R-103 overhead contain NH_3, H_2S, SO_2, CO_2, inerts, sulfur vapors, and entrained PEG. These are quenched with boiler feedwater and sent to thiosulfate drum V-102. Water quenching lowers the temperature to the dew point of the gases. SO_2 reacts with sulfur in the presence of NH_3 to give thiosulfates, which are dissolved in water. This solution, also containing traces of PEG, is sent back by P-105 to the gas line coming from R-102. The temperature being higher than 190°F, the thiosulfates are cracked into gaseous SO_2, NH_3, and sulfur.

The off gases from V-102 at approximately 200°F are cooled in exchanger E-102. Most of the water is condensed and collected in drum V-103 with some ammonia, H_2S, and CO_2 dissolved in it.

The top gases from V-103 are washed in drum V-105 by bubbling contact with brine from the ammonia scrubber bottom to further reduce pollutant release into atmosphere.

The following reaction takes place:

$$NH_4HSO_3 + NH_3 = (NH_4)_2SO_3$$

$$2(NH_4)_2SO_3 + 2NH_4HSO_3 + 2H_2S = 3(NH_4)_2S_2O_3 + 3H_2O$$

The operating conditions and feed composition of a flue gas desulfurization unit are shown in Table 8-12. The feed and product compositions and utility consumption are shown in Tables 8-13 to 8-15.

AMINE TREATMENT

The objective of an amine treatment unit is to remove H_2S, CO_2, and mercaptan compounds from various gas streams, such as recycled gas in hydrotreating and hydrocracking processes, hydrogen plant feed, and fuel gas systems. The H_2S recovered is used as feed for the sulfur recovery unit.

Hydrogen sulfide is removed from a gas stream by contact with an aqueous solution (18 wt%) of monoethanol amine (MEA) or diethanol amine (DEA) according to the following reaction:

$$\begin{matrix} OH{-}CH_2{-}CH_2 \\ OH{-}CH_2{-}CH_2 \end{matrix} \!\!> NH + H_2S \rightleftharpoons \begin{matrix} OH{-}CH_2{-}CH_2 \\ OH{-}CH_2{-}CH_2 \end{matrix} \!\!> NH_2^+ HS^- + HEAT$$

LEAN DEA   WEAK ACID   RICH DEA (WATER SOLUBLE SALT)

The amount of H_2S that can react depends on the operating conditions. Low temperature, high pressure, and high concentration of H_2S in the H_2S absorber favor the reaction.

The temperature of the lean amine solution to the H_2S absorber is controlled by a temperature controller on E-101. The temperature of lean amine is maintained at least 10°F above H_2S feed to the absorber to

Table 8-12
Flue Gas Desulfurization Operating Conditions

OPERATING VARIABLE	UNITS	
SULFITE EVAPORATOR		
TEMPERATURE	°F	300
PRESSURE	psia	30
SULFATE REDUCTION REACTOR		
TEMPERATURE	°F	660
PRESSURE	psia	32
SULFUR CONVERSION REACTOR		
TEMPERATURE	°F	300
PRESSURE	psia	18.4
SOLVENT (POLYETYLENE GLYCOL), FOR SULFUR CONVERSION REACTOR		
MOLECULAR WEIGHT		380–420
FREEZING POINT	°F	39–46
SPECIFIC GRAVITY		1.125
VISCOSITY	122°F, cst	28
	212°F, cst	6.5–8.2
	275°F, cst	3.9
SULFUR SOLUBILITY		
	68°F, gm/liter	2–10
	248°F, gm/liter	27
	248°F, gm/liter	35
SO_3 CONVERSION REACTOR CATALYST		CO-MO ON ALUMINA BASE
TEMPERATURE	°F	660
PRESSURE	psia	31

Table 8-13
Utility Consumption, *per Tons Sulfur Recovered*

UTILITY	UNITS	AMMONIA SCRUBBING	CONVERSION	TOTAL
ELECTRICITY	kWhr	1883	974	2857
FUEL, mm Btu	mmBtu	14.2	41.2	55.4
STEAM	mmBtu	0	−36.7	−36.7
DISTILLED WATER	MIG	0	0.057	0.057
COOLING WATER	MIG	0	2.86	2.86
PROCESS WATER	MIG	0.095	0	0.095

Table 8-14
Feed and Product Properties

PROPERTY	UNITS	FEED FLUE GAS	TREATED FLUE GAS	SULFUR
SO_2	ppm	1600	200	
SO_3	ppm	75	0	
AMMONIA LOSSES	ppm		10	
CARBONILE MATTER	ppm			<1000
SULFUR PURITY	Wt%			99.7
COLOR				YELLOW

prevent condensation of gas, which results in foaming in the absorber or hydrocarbon contamination of the H_2S product.

Rich amine solution is fed to a flash and surge drum. The object of the flash drum is to separate liquid hydrocarbons and dissolved gases from rich amine solution. A hydrocarbon skim is provided to remove any liquid hydrocarbons. This prevents foaming in the amine regenerator and contamination of the sulfur plant where the sulfur plant catalyst is coked. The drum also serves as surge volume for circulating the amine solution. Pressure in this drum is maintained as low as possible to release hydrocarbons from the rich amine solution.

H_2S is removed from the amine solution in an amine regenerator, where high temperature and low pressure are used to strip H_2S from amine solution.

At low temperature (<180°F), aqueous amine solutions are not too corrosive to carbon steel. Organic acids and decomposition products of DEA contribute to corrosion where surface temperatures are high enough to decompose the salts and release acid.

Table 8-15
Product Composition

COMPOSITION	
AMMONIA SOLUTION (RECYCLE) TO AMMONIA SCRUBBER, WT%	
$(NH4)_2CO_3$	18.9
NH_4OH	2.2
$(NH4)_2S$	0.3
H_2O	78.6
SULFITIC BRINE FROM AMMONIA SCRUBBER, WT%	
NH_4HSO_3	17.39
$(NH_4)_2SO_3$	12.12
$(NH_4)_2S_2O_3$	13.40
$(NH_4)_2SO_4$	7.10
H_2O	49.99
TOTAL	100.00
TREATED VENT GAS, VOL%	
H_2S	0.47
H_2O	11.70
NH_3	0.12
CO_2	22.20
UNCONDENSABLES	65.51
TOTAL	100.00

In amine plants, provisions are made to allow additions of chemicals such as sodium carbonate to neutralize acids and reduce acid-type corrosion. Sometimes, a DEA solution can promote severe H_2S corrosion. A filming amine corrosion inhibitor is usually added to prevent contact between corrosive material and metal.

THE AMINE TREATMENT PROCESS

Most refineries have more than one amine contactor, one for low pressure H_2S contaminated gases for treating refinery fuel gas and another for treating high pressure recycle gas from units such as hydrocrackers and hydrotreaters.

In LP absorber V-104, the gas feed enters the absorber at the bottom and the circulating amine solution enters the column at top (see Figure 8-4). The amine absorber has approximately 24 plates or a packed column of equivalent length. An aqueous amine solution (18%) enters the column at top, while

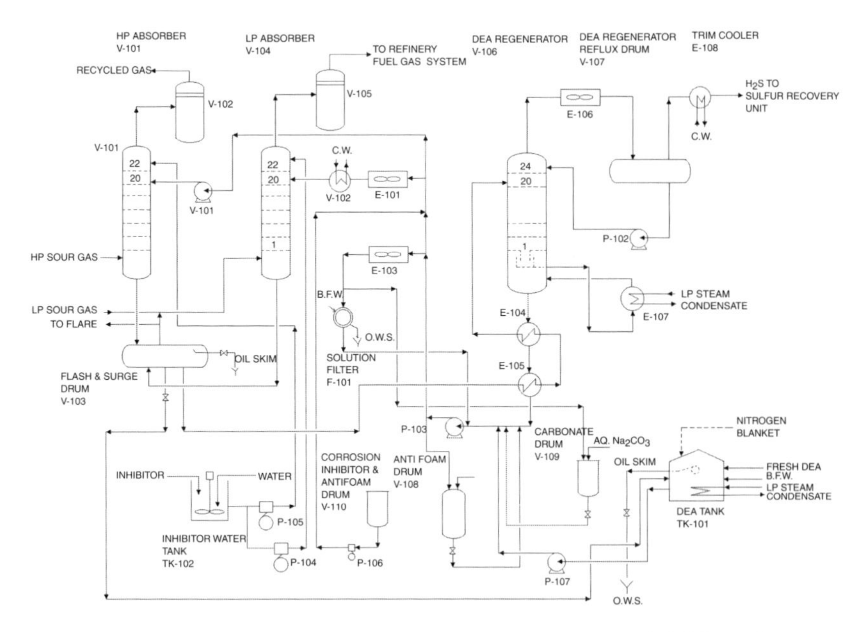

Figure 8-4. Amine treatment unit. B.F.W. = boiler feedwater; O.W.S. = oily water sewer.

sour gas enters the column at the bottom. Treated gas leaves the column at the top after passing through knockout drum V-105 and is sent out of battery limits. Rich amine solution is collected in flash and surge drum V-103. HP sour gas is similarly treated in column V-101 with 24 plates.

Rich amine from both columns V101 and V-104 is collected in V-103, which acts both as a flash and a surge drum. Liquid from V-103 exchanges heat with regenerator column V-104 bottoms in E-104 and E-105 and enters the amine regenerator column V-104 at the top. The regenerator column has 24 plates and a steam reboiler. In this column, acid gases, such as H_2S and CO_2, are separated as the overhead stream while the regenerated amine solution is pumped with recirculation pump P-103 to high- and low-pressure absorber columns V-101 and V-104. Acid gas from amine regeneration is sent to the sulfur recovery plant as feed. Liquid collected after condensation of V-106 vapor in E-106 is collected in V-107 and sent back to the column as reflux.

The amine solution is cooled in exchangers E-101 and E-102 before entering contactor V-104. The circulating amine solution is dosed with sodium carbonate and antifoaming additives to prevent foaming and corrosion in the contactors.

Table 8-16
Amine Treatment Unit Operating Conditions

OPERATING VARIABLE	UNITS	
LOW-PRESSURE AMINE ABSORBER		
INCOMING LEAN DIETHANOL AMINE TEMPERATURE	°F	150
INCOMING SOUR GAS TEMPERATURE	°F	140
ABSORBER BOTTOM PRESSURE	psig	27
NUMBER OF PLATES IN ABSORBER		24
HIGH-PRESSURE AMINE ABSORBER		
INCOMING LEAN DEA TEMPERATURE	°F	150
INCOMING SOUR GAS TEMPERATURE	°F	140
RICH AMINE LOADING	MOLES H_2S/ MOLE DEA	0.35–0.45
ABSORBER BOTTOM PRESSURE	psig	2410
NUMBER OF PLATES		24
DEA REGENERATOR		
REGENERATOR COLUMN TOP TEMPERATURE	°F	220
REGENERATOR COLUMN TOP PRESSURE	psig	13
REFLUX DRUM TEMPERATURE	°F	140
REFLUX DRUM PRESSURE	psig	9
REGENERATOR FEED TEMPERATURE	°F	215
REGENERATOR COLUMN BOTTOM TEMP.	°F	253
REGENERATOR COLUMN BOTTOM PRESSURE	psig	14

A small, lean amine slipstream is sent to amine filter F-101 to remove undesirable salts that contaminate the system. The slipstream filtration system is provided to maintain the cleanliness of the circulating amine.

Makeup amine, received in drums, is emptied in an amine sump and diluted with water to about an 18% solution. The aqueous amine solution is pumped to amine storage tank TK-101. The tank is blanketed with nitrogen gas to prevent air oxidation of the amine. Fresh amine is pumped into the circulating amine system as required.

Table 8-17
Amine Unit Utility Consumption, *(per Ton Feed)*

UTILITY	UNITS	CONSUMPTION
FUEL	mmBtu	
POWER	kWhr	13
STEAM	mmBtu	2.2
COOLING WATER	MIG	2.2
DISTILLED WATER	MIG	0.003

The operating conditions and feed composition of an amine treatment unit are shown in Table 8-16. The properties of amine, feed and product, and utility consumption are shown in Tables 8-17 to 8-19.

Table 8-18
Feed and Product Composition

COMPONENTS	FEED, mol%	TREATED GAS, mol%	ACID GAS, mol%
H_2	11.24	14.97	
H_2S	24.89	0.00	91.45
C_1	19.07	25.39	0.54
C_2	11.02	14.68	0.30
C_3	12.99	17.30	0.25
IC_4	5.15	6.86	0.05
NC_4	13.46	17.92	
IC_5	0.06	0.08	
NC_5	0.32	0.43	
C_{6+}	0.41	0.54	
H_2O	1.37	1.83	7.42
TOTAL	100.00	100.00	100.00

Table 8-19
Typical Properties of Amines*

PROPERTY	UNITS	MONOETHANOL AMINE	DIETHNANOL AMINE
FORMULA		$H_2NCH_2CH_2OH$	$HN(CH_2CH_2OH)_2$
MOLECULAR WEIGHT		61.08	105.14
BOILING POINT	760 mm Hg, °F	338.7	498.2
	50 mm Hg, °F	213.8	359.6
	10 mm Hg, °F	159.8	302
FREEZING POINT	°F	50.9	82.4
SPECIFIC GRAVITY	68°F	1.017	1.092
VISCOSITY	CP, 86°F	16.2	380
REFRACTIVE INDEX	n_D^{30}	1.4539	1.4747(1)
SURFACE TENSION	dynes/cm	48.3(4)	48.5(5)
FLASH POINT	°F	205	375
HEAT OF VAPORIZATION	Btu/lb	390(2)	300(3)
CRITICAL TEMPERATURE	°F		827.8
CRITICAL PRESSURE	Atm		32.3

(1) AT 30°C. (2) AT 212°F. (3) AT 302°F. (4) AT 77°F. (5) AT 86°F.
*DOW CHEMICAL COMPANY DATA.

CHAPTER NINE

Refinery Water Systems

Water is used in an oil refinery for the following purposes:

- Cooling.
- Steam generation.
- Domestic and sanitation purposes.
- Washing products.
- Flushing equipment, pipelines, and hydrotests.
- Fire fighting.

Water usage in the refinery also generates large volumes of water contaminated with oil and other chemical impurities. Part of this water can be reused after treatment, while the remainder must be discharged to sea or other water bodies. But, before this can be done, the aqueous effluent must be treated to remove oil, chemical, biological, and other impurities, so that the impact on the environment is minimal. The principle wastewater streams that require treatment are oily water effluents, biological waste generated by the refinery workforce, and sour water streams containing dissolved H_2S and NH_3.

COOLING WATER SYSTEM

Refining operations are conducted at elevated temperatures. In a rough overall sense, a refinery must be in heat balance. All heat added in the forms of fuel burned, steam consumed, or coke burned must be removed by one of the various cooling systems. Water cooling is one such system. The others are air cooling and heat exchange with other streams. Cooling accounts for about 90% of the total refinery water requirements. Approximate cooling water requirements of a refinery can be estimated as a function of refinery complexity.[1]

Water in a refinery's cooling system either travels through the system once or is recirculated. In a once-through system, pumps suction water from a source, such as a sea, river, or lake, and deliver it to process units or other water users within the refinery. After passing through the cooling equipment, the hot cooling water is conducted to a point of disposal through a pressure system of piping or through a gravity flow system.

In recirculated systems, pumps suction water from a cooling tower basin and deliver it to cooling equipment. After passing through water user equipment, the hot cooling water is discharged through a pressure return system to the top of the cooling tower. The water cooling system includes heat exchangers, pumping equipment, distribution piping, water intake stations, and cooling towers.

In some refineries in coastal areas, warm water (at approximately 113°F) returning from the process heat exchangers and other equipment is first cooled by heat exchange with sea water in sea water exchangers then sent back to the cooling tower for further cooling (Figure 9-1). The advantage of this system is that the heat load on the cooling tower is reduced and corrosive sea water do not come in contact with process equipment. Good, soft water, containing very little dissolved solids, salts, and iron, allows the use of carbon steel for coolers and condensers. Sea water, on the other hand, contains large amounts of dissolved salts (Table 9-1), is a good conductor of electric current, and causes severe corrosion of metals. For example, carbon steel registers a corrosion penetration rate of more than 2 mm/yr in sea water. Nonferrous materials, such as aluminum, brass, and titanium, have to be used, in the case of direct sea water cooling.

SEA WATER COOLING SYSTEM

A typical direct sea water cooling system is shown in Figure 9-2. Because of the large volume of water intake for cooling, water is pumped from the sea to an inlet sump through a battery of low lift pumps. Sea water next flows through a system of bar screens, scrapper screens, and rotary screens to the suction of the high lift pump manifold. The screen system prevents entry of marine life, seaweed, algae, and the like into pump suction. To prevent growth of algae and fungi and suppress microbial activity, chlorine is injected (0.5–1.0 ppm) into sea water before it enters the suction of the high lift pump battery. As sea water is very corrosive, reinforced concrete pipes are used for all piping greater than 30 in. diameter.

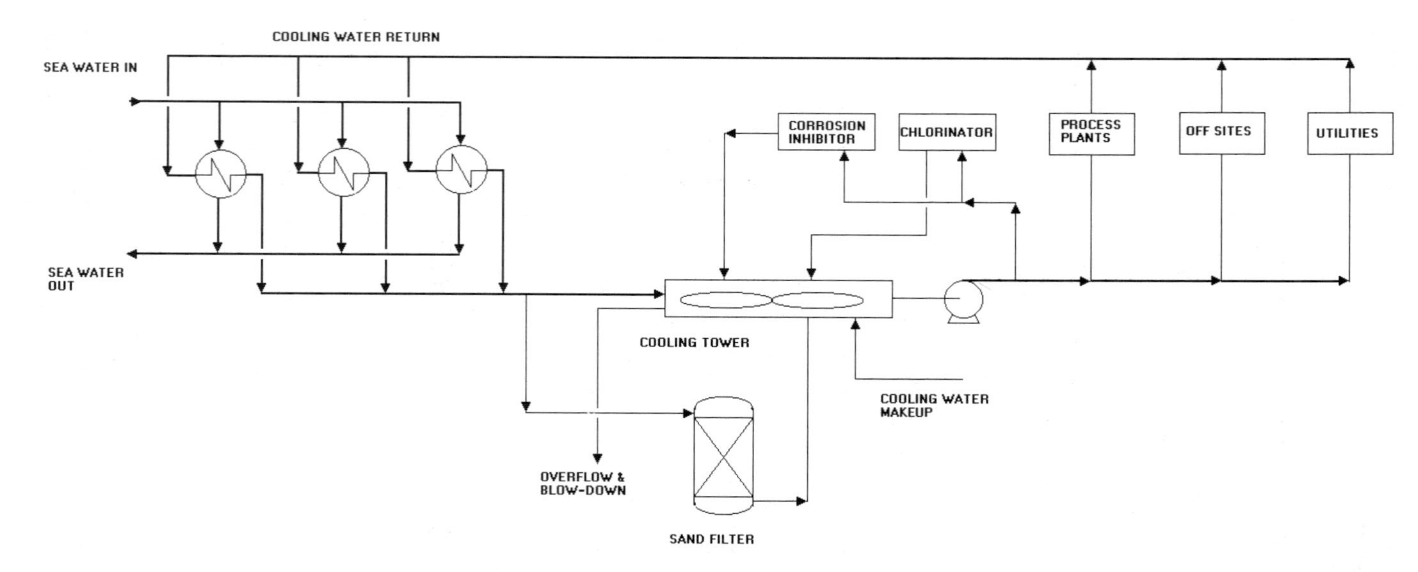

Figure 9-1. Refinery cooling water system.

Table 9-1
Typical Sea Water Analysis

PARAMETER	UNITS	
TOTAL DISSOLVED SOLIDS	ppm	49,700–51,200
NACl	ppm	39,000–41,000
TOTAL HARDNESS		8,200
ALKALINITY	AS $CACO_3$	114
DENSITY	kg/m^3	1,030
TEMPERATURE	°F	90
VISCOSITY	CP	0.76

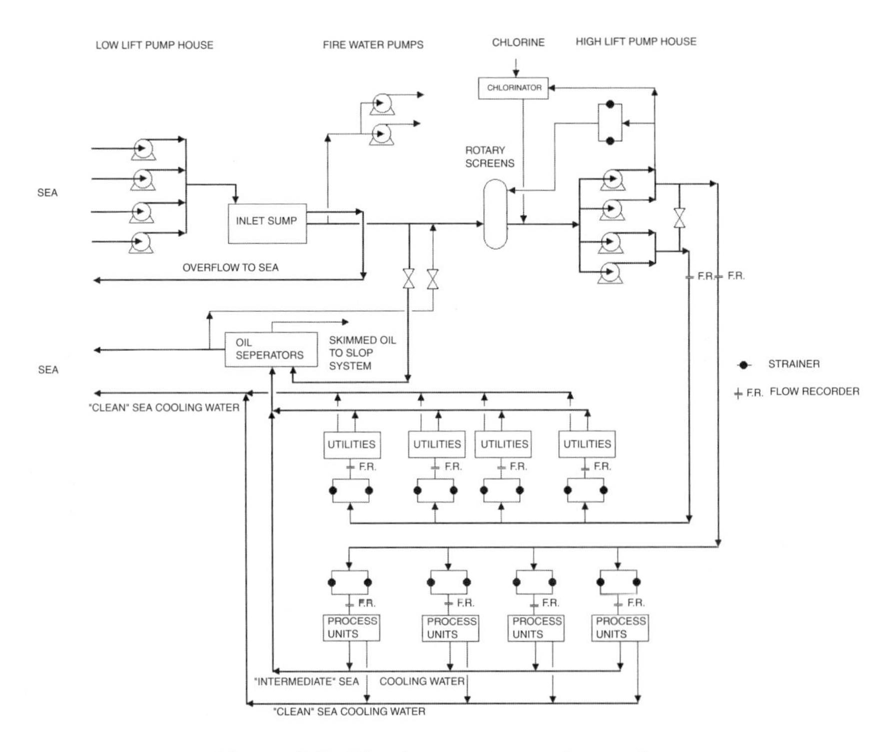

Figure 9-2. Direct sea water cooling system.

Below 28 in. diameter, cement lined steel pipes are used for sea water service. The battery of high lift pumps supplies sea water to a sea water manifold running throughout the refinery, from which individual process units and utilities tap their cooling water supply. To prevent any grit, debris, or scale from entering the exchangers, strainers are placed at the entry of every process or utility unit. Warm sea water coming out of an exchanger is segregated into two manifolds: a "clean" sea cooling water and an "intermediate" sea cooling water.

Intermediate, or oil-contaminated, sea water, by definition, is that sea water used in service where the oil side of exchanger is operating at 55 psig or greater. All contaminated sea water may be routed through the intermediate sea cooling water manifold to a battery of corrugated plate interceptor separators before discharging it to sea. If the oil side of the exchanger is operating at less than 55 psig, all such returning sea water is routed to the clean sea cooling water manifold and discharged directly to the sea.

COOLING TOWERS

The cooling towers function by direct removal of heat from water by air flow and vaporizing a portion of water. Both forms of cooling are accomplished by a counterflow of air and water. The towers are constructed of wood, metal, or concrete with wood or plastic packing for distribution of the water flow. A portion of water passing over the cooling tower is vaporized. Any solid this water contains is left behind and increases the concentration of solids in water. To limit the concentration of solids and prevent their deposit on cooling surfaces, it is necessary to blow down a certain percentage of circulating water. Further water loss occurs when water drifts off the tower in the wind, called *drift* or *windage loss*.

The initial step in the design of cooling water systems is to determine the design temperature and system capacity. The system capacity varies with design temperature as limited by process conditions. The usual cooling range is between 25 and 30°F. The inlet temperature of water to cooling equipment is established by ambient conditions, generally in the range 75–86°F, and the outlet temperature is in the range 104–114°F. The type and quality of water set the outlet water temperature. The maximum temperature of water in the heat exchangers must be limited to prevent corrosion and deposit of solids.

Cooling tower losses are usually evaluated as follows:

1. Evaporation losses are approximately 1% of tower throughput for each 10°F of cooling tower temperature differential.
2. Drift loss is limited in the design of the cooling tower to 0.2% cooling water throughput.

3. Miscellaneous liquid loss is assumed to equal 15% of evaporation losses minus the drift losses.
4. Blow-down is determined on assumption that makeup water can be concentrated five to six times in the cooling tower. Blow-down losses can be estimated approximately by the following empirical relationship:

$$B = \frac{10.7 \times \Delta T \times R}{100{,}000}$$

where

B = blow-down rate, gpm;
ΔT = temperature differential, °F;
R = circulation rate, gpm.

BOILER FEEDWATER SYSTEM

Municipal water feed to the refinery must be upgraded to boiler feedwater quality by further treatment (see Figure 9-3). Boiler feedwater is used as makeup to fired and nonfired steam generators located in the utility plants and process units. Boiler feedwater is also used for process purposes.

Fresh water, as received from an outside supply, may not have enough pressure to fill tank TK-101. Booster pump P-101 is used to elevate the water supply pressure. Feedwater is stored in tank TK-101, holding approximately 72-hr supply. Water is pumped by P-102 to water filter V-101, which removes most entrained solids. Filter aids are added at the discharge of P-102, using a dozing pump to facilitate removal of suspended solids. Filtered water next passes through a bed of cation and anion resins in V-102 and V-103, which remove all anions and cations in the water by an ion exchange process. A high-purity water, which is essentially free of salts, is produced and stored in tank TK-102. Demineralizer beds are exhausted after 3–7 days operation, and they are regenerated by treatment with dilute sulfuric acid (cation resins) and caustic soda (anion resins). Each demineralizer bed is equipped with automatic regeneration facility. The typical feedwater and BFW water qualities are shown in Table 9-2. The demineralized water is supplied to all low-pressure deaerators as makeup to LP boiler water system.

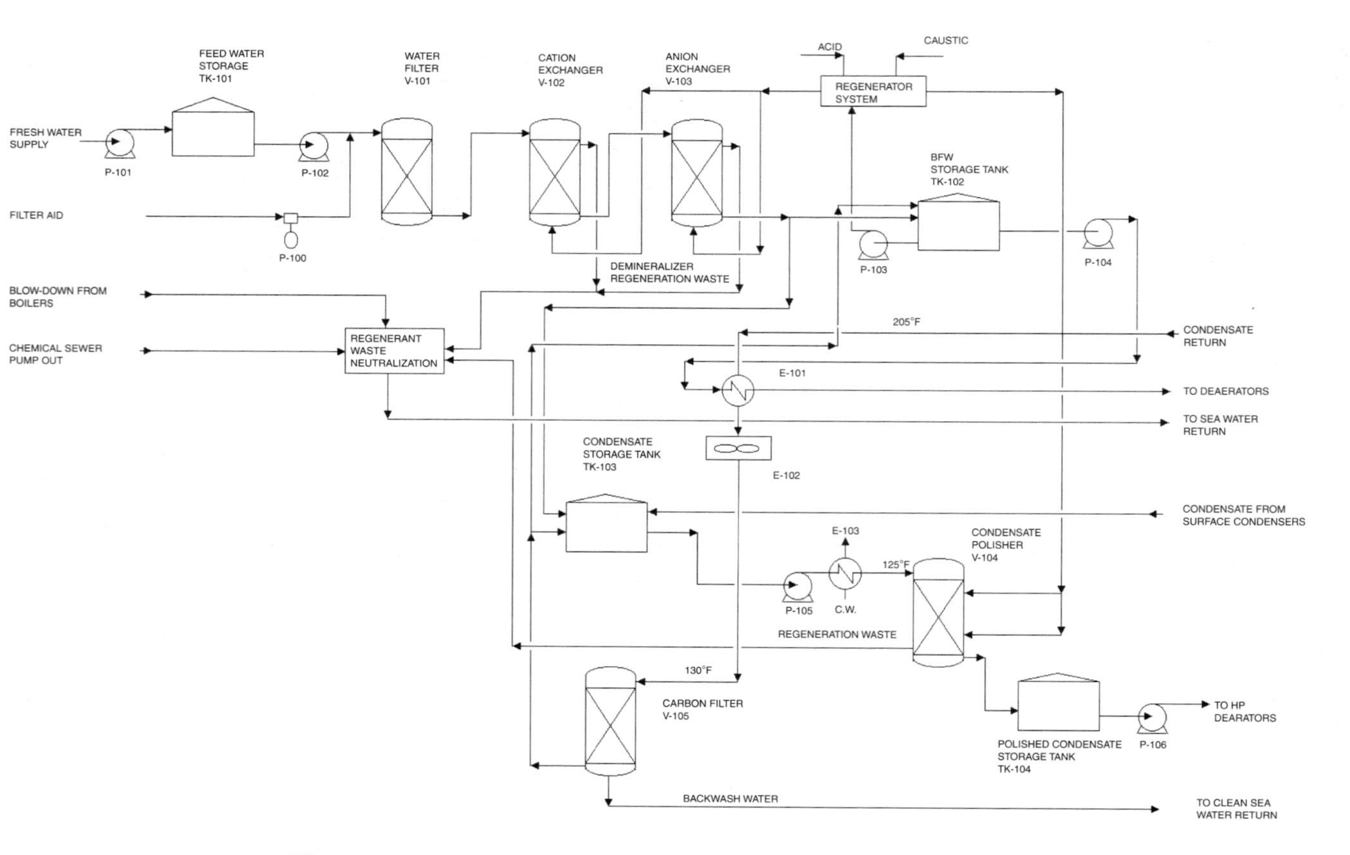

Figure 9-3. Treatment for boiler feedwater (BFW). HP = high pressure.

Table 9-2
Properties of Various Water Streams

	UNITS	FRESH WATER[1]	DISTILLED WATER[2]	LP BOILER FEEDWATER[3]	CONDENSATE RETURN[4]	POLISHED CONDENSATE[5]
CONSTITUENTS	AS					
CALCIUM	$CaCo_3$	57	10			
MAGNESIUM	$CaCo_3$	40				
SODIUM	$CaCo_3$	77	40			
POTASSIUM	$CaCo_3$	2				
AMMONIA	$CaCo_3$	0.2	0.04			
TOTAL ANIONS						
BICARBONATES	$CaCo_3$	9	5			
SULFATE	$CaCo_3$	81	10			
CHLORIDE	$CaCo_3$	87	35			
FLUORIDE	$CaCo_3$	0.4				
ALKALINITY	$CaCo_3$	9	5		10	<1
TOTAL HARDNESS	$CaCo_3$	97	10	<0.3	1	<0.02
pH		7.2	6.3–6.7			
SILICA	SiO_2	1.4	0.05		1	
IRON	Fe	0.3	<1			
COPPER	Cu		<1			<0.015
TURBIDITY	jtu	3	4	<1	5	<1
TDS		238	60	<30	10	<0.5
SPECIFIC CONDUCTANCE	umho/cm	400	100		20	<1

NOTES:
[1]WATER RECEIVED FROM MUNICIPAL SUPPLY.
[2]WATER FROM A DESALINATION PLANT.
[3]DEMINERALIZED WATER FEED TO LOW-PRESSURE BOILERS.
[4]UNCONTAMINATED CONDENSATE RETURNED FROM THE PROCESSING UNITS.
[5]POLISHED CONDENSATE FEED TO HP BOILERS.
LP = LOW PRESSURE; TDS = TOTAL DISSOLVED SOLIDS; JTU = JACKSON TURBRIDITY UNITS; ALL DISSOLVED SALT CONCENTRATIONS IN PPMW.

CONDENSATE POLISHER

Condensate is returned to the water treatment system from two sources:

1. Steam condensate condensed in surface condensers is considered clean and returned directly to condensate storage tank TK-103. Conductivity monitoring is provided on the surface condenser condensate header. This measurement detects any saltwater leakage into the condensate that may occur due to condenser failure. To save as much water as possible, two set points are provided on the conductivity controller. If the conductivity exceeds 100 micro-mho, the surface condenser condensate is diverted to a utility water storage tank. If a severe leak occurs and the condensate conductivity exceeds 500 micro-mho, the condensate is considered too contaminated to save and is diverted to a sewer.
2. Condensate is also collected from refinery's process equipment and returned to the water treatment system. This condensate is considered potentially contaminated with oil. The condensate passes through an activate carbon filter V-105 before entering condensate storage tank TK-103. The carbon filters are typically designed to process feed with 20 ppm oil and reduce it to less than 0.3 ppm oil in the filtered condensate. The condensate is next passed through condensate polisher V-103, which removes any dissolved salts contamination picked up.

The process condensate is returned to the water treatment unit at approximately 205°F. Since the ion exchange resin in condensate polisher V-104 is extremely temperature sensitive, this stream is cooled to 130°F. BFW heat exchangers E-101 and E-102 by exchanging heat with LP boiler feedwater.

During the operation of the steam system, the demand for HP boiler feedwater exceeds the returned condensate from all sources. To make up for this shortfall, demineralized water from BFW storage tank TK-102 is used. The design flow rate for this service is dictated by the loss of condensate.

Polished condensate is supplied as makeup to all high-pressure dearators as HP boiler feedwater. In addition, the process condensate and surface condenser condensate are returned from the steam system as feed to condensate polisher system.

NEUTRALIZATION

The neutralization system is designed to collect the acid and caustic regeneration waste prior to discharge to the sea or another outlet. Facilities are provided to add acid and caustic to produce a neutral wastewater stream. The neutralization basin has two separate chambers, each designed to hold regeneration waste from three regenerations of cation exchangers, three regenerations of anion exchangers, and one regeneration of condensate polisher. Also used in neutralization are a 98% sulfuric acid and 50% caustic solution for ion exchange regeneration.

UTILITY WATER SYSTEM

As shown in Figure 9-4, the utility water distribution system comprises the following subsystems: potable water and irrigation water makeup, utility water distribution stations, makeup water for cooling towers, and fire water makeup.

POTABLE WATER

Both potable and irrigation water are supplied through the utility water distribution system in the refinery. The refinery may draw its potable water supply from a municipal water supply or a desalination plant in coastal refineries. For supply to refinery drinking water and safety showers, further treatment is required. The treatment consists of passing through a bed of activated carbon in the form of extrudates and supported on a bed of graded quartz. Filtered water is next cooled to 85°F in a water cooler and chlorinated. The treated water must have a residual chlorine content of no less than 0.5 mg/liter.

In case the refinery potable water supply comes from a desalination plant, the total dissolved solids (TDS) may be very low, on the order of 2 mg/liter. The water must be chemically conditioned to make it potable and increase the TDS to potable water requirements (80 mg/liter minimum). Chemical conditioning consists of dosing measured amounts of calcium chloride, sodium bicarbonate, and sodium carbonate into the feedwater stream. Typical dosage of chemicals are these:

Calcium chloride: 28 mg/liter
Sodium bicarbonate: 62 mg/liter
Sodium carbonte: 12 mg/liter

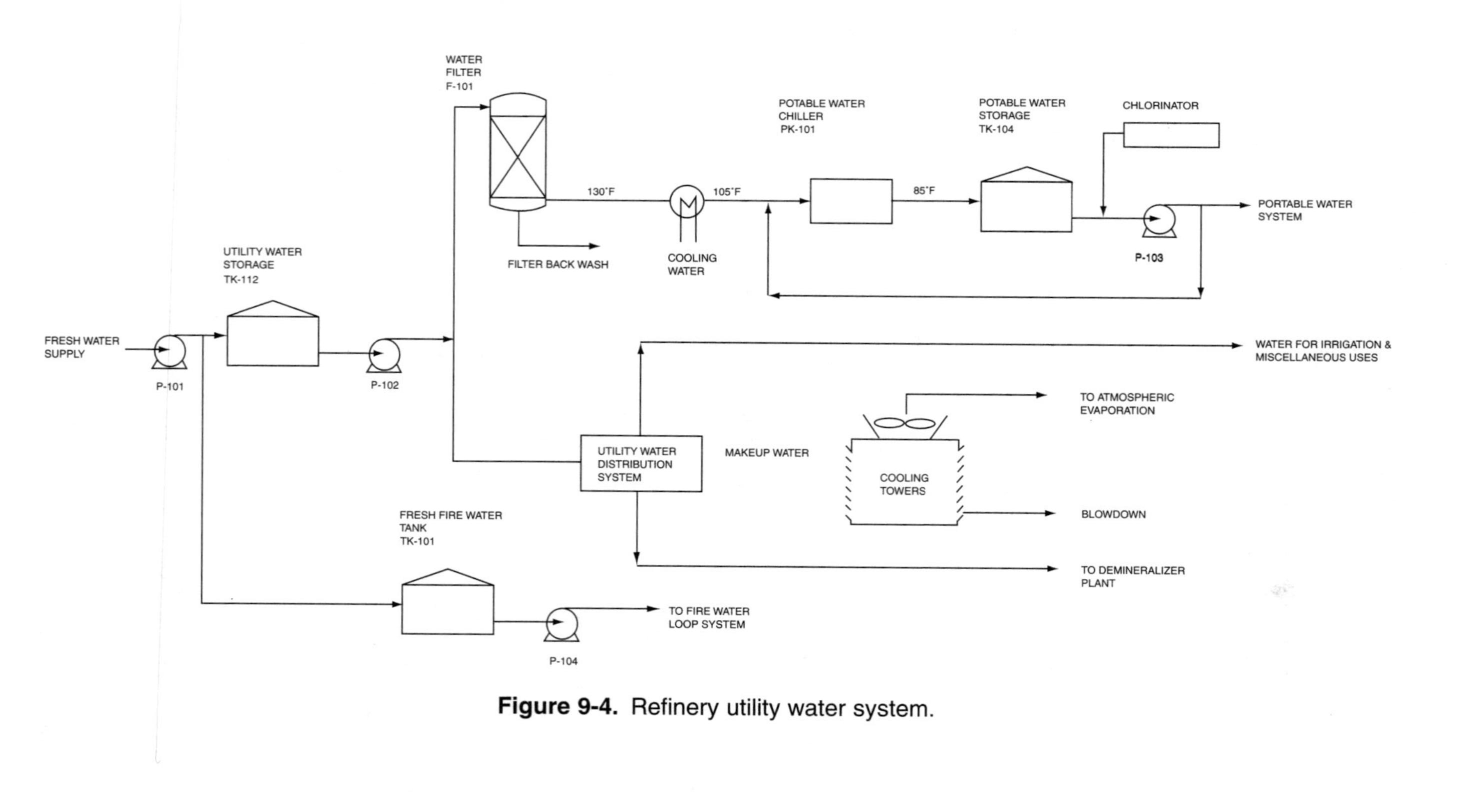

Figure 9-4. Refinery utility water system.

Table 9-3
Potable Water Quality Requirements

PARAMETER	UNITS	
TOTAL DISSOLVED SOLIDS	mg/liter, minimum	80
ALKALINITY	mg/liter as $CaCO_3$	60–80
pH		APPROXIMATELY 8.7
CHLORINE, RESIDUAL	mg/liter, minimum	0.5

If a refinery has an assured supply of carbon dioxide, an alternative conditioning treatment is possible. In this case, carbon dioxide is directly injected into feedwater, and this water is passed through a bed of limestone, where calcium bicarbonate is formed. Excess CO_2 is necessary for the correct hardness level. The remaining CO_2 is neutralized with a nonlimestone-based alkali (caustic soda) to avoid forming turbidity. Typical potable water quality requirements are shown in Table 9-3. The average potable and irrigation water demand is estimated at 35–75 gal/day per person working in the refinery, the higher figures being for hotter climates.

The instantaneous demand for potable water is based on two criteria. The first is the computed peak demand resulting from the number of potable water fixtures in the refinery. The second consideration is the simultaneous use of safety showers. An average safety shower requires water flow of approximately 65 gal/minute. The designed irrigation water usage is generally based on approximately 3 in. water per week applied to the irrigated area of the refinery.

FIRE WATER SYSTEM

Many hydrocarbon processing plants are located along waterways, so that availability of fire water is essentially guaranteed. Similarly, coastal refineries may use sea water as backup fire water in an emergency. If, however, fire water is obtained from municipal sources, storage of the water would be required, for it is unlikely that a municipal water supply system would permit the volumes of draw-down required to fight a fire. The wide variety and varying intensity of fires possible in hydrocarbon processing facilities make precise calculation of fire water requirements difficult. Instead, fire water usage in case of actual fires from historical data of the industries is relied on.

The size of the storage tank should be sufficient to provide a 4–6 hr supply at the estimated maximum fire water rate. A good general figure for estimating fire water is 2500–3000 gpm (gallons per minute) in case of an actual fire. Another important consideration is system pressure. The knowledge of the system capacity as well as the suggested layout of the hydrants, monitors, and water spray systems are the basic input required to determine the size and configuration of the distribution system. The other major input is the residual pressure requirements. The minimum pressure desirable at the extremity of the system is 70 psig, with 100 psig required anywhere in the process area. The maximum pressure that can be tolerated on a handheld hose is 100 psig nozzle pressure and that is a consideration in setting maximum system pressure.

REFINERY WASTEWATER TREATMENT

During the processing of oil in the refinery, large volumes of water flow through the processing units in the form of cooling water, process water, steam, equipment washings, unit hydrotests, and the like; and these streams pick up small quantities of hydrocarbons or oil from equipment and spillage. Oil-contaminated storm water flows from the curbed areas of the refinery add to this load. The objective of the treatment is to separate oil from the wastewater before this water is allowed to be discharged into the sea or any other water body, with minimum negative environmental impact. Typical tolerance limits for refinery effluent discharged into marine coastal areas is shown in Tables 9-4 and 9-5.

Water streams that need treatment is shown in Figure 9-5. Contaminated drainage and storm flows from process areas are routed to retention basins. Storm water is retained for the removal of floating oil, then discharged to the sea with no further treatment. Incidental contaminated drainage is retained for removal of floating oil and pumped to the API oil/water separator system for additional treatment.

Retention basins are designed to collect all water transported to the contaminated sewer network. This includes contaminated storm and fire water flows as well as continuous nuisance flows from daily maintenance operations throughout the refinery. The normal operation of the basins is to provide preliminary oil removal from nuisance flows, using the belt oil skimmers, and act as a sump preparatory to pumping of these flows to API separators.

Table 9-4
Liquid Effluent Discharge to Sea*

PARAMETER	UNIT	REFINERY AVERAGE
TEMPERATURE	°C	45
pH	S.U	5.5–9.0
CHEMICAL OXYGEN DEMAND	mg/l	200
BIOLOGICAL OXYGEN DEMAND	mg/l	30
TOTAL SUSPENDED SOLIDS (TSS)	mg/l	45
OIL/GREASE	mg/l	10
PHENOLS	mg/l	1
SULFIDES	mg/l	0.5
NITROGEN AS AMMONIA	mg/l	10
PHOSPHATES – INORGANIC	mg/l	0.5
CYNIDES	mg/l	1
IRON	mg/l	2
FREE CHLORINE	mg/l	0.6
CHROMIUM AS CrO_4	mg/l	0.3
ZINC	mg/l	0.001–0.05

*TYPICAL REFINERY STANDARDS.

Table 9-5
U.S. Refineries (EPA) Standards*

PARAMETER	UNITS	AVERAGE	MAXIMUM
TEMPERATURE	°C	35	37
pH			6–9
CHEMICAL OXYGEN DEMAND	KG/DAY	2371	4571
BIOLOGICAL OXYGEN DEMAND	KG/DAY	338	610
TOTAL SUSPENDED SOLIDS (TSS)	KG/DAY	270	427
OIL/GREASE	KG/DAY	99	185
PHENOLS	KG/DAY	2.1	4.4
SULFIDES	KG/DAY	1.8	3.8
NITROGEN AS AMMONIA	KG/DAY	136	273
PHOSPHATES – INORGANIC			
CYNIDES			
IRON			
FREE CHLORINE			
CHROMIUM AS CrO_4	KG/DAY	3.3	9.5
ZINC	KG/DAY	1.7	3.8

*FOR A 180,000 BPD REFINERY ON THE U.S. GULF COAST.

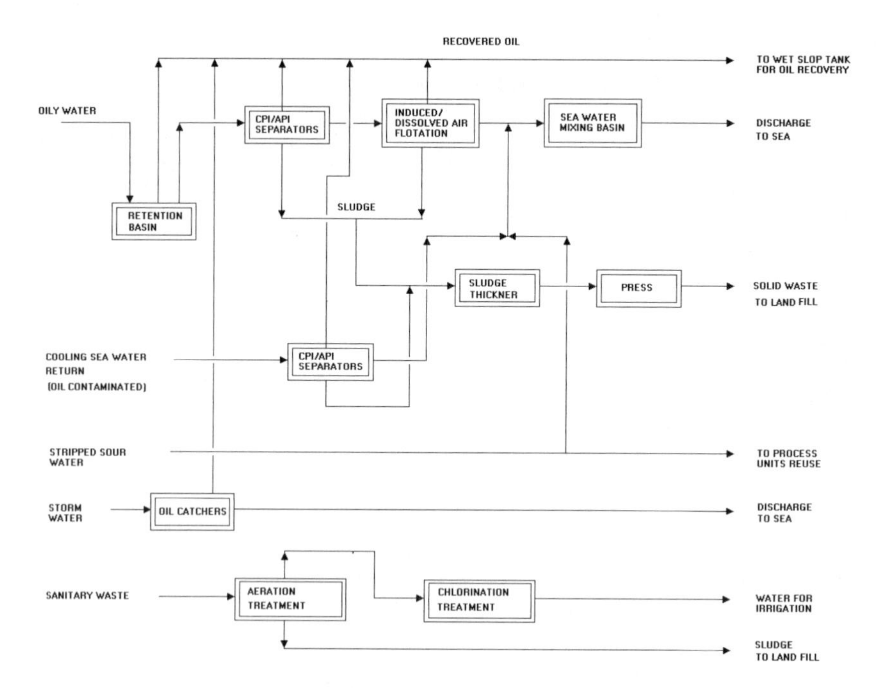

Figure 9-5. Refinery aqueous effluent treatment system. CPI = corrugated plate interceptor.

Flows in excess of nuisance flows, such as contaminated rainwater and fire control runoff, is also collected in the retention basins. This flow is retained for sufficient time for preliminary oil removal using belt and rope skimmers. After this partial deoiling, the water is discharged to the sea.

In coastal refineries, the saline wastewater consists of potentially oil-contaminated once-through sea cooling water, boiler blow-down, and cooling tower blow-down. Oil-contaminated sea water is typically sea water used where the oil side of the exchanger is operating at 55 psig or greater. Potentially oil-contaminated sea water is passed through a battery of API separators. The effluent is combined with uncontaminated once-through sea cooling water in a sea water mixing basin prior to discharge to sea.

Oily wastewater generated in the refinery is segregated into two streams: one low (160–250 ppm) and one high in oil content (>250 ppm). The low-oil stream is made up primarily of stripped sour water and desalter underflow. This stream requires no further treatment and is routed directly to mixing basin, where it is combined with sea cooling water prior to discharge to sea.

The high-oil stream originates from the oily water sewer, oily water pumped out of retention basins, and desalter underflows. The oil in the desalter underflows is in two forms: emulsified and free oil. All potentially high-oil-content streams are treated for the removal of oil using a corrugated plate interceptor type API oil/water separators followed by dissolved air flotation (DAF) oil/water separators. Effluent from DAF units is combined with nonoily wastewater streams and returning cooling water in the sea water mixing basin and discharged to the sea.

The sanitary sewage of the refinery is collected and treated in an aeration unit, which reduces the BOD (biological oxygen demand) and COD (chemical oxygen demand) of the influent. The biological sludge is separated and clarified aqueous effluent is chlorinated to reduce coliform bacteria before discharging this water for irrigation purposes. The sludge separated is disposed of as landfill.

TREATMENT OF OILY WATER

Water from desalters is cooled from 225°F to approximately 130°F in exchanger E-101 to prevent thermal damage to CPI oil/water separator plates (see Figure 9-6). The cool effluent next joins all other potentially contaminated water streams for removal of oil in CPI-type API separators CPI-101A and CPI-101B. CPI separators remove most of the free oil and some of the settleable solids associated with influent streams (Table 9-6). They do not reduce emulsified oil, H_2S, or NH_3. Oil removed from the CPI units is sent to a wet slop oil tank for oil recovery by pump P-101. The effluent water from CPIs is pumped to DAF surge tank TK-101. The oily sludge that settles to the bottom of the CPI units is drawn off periodically by a vacuum truck and taken to a sludge dump.

Waste treatment processes require a relatively steady-state operation to function optimally. Normal and emergency flows fluctuate dramatically throughout a normal operating day. These flows vary from periods of little or no flow to periods of flow approaching the design capacities of the lines. Surge tank TK-101 is therefore located prior to dissolved air flotation units to minimize erratic peak of flows and ensure optimum operation.

DAF units DAF-101A and DAF-101B treat all wastewater from the DAF surge tank. Wastewater is drawn from the surge tank using two pumps operating simultaneously. The flow from the tank is controlled using a manually operated control valve. The DAF unit reduces the oil

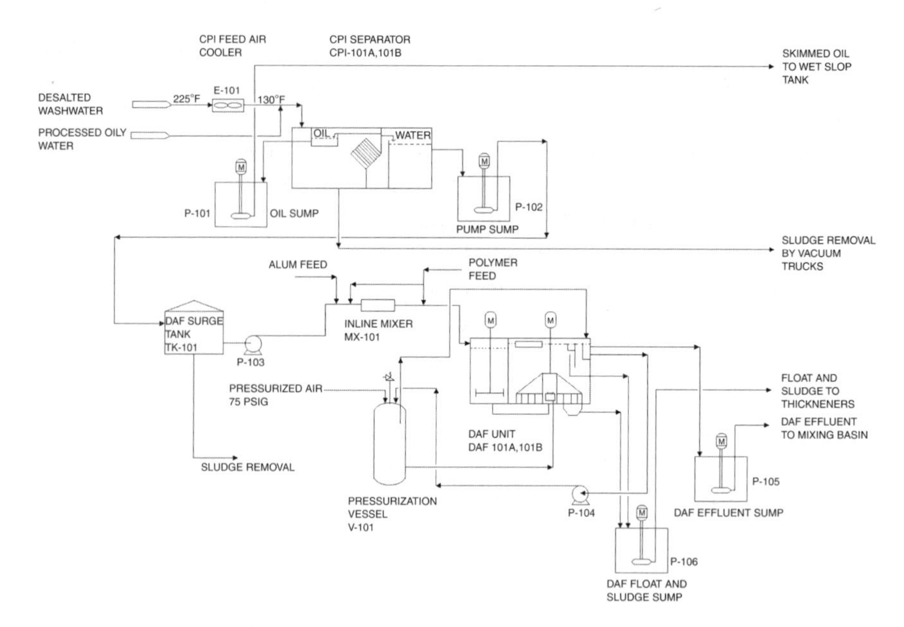

Figure 9-6. Primary oil separation from water. M = motor (pump driver).

Table 9-6
Operating Data of an API Separator (CPI type)

PARAMETERS	UNITS	INFLUENT	EFFLUENT
DESALTED WASHWATER FEED			
SUSPENDED SOILDS	ppm	500	500
TOTAL SOLIDS	ppm	750	750
FREE OIL	ppm	250	20
EMULSIFIED OIL	ppm	160	160
TOTAL OIL	ppm	410	180
COD	ppm	1000	500
BOD	ppm	620	310
PHENOLS	ppm	3	3
SULFIDES AS H_2S	ppm	10	10
NITROGEN AS AMMONIA	ppm	35	35
TEMPERATURE	°F	225	130
OILY WATER FEED			
TOTAL OIL	ppm	2200	190

content of wastewater (emulsified plus free oil) to about 15 ppm. De-emulsifying agents, such as polyelectrolytes or alum, are added to DAF influent in inline mixer MX-101 to increase the oil removal efficiency of each DAF unit. Typical operating conditions and effluent qualities from a DAF unit are shown in Tables 9-7 and 9-8. Three DAF units are normally provided, each sized with 50% of the design capacity, to permit routine maintenance and cope with wide fluctuations in flow.

Sludge management consists of mechanical thickening, dewatering, and sludge storage. The sludge originates from the CPI system and DAF

Table 9-7
DAF Unit Influent and Effluent Quality

PARAMETERS	UNITS	INFLUENT	EFFLUENT
SUSPENDED SOILDS	ppm	<500	20
FREE OIL	ppm	50	3
EMULSIFIED OIL	ppm	160	10
TOTAL OIL	ppm	210	13
COD	ppm	640	300
BOD	ppm	320	150
PHENOLS	ppm	1.8	1.8
SULFIDES AS H_2S	ppm	10	8
NITROGEN AS AMMONIA	ppm	35	15
pH		6.5–7.5	6.0–7.0

Table 9-8
DAF Unit Operating Conditions

PARAMETERS	UNITS	
FLOW RATE/UNIT	gpm	240
RECYCLE RATE, % FEED		50
AIR TO SOLID RATIO		0.1–0.3
AIR PRESSURE	psig	75
AIR FLOW RATE	scfm	2.0–5.0
FLOTATION DETENTION TIME	min	40
HYDRAULIC LOADING	gpm/ft^2	0.5–2.0
COAGULATING CHEMICAL		
ALUM (10% SOLUTION) DOSAGE	ppm	10–20
FLOCCULATING CHEMICAL		
POLYMER SOLUTION	ppm	0.5

units. The sludge contains considerable water, oil, and some solids, which may be considered hazardous or toxic. Thickening is done in a thickener, followed by dewatering by a belt press to reduce sludge volume. The decanted or released water is returned to CPI system. The dewatered sludge, rich in solids and oil, is conveyed to a sludge storage area by conveyer belt to await disposal as landfill. Polyelectrolytes are added to the sludge as thickener and in the belt press influent to increase solid removal efficiency for both thickener and belt presses.

Effluent from the DAF units is combined with nonoily wastewater streams and cooling water return in a mixing basin and discharged to the sea or other water body.

WET SLOP OIL SYSTEM

The slop oil streams from sour water stripper, wastewater treatment, the oil terminal, and such contain large quantities of water, which need to be separated, using heating, settling, and by adding oil/water-emulsion-breaking chemicals (see Figure 9-7).

Wet slop treatment is a batch operation. All the wet slop produced is pumped to two wet slop tanks, T-101A and T-101B. Each tank is provided with a floating roof and a swing arm suction line. The tanks are in service alternately. When a tank gets full, its contents are heated and mixed by pumping the liquid through wet slop heater E-101 and back to the tank. LP steam is used to raise and maintain the wet slop temperature to 120°F. To

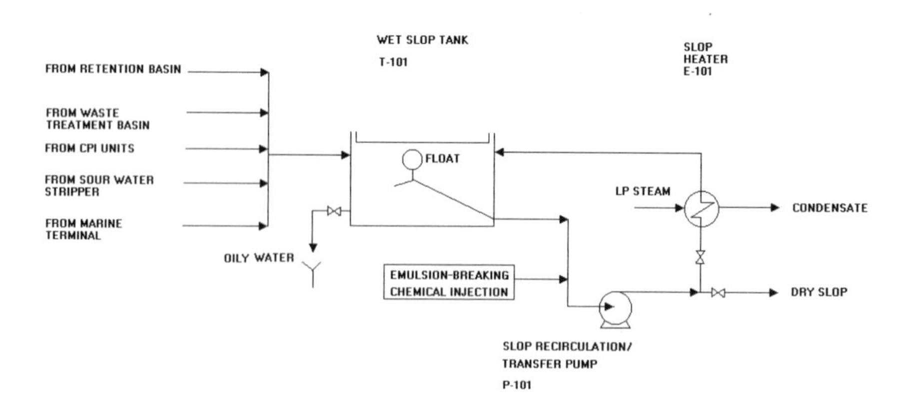

Figure 9-7. Wet slop system. LP = low pressure.

facilitate the breaking up of oil/water emulsions, an emulsion-breaking chemical is added to the wet slop oil during pump/mix cycle.

The water-free slop oil is pumped to a dry slop oil tank. The oil-free water is drained to oily water sewer and returned to the CPI system. The entire operation, consisting of collection, heating, and transfer, is a manual batch type operation.

TREATMENT OF SANITARY SEWAGE

All sanitary sewage in the refinery is collected in a common sewage lift station, from which it is pumped to equalization tank TK-101 (see Figure 9-8). The sewage equalization tank is sized for approximately 6 hr retention of normal flows plus capacity to accommodate surges associated with peak flows from the lift station. The equalization tank is aerated to prevent the sanitary wastewater from becoming septic. The sewage is treated in two aeration tanks TK-103, with each unit having the capacity to treat approximately 75% of daily refinery sewage production. The design thus provides flexibility to allow normal maintenance while maintaining the required degree of treatment.

The aeration treatment reduces both the BOD and suspended solids loading down to 30 mg/liter. The effluent stream from the aeration units is chlorinated in TK-105 to reduce the coliform bacteria content to acceptable level. The treated effluent is discharged to the sea. Operating conditions and quality of effluent after treatment are shown in Table 9-9.

During the aeration process, biological sludge, approximately 9% of the feed volume, is produced. This sludge is removed from sludge digester tank TK-102 by vacuum trucks and disposed of off plot.

SOUR WATER TREATMENT

Sour water, rich in NH_3 and H_2S content, is produced from refinery crude distillation units, hydrocracker, hydrodesulfurization units, gas handling, amine regenerators, delayed cokers, sulfur plant tail gas treating units, and the like. Sour water is produced by the steam stripping operation during crude distillation hydrotreating and hydrocracking operations. During most hydrocracking and hydrotreating operations, NH_3 and H_2S

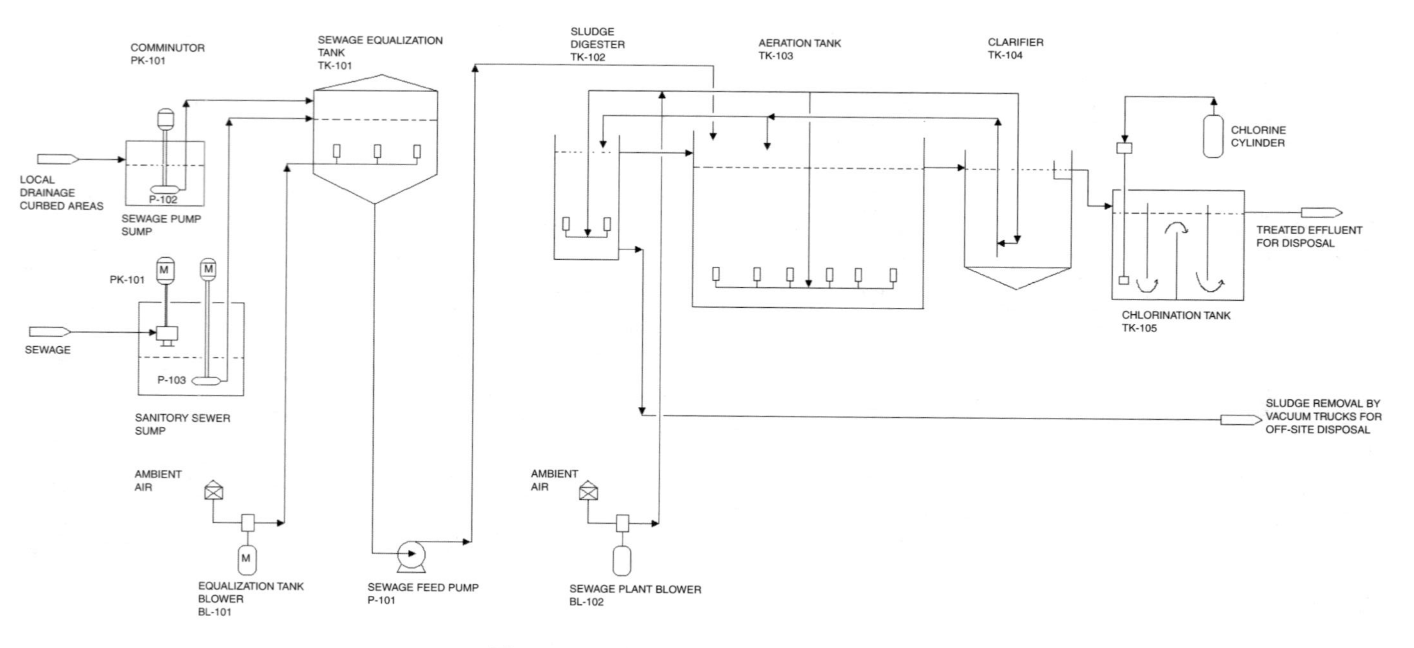

Figure 9-8. Sewage treatment system.

Table 9-9
Sanitary Sewage Treatment (Aeration Unit) Operating Conditions and Stream Quality

PARAMETERS	UNITS	INFLUENT	TREATED EFFLUENT
FLOW RATE/UNIT	gpd	60,000	
SUSPENDED SOLIDS	ppm	<500	30
BOD	ppm	300	30
COD	ppm	1000	30
TOTAL OIL	ppm	150	0
TOTAL NITROGEN AS NH_3	ppm	80	0
pH		6.5–7.5	6.5–8.0
TEMPERATURE	°F	90–95	90–95
DETENTION TIME	hr	22	
vol/lb BOD	ft^3	48.5	
AIR SUPPLIED, lb/BOD REMOVED	SCF	105	
RESIDUAL OXYGEN	ppm	2	>2
CHLORINE RESIDUAL	ppm	0.5–1.0	

are formed in the reactors due to elimination of sulfur and nitrogen in the feed. These two react to form solid ammonium sulfide salts. Ammoniacal condensate injection is required in the reactor effluent ahead of HP separation to prevent deposit of solid ammonium sulfide salts in pipes. Deposit is prevented by dissolving most of the ammonia and approximately an equal quantity of H_2S in water, followed by removing the resulting aqueous solution from the system in the form of sour water. Also in hydrotreating plants, steam stripping of desulfurized liquid streams, to remove H_2S, produces sour condensate containing dissolved H_2S.

Sour water coming from various refinery units has a very high concentration of H_2S and NH_3, typical values being 10,000–20,000 and 5000–10,000 ppmw, respectively. This water also contains some suspended and dissolved hydrocarbons. The composition of sour water from a distillate hydrocracker unit is shown in Table 9-10. Typically a 200 MBPD refinery of medium complexity may produce 1000 gpm sour water. After removing NH_3, H_2S, and small amount of oil, this water can be reused.

Table 9-10
Sour Water Composition

SOURCE	SOUR WATER, GPM	NH_3, PPMW	H_2S, PPMW	OIL, PPMW
HP DRUM (2585 psig)	80.4	12750	20400	—
LP DRUM (500 psig)	8.1	6800	25500	10
STRIPPER OVERHEAD DRUM	30.3	850	3400	5
TOTAL	118.8			

NOTES:
BASIS 44,000 BPSD HYDROCRACKER UNIT.
FEED SULFUR = 2.6 WT%, NITROGEN 695 ppmw.

SOUR WATER STRIPPING UNIT

Ammoniacal water streams from various refinery units are mixed and cooled with cooling water in E-101 to approximately 120°F (see Figure 9-9). The cooled effluent is degassed in V-101. The gases evolved are sent to flare. The degassed foul water is pumped by P-101 to tank TK-101 and kept under fuel gas blanket. Sufficient detention time is allowed in this tank to separate out traces of oil in the water. Separated oil is skimmed off with a swing arm suction line and sent to the slop oil system. The water is pumped to steam stripper column V-103, with about 40 sieve plates.

Ammonia and H_2S are stripped from this water in column V-103, with 40 plates. The column has a thermosyphon reboiler E-104, heated with 150 psig steam. Feed before entering the column is heated to approximately 205°F by heat exchange with the V-103 bottom stream in E-102 and feed enters the column V-102 at the 30th plate. To provide reflux to the column, a sidestream is withdrawn from the 34th plate, cooled to 150°F from 195°F and returned to the top plate of the column. Column top pressure is maintained at 15 psig. The top vapor distillate containing most of H_2S and NH_3 is sent to sulfur plant or flare. The bottom product, stripped water at approximately 262°F, is first cooled to 175°F by heat exchange with incoming column feed. It is next cooled in air cooled exchanger E-105 then in trim cooler E-106 to 105°F before sending the stream outside the battery limit.

The operating conditions of a sour water stripping unit are shown in Table 9-11. The unit utility consumption and yields are shown in Tables 9-12 and 9-13.

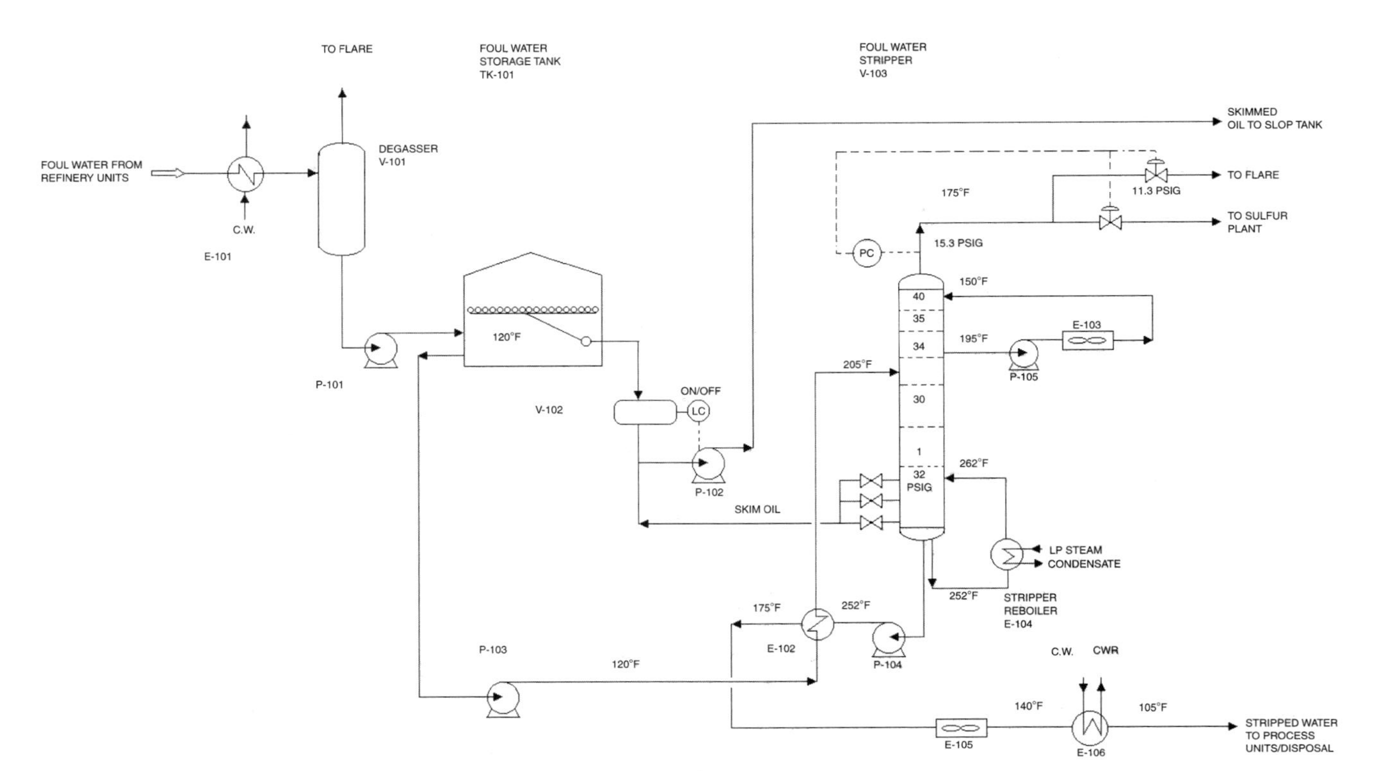

Figure 9-9. Sour water supply. C.W. = cooling water; LP = low pressure; CWR = cooling water return.

Table 9-11
Sour Water Stripper Operating Conditions

PARAMETER	UNIT	
STRIPPER FEED TEMPERATURE	°F	205
STRIPPER TOP TEMPERATURE	°F	175
STRIPPER TOP PRESSURE	psig	15.3
STRIPPER REBOILER TEMPERATURE	°F	262
STRIPPER BOTTOM PRESSURE	psig	32.0
TOTAL NUMBER OF PLATES IN STRIPPER		40
NUMBER OF PLATES ABOVE FEED PLATE		10
REFLUX RATIO, REFLUX/FEED		3.13
REFLUX TEMPERATURE	°F	150
STRIPPED WATER DISCHARGE TEMPERATURE	°F	105

Table 9-12
Sour Water Stripper Utility Consumption per Ton Feed

UTILITY	UNITS	CONSUMPTION
STEAM	mmBtu	0.345
ELECTRICITY	kWhr	2.73
COOLING WATER	MIG*	0.135

*MIG = 1000 IMPERIAL GALLONS.

EXAMPLE 9-1

A 200-mbpcd refinery requires 90,000 gpm sea water for cooling. Sea water is supplied through a 72 in. (internal diameter 1800 mm) pipe from a water intake pit located 4 km from the refinery battery limits. Sea water is to be supplied at a pressure of 75 psi to the manifold, located at an elevation of 22 ft from the intake pit. Determine the power consumption for pumping sea water supply to the refinery. Assume a surface roughness factor of 100 for the pipe. Sea water properties can be assumed as follows:

$$\text{Temperature} = 90^{\circ}\text{F}$$

$$\text{Density} = 1.030\,\text{kg/m}^3$$

$$\text{Viscosity} = 0.76\,\text{cp}$$

Table 9-13
Material Balance of a 500 gpm Sour Water Stripper Unit

	UNITS	SOUR WATER FEED	TREATED WATER	SIDE REFLUX	SOUR GAS TO SULFUR PLANT
STREAM					
WATER	lb mol/hr	13620.90	13562.50	33408.50	58.40
H_2S	lb mol/hr	97.50	0.06	2549.20	97.44
NH_3	lb mol/hr	98.70	0.29	5563.40	98.41
TOTAL	lb mol/hr	13817.10	13562.85	41521.10	254.25
TOTAL FEED	lb/hr	250190.00	244730.00	782880.00	6050.00
H_2S	ppmw	13251	0	(11.07 Wt%)	(54.87 Wt%)
NH_3	ppmw	6707	20	(12.08 Wt%)	(27.71 Wt%)
mol Wt		18.1	18	18.9	23.8

The first step is to determine the pressure drop in the pipe, using William-Hazen's equation as follows:

$$\Delta P = \frac{10.67 \times Q^{1.85} \times \rho \times L}{C^{1.85} \times D^{4.87}} \times 10^{-4}$$

where

$\Delta P =$ pressure drop, kg/cm^2;
$Q =$ flow rate, m^3/sec;
$\rho =$ density of fluid, kg/m^3;
$L =$ equivalent of pipe, m;
$C =$ pipe roughness factor;
$D =$ internal diameter of pipe, m.

Here

$Q = 5.68;$
$L = 4000;$
$C = 100;$
$\rho = 1030;$
$D = 1.8.$

$$\begin{aligned}\Delta P &= \frac{10.67 \times (5.68)^{1.85} \times 1030 \times 4000 \times (10)^{-4}}{100^{1.85} \times (1.8)^{4.87}} \\ &= 1.245\ \text{kg/cm}^2 \\ &= 17.71\ \text{psi}\end{aligned}$$

$$\begin{aligned}\Delta P\ (\text{feet of fluid}) &= \frac{(\Delta P \text{ in psi})}{(0.434 \times \text{specific gravity})} \\ &= 39.61\ \text{feet}\end{aligned}$$

Total head required = discharge pressure + vertical discharge elevation + frictional ΔP, as follows:

	psi	WATER, ft
DISCHARGE PRESSURE	75.00	167.80
VERTICAL ELEVATION	9.83	22.00
FRICTIONAL LOSSES	17.71	39.61
TOTAL PUMP HEAD	102.54	229.41

$$\text{Brake Horse Power (BHP)} = \frac{\text{gpm} \times \Delta H \times \text{specific gravity}}{3960 \times \eta}$$

$$\text{kW power} = \text{BHP} \times 0.746$$

where

gpm = flow rate, gal/min;
ΔH = head, ft of fluid;
η = pump efficiency

Assuming a pump efficiency of 70%,

$$\text{Pump BHP} = \frac{90{,}000 \times 229.41 \times 1.03}{3960 \times 0.70} = 7671.8$$

$$\text{Pump kW power} = 7670.5 \times 0.746 = 5{,}723$$

NOTE

1. *Oil and Gas Journal* (June 21, 1976).

CHAPTER TEN

Refinery Off-Site Facilities and Utility Systems

REFINERY TANKAGE

Tanks are required in refineries for storage of crude, blend stocks, and products for shipping. Tanks are also required as charge tanks for many secondary processing units. In a refinery, the cost of tanks alone roughly equals that of all the process units in it.

From an operational point of view, refinery tankage can be divided into the following categories: crude storage tanks, charge tanks for secondary processing units, base stock or component storage tanks for product blending, and shipping tanks. Tanks used in various utility services are not considered here.

CRUDE STORAGE

From the viewpoint of operational safety, a refinery located close to an oilfield or pipeline terminal requires a minimum inventory equivalent to 16 hr throughput. This includes a 4-hr allowance for settling, sampling, and testing. The total minimum operating inventory comprises the following elements: crude oil working stock (= 16 hr crude throughput), unavailable tank heels (of all crude tanks in operation), and line content. In practice, a storage capacity equivalent to 48 hr crude throughput is generally considered adequate for such refineries.

For coastal refineries, which are supplied crude by marine tankers, the crude storage capacity is dictated by the parcel size and frequency of crude tankers. Normally, storage capacity equivalent to 15 days throughput is considered adequate.

CHARGE TANKS

Charge tanks are required for most secondary processing units. Charge tanks are required as surge tanks, to insulate the unit from any temporary upset upstream of the unit from which it is receiving feed. These are essentially low-capacity tanks with a working capacity of approximately 3–4 days operation of the unit it is feeding. For example, if a refinery has a naphtha unifiner, cat reformer, diesel hydrodesulfurizer, and hydrocracker, each unit may have a charge tank. Any upset in the crude unit is prevented from affecting operation of these units. Crude storage tanks can be considered charge tanks for crude distillation units.

BASE STOCK TANKS

Product received from any single unit of the refinery may not necessarily be a finished product. It can be feedstock for another processing unit or a blend stock for a product or group of products. For example, reformate from the cat reformer unit may be a base or blend stock for all the gasoline grades being produced by the refinery. If a tank is utilized to store an intermediate stock from a refinery unit for product blending, it is called a *base stock tank*. In refineries with a continuous on-line blending system, base stock tanks also serve as shipping tanks.

SHIPPING TANKS

Various base stocks are blended to yield finished products. Finished products, after blending and testing in the refinery laboratory, are stored in shipping tanks. The shipping tanks constitute the major part of tankage in a refinery, if batch blending methods are employed. The working storage capacity required for any product group is determined from the following factors: production rate, maximum parcel size, number of different product grades in a product group, and product loading rates.

The production rate, in tons or barrels per day, for all product groups (naphtha, gasoline, kerosene, diesel, and fuel oil) can be determined from the overall material balance of the refinery.

The maximum parcel size is the quantity, in tons or barrels, lifted from the terminal by a single ship at a time. It depends on the dwt (dead weight tons) of the tanker arriving to lift products and storage capacity available for that product in the refinery. All refinery terminals have norms for the maximum allowable loading time for different-sized tankers, within which loading must be completed.

SHIPPING TERMINALS AND SEA LINES

Blended products are stored in shipping tanks away from the process area. From shipping tanks, the products are pumped by high-capacity shipping pumps through dedicated sea lines to loading arms at the piers and finally by hoses to individual tankers. The terminal product transfer lines sizes and pumping capacities are designed to achieve the required loading rates for different product groups. For example, if the maximum parcel size of naphtha is 80,000 tons, the loading pumps be able to pump the product from a number of shipping tanks to the ship within a reasonable time. A 30-hr loading time may be allowed for a 80,000 dwt tanker, requiring a loading rate of 2700 tons per hour. Also, the sea lines must be suitably sized to take this product rate within the permissible pressure drop. The sea line required for this service may be 16–20 in. in diameter. Marine terminals usually have a number of berths. Each berth, in turn, has restrictions as to the minimum and maximum dwt of the tanker it can accept. Specifications of a typical marine terminal serving a 250 mbpd refinery are shown in Table 10-1. Each of the berths may not have connection to all sea lines. Normally, each product group has at least one dedicated sea line.

Facilities available at the marine terminal (maximum size of the tanker accepted, tanker loading rates, etc.) decide the maximum parcel size and capacity of shipping tanks required for a given production rate of the refinery.

The working capacity of base stock and shipping tanks is computed taking into account the following components:

- 10 days product storage.
- Contingency equivalent to 10 days production of the refinery.
- One maximum size parcel for each product group.
- Any fixed requirement for each product group.

Contingency storage capacity is required for unforeseen situations, such as slippage of vessels causing lifting delays, disruption of export schedule due to bad weather conditions, or scheduled and unscheduled shutdowns of key refinery units. Working capacity also takes into account

Table 10-1
Specifications of a Typical Refinery Marine Terminal

PRODUCT	SHIPPING LINE 1	2	3	4	5	6
NAPHTHA	X					
MOGAS		X				
ATK/KEROSENE			X			
DIESEL				X	X	
FUEL OIL						X
MAXIMUM PUMPING RATE 000' tons/hr	2.22	2.22	2.22	2.22	2.22	2.22

BERTH	dwt RESTRICTIONS MIN.	MAX.	SHIPPING LINE CONNECTIONS
1	10	160	1, 3, 4, 5, 6
2	10	50	1, 3, 4, 5, 6
3	10	160	1, 2, 3, 4, 5, 6
4	10	50	1, 2, 3, 4, 5, 6
5	10	50	1, 2, 3, 4, 5, 6

loss of storage capacity due to release of tanks for scheduled and unscheduled maintenance.

REFINERY TANKAGE ESTIMATION

For estimating the tankage requirements for a refinery, the following information is required:

1. Refinery process unit capacities, in tons and barrels per day.
2. Refinery material balance and product blending schedule. (Figure 10-1).
3. Rundown temperature of all streams flowing to the base stock and shipping tanks (Table 10-2) to determine product density.
4. Information on the refinery marine terminal product loading facilities; that is, the available sea lines, product pumping rates, and connections to various berths (Table 10-1). A refinery maintains

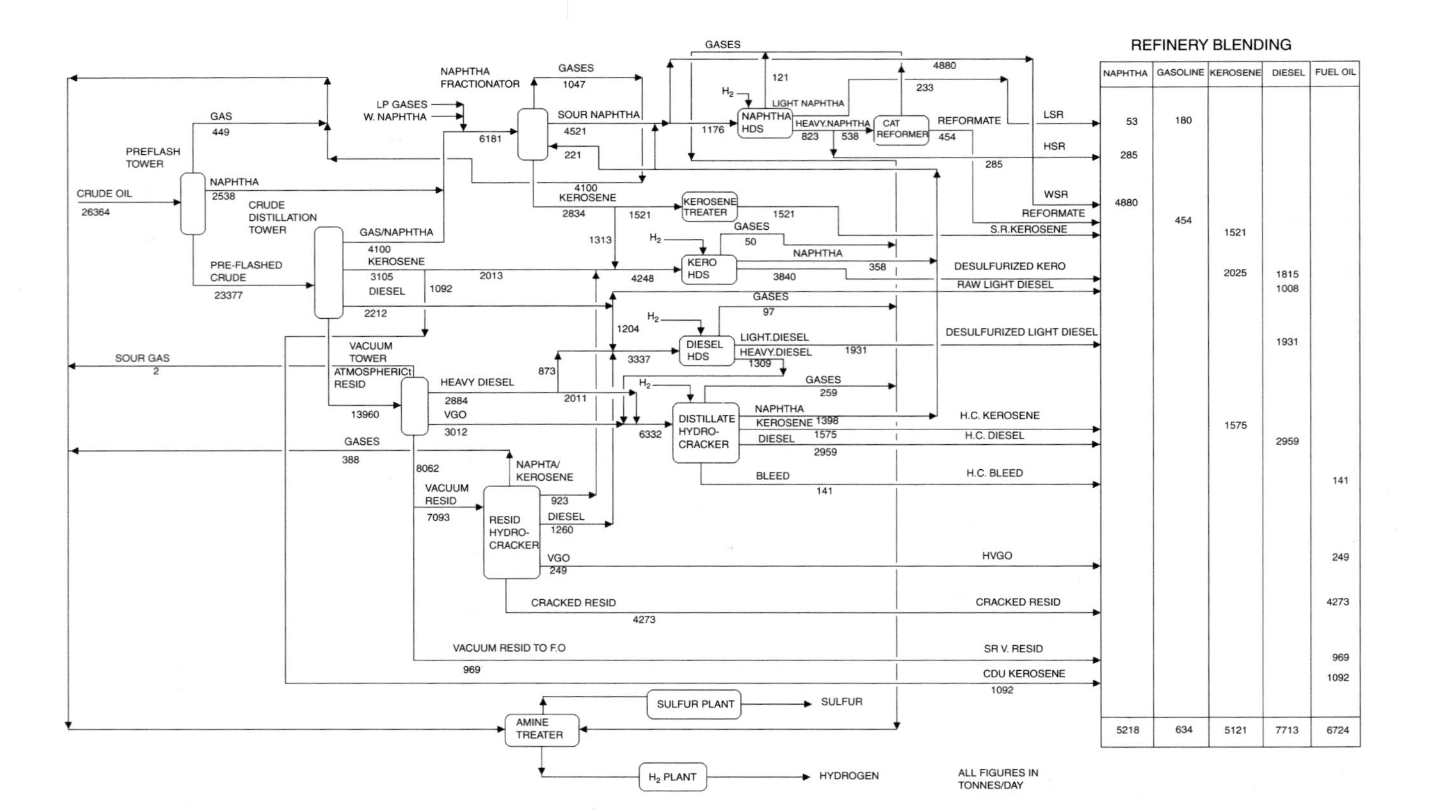

Figure 10-1. Refinery block flow diagram and blending schedule. SR = straight run; LSR = light straight run; HSR heavy straight run; WSR = whole straight run; VGO = vacuum gas oil; HVGO = heavy vacuum gas oil; CDU = crude distillation unit; H.C. = hydrocracker; SR.V.Resid = Straight Run Vacuum Resid.

Table 10-2
Storage Temperatures of Refinery Streams

STOCK	TYPE	NORMAL FLASH POINT, °F	MINIMUM FLASH POINT, °F	NORMAL RVP, psi	MAXIMUM RVP, psi	NORMAL STORAGE TEMPERATURE, °F
CRUDE OIL	FLOATING			6.0		115
FULL RANGE NAPHTHA				7–9.5		105
CAT REFORMER CHARGE	FLOATING			1.5		100
REFORMATE	FLOATING			8–9	9.5	100
UNLEADED GASOLINE	FLOATING			8–10	10.5	100
KEROSENE UNIFINER CHARGE	FLOATING					125
KEROSENE/DIESEL	FIXED	100–160	100/140			110
DIESEL	FIXED	150–200	140			125
MARINE DIESEL OIL	FIXED	150–180	130			110
DIESEL HDS CHARGE	FLOATING	>200	200			160
RAW LIGHT DIESEL	FLOATING	150–230	150			140
DESULFURIZED GAS OIL	FIXED	>200	150			125
HYDROCRACKER CHARGE/HVGO	FLOATING	>200	200			175
FUEL OIL	FIXED	150–230	150			155
RESID HYDROCRACKER CHARGE/V.R	FIXED	>200	200			490
SLOP	FIXED			0–15		110

HVGO = HEAVY VACUUM GAS OIL; VR = VACUUM RESID.

historic data of previous years on the average and maximum size of tankers calling on the terminal to take product, by product group (naphtha, kerosene, diesel, fuel oil. etc.), which serves as the basis for future projections (Table 10-3).

An example for determination of the working capacity of refinery tankage follows.

EXAMPLE 10-1

A coastal refinery processes 180,000 barrels per day of 30.6 API Middle Eastern crude. The capacity of various processing units in mbpd and tonnes/day are shown in Table 10-4. Figure 10-1 shows the overall material balance and product blending schedule of the refinery. Table 10-2 shows the rundown temperatures of base stocks and blended products. Table 10-5 shows the facilities available at the refinery marine terminal. Table 10-1 shows the historic for the distribution of tanker fleet arriving at the terminal for product removal.

The refinery employs a batch blending system. The crude is received into the refinery tankage from the pipeline from oil fields. We want to estimate the tankage requirement for storing crude oil, charge tank capacities, and shipping tankage capacities for the product slate of this refinery.

SOLUTION

Crude Oil and Charge Tanks Capacities

Estimated crude oil storage requirements are based on 48-hr crude storage. Various processing units charge tanks are based on 3–4 days of feed storage, depending on the process requirements of the unit (Table 10-6).

Base Stocks and Shipping Tanks

Table 10-5 shows the calculations for estimating the shipping tankage required for every product group. Base stock tankage capacity is also considered a part of shipping tankage capacity in a given product group. As discussed earlier, operational capacity is based on 20 days production (10 days production and 10 days contingency) plus one maximum parcel

Table 10-3
Distribution of Tanker Fleet

TANKER DWT RANGE, 1000 TONS			PERCENT OF CARGO VOLUME				
MINIMUM	MAXIMUM	AVERAGE	NAPHTHA	KEROSENE	MOGAS	DIESEL	FUEL OIL
10	20	15	0	0	0	1	0
20	30	25	2	10	10	10	4
30	40	35	31	47	46	42	4
40	60	50	10	28	18	24	6
80	100	90	37	0	0	4	33
100	130	115	11	3	0	3	28
130	160	145	0	0	0	1	12
160	190	175	0	6	0	4	2
190	220	205	0	1	0	0	0
AVERAGE PARCEL SIZE, 1000 TONS			53	44	38	45	86

Table 10-4
Refinery Unit Capacities

UNIT	000' bpsd	000' TONS/DAY
CRUDE DISTILLATION	180.00	24.98
VACUUM DISTILLATION	85.00	13.23
NAPHTHA UNIFINER	36.00	4.16
CATALYTIC REFORMER	15.00	1.92
KEROSENE UNIFINER	31.50	4.02
DIESEL UNIFINER	22.50	3.16
KEROSENE MEROX	15.00	1.90
DISTILLATE HYDROCRACKER	41.30	6.00
RESID HYDROCRACKER	41.20	6.72

BPSD = BARRELS PER STREAM DAY.

size. To the capacity so estimated is added any fixed product requirements for a product group; for example, tankage required for local marketing, bunker sales, or meeting seasonal requirements for a particular grade.

PRODUCT BLENDING SYSTEM

The process units produce various product components and base stocks, which must be combined or blended, sometimes with suitable additives, to manufacture finished products. These finished products are generally grouped into the broad categories of gasoline, kerosene, diesel and fuel oil, and so forth.

Different methods of product blending are used to suit variation in the type of product, available components, operating procedures, shipping and marketing requirements, and available storage facilities. Blending methods normally employed include batch blending, partial in-line blending, and continuous in-line blending.

Petroleum products are shipped in bulk using pipelines, marine tankers, and occasionally road or rail facilities.

BATCH BLENDING

In batch blending, the components of a product are added together in a tank, one by one or in partial combination (see Figure 10-2). The materials are then mixed until a homogenous product is obtained.

Table 10-5
Estimation of Product Tankage Requirements

PRODUCT	PRODUCT TYPE	PRODUCTION RATE[2] TONS/DAY	OPERATIONAL REQUIREMENTS[1] 000' TONS	PARCEL SIZE[3] 000' TONS	FIXED[4] REQUIREMENTS 000' TONS	TOTAL CAPACITY	
						000' TONS	000' BARRELS
NAPHTHA	LSR	848.0	17.0				
NAPHTHA	WSR	1998.6	40.0				
NAPHTHA	GROUP TOTAL	2846.6	56.9	80	11.0	147.9	1321
REFORMATE	98 RON	0.0	–				
REFORMATE	96 RON	69.6	1.4				
MOGAS 90	90 RON	289.5	5.8				
MOGAS 98	98 RON	1179.6					
MOGAS	GROUP TOTAL	1538.7	30.8	35	12.0	77.8	676
KEROSENE	DPK	1005.5	20.1				
KEROSENE	JP-5	805.2	16.1				
KEROSENE	GROUP TOTAL	1810.7	36.2	80	108.0	224.2	1771
DIESEL	0.2% S	6494.1	129.9				
	53 CETANE	2494.1					
DIESEL	GROUP TOTAL	8988.2	179.8	80	0.0	259.8	1933
MARINE DIESEL		615.9	12.3				
MARINE DIESEL	GROUP TOTAL	615.9	12.3	80	70.0	162.3	1152
FUEL OIL	3.5% S, 380 CST	3995.7	79.9				
1% SULFUR FUEL OIL	3.5% S, CRACKED	2585.7	51.7				
FUEL OIL	GROUP TOTAL	6581.4	131.6	100	80.0	311.6	2029
TOTAL PRODUCTION		22381.5	447.6			1183.6	8882

NOTES:
[1]OPERATIONAL REQUIREMENTS EQUIVALENT TO 20 DAYS PRODUCTION.
[2]CORRESPONDING TO REFINERY CRUDE THROUGHPUT OF 180 MBPSD.
[3]MAXIMUM PARCEL SIZE DEPENDENT ON TANKER SIZE FOR PRODUCT LIFTING, CAPACITIES OF SHIPPING TANKS, AND LOADING RATES OF THE TERMINAL.
[4]FIXED REQUIREMENTS CORRESPOND TO STORAGE CAPACITY REQUIRED FOR SPECIFIC PRODUCT GRADES TO MEET SEASONAL REQUIREMENTS, LOCAL MARKETING, OR OTHER SPECIFIC USES.

Table 10-6
Crude Storage and Units' Charge Tank Capacities

UNITS	STORAGE TIME hr	WORKING CAPACITY 000' TONS	000' BBLS
CRUDE STORAGE	26	27.05	195
NAPHTHA UNIFINER	72	12.47	108
CATALYTIC REFORMER	96	7.67	60
KEROSENE UNIFINER	72	12.07	94.5
DIESEL UNIFINER	72	9.48	67.5
DISTILLATE HYDROCRACKER	72	17.99	123.9
RESID HYDROCRACKER	96	26.88	164.8

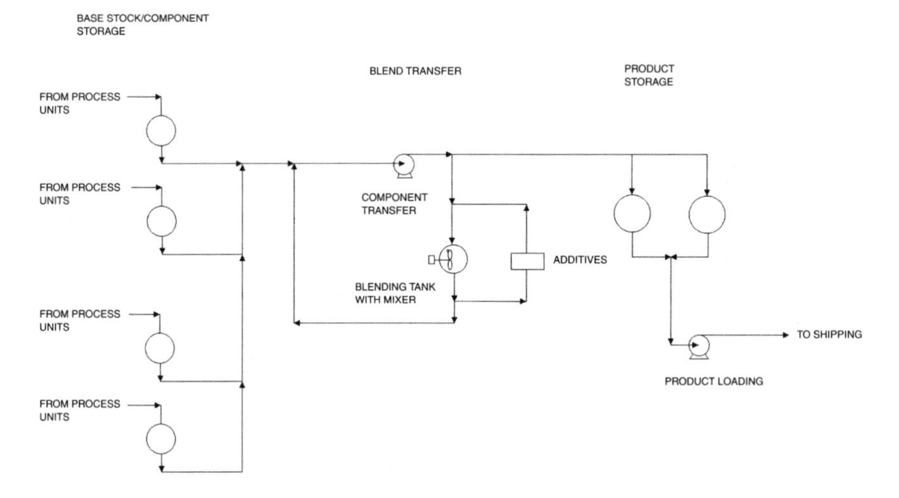

Figure 10-2. Batch blending system.

The components are run from process units to "base stocks" tanks. Each component stream is next pumped separately from base stock tanks into a blending tank, and the tank is gauged after each addition. Additives such as gasoline dyes, TEL (tetraethyl lead, a gasoline additive for increasing octane number of gasoline) kerosene antistatic or antiicing agents, diesel flow improvers, and the like are then added and mixed thoroughly. After laboratory analysis, the blended product is pumped to storage or shipping tanks.

When butane is blended into gasoline, the batch procedure is varied to bring butane into the tank after line-blending it with another component. In this manner, the butane blended is dissolved in the heavier material with less butane waste and without the fire hazards introduced by static electricity generated by in-tank blending of butane.

Storage tanks are required for base stocks and finished product shipping. The required capacity can be estimated depending on the refinery production rate and shipping parcel size. The capacity of mixing or blending tanks is not included in the total storage capacity. These tanks are usually of sufficient capacity to allow blending two to three days product per batch.

Components are transferred to the mixing tank by one or more pumps. A pump of relatively high capacity is required because each component must be transferred and gauged separately. In general, each component transfer pump is large enough to fill the blending tank in approximately 3 hr.

The blended material must be mixed thoroughly by propeller mixers, in-tank jet mixers, or recirculation pumps and piping. For large tanks, multiple mixers are required. For blending by in-tank jet mixers, a pump is necessary. The pump takes suction from the blend tank and discharges it, through suitable piping, back to the tank.

If the discharge into the tank is through a distribution spider or a swing line, a large-capacity pump is necessary. The mixing is considered complete when the entire contents of the tank has been pumped through the pump at least once.

A jet mixer consists of a nozzle instead of a spider or swing line. This nozzle is directed upward from the bottom of the tank at an angle. The high velocity of the jet stream induces circulation of the entire contents of the tank. The pump used with the in-tank jet mixer can be a low-capacity, high-head pump. After the required mixing hours, samples of the blended product are drawn from the top, middle, and bottom of the blending tank, and the three samples analyzed in the laboratory. If the three analyses are identical, the blending is considered satisfactory and the blended material can be transferred to the shipping tank. If the three analysis results are not identical, more mixing time is allowed until the tank contents are uniformly mixed.

Batch blending is most adaptable to use in small refineries, in which a limited variety of blends are to be produced. In a refinery, the cost of extra blending tanks, pumps, and related equipment may not be as large as the cost of instrumentation and equipment needed for in-line blending; and for this reason, many large refineries continue to use the batch blending system because of its ease and flexibility of its operation.

PARTIAL IN-LINE BLENDING

Partial in-line blending is accomplished by adding together product components simultaneously in a pipeline at approximately the desired ratio without necessarily obtaining a finished specification product (see Figure 10-3). Final adjustments and additions are required, based on laboratory tests, to obtain the specification product. In partial in-line blending, the mixing is required only for final adjustment. Additives, if any, such as dye for gasoline, are added as a batch into the blending header during the final stages of the blend or final adjustment stage.

The required components are pumped simultaneously from each base stock tank through the appropriate flow controller into a blending header. The component is mixed by turbulence in the header, as combined components flow to the finished product storage tank. Additives are introduced into the blend by a bypass stream with a suitable booster pump to an aductor or by a proportioning pump delivering a premix.

For partial in-line blending, an individual pump is required for each component. The capacity of the pump must be established to permit simultaneous pumping and delivery of one day's blend to product tanks within a reasonable time. The usual practice is to complete a blending operation within about 6 hours. The quantity of each component of a blend must be proportioned by the use of a flow meter and control valve.

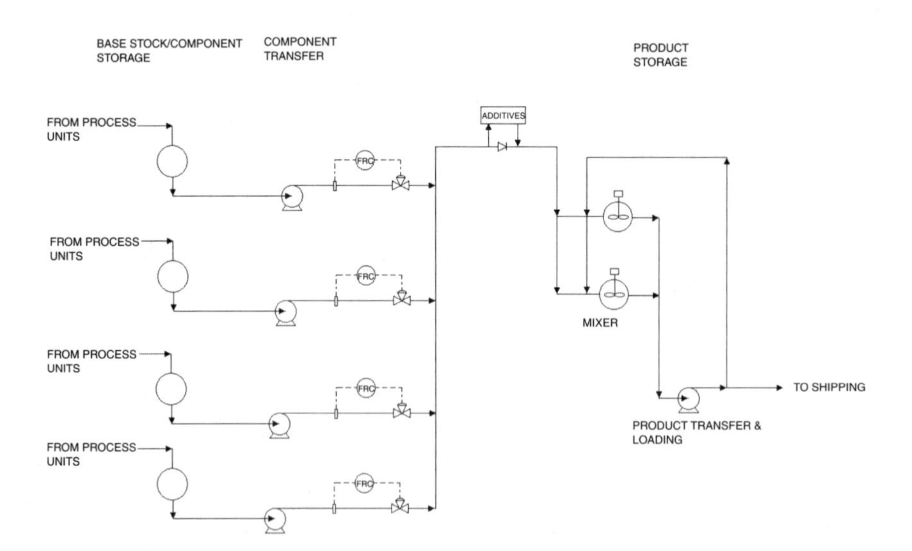

Figure 10-3. Partial in-line blending system. FRC = flow recorder and controller.

Flow controllers are set to a predetermined rate and flow is recorded. Flow meters used for partial in-line blending need not be extremely accurate. Accuracy ranges of 5% as attained with orifice meter are suitable. Mixers are required in final storage tanks for correction of blends by addition of components.

Partial in-line blending is suitable for moderate-sized refineries, where the cost of blend tanks would be excessive and blending time must be minimized. Blending time is substantially reduced because of the following:

1. Simultaneous pumping of components instead of consecutive pumping, as is the case in batch blending.
2. Reduction of overall mixing time.
3. Elimination of multiple gauging operations.

Many refineries prefer partial in-line blending as an initial installation that can be adapted to continuous blending in the future with minimal change in pumps, piping, and tankage. The disadvantages are the cost of additional meters, flow controllers, and pumps and high maintenance costs of the instruments.

CONTINUOUS IN-LINE BLENDING

In continuous in-line blending, all components of a product and all additives are blended in a pipeline simultaneously, with such accuracy that, at any given moment, the finished specification product may be obtained directly from the line (see Figure 10-4). As a result of the accuracy and safeguards included in the system, no provision is necessary for reblending or correction of blends.

Each base stock component is stored in two tanks. Samples of the component are test blended in the plant laboratory, and blends are analyzed to determine the most suitable proportion of components and additives for a specified product. The required components and additives are then pumped simultaneously, at a controlled rate, into a blending header. Various methods of controlling individual flow rates with interlock provisions have been used to ensure delivery of only the specified material.

The product can be sent to final storage, delivered to the product pipeline for transmission to a remote terminal, or loaded directly onto marine tankers.

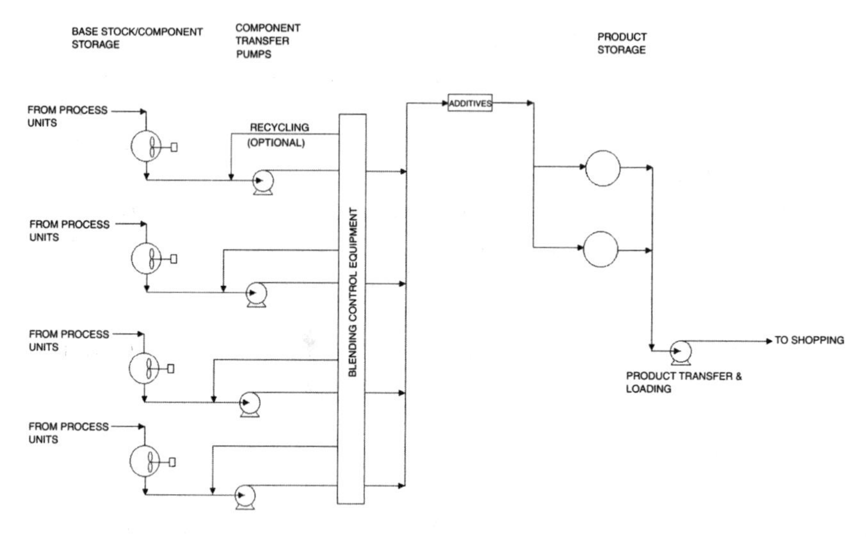

Figure 10-4. Continuous in-line blending system.

Storage tanks are required for components and most additives. Storage of the blended product is required only to suit shipping methods. Thus, the greatest proportion of storage can be in the form of components or base stocks, with minimum finished product storage.

An individual pump is required for each component. Pumps are also required for dyes and additives preparation and delivery. Dyes and additives are stored in solution form and added by proportioning pumps.

The quantity of each component of a blend must be accurately delivered. The recording flow meters and flow control valves used to proportion components are similar to those used for partial in-line blending, but a greater degree of accuracy is necessary. An accuracy of 0.25% or better is expected. Orifice meters are unsuitable for use in continuous in-line blenders. Positive displacement meters, venturi tubes, or velocity meters are generally used.

To ensure continued accuracy of the blends under varying operating conditions, the blending equipment is designed to provide for adjustment of individual component flow in proportion to total flow. Failure of the system to readjust should result in complete shutdown of the blending operations by means of flow stoppage or recycle.

Two types of blending controls are used to adjust component flows to desired rates: a mechanical system or an electronic system. In the mechanical system, the rotary motion generated by component flow in the meter is matched through a differential gear device against preset rotary motion. When the metered rate differs from the preset rate, pneumatic or electronic controls are actuated to adjust the flow control valve and change the flow rate to the desired value.

In the electronic system, electronic pulses are generated as a result of fluid flow through the meter. These are matched against preset pulses generated by an electronic device. Differences in pulse rate are detected by a digital totalizer, which feeds back a signal to adjust the flow control valve and adjust the flow to desired value.

To ensure the accuracy of the blend, it is necessary to calibrate meters frequently. One method of meter calibration is to remove the meter from the system and replace it with a calibrated space meter. This requires use of a prover tank or similar device, which can also be used to calibrate product shipping meters.

As a result of work involved in replacing a large numbers of meters frequently, many refineries that use continuous in-line blending system use a metering loop in the blending system. Such a loop is basically a pipe of known length and, therefore, known volume. The flowing fluid forces a "pig" through the loop, passing control points from which signal are sent for comparison. Loops are generally made of 600–1000 feet of 8-in, diameter pipe to ensure sufficient accuracy with electronic controls and digital totalizers.

Continuous in-line blending is best for large refineries that make several grades of products. When products can be transported directly to a pipeline or bulk transport, such as tankers, continuous in-line blending is economically more attractive than the batch methods because of the following factors:

1. Reduced blending time.
2. Minimum finished product storage, since components are stored and blended as required.
3. Increased blending accuracy with minimum "give away" on quality.
4. Reduction in loss through weathering of the finished product.
5. Minimum operating personnel.

However, the following disadvantages are associated with a continuous in-line blending system:

1. When products are transferred directly to a pipeline or bulk transport, a complete blender is required for each product, which must be

loaded simultaneously. For example, if a tanker is to be loaded with two grades of gasoline simultaneously, two blenders are necessary; otherwise, the advantage of reduced product tankage cannot be realized.

2. There is extreme difficulty in correcting errors, if they occur. As a result of safeguards built into the system, the only possible errors are human errors.
3. High initial investment and high maintenance cost of instruments.

REFINERY FLARE SYSTEM

Flaring is used as an effective and safe method of hydrocarbon vapor disposal whenever excess hydrocarbons must be controlled due to equipment failure or major emergencies, such as instrument malfunction, power failure, or plant fire. Before the utilization of flares, the gaseous waste streams were vented directly to atmosphere. This created two problems. The first is personnel and neighborhood safety. There was always the possibility of combustible vapor reaching the ground in sufficient concentration to be ignited. The result would be an explosion and fire, with devastating consequences for the operating personnel. The second problem was that the hydrocarbon discharges produced a major strain on the environment, and their emissions had to be controlled. Flaring provided a good solution to these problems but not without its own problems of heat, smoke, light, and noise. By burning the hydrocarbon vapor, the pollutants are converted to safe, less-noxious components: carbon dioxide and water vapor. The combustion process, however, generates a lot of heat, and thermal radiation from the flame must be reduced to levels safe for nearby operating personnel and equipment.

FLARE SYSTEM DESIGN

The flare system is designed to provide safe receipt and disposal of combustible, toxic gases and vapors released from process equipment during normal operation and during upset conditions. The safe disposal is achieved by knocking down the heavy ends and condensables in the flare knockout drum and burning the gases through an elevated stack. A controlled amount of steam is used to ensure smokeless burning and easy dispersion of combustion products. The process helps maintain an acceptable level of pollution at ground level.

A flare facility, particularly the flare burner, must have a stable flame capable of burning the hydrocarbon vapor released during a major operational failure. Also, the vapor must be sufficiently free from liquid droplets before entering the stack. Smoke is minimized by the injection of steam into the flame. The stack is located remote enough from operating units to provide safety for operating personnel and equipment. The flare system is purged with inert gas to prevent flame flashback.

BURNER DIAMETER

A flare stack, particularly the flare burner, must be of a diameter suitable to maintain a stable flame and prevent a blowout should there be a major failure. Flame blowout occurs when vapor exit velocities are as high as 20–30% of sonic velocity of the stack vapor.

The diameter of the flare burner can be determined from the following relationship:

$$d^2 = \frac{W}{1370}\sqrt{\frac{T}{M}}$$

where

D = diameter of the flare tip, in.;
W = mass flow, lb/hr;
T = temperature of vapor, °Rankine;
M = molecular weight of vapor.

This equation is based on specific heat ratio of $K = 1.2$, which is generally true for most hydrocarbon vapors. For vapors of significantly different values of K, the stack diameter obtained above is multiplied by $1.25\,K^{-0.25}$.

KNOCKOUT DRUM

A knockout drum in a flare system is used to prevent the hazards associated with burning liquid droplets escaping from flare stack. Therefore, the drum must be of sufficient diameter to effect the desired liquid/vapor separation. The drum diameter can be obtained from the following empirical relation (for a horizontal knockout drum to separate drops of up to 400 micron particle size):

$$D^2 = \frac{W}{9900}\sqrt{\frac{T}{M}}$$

where

D = diameter of the drum, ft;
W = mass flow rate, lb/hr;
T = temperature of vapor, °Rankine;
M = molecular weight of the vapor.

This equation is applicable to a single-flow knockout drum. Split-flow drums, where vapor enters in the middle of the drum and leaves it at both ends, have twice the capacity of single-flow drums. The diameter of split-flow drums is, therefore, 0.7 times the diameter obtained from this equation. The equation is based on an empty drum. An increase of diameter may be required if the drum is to hold a large volume of liquid. Vertical drums usually require a larger diameter, about 1.4 times the diameter for single flow drums. The length/diameter ratio for a horizontal drum is between 3 and 4. Also, the diameter of the knockout drum is three to four times the stack diameter.

HEIGHT AND LOCATION OF THE FLARE STACK

In designing a flare, the effect of heat radiation on operating personnel and equipment is considered. The selection of height and location of flare is done on the basis of safety for operating personnel and equipment. The effect of heat radiation on human beings is as follows:

HEAT INTENSITY, Btu/(hr)(ft^2)	PAIN THRESHHOLD	BLISTERING
2000	8 sec	20 sec
5300	—	5 sec

With a heat intensity of 2000 Btu/(hr)(ft^2), which is six times the solar radiation, the pain threshhold is 8 sec. Therefore, if time is to be allowed for a person to run to safety, the individual should not be subjected to a heat intensity of higher than about 1500 Btu/(hr)(ft^2) in the event of a major refinery failure (Figure 10-5(c)). A stack of sufficient height can be selected to satisfy this condition. Radiation levels used for design are

Service	**Btu/hr(ft^2)**
Equipment protection	3000
Personnel, short-time exposure	1500
Personnel, continuous exposure	440

Solar radiation adds to the calculated flame radiation. Typical values are 200–300 Btu/hr(ft^2).

Thermal radiation is of prime concern in flare design. Thermal radiation must be calculated to avoid dangerous exposure to personnel, equipment, and the surrounding area (trees, grass, etc.). The following calculation procedure is a convenient way to find the height of the flare stack and intensity of radiation at different locations.

The heat intensity is given by the following relation:[1]

$$I = \frac{\text{flow} \times \text{NHV} \times \varepsilon}{4\pi R^2}$$

where

I = radiation intensity at point X, Btu/hr (ft^2);

flow = gas flow rate, lb/hr;

NHV = net heating value of flare gas, Btu/lb;

ε = emissivity;

R = distance from flame center to point X.

This equation has been found quite accurate for distances as close to the flame as one flame length. The preceding equation is perfectly valid so long as proper values of emissivity are inserted. Emissivity is considered a fuel property alone. Emissivity of some common gases follow:

GAS	EMISSIVITY
CARBON MONOXIDE	0.075
HYDROGEN	0.075
METHANE	0.10
HYDROGEN SULFIDE	0.070
PROPANE	0.11
BUTANE	0.12
ETHYLENE	0.12
PROPYLENE	0.13

To calculate the intensity of radiation at different locations, it is necessary to determine the length of flame and its angle with respect to stack. A convenient expression to estimate the length of the flame is as follows:[2]

$$L_f = 10 \times D \times \left(\frac{\Delta P}{55}\right)^{0.5}$$

where

$$L_f = \text{length of flame, ft;}$$
$$\Delta P = \text{pressure drop, in. water;}$$
$$D = \text{tip diameter, in.}$$

The center of the flame is assumed to be located a distance equal to one third the length of the flame from the tip.

The angle of the flame results from the vectorial addition of the velocity of the wind and the gas exit velocity (Figure 10-5B):

$$\theta = \tan^{-1}\left(\frac{V_{\text{wind}}}{V_{\text{exit}}}\right)$$

$$V_{\text{exit}} \approx 550\left(\frac{\Delta P}{55}\right)^{0.5} \text{ ft/sec}$$

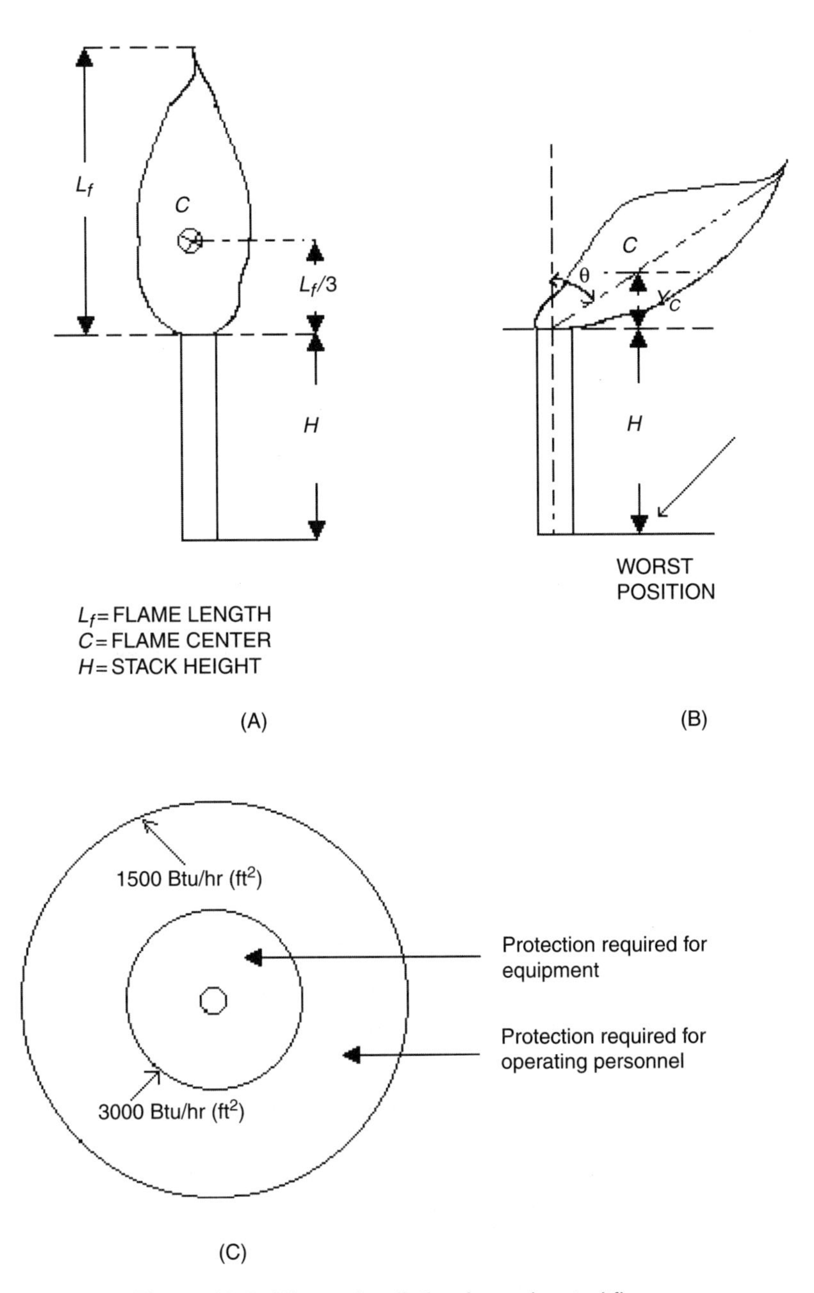

Figure 10-5. Thermal radiation from elevated flares.

and the coordinates of the flame center (located at $L_f/3$), with respect to the flare tip, are

$$X_c = \left(\frac{L_f}{3}\right)\cos\theta$$

$$Y_c = \left(\frac{L_f}{3}\right)\sin\theta$$

The distance from any point on the ground level to the center of the flame is

$$R = \sqrt{(X - X_c)^2 + (Y - Y_c)^2}$$

These equation allow you to determine the radiation intensity over any location.

The worst position for a given gas flow and wind velocity is vertically below the center of the flame (Figure 10-5(B)). For this location,

$$R = [(H + Y_c)^2]^{0.5}$$

or

$$R = H + Y_c$$

$$H = R - Y_c$$

This method assumes that there is no effect of wind on the flame length, which is not true at high wind velocities.

EXPLOSION HAZARD

An explosion could occur in a flare system if the oxygen concentration reaches the lower explosive limit for the gas or vapor contained in the system. Basically, two methods are used to minimize air concentration in the flare system: purging and sealing.

Purging is required during start-up to initially displace air from the flare system or when waste gas flow rate decreases below a certain critical level or stops completely. Any gas can be used, provided it contains no oxygen.

Sealing involves two aspects: Sealing the piping header up to the base of the stack is generally accomplished by a liquid seal drum. The stack is protected by a sealing device placed, ideally, as close to the tip of the flare as possible. The seal should be able to prevent reverse flow due to flow oscillations and thermal contractions.

LIQUID CARRYOVER

Flammable liquid droplets greater than 150 microns should be stopped before reaching the tip of the flare; otherwise, they can be dangerously projected over the surrounding area as a burning rain. During normal flaring conditions, which are usually low flow rates, a simple settling-chamber-type knockout drum is completely effective in removing the liquid droplets. However, during emergency flaring conditions, drums of excessively large diameter are required to knock out liquid particles.

POLLUTION CONTROL

Sometimes, flared gas or its combustion products could be toxic. In these cases, the stack is designed to ensure ground-level concentration below the safe limit, even for the worst possible situation—that is, flame out. Typical norms for emission of sulfur dioxide, particulate matter, carbon monoxide, and so forth, with which refineries must comply by choosing the appropriate technology for processing units, sulfur recovery, refinery fuel, and stack height are shown in Table 10-7.

Igniter reliability contributes to minimizing possible release of unburned hydrocarbons. The flare tip and the pilot must be capable of sustaining a stable flame and positive ignition even under the most severe winds. This capability is achieved by special devices, such as flare and pilot windshields. The highest reliability is obtained from flame front igniters, which automatically adjust for changes in the properties of the igniter gas/air mixture.

SMOKELESS FLARE

Smokeless flaring is based on the principle of increasing the burning rate by the injection of steam into a flame, creation of turbulence in the

Table 10-7
Ambient Air Quality Standards

PARAMETER	TYPICAL REFINERY STANDARDS	U.S. STANDARDS
SULFUR DIOXIDE	0.03 ppm ANNUAL ARITHMETIC MEAN	0.03 ppm ANNUAL ARITHMETIC MEAN
	0.14 ppm MAX 24-hr CONCENTRATION	0.14 ppm MAX 24-hr CONCENTRATION
PARTICULATE MATTER	0.075 mg/m^3 ANNUAL ARITHMETIC MEAN	0.05 mg/m^3 <10 MICROGRAM SIZE ANNUAL ARITHMETIC MEAN
	0.286 mg/m^3 24-hr CONCENTRATION ONCE/yr	0.15 mg/m^{3}MICROGRAM SIZE 24-hr AVERAGE
CARBON MONOXIDE	9 ppm 8-hr AVERAGE CONCENTRATION ONCE/yr	9 ppm 8-hr AVERAGE CONCENTRATION ONCE/yr
	35 ppm 1-hr, AVERAGE CONCENTRATION, ONCE/yr	35 ppm 1-hr, AVERAGE CONCENTRATION ONCE/yr
OZONE	0.08 ppm, 1-hr TOTAL/yr	0.12 ppm, 1-hr TOTAL/yr
NITROGEN DOXIDE	0.05 ppm, ANNUAL ARITHMETIC MEAN	0.053 ppm, ANNUAL ARITHMETIC MEAN
	0.25 ppm MAX, 24-hr CONCENTRATION	
LEAD	10 MICROGRAM/m^3, 30-DAY AVERAGE CONCENTRATION	1.5 MICROGRAM/m^3, QUARTERLY ARITHMETIC MEAN
VOLATILE ORGANIC COMPOUNDS	0.1 ppm FOR 1 hr	NO STANDARD

reacting gases, and the inspiration of air, thereby reducing the formation of soot. Soot formation is also reduced by a water/gas reaction:

$$C + H_2O = CO + H_2$$

The tendency of the hydrocarbon vapor to smoke when burned depends on its molecular structure, degree of unsaturation, and molecular weight. The products of combustion are mainly steam and carbon dioxide. The higher the molecular weight of the hydrocarbon, the lower is the ratio of steam to carbon dioxide and greater is the tendency to smoke. The rate of steam to the flare is automatically controlled to avoid excessive steam usage and ensure continued smokeless operation. Typical values for steam or fuel gas are 0.15–0.5 lb/lb hydrocarbon flow

Smokeless operation is obtained by proper flare tip design and optimum use of mechanical energy to induce air mixing with waste gas. The mixing energy can be obtained from the same gas when sufficient pressure is available or from steam.

However, steam is not the only medium that can be used for smokeless flaring. In situations where steam is unavailable, for example, around an offshore rig, other methods must be employed to achieve smokeless flaring.

The use of high pressure assist gas produces smokeless operation but has several drawbacks: the increased thermal radiation due to addition of the heating value of the high pressure gas and a waste of energy. Gas-assisted flares require approximately 0.15–0.3 lb high pressure gas per pound of waste flare gas.

Direct injection of water, sprayed into the flare, also eliminates smoke. The amount of water required depends on the degree of atomization of water stream. Approximately 1–2 lb water are required for each pound of hydrocarbon vapor. If the molecular weight of the flare gas increases, the amount of water injected also must increase. The degree of atomization of water is an important variable. With a coarse spray, the amount of water required can increase tenfold. Of course, most of this water falls through the flame without being utilized.

FIRED OR ENDOTHERMIC FLARES

Fired or endothermic flares are used for low heat waste streams, such as sulfur plant tail gas, ammonia vapor, and the like. Whenever the heat content of waste gas is below 150 Btu/ft^3, then a fired flare with high energy assist gas is required for complete combustion.

Certain gases, like ammonia, though having a relatively high calorific value of 365 Btu/ft^3, still require assist gas to increase the heat content and ensure complete combustion while minimizing NO_2 production. This is because the fuel-bound nitrogen has a quenching effect on the flare flame and can generate NO_x.

The basic design of fired or endothermic flare depends on the required amount and available pressure of the assist gas. Small flares or flares requiring small amounts of assist gas use inspirating incinerator burners to oxidize the waste gases. For large flares, flares with rings and center injection are used to supply the gas to produce turbulence mixing.

Thermal radiation of the fired flares is quite different from conventional hydrocarbon flares because of the very low heat content and flame temperature. The flame length is approximately 10% shorter than a conventional hydrocarbon flare, while flame emissivity is from 20 to 40% lower.

GROUND FLARES

Enclosed ground flares conceal the flame and provide smokeless operation without steam injection. By eliminating steam, one source of noise is completely removed. Combustion noise is also reduced by using many small burners and many small individual flames. The combustor is lined with an acoustically absorbent high-temperature ceramics to reduce the combustion noise. The combustion air inlet is acoustically shrouded and baffled to reduce noise outside the unit. The main disadvantages of ground flares are the larger ground area required and high initial cost. However, they provide the best overall control of noise and combustion problems and offer best solution for performance and reliability.

A refinery may have both elevated and ground flares (Figure 10-6). In this case, safe disposal of a normal load is achieved by knocking down the heavy ends and condensables in the flare knockout drum and burning the gases in an elevated stack. A controlled amount of steam is used to ensure smokeless burning. A water seal is used to prevent flame flashback.

In emergency loads, the high rate of gas flowing through the common header causes the flow to be partly diverted to a ground flare and, after passing through a ground flare water seal, are burned off in the ground flare.

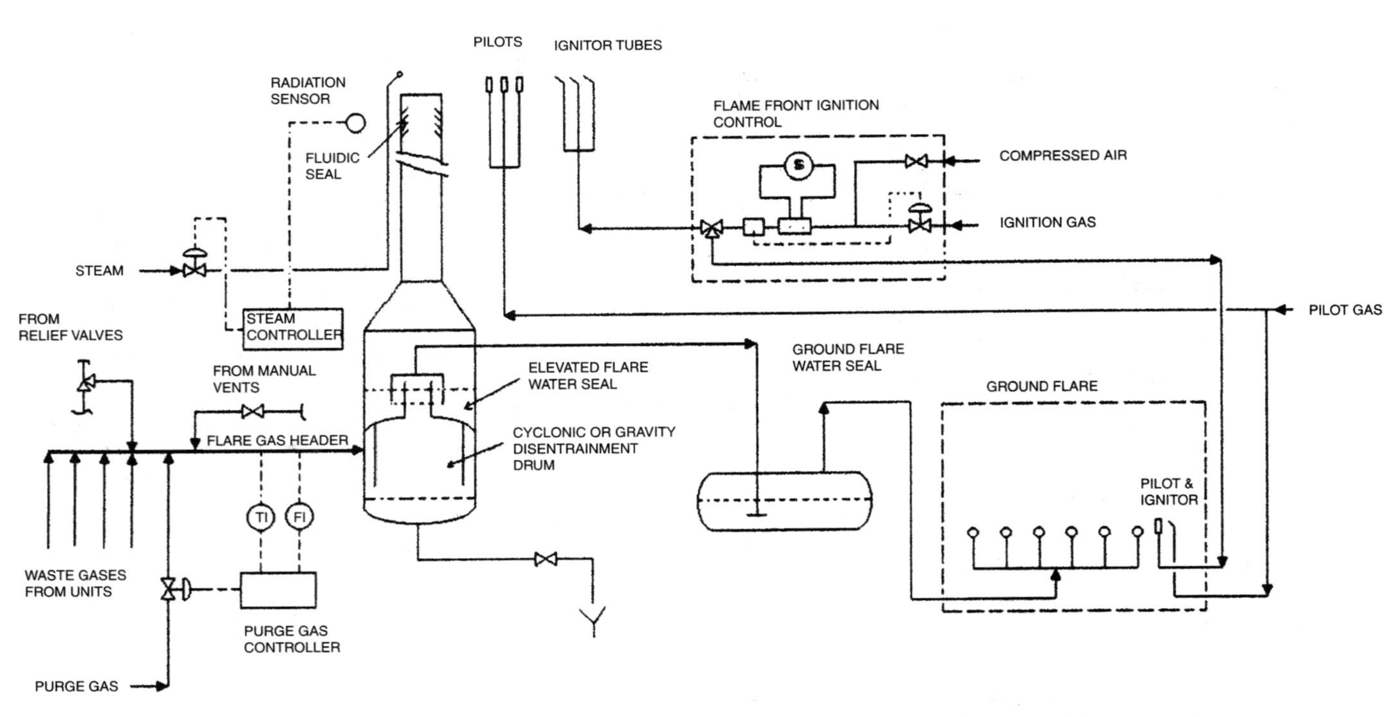

Figure 10-6. Flare system with elevated and ground flares. TI = temperature indicator; FI = flow indicator.

REFINERY STEAM SYSTEM

Roughly 10% of the crude throughput of the refinery ends up as refinery fuel. Steam generation alone accounts for about one third of the refinery total fuel consumption. The generation and distribution of steam and electrical power constitute a major part of a refinery's utility system.

There are three broad categories of usage of steam in the refineries:

1. *Heating loads.* These are usually well defined. Included in this category is steam for heat exchangers, reboilers, steam tracings, and off-site and utility plant heating. Variation in feedstock quality and throughput can sometimes result in maximum and minimum loads.
2. *"Process" or open steam loads.* These loads include fractionators, stripping steam, vacuum jet ejectors, spargers, smokeless flares, fuel atomizing steam, and the like. Included in this category could also be heating steam in remote areas where condensate return is not justified.
3. *Power loads.* These loads are represented by turbine drivers on pumps, compressors, generators, and so forth. Loads include turbines driven for reasons of reliability, economy, control, emergency coverage, and so on. These can be termed *fixed requirements.*

STEAM GENERATION AND DISTRIBUTION

Steam for refinery use can be generated from one or more of these following sources: fired steam or unfired steam generators, turbine exhaust or extraction. Figure 10-7 shows the general design principles for establishing refinery steam systems. The steam system consists of steam generators and a distribution network at different pressure levels for process and utility requirements.

The feed is a mixture of condensate and de-mineralized water which is deaerated before being fed to the boilers. The product is high-pressure steam, typically at 900 psig, depending on the end use of the steam. Many process units, such as hydrogen and sulfur plants, have waste heat boilers; and these also generate high-pressure steam. Only boiler feedwater is supplied to these unfired steam generators. The product HP (900 psig) steam is sent to an HP steam header.

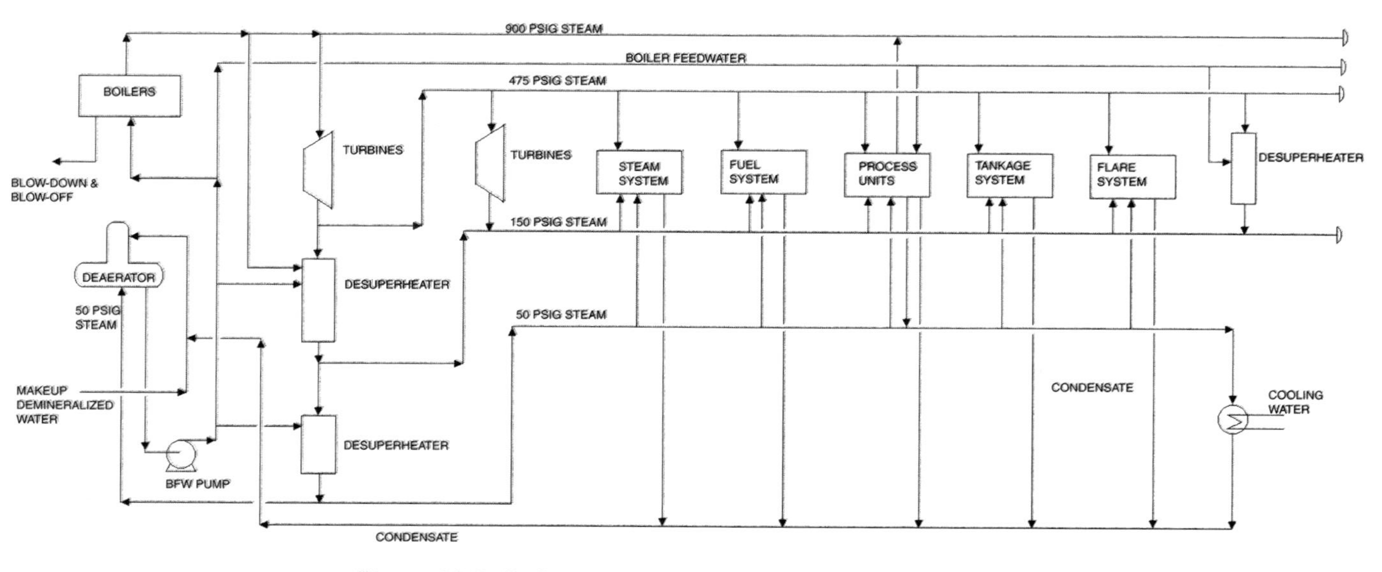

Figure 10-7. Refinery steam system. BFW = boiler feedwater.

Typically four types of headers are used for steam distribution in a refinery. In refineries with in-house power generation (Figure 10-8), steam is generated at a higher pressure of approximately 1450 psig instead of 900 psig, which increases the efficiency of power generation and thus lowers per unit cost of power produced. The four headers are as follows:

1. *900 psig*. Steam is supplied to the 900 psig header from the boiler plant. Some process units, such as hydrogen and sulfur plants, have waste heat boilers that also generate steam at 900 psig. The 900 psig steam is consumed by steam turbines to drive recycle compressors, charge pumps, and the like for many hydroprocessing units of the refinery and as makeup to 475-psig header through a reducing or desuperheating station.
2. *475 psig*. Sources of 475 psig steam are the exhaust steam of turbines and makeup from the 900 psig steam header. The 475 psig steam is consumed in steam turbines, reboilers, and certain process plants, such as hydrogen. All excess 475 psig steam is reduced and desuperheated to a 150 psig header.
3. *150 psig*. Sources of 150 psig steam are steam turbine exhaust, waste heat boilers in the process units, and makeup from 475 psig steam. The consumers are process stripping, reboilers, and steam tracing. All excess 150 psig steam is reduced and desuperheated to 50 psig steam header.

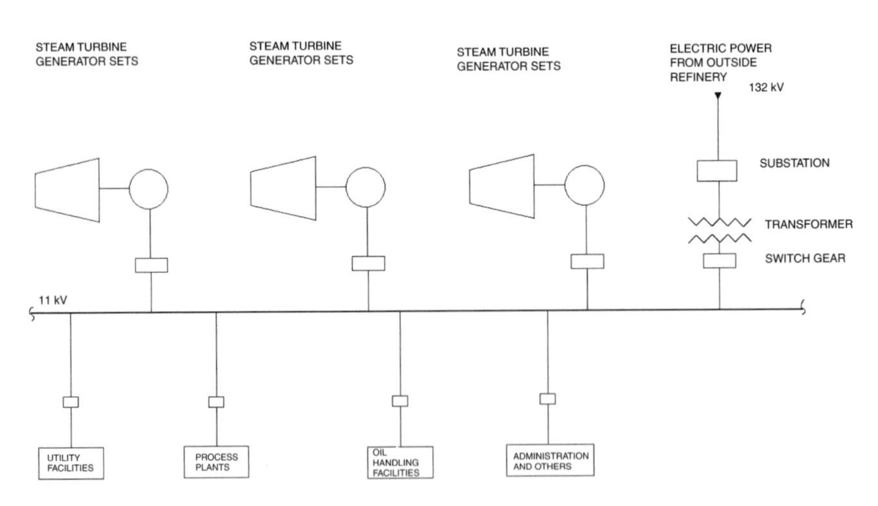

Figure 10-8. Refinery electrical system.

4. *50 psig*. Sources of 50 psig steam are steam turbine exhaust, waste heat boilers in certain process plants (sulfur plant condensers), boiler plant FD fans, and from 150 psig steam reduction and desuperheating. The 50 psig steam is consumed for process steam stripping, reboilers, and steam tracing. All excess 50 psig steam is condensed and returned to the reboiler feedwater.

NORMAL AND PEAK STEAM DEMAND

Normal steam demand at different pressure levels can be estimated by adding together the demand of the individual process units of the refinery. Steam demand for off-site areas is similarly estimated. To this is added the following:

1. Demand of steam for major equipment, which must be steam driven.
2. The quantities of steam generated by waste heat boilers (negative demand).
3. The exhaust steam demand for heating the building and off-site areas.
4. The exhaust steam demand for atomizing fuel and for deaerators.

In typical refineries, where most of the power is purchased and motors are backed up with turbines, the major peak load occurs during power failure. The other intermittent and emergency load may be for unit start-up. In colder climates, the winter heating load frequently overrides other factors, and the possibility of a winter power failure often fixes the maximum steam requirement. Elsewhere, the maximum boiler steam requirement might be set for peak start-up demand, a turnaround in which a major waste heat steam generator is down.

BOILER CAPACITY

Steam is preferably generated in multiple steam generators instead of one or two large-capacity boilers, and the boilers have a total installed capacity no less than 133% of normal requirements. All steam boilers are considered to be operating at partial load to supply normal refinery requirements. Multiple boiler units have more flexibility to match demand with supply of steam in the refinery due to variations in throughput or shutdown of any processing unit.

The total installed capacity must also equal at least the emergency steam requirements of the refinery. An additional factor to be considered is shutdown of the boilers. When a unit is out of service, the remaining units should be capable of fulfilling the process and off-site steam requirements of the refinery without the need for any emergency shutdown of any process unit.

The design conditions for an individual boiler and associated piping and auxiliaries should equal the rated capacity of the boiler. The emergency design conditions for each boiler should be equal to 110% of the rated capacity of the boiler, with pressure and temperature assumed equal to those at design conditions. The design pressure of boilers is 10% greater than the maximum operating pressure but no less than 25 psi in excess of this pressure.

COST OF STEAM GENERATION

The cost of steam generation is basically the cost of energy contained in steam. Steam boilers consume a the large percent of refinery total fuel consumption. Typical utility consumption per metric tonne of HP steam raised is presented in Table 10-8.

In addition to fuel, power, boiler feedwater, and cooling water, steam boilers require minor amounts of certain chemicals (caustic soda, soda

Table 10-8
Utility Consumption for Steam Boilers per Tonne HP (900 psig) Steam

UTILITY	UNITS	CONSUMPTION
FUEL	mmBtu	3.319
STEAM	mmBtu	0.176
ELECTRICITY	kWhr	4.848
DEMINERALIZED WATER	MIG*	0.088
COOLING WATER	MIG*	0.018
ASSUMPTIONS		
HEAT CONTENT OF BOILER FEEDWATER 20 psi AND AT 250°F	mmBtu	0.4408
HEAT CONTENT OF BOILER STEAM AT 900 psig AND AT 810°F	mmBtu	3.0856
HEATING EFFICIENCY OF BOILER	%	85
COST OF CHEMICALS EXCLUDED		

*MIG = 1000 IMPERIAL GALLONS.

ash, sodium phosphates, and hydrazine). The cost of chemicals, although small compared to energy cost, nevertheless must be taken into consideration in estimating the cost per tonne of steam.

The cost of steam at lower pressures is taken equal to high-pressure steam if lower-pressure steam is generated simply by pressure reduction of high-pressure steam. If lower-pressure steam is obtained through power turbines, the marginal value of steam equals high-pressure steam less the cost of power generated.

REFINERY FUEL SYSTEM

Heating may be required in the refinery at a number of places for various process applications and steam generation in its utility plant. Burning fuel provides the necessary heat. The refinery fuel system includes facilities for the collection, preparation, and distribution of fuel to users. The commonly used refinery fuels are fuel oil and gas. On average, for every 100 barrels of crude processed in the refinery, 10 barrels are used as refinery fuel.

FUEL SELECTION

Fuel gas and residual fuel are the most commonly used fuels in the refinery. Other refinery products of low monetary value, such as heavy pitch residues, visbreaker tars, FCCU decanted oil, vacuum tower bottoms from certain crudes, lube extracts, and waxes are also used as fuel in the refinery itself. The majority of these materials would be difficult to blend in a commercial fuel of acceptable specifications because of high viscosity, chemical aggressiveness, high contaminant level (sulfur, metals, etc.), and associated environmental problems in their use.

Gaseous streams diverted to the refinery as fuel are those gases that cannot be processed to saleable products economically. These include H_2, CH_4, C_2H_6, and frequently C_3 and C_4 gases. In a refinery of average complexity, approximately two thirds of refinery fuel requirements may come from refinery gases. The rest of the fuel requirement is made up from a natural gas supply, if available, or residual fuel oil produced in the refinery.

In many refineries, both gaseous and liquid fuels are used simultaneously. Furnaces and boilers are equipped with combination burners, suitable for both gas and oil firing. Exception to this are certain refinery units, such as cat reforming, where only gas firing is permissible because of the need for precise temperature control. Furnaces that are operated when no gaseous fuel is available as during refinery start up must be equipped with oil burning capability.

REFINERY FUEL OIL SYSTEM

The purpose of refinery fuel oil system is to ensure a constant, regular supply of oil to burners of steam boilers and to process furnaces. The system includes facilities for storage, pumping, heating, and distribution of oil at suitable pressure and viscosities so that the atomization and burning of oil is possible. A typical refinery home fuel system is shown in Figure 10-9. The aim of the fuel oil system design is that operational changes to one furnace will not cause fluctuation of supply to another portion of the refinery.

Storage is an essential need, which must be met with availability of at least 5 days supply at normal firing rate of the furnaces and boilers, which are usually oil fired. This may require one or more tanks for storage. If the refinery fuel is obtained from more than one source, it may be necessary to store the blended material, too, and so more than one tank may be required. It is often possible to line blend the various fuel components as these leave their units without the need for intermediate storage. Precise blending control is not required. Oil is delivered to the burner at a pressure of approximately 100 psig for control and atomization. To

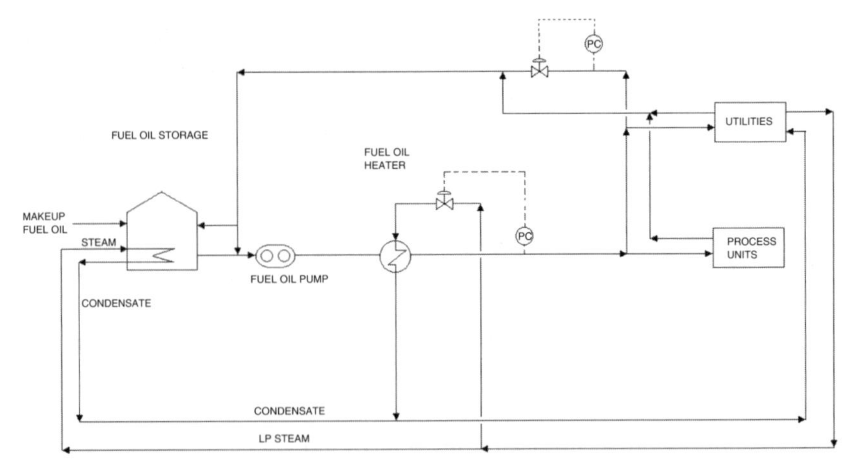

Figure 10-9. Refinery home fuel (liquid) system. LP = low pressure; PC = pressure controller.

provide this pressure at the burner and allowing for pressure drop in the lines, oil is pumped from storage at 120–150 psig.

The temperature at which fuel oil is discharged from fuel oil storage varies with the material being pumped. For proper atomization, the temperature at which oil is being pumped should be sufficient to lower the viscosity of oil to 30–40 centistokes.

The design capacity of fuel oil heating and pumping equipment usually equals 125% of the design requirement of the plant, assuming simultaneous firing at the design rate of all fuel oil burning furnaces and boilers. This allows for 25% recirculation of oil. Rotary pumps are used, one driven by electric motor and the other by steam turbine. Heaters used are steam heaters. Relief valves are located on the discharge side of the pumps and on oil heaters. Relief valve discharge is piped back to storage tank. Strainers are provided at the pump inlet and outlet to strain out any precipitated carbon particles. Piping from the suction line to the fuel oil pumping equipment is sized for a pressure drop not exceeding 3 psi/1000 ft. The discharge piping supplying burners and recirculation piping returning to storage have protective heating. Separate nozzles are provided on storage tanks for makeup fuel and for recirculation and withdrawal of oil.

REFINERY FUEL GAS SYSTEM

The refinery fuel gas system is designed to supply fuel gas to steam boilers, process furnaces, gas engines, and gas turbines at a regulated pressure and reasonably constant heating value. The system includes mixing drum controls and distribution piping. Wherever necessary, a stand by storage of liquefied petroleum gas (LPG) is included. A typical refinery fuel gas system is shown in Figure 10-10.

The mixing gas drum is typically operated at 30 to 40 psig, which allows for delivery of gas at burners approximately 15–20 psig. For refineries with an operating gas turbine, the drum pressure is much higher, at 125 psig. The mixing vessel is fitted with mixing baffles and a steam coil for vaporizing any liquid carryover or LP gases. Refinery gas burners are designed for a particular heating value and gas density. The composition of gas to the gas mixing drum is so adjusted that large changes in calorific value or density of gas to various refinery furnaces are avoided. For example, it may not be possible to replace refinery gases with high calorific value and density totally with natural gas with lower calorific value and density without affecting performance of the furnaces using them.

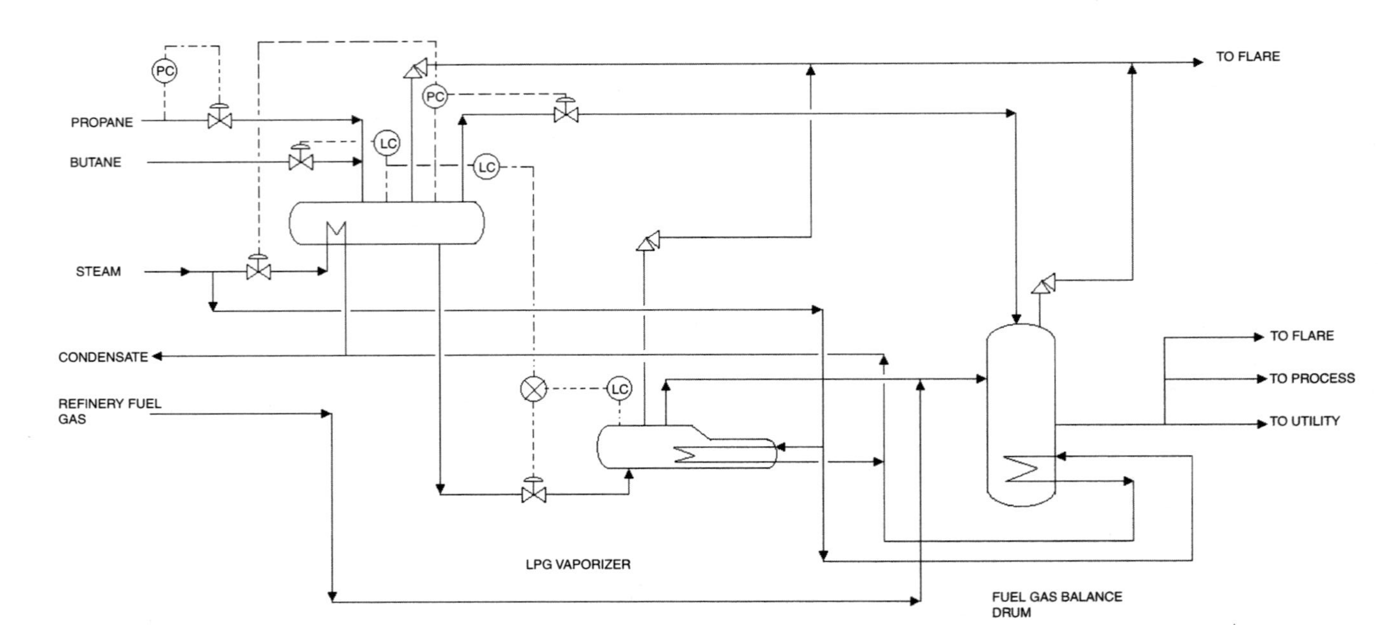

Figure 10-10. Refinery home fuel (gas) system.

NOTES

1. T. A. Brzustowski and E. C. Sommer, Jr. "Predicting Radiant Heating from Flares," *Proceedings of the Division of Refining, A.P.I.* 53 (1973), pp. 865–893.
2. J. F. Straitz and R. J. Altube. "Flare Design and Operation." National Airoil Burner Company (USA) bulletin.

CHAPTER ELEVEN

Product Blending

GASOLINE OCTANE BLENDING

The research (RON) and motor (MON) octane numbers of a gasoline blend can be estimated using the following equations:[1]

$$R = R_1 + C_1 \times (R_2 - R_1 \times J_X) + C_2 \times (O_1 - O_2) + C_3(A_1 - A_2) \tag{11-1}$$

where

R = research octane numbers of blend;
R_0 = research octane number of each component;
R_1 = volume average of octane number;
R_2 = volume average of product of R_0 and J;
J_X = volume average of sensitivity;
O_1 = volume average of squared olefin content;
O_2 = square of volume average olefin content;
A_1 = volume average of squared aromatic content;
A_2 = square of volume average aromatic content.

and

$$M = M_1 + D_1 \times (M_2 - M_1 \times J_X) + D_2(O_1 - O_2) + D_3\left[\frac{(A_1 - A_2)}{100}\right]^2 \tag{11-2}$$

where

M = motor octane of the blend;
M_0 = motor octane of each component;
M_1 = volume average motor octane;
M_2 = volume average of product of M_0 and J.

These equations represent straight-line blending with three correction terms added to account for the blending deviations normally experienced with gasoline blends. The first term (sensitivity function) is a correction for the blending deviation arising because octane numbers are determined at compression ratio different from those at which the blends are rated. The second (function of olefin content) and third terms (function of aromatic content) are corrections that reflect the influence of chemical interaction of the components in the blend.

The coefficients to be used in octane blending follow. The RON equation coefficients are

$C_1 = 0.04307$
$C_2 = 0.00061$
$C_3 = -0.00046$

The MON equation coefficients are

$D_1 = 0.04450$
$D_2 = 0.00081$
$D_3 = -0.00645$

These coefficients are derived from the regression analysis of RON and MON data of actual laboratory blends.

EXAMPLE 11-1

Determination of the RON and MON of a gasoline blend are done with the help of a spreadsheet program, assuming the RON, MON, aromatic, and olefin content of all blend components are available. Sample data on the properties of gasoline blend components such as the RON, MON, aromatic, and olefin content of the blend components and calculations for computing the RON and MON of the blend by equations (11-1) and (11-2) are shown in Tables 11-1 and 11-2.

GASOLINE BLENDING BY THE INTERACTION COEFFICIENT METHOD[2]

In a given refinery, where the maximum number of gasoline blend components and their properties are known, it is possible to develop an accurate blending spreadsheet program, using the individual blendstock properties and the binary blend interaction coefficients. The only additional laboratory work required is the determination of the properties: RON, MON, and ASTM distillation of all possible binary blends for a given number of blend components.

Interaction coefficients are determined for all binary blends and used in the model to accurately predict the properties of any gasoline blend of these components.

BLENDING ALGORITHM

A study of the gasoline blending data has shown that nonlinear gasoline blending behavior can be described by an equation of the following type:

$$P_{CALC} = P_{VOL} + I_{(1,2)} \times X_1 \times X_2 + I_{(1,3)} X_1 \times X_3 + \cdots + I_{(8,9)} \times X_8 \times X_9 \quad (11\text{-}3)$$

where

P_{CALC} = calculated property;
P_{VOL} = volumetric weighted average property;
$I_{(1,2)} \ldots I_{(8,9)}$ = the component interaction coefficients;
$X_1 \ldots X_9$ = volume fraction of each component.

The interaction coefficient for a binary blend can be calculated as follows:

$$I_{(A,B)} = \frac{(P_{ACTUAL} - P_{VOL})}{(V_A \times V_B)} \quad (11\text{-}4)$$

Table 11-1
Research Octane Blending

COMPONENT	VOL % (1)	RON, R (2)	MON (3)	OLEFIN VOL % (4)	OLEF^2 (5)	AROMATIC VOL % (6)	AROM^2 (7)	SENSITIVITY, J (8)	R*J (9)	VOUMETRIC, R*J (10)
LSR	0.000	55.2	55.0	0.10	0.01	3.50	12.25	0.20	11.04	0.00
REF 90	0.200	90.7	82.5	0.70	0.49	43.00	1849.00	8.20	743.74	148.75
REF 95	0.000	95.0	85.2	3.00	9.00	47.30	2237.29	9.80	931.00	0.00
REF 97	0.000	97.2	87.1	0.60	0.36	52.90	2798.41	10.10	981.72	0.00
LCN	0.410	91.5	79.0	44.60	1989.16	6.60	43.56	12.50	1143.75	468.94
MCN	0.140	84.0	75.9	39.00	1521.00	13.30	176.89	8.10	680.40	95.26
VBU NAPH	0.060	63.4	59.8	26.80	718.24	6.50	42.25	3.60	228.24	13.69
POLY	0.150	97.5	82.9	94.10	8854.81	0.70	0.49	14.60	1423.50	213.53
BUTANE	0.040	93.0	91.0	0.00	0.00	0.00	0.00	2.00	186.00	7.44
MTBE	0.000	110.0	101.0	0.00	0.00	0.00	0.00	9.00	990.00	0.00
VOL AVG	1.000	89.56	79.18	1568.87	2399.91	186.68	415.03	10.39		947.60

NOTES:
BLEND MON = 90.72
COLUMNS ARE IN PARENTHESES

2 = RON OF BLEND COMPONENTS.
3 = MOTOR OCTANE OF BLEND COMPONENTS.
5 = SQUARE OF OLEFIN CONTENT (4).
7 = SQUARE OF AROMATIC CONTENT (6).
8 = SENSITIVITY (RON – MON) OF COMPONENTS
9 = COL 8 × COL 1.
10 = COL 9 × COL 1.

Table 11-2
Motor Octane Blending

COMPONENT	VOL %, (1)	RON, (2)	MON, (3)	OLEFIN M,VOL %, (4)	OLEF^2, (5)	AROMATIC VOL %, (6)	AROM^2, (7)	SENSITIVITY, J, (8)	R*J, (9)	VOUMETRIC, R*J, (10)
LSR	0.000	55.2	55.0	0.10	0.01	3.50	12.25	0.20	11.00	0.00
REF 90	0.200	90.7	82.5	0.70	0.49	43.00	1849.00	8.20	676.50	135.30
REF 95	0.000	95.0	85.2	3.00	9.00	47.30	2237.29	9.80	834.96	0.00
REF 97	0.000	97.2	87.1	0.60	0.36	52.90	2798.41	10.10	879.71	0.00
LCN	0.410	91.5	79.0	44.60	1989.16	6.60	43.56	12.50	987.50	404.88
MCN	0.140	84.0	75.9	39.00	1521.00	13.30	176.89	8.10	614.79	86.07
VBU NAPH	0.060	63.4	59.8	26.80	718.24	6.50	42.25	3.60	215.28	12.92
POLY	0.150	97.5	82.9	94.10	8854.81	0.70	0.49	14.60	1210.34	181.55
BUTANE	0.040	93.0	91.0	0.00	0.00	0.00	0.00	2.00	182.00	7.28
MTBE	0.000	110.0	101.0	0.00	0.00	0.00	0.00	9.00	909.00	0.00
VOL AVG	1.000	89.56	79.18	1568.87	2399.91	186.68	415.03	10.39		827.99

NOTES:
BLEND MON = 80.07
COLUMNS ARE IN PARENTHESES

2 = RON OF BLEND COMPONENTS.
3 = MOTOR OCTANE OF BLEND COMPONENTS.
5 = SQUARE OF OLEFIN CONTENT (4).
7 = SQUARE OF AROMATIC CONTENT (6).
8 = SENSITIVITY (RON – MON) OF COMPONENTS
9 = COL 8 × COL 3.
10 = COL 9 × COL 1.

where

$I_{(A,B)}$ = interaction coefficient of components A and B;
P_{ACTUAL} = property of the blend, as determined in the laboratory;
P_{VOL} = volumetric weighted average property of the blend;
V_A, V_B = volume fractions of components A and B.

In this model, the concept of a blending interactions coefficient is considered and a spreadsheet model developed to predict the octane and volatility of the multicomponent blend.

EXAMPLE 11-2

An example of the interaction coefficient method spreadsheet for a multicomponent blend follows. We want to determine the RON, MON, and ASTM distillation of a blend of these components:

FCC light naphtha (LCN)
FCC medium naphtha (MCN)
Light straight run (LSR)
Polymer gasoline (POLY)
Reformate 95 RON (REF 95)
Reformate 97 RON (REF 97)

The following properties are determined in the laboratory for each of the blend components and for all possible binary blends: RON, MON, and ASTM distillation; percent evaporated at 150°, 195°, 250°, and 375°F.

The number of binary blend components determines how many binary blends are possible. Hence, for a six-component blend, 15 binary blends are possible. All the 15 binary blends are made in the laboratory and their properties determined for computing the interaction coefficient for each binary blend. Once the binary interaction coefficients are known, the properties for any blend composition can be determined by means of the blending equation. The calculations are facilitated by means of a spreadsheet program.

The properties of the pure components and all binary blends interaction coefficients are shown in Tables 11-3 to 11-6. To calculate the properties (RON, MON, distillation) of the blend, the blend composition is entered in Table 11-6 and the blend properties are read from Table 11-7.

Table 11-3
Blend Component Properties

COMPONENT		LCN	MCN	LSR	POLY	REF 97	BUTANE
DENSITY		0.70	0.75	0.68	0.73	0.79	0.57
VAP. PRESSURE	psia	9.1	2.6	9.7	9.1	6.7	65.3
ASTM DISTILLATION							
VOL% EVAPORATED							
IBP	°C	37.0	53.0	36.0	33.0	38.0	0.0
@ 60°C		24.5	0.0	31.0	8.0	8.0	100.0
@ 65°C		32.5	0.5	41.0	10.0	11.0	100.0
@ 80°C		51.0	4.0	66.0	14.5	18.5	100.0
@120°C		88.0	55.0	100.0	50.0	40.0	100.0
@190°C		100.0	100.0	100.0	91.5	98.5	100.0
FBP, °C		158.0	153.0	117.0	230.0	191.0	0.0
FIA ANALYSIS							
SATURATES	Vol %	48.8	47.7	96.4	5.2	46.5	0.0
OLEFINS	Vol %	44.6	39.0	0.1	94.1	0.6	0.0
AROMATICS	Vol %	6.6	13.3	3.5	0.7	52.9	0.0
SULFUR	% W/W	0.0	0.0	0.0	0.0	0.0	
OCTANE NUMBER							
RON		89.6	84.0	55.2	97.5	97.2	87.1
MON		78.5	75.9	55.0	82.9	87.1	87.0

ASTM DISTILLATION BLENDING

Two methods are available for estimating the ASTM distillation of a blend: Edmister's method and empirical correlation.

EDMISTER'S METHOD

ASTM distillation is converted to the true boiling point (TBP) distillation using Edmister's correlation. The blend TBP can be determined simply by adding together the volumes contributed by all the components at a chosen temperature, dividing by the total volume, and plotting a temperature vs. percent distillation chart. The TBP distillation vs. temperature graph can be converted back into ASTM distillation again, by using Edmister's correlation in the reverse.

This procedure is not very accurate and the blend can be off by as much as 8–10°C. The inaccuracy can be attributed to the inadequacy of

Table 11-4
Quality of Binary Blends and Interaction Coefficients

COMPONENT PAIR	LCN MCN (1)	LCN LSR (2)	LCN POLY (3)	LCN REF 97 (4)	LCN BUTANE (5)	MCN LSR (6)	MCN POLY (7)	MCN REF 97 (8)	MCN BUTANE (9)	LSR POLY (10)	LSR REF 97 (11)	LSR BUTANE (12)	POLY REF 97 (13)	POLY BUTANE (14)	REF 97 BUTANE (15)
COMPONENT A, VOL%	0.5	0.5	0.5	0.5	0.8	0.5	0.5	0.5	0.7	0.5	0.5	0.8	0.5	0.80	0.75
COMPONENT B, VOL%	0.5	0.5	0.5	0.5	0.2	0.5	0.5	0.5	0.3	0.5	0.5	0.2	0.5	0.2	0.25
VAPOR PRESSURE bar	6.2	9.61	9.15	7.9	22.01	6.67	5.89	4.96	24.80	9.92	8.68	22.48	8.06	22.94	23.87
ASTM DISTIL															
VOL% EVAPORATED															
@ 60°C	6.0	29.0	15.0	14.5	44.5	7.5	2.5	3.0	32.5	16.0	17.0	49.5	8.0	28.0	36.0
@ 65°C	10.0	37.0	20.0	19.5	48.5	13.0	3.5	4.5	34.0	20.0	23.0	54.5	10.0	29.0	37.5
@ 80°C	27.0	58.5	32.0	32.5	61.0	32.0	8.5	11.5	37.5	35.0	38.0	72.5	16.0	32.5	41.5
@ 120°C	73.0	94.0	72.0	63.5	90.0	78.0	52.0	46.0	67.5	75.5	70.0	100.0	44.5	60.5	55.5
@ 190°C	100.0	100.0	96.5	100.0	100.0	100.0	95.5	100.0	100.0	95.0	100.0	100.0	95.5	93.0	100.0
RON	88.0	75.6	95.0	93.4	94.0	72.0	95.4	91.7	91.0	85.8	80.2	65.0	97.8	92.8	96.8
RON @ 0.40 TEL	94.7	86.7	99.1	98.9	98.4	83.2	98.1	96.9	96.3	92.8	88.3	78.2	101.6	100.8	102.1
RON @ 0.84 TEL	96.8	90.4	100.5	100.2	100.0	87.6	99.3	98.4	98.0	95.2	91.4	83.6	102.7	103.5	104.5
MON	81.2	71.5	83.1	83.8	83.1	67.5	82.8	84.0	82.3	76.8	75.2	64.8	85.7	85.0	86.6
MON @ 0.40 TEL	84.0	80.5	85.6	87.8	87.3	78.4	85.4	87.0	87.1	83.0	82.8	77.2	88.5	87.2	94.1
MON @ 0.84 TEL	85.2	83.4	86.1	89.0	88.8	82.3	86.1	88.1	89.2	85.2	85.8	82.3	89.4	88.0	97.0
COEFFICIENTS															
VAPOR PRESSURE, psia	1.400	0.840	0.200	0.000	10.438	2.080	0.160	1.240	16.143	2.080	1.920	10.375	0.640	16.250	13.440
ASTM DISTL.															
VOL% EVAPORATED															
@ 60°C	−25.000	5.000	−5.000	−7.000	30.625	−32.000	−6.000	−4.000	11.905	−14.000	−10.000	29.375	0.000	10.000	26.667
@ 65°C	−26.000	1.000	−5.000	−9.000	15.625	−31.000	−7.000	−5.000	17.381	−22.000	−12.000	10.625	−2.000	6.250	22.667
@ 80°C	−2.000	0.000	−3.000	−9.000	1.250	−12.000	−3.000	1.000	22.381	−21.000	−17.000	−1.875	−2.000	5.625	14.000
@ 120°C	6.000	0.000	12.000	−2.000	−2.500	2.000	−2.000	−6.000	−4.762	2.000	0.000	0.000	−2.000	3.125	2.667
@ 190°C	0.000	0.000	3.000	3.000	0.000	0.000	−1.000	3.000	0.000	−3.000	3.000	0.000	2.000	−16.375	11.333
MON	16.000	19.000	9.60	4.000	18.12	8.20	13.600	10.000	14.619	31.400	16.600	21.250	2.800	8.000	−2.533

NOTE: COLUMN NUMBERS ARE IN PARENTHESES.

Table 11-5
Weighted Coefficients

COMPONENT PAIR	LCN MCN (1)	LCN LSR (2)	LCN POLY (3)	LCN REF 97 (4)	LCN BUTANE (5)	MCN LSR (6)	MCN POLY (7)	MCN REF 97 (8)	MCN BUTANE (9)	LSR POLY (10)	LSR REF 97 (11)	LSR BUTANE (12)	POLY REF 97 (13)	POLY BUTANE (14)	REF 97 BUTANE (15)	TOTAL INTERACTION COEFFICIENT (16)	VOL. AVG QUALITY (17)	ESTIMATED QUALITY (18)
VAPOR PRESSURE	0.0420	0.0126	0.0072	0.0000	0.0939	0.0104	0.0019	0.0496	0.0484	0.0125	0.0384	0.0156	0.0307	0.0585	0.0484	0.4701	9.206	9.6761
ASTM 60°C	−0.7500	0.0750	−0.1800	−0.8400	0.2756	−0.1600	−0.0720	−0.1600	0.0357	−0.0840	−0.2000	0.0441	0.0000	0.0360	0.3200	−1.6596	16.06	14.4004
65°	−0.7800	0.0150	−0.1800	−1.0800	0.1406	−0.1550	−0.0840	−0.2000	0.0521	−0.1320	−0.2400	0.0159	−0.0960	0.0225	0.2720	−2.4288	20.45	18.0212
80°	−0.0600	0.0000	−0.1080	−1.0800	0.0112	−0.0600	−0.0360	0.0400	0.0671	−0.1260	−0.3400	−0.0028	−0.0960	0.0203	0.1680	−1.6022	31.14	29.5378
120°	0.1800	0.0000	0.4320	−0.2400	−0.0225	0.0100	−0.0240	−0.2400	−0.0143	0.0120	0.0000	0.0000	−0.0960	0.0113	0.0320	0.0405	61.9	61.9405
190°	0.0000	0.0000	0.1080	0.3600	0.0000	0.0000	−0.0120	0.1200	0.0000	−0.0180	0.0600	0.0000	0.0960	−0.0045	0.0720	0.7815	98.38	99.1615
RON	0.1440	0.1920	0.2088	0.0000	0.2756	0.0480	0.2232	0.1760	0.0867	0.2268	0.3200	0.0321	0.0864	−0.0590	0.1360	2.0967	91.233	93.3297
MON	0.4800	0.2850	0.3456	0.4800	0.1631	0.0410	0.1632	0.4000	0.0439	0.1884	0.3320	0.0319	0.1344	0.0288	−0.0304	3.0869	81.288	84.3749

Table 11-6
Blend Composition

BLEND	VOL%
LCN	0.30
MCN	0.10
LSR	0.05
POLY	0.12
REF 97	0.40
BUTANE	0.03
TOTAL	1.00

Table 11-7
Blend Results by Interaction Coefficient Method

VAPOR PRESSURE	psia	9.7
ASTM		
DISTILLATION		
60°	°C	14.4
65°		18.0
80°		29.5
120°		61.9
190°		99.2
RON CLEAR		93.3
MON CLEAR		84.4

Edmister's correlation, particularly in converting ASTM distillation to TBP distillation.

GRAPHICAL SUMMATION METHOD

An empirical method is described for estimating ASTM distillation of a blend from its composition and ASTM distillation temperature of blend components. This method is used for the following calculations: estimate of the initial boiling point (IBP), 10%, 20–90% points and the estimation of the ASTM end point.

Determination of ASTM IBP, 10%, 20–90% Points of Blend

This method is applicable to blends containing distillate stocks having an ASTM initial boiling point higher than 85°F and an ASTM end point lower than 700°F. It is based on the observation that a straight summation line can be drawn through an ASTM distillation point of a blend.

The slope of this line is such that the sum of the proportions of each blend component corresponds to its intersection with ASTM distillation curve. For TBP distillation, the summation lines are parallel to volume percent axis on an ASTM distillation plot. ASTM summation lines slope due to poor fractionation of ASTM distillation, and the slope varies according to distillation end point. The slope to be used follows:

DISTILLATION POINT	SLOPE OF SUMMATION LINE, °F
IBP	−180°F PER 100% DISTILLED
10%	−180°F PER 100% DISTILLED
20%	−100°F PER 100% DISTILLED
30%	−80°F PER 100% DISTILLED
40%	−50°F PER 70% DISTILLED
50–90%	−20°F PER 70% DISTILLED

ASTM 10–90% Points

ASTM distillation curves are drawn for each blend component, with the temperature on the vertical axis and the volume percent distilled on the horizontal axis. Distillation must be on a consistent basis for all components; that is, either percent evaporated or percent recovered.

A guess is made on the temperature at which a given proportion of the blend is distilled, and the corresponding point is marked on the graph. A summation line of specified slope is drawn through the point. The vol% distilled is read off vertically below the intercept of the summation line and ASTM distillation curve of the each component (Figure 11-1). The sums for all blend components should equal the proportion of blend originally estimated. If not, a new guess of temperature at which the specified proportion of blend is distilled is made and the procedure repeated. If the second estimate also does not give the required result, an interpolation is made between the earlier determinations.

Initial Boiling Point

This method is identical to 10–90% points, except that the distillation curves for the components are extrapolated to −1.4%. Therefore, −1.4%

becomes zero of the modified scale and 10% becomes 11.4%. The volume distilled at 1.4% is next calculated by previous procedure to give the IBP of the ASTM curve.

EXAMPLE 11-3

Calculate the IBP and 10–90% points of a blend of FCC naphtha (50% volume), coker naphtha (16% volume), and cat reformate (34% volume) with the following ASTM distillation:

VOL%	FCC NAPHTHA, °F	COKER NAPHTHA, °F	CAT REFORMATE, °F
IBP	97	111	111
5	115	138	131
10	120	144	156
20	126	151	185
30	136	160	205
40	146	170	226
50	156	180	246
60	177	195	260
70	198	210	274
80	218	226	288
90	239	241	302
95	253	253	320
EP	295	289	356

The IBP of the blend, is calculated as follows. Assume that the IBP represents 1.4% distilled instead of 0% and modify the preceding data as follows:

VOL%	FCC NAPHTHA, °F	COKER NAPHTHA, °F	CAT REFORMATE, °F
1.4	97	111	111
6.4	115	138	131
11.4	120	144	156

Now draw the ASTM distillation curves with percent distilled on the X- axis and distillation temperature on Y-axis. Read off the temperature at which 1.4% volume is distilled off. Assuming IBP (1.4% distilled) at 100°F,

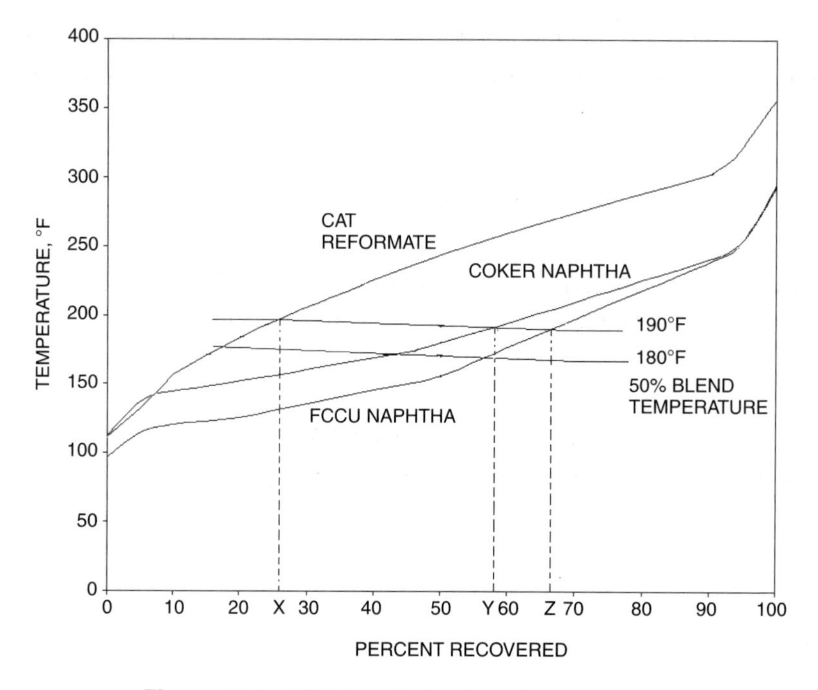

Figure 11-1. ASTM distillation blending procedure.

BLEND COMPONENT	VOL% AT 100°F	% IN BLEND	TOTAL
FCC NAPHTHA	1.4	50.0	0.7
COKER NAPHTHA	0	16.0	0
CAT REFORMATE	0	34.0	0
TOTAL		100.0	0.7

As the percent distilled is less than 1.4%, next assume a higher IPB temperature, at 110°F, and repeat the procedure:

BLEND COMPONENT	VOL% AT 100°F	% IN BLEND	TOTAL
FCC NAPHTHA	4.25	50.0	2.13
COKER NAPHTHA	0	16.0	0
CAT REFORMATE	0	34.0	0
TOTAL		100.0	2.13

Therefore, the blend distilled at 110°F is 2.13 vol%. By interpolation between these two values (0.7% and 2.13%), we determine the temperature at which 1.4% of the blend is distilled off; that is, 105°F. Calculations for ASTM 10–90% points are shown in Table 11-8.

ASTM End Point of Blend

The end point of a two-component blend is a function of the end point, proportion of blend component, and slope of the tail of the distillation curve of the higher-boiling component.

Procedure

From Table 11-9, read off the factor relating the difference between the end point of the components and their proportion in the blend. From Table 11-10, read off the factor relating slope of the tail of the higher-boiling component and its proportion in the blend. Add the product of these two factors to the end point of the lower-boiling component, and the result is the predicted end point of the blend.

Multicomponent blends are calculated as though the final blend were the result of a series of binary blends, starting with lowest-end-point component and successively adding higher-end-point components. This procedure is elaborated in Table 11-11.

VAPOR LOCK PROTECTION TEMPERATURE

When the volatility of gasoline is too high or when high temperatures or low pressure conditions prevail, bubbles of vapor can form at critical points in the fuel systems. This prevents adequate supply of fuel to the engine by preventing the fuel pump from operating because of low or negative suction pressure.

Vapor lock has a number of unwelcome effects, such as difficulty restarting a hot engine, uneven running, and reduced power output at high speed.

Vapor lock is influenced by the volatility characteristics of the fuel. The degree to which a fuel is liable to produce vapor lock depends mainly on the front-end volatility of the fuel blend.

Table 11-8
Calculation of the ASTM Distillation of the Blend

ASTM DISTILLATION	FCC NAPHTHA, % DISTILLED	COKER GASOLINE, % DISTILLED	CAT REFORMATE, % DISTILLED	FCC NAPHTHA, % DISTILLED × VOL%	COKER GASOLINE, % DISTILLED × VOL%	CAT REFORMATE, % DISTILLED × VOL%	BLEND, % DISTILLED	ASTM BLEND TEMPERATURE, °F INTERPOLATED
ASSUME IBP = 100°F	1.40	0.00	0.00	0.70	0.00	0.00	0.70	105
ASSUME IBP = 110°F	4.25	0.00	0.00	2.13	0.00	0.00	2.13	
ASSUME 10% = 120°F	10.00	2.00	2.00	5.00	0.32	0.68	6.00	125
ASSUME 10% = 130°F	25.00	4.00	4.00	12.50	0.64	1.36	14.50	
ASSUME 30% = 150°F	42.50	20.00	8.00	21.25	3.20	2.72	27.17	154
ASSUME 30% = 160°F	51.00	30.00	12.50	25.50	4.80	4.25	34.55	
ASSUME 50% = 180°F	62.00	50.00	20.00	31.00	8.00	6.80	45.80	187
ASSUME 50% = 190°F	67.50	57.50	26.00	33.75	9.20	8.84	51.79	
ASSUME 70% = 220°F	81.50	78.50	41.50	40.75	12.56	14.11	67.42	226
ASSUME 70% = 230°F	86.00	83.00	46.00	43.00	13.28	15.64	71.92	
ASSUME 90% = 270°F	97.50	96.00	70.00	48.75	15.36	23.80	87.91	276
ASSUME 90% = 280°F	99.00	97.50	77.50	49.50	15.60	26.35	91.45	

NOTE:
IBP IS THE TEMPERATUTRE ON MODIFIED SCALE WITH ORIGIN SHIFTED TO −1.4% VOLUME, WHERE 1.4% BLEND DISTILLS OFF.

Table 11-9
ASTM End-Point Coefficients, High-End-Point Component in Blend

°F	PROPORTION OF HIGH-END-POINT COMPONENT IN BINARY BLEND, Δ										
	5	10	15	20	25	30	40	50	60	70	80
5	0	1	1	1	1	2	2	3	3	4	4
10	1	2	2	3	3	4	5	6	6	7	8
15	1	2	3	4	5	6	8	8	10	11	12
20	2	3	4	6	7	8	10	12	14	15	15
25	2	4	6	8	9	11	13	15	17	20	22
30	3	5	7	9	11	13	16	19	21	24	26
35	3	6	8	11	13	16	20	23	26	28	31
40	4	8	10	13	16	19	23	27	30	33	36
45	4	9	12	16	19	22	26	31	34	38	40
50	5	10	14	18	22	25	30	35	39	42	45
55	6	12	16	20	25	28	34	40	44	47	50
60	6	13	18	23	28	32	39	44	48	52	55
65	7	15	20	26	32	36	43	49	53	57	60
70	8	17	23	29	35	40	48	54	58	62	65
75	9	19	26	33	39	44	52	59	63	67	70
80	10	21	29	36	43	48	57	64	68	72	75
85	11	23	32	40	47	52	62	69	73	77	80
90	12	26	36	44	51	57	67	74	78	82	85
95	14	29	40	48	56	61	72	79	84	87	90
100	15	32	44	53	61	66	77	84	89	92	95
105	17	35	48	58	66	71	82	89	94	98	100
110	19	38	53	63	71	76	87	94	99	103	106
115	21	42	58	68	76	82	92	99	104	108	111
120	23	46	63	73	81	87	97	104	109	113	116
125	26	51	68	78	87	92	102	110	114	118	121
130	29	56	74	84	92	98	108	115	120	123	126
135	33	62	79	90	98	103	113	120	125	128	132
140	38	69	85	96	104	109	118	125	130	133	137
145	45	77	91	101	109	114	124	130	135	138	142

NOTE:
DELTA IS THE DIFFERENCE IN THE END POINTS OF COMPONENTS.

The vapor lock protection temperature (VLPT) of a gasoline blend is the temperature at which a certain fixed vapor/liquid ratio (usually 20 or more) exists. Although a number of different indices exist, they are equally valid for predicting the susceptibility of the fuel to cause vapor lock problems.

Table 11-10

ASTM End-Point Coefficients, Difference between 90% Point and End Point of Higher-Boiling Component

	% HIGHER END POINT STOCK IN BLEND										
DIFFERENCE	5	10	15	20	25	30	40	50	60	70	80
18	5.50	3.18	2.38	2.08	1.90	1.75	1.54	1.42	1.33	1.28	1.24
20	4.80	2.70	2.05	1.80	1.65	1.54	1.40	1.29	1.20	1.17	1.15
25	3.30	2.23	1.74	1.55	1.43	1.34	1.24	1.18	1.13	1.10	1.08
30	2.55	1.92	1.55	1.42	1.32	1.26	1.17	1.12	1.08	1.05	1.04
35	2.04	1.65	1.40	1.30	1.24	1.20	1.12	1.08	1.05	1.02	1.01
40	1.68	1.44	1.28	1.22	1.17	1.14	1.08	1.05	1.03	1.01	1.00
45	1.40	1.27	1.18	1.14	1.11	1.09	1.05	1.03	1.01	1.00	1.00
50	1.18	1.13	1.09	1.07	1.05	1.04	1.02	1.01	1.00	1.00	1.00
55	1.00	1.00	1.00	1.00	1.00	1.00	1.00	1.00	1.00	1.00	1.00
60	0.88	0.90	0.92	0.94	0.95	0.96	0.97	0.98	0.99	0.99	1.00
65	0.78	0.82	0.85	0.88	0.90	0.91	0.93	0.95	0.97	0.98	0.99
70	0.70	0.74	0.78	0.82	0.85	0.87	0.90	0.93	0.95	0.97	0.98
75	0.62	0.67	0.72	0.76	0.80	0.82	0.86	0.90	0.93	0.95	0.97
80	0.55	0.60	0.66	0.71	0.75	0.73	0.83	0.87	0.91	0.93	0.96
85	0.49	0.55	0.61	0.66	0.70	0.74	0.79	0.84	0.88	0.91	0.94
90	0.44	0.50	0.56	0.61	0.65	0.69	0.75	0.80	0.84	0.88	0.91
95	0.39	0.46	0.51	0.56	0.60	0.64	0.71	0.76	0.80	0.84	0.87
100	0.34	0.42	0.47	0.51	0.56	0.60	0.66	0.72	0.76	0.80	0.83

Table 11-11
ASTM End Point of Blend

Calculate the end point
of the following blend:
FCC naphtha = 50%
Coker naphtha = 16%
Cat reformate = 34%
ASTM distillation of the blend
components are as per earlier example

(1) CONSIDER A BINARY BLEND OF FCC NAPHTHA AND COKER NAPHTHA.

	FCC NAPHTHA		COKER NAPHTHA	DELTA
ASTM END POINT, °F	295		289	6
% FCC GASOLINE IN BLEND		=	0.758	
REFERRING TO TABLE 11-9, FACTOR 1		=	4	
DIFFERENCE BETWEEN END POINT AND 90% POINT (HIGHER-END POINT COMPONENT)		=	(295 − 239)	
		=	56	
REFERRING TO TABLE 11-10, FACTOR 2		=	1	
FACTOR 1 × FACTOR 2		=	4	
THEREFORE END POINT OF BINARY		=	(289 + 4)	
FCC NAPHTHA AND COKER NAPHTHA		=	293°F	
(2) NEXT CONSIDER A BLEND OF ABOVE BINARY AND CAT REFORMER NAPHTHA				
DIFFERENCE IN END POINTS		=	356 − 293	
(FCC + COKER) BINARY AND C. REFORMATE NAPHTHA		=	63°F	
PROPORTION OF C. REFORMATE IN BLEND		=	34%	
REFERRING AGAIN TO TABLES 11–9 AND 11–10				
FACTOR 1		=	40	
FACTOR 2		=	1.16	
FACTOR 1 × FACTOR 2		=	46.4	
END POINT OF THE BLEND		=	(293 + 46.4)	
(FCC Naphtha + Coker Naphtha + Cat Reformate)		=	339.4°F	

Jenkin Equation[3]

VLPT, T_{20} is the temperature at which vapor/liquid ratio is 20. VLPT is expressed as a function of the RVP (Reid Vapor Pressure) and the ASTM 10 and 50% points:

$$\text{VLPT} = 52.47 - 0.33 \times (\text{RVP}) + 0.2 \times (10\% \text{ point}) + 0.17 \times (50\% \text{ point})$$

where

$$\text{VLPT} = \text{temperature, °C;}$$
$$\text{RVP} = \text{RVP, kPa;}$$
$$10\%, 50\% = \text{ASTM distillation point, °C.}$$

Acceptable VLPT numbers depend on the maximum ambient temperature of the area where the gasoline is designed to be used. For example, if the maximum summer temperature touches 50°C at any location, the VLPT must be more than 50°C by reducing the lower-volatility blend components in the gasoline formulation.

VISCOSITY BLENDING

The viscosities of petroleum fractions do not blend linearly, and viscosity blending is done with the help of blending indices. Table 11-12 presents volume blending indices, and Table 11-13 presents weight blending indices at 122°F.

EXAMPLE 11-4

Determine the amount of cutter stock need to blend vacuum resid with a kinematic viscosity of 80,000 cst at 122°F to finished fuel oil with viscosity of 180 centistokes at 122°F. The cutter stock viscosity is 8.0 cst.

To estimate the cutter requirements, determine the viscosity blend indices for vacuum residuum, cutter stock and finished fuel oil from the viscosity blend indices table and blend these values linearly. We see from the table on page 330 that 42.7% cutter stock is required to reduce the final blend viscosity to 175 centistokes.

Table 11-12
Viscosity Blending Indices

cst	0	0.1	0.2	0.3	0.4	0.5	0.6	0.7	0.8	0.9
0	0.0	−1447.4	−992.1	−787.7	−658.0	−564.0	−490.9	−431.2	−381.0	−337.9
1	−300.0	−270.9	−246.6	−222.2	−199.4	−179.1	−161.2	−145.5	−131.5	−119.0
2	−107.6	−97.2	−86.6	−78.6	−70.3	−62.5	−55.1	−48.1	−41.4	−35.0
3	−28.9	−23.1	−17.5	−12.1	−6.9	−1.8	3.1	7.8	12.4	16.8
4	21.2	25.4	29.5	33.5	37.5	41.3	45.0	48.7	55.8	59.3
5	59.3	62.7	66.0	69.2	72.4	75.6	78.7	81.4	84.7	87.7
6	90.6	93.4	96.2	99.0	103.7	105.8	108.0	110.1	112.1	114.2
7	116.2	118.2	120.1	122.0	123.9	125.8	127.7	129.5	131.3	133.1
8	134.8	136.6	138.3	140.0	141.7	143.3	144.9	146.5	148.8	149.7
9	151.3	152.8	154.3	155.8	157.3	158.8	160.2	161.6	163.0	164.4
	0	**1**	**2**	**3**	**4**	**5**	**6**	**7**	**8**	**9**
10	165.4	178.9	190.7	201.4	211.2	220.3	228.6	236.3	243.4	250.2
20	256.5	262.4	268.0	273.3	278.3	283.0	287.5	291.8	296.0	299.9
30	303.7	307.3	310.7	314.1	317.3	320.4	323.4	326.3	329.0	331.7
40	334.4	336.9	339.3	341.7	344.0	346.3	348.5	350.6	352.7	354.7
50	356.7	358.6	360.5	362.3	364.1	365.9	367.5	369.2	370.8	372.4
60	374.0	375.5	377.0	378.5	379.9	381.3	382.7	384.1	385.4	386.7
70	388.0	389.3	390.5	391.8	393.0	394.1	395.3	396.5	397.6	398.7
80	399.8	400.9	401.9	403.0	404.0	405.0	406.0	407.0	407.9	408.9
90	409.8	410.8	411.7	412.6	413.5	414.4	415.2	416.1	416.9	417.8
	0	**10**	**20**	**30**	**40**	**50**	**60**	**70**	**80**	**90**
100	418.6	426.4	433.3	439.6	445.3	450.5	455.3	459.8	464.0	467.9
200	471.5	475.0	478.2	481.3	484.2	487.0	489.7	492.2	494.6	496.9
300	499.2	501.3	503.4	505.4	507.3	509.1	511.0	512.7	514.4	516.0
400	517.6	519.1	520.6	522.0	523.5	524.8	526.2	527.5	528.8	530.0
500	531.2	532.4	533.6	534.7	535.8	536.9	538.0	539.0	540.0	541.0
600	542.0	542.9	543.9	544.8	545.7	546.6	547.5	548.4	549.2	550.0
700	550.9	551.7	552.5	553.2	554.0	554.8	555.5	556.2	557.0	557.7

Table 11-12
Continued

cst	0	10	20	30	40	50	60	70	80	90
800	558.4	559.1	559.7	560.4	561.1	561.7	562.4	563.0	563.6	563.3
900	564.9	565.5	566.1	566.6	567.2	567.8	568.4	568.9	569.5	570.0
	0	**100**	**200**	**300**	**400**	**500**	**600**	**700**	**800**	**900**
1000	570.6	575.7	580.3	584.4	588.3	591.8	595.0	598.1	600.9	603.6
2000	606.1	608.5	610.7	612.9	614.9	619.9	618.7	620.5	622.2	623.8
3000	625.4	626.9	628.4	629.8	631.2	632.5	633.8	635.0	636.2	637.4
4000	638.6	639.7	640.7	641.8	642.8	643.8	644.8	645.7	646.6	647.5
5000	648.4	649.3	650.1	651.0	651.8	653.0	653.4	654.1	654.9	655.6
6000	656.3	657.0	657.7	658.4	659.1	659.7	660.4	661.0	661.6	662.3
7000	662.9	663.5	664.0	664.6	665.2	665.8	666.3	666.9	667.4	667.9
8000	668.4	669.0	669.5	670.0	670.5	671.0	671.4	671.9	672.4	672.8
9000	673.3	673.8	674.2	674.6	675.1	675.5	675.9	676.4	676.8	677.2
	0	**1000**	**2000**	**3000**	**4000**	**5000**	**6000**	**7000**	**8000**	**9000**
10000	677.6	681.4	684.9	688.1	691.0	693.7	696.2	698.5	700.7	702.8
20000	704.7	706.7	708.3	710.0	711.6	713.1	714.6	715.9	717.2	718.6
30000	719.8	721.0	722.2	723.3	724.3	725.4	726.4	727.4	728.3	729.3
40000	730.2	731.0	731.9	732.7	733.5	734.3	735.1	735.8	736.6	737.3
50000	738.0	738.7	739.4	740.0	740.7	741.3	742.0	742.6	743.2	743.8
60000	744.3	744.9	745.5	746.0	746.5	747.1	747.6	748.1	748.6	749.1
70000	749.6	750.1	750.5	751.0	751.5	751.9	752.4	752.8	753.2	753.7
80000	754.1	754.5	754.9	755.3	755.7	756.1	756.5	756.9	757.3	757.7
90000	758.0	758.4	758.8	759.1	759.5	759.8	760.2	760.5	760.8	761.2
100000	761.5									

NOTE:
VISCOSITIES OF PETROLEUM PRODUCTS DO NOT BLEND LINEARLY ON VOLUME OR WEIGHT BASIS. BLENDING INDICES ARE THEREFORE EMPLOYED. INDICES FOR VOLUMETRIC BLENDING ARE PRESENTED IN THE TABLE. THE UNITS OF KINEMATIC VISCOSITY ARE IN CENTISTOKES.

Table 11-13
Viscosity Blending Indices, Weight Basis

cst	0	0.1	0.2	0.3	0.4	0.5	0.6	0.7	0.8	0.9
1	3.25	4.53	5.65	6.64	7.52	8.32	9.04	9.70	10.31	10.88
2	11.40	11.89	12.34	12.77	13.17	13.55	13.91	14.25	14.57	14.88
3	15.17	15.45	15.72	15.98	16.22	16.46	16.69	16.91	17.12	17.32
4	17.52	17.71	17.89	18.07	18.24	18.41	18.57	18.73	18.88	19.03
5	19.17	19.31	19.45	19.58	19.71	19.84	19.97	20.09	20.20	20.32
6	20.43	20.54	20.65	20.76	20.86	20.96	21.06	21.16	21.25	21.35
7	21.44	21.53	21.62	21.70	21.79	21.87	21.95	22.03	22.11	22.19
8	22.27	22.34	22.42	22.49	22.56	22.63	22.70	22.77	22.84	22.90
9	22.97	23.03	23.10	23.16	23.22	23.28	23.34	23.40	23.46	23.52
	0	**1**	**2**	**3**	**4**	**5**	**6**	**7**	**8**	**9**
10	23.57	24.11	24.58	25.00	25.38	25.73	26.05	26.35	26.62	26.87
20	27.11	27.33	27.54	27.74	27.93	28.11	28.28	28.44	28.59	28.74
30	28.88	29.01	29.14	29.27	29.39	29.50	29.61	29.72	29.83	29.93
40	30.03	30.12	30.21	30.30	30.39	30.47	30.55	30.63	30.71	30.79
50	30.86	30.93	31.00	31.07	31.14	31.20	31.27	31.33	31.39	31.45
60	31.51	31.57	31.62	31.68	31.73	31.79	31.84	31.89	31.94	31.99
70	32.04	32.09	32.13	32.18	32.23	32.27	32.31	32.36	32.40	32.44
80	32.48	32.52	32.56	32.60	32.64	32.68	32.72	32.76	32.79	32.83
90	32.86	32.90	32.93	32.97	33.00	33.04	33.07	33.10	33.13	33.17
	0	**10**	**20**	**30**	**40**	**50**	**60**	**70**	**80**	**90**
100	33.20	33.49	33.76	34.00	34.21	34.41	34.60	34.77	34.93	35.08
200	35.22	35.35	35.48	35.60	35.71	35.82	35.92	36.02	36.11	36.20
300	36.29	36.37	36.45	36.53	36.60	36.67	36.74	36.81	36.88	36.94
400	37.00	37.06	37.12	37.18	37.23	37.28	37.34	37.39	37.44	37.48
500	37.53	37.58	37.62	37.67	37.71	37.75	37.79	37.83	37.87	37.91
600	37.95	37.99	38.03	38.06	38.10	38.13	38.17	38.20	38.23	38.27
700	38.30	38.33	38.36	38.39	38.42	38.45	38.48	38.51	38.54	38.56
800	38.59	38.62	38.64	38.67	38.70	38.72	38.75	38.77	38.80	38.82
900	38.84	38.87	38.89	38.91	38.94	38.96	38.98	39.00	39.02	39.05
	0	**100**	**200**	**300**	**400**	**500**	**600**	**700**	**800**	**900**
1000	39.07	39.27	39.45	39.61	39.76	39.90	40.02	40.14	40.25	40.36
2000	40.46	40.55	40.64	40.72	40.80	40.88	40.95	41.02	41.09	41.15
3000	41.21	41.27	41.33	41.38	41.44	41.49	41.54	41.59	41.63	41.68
4000	41.72	41.77	41.81	41.85	41.89	41.93	41.97	42.00	42.04	42.08
5000	42.11	42.14	42.18	42.21	42.24	42.27	42.30	42.33	42.36	42.39
6000	42.42	42.45	42.47	42.50	42.53	42.55	42.58	42.60	42.63	42.65
7000	42.67	42.70	42.72	42.74	42.76	42.79	42.81	42.83	42.85	42.87
8000	42.89	42.91	42.93	42.95	42.97	42.99	43.01	43.03	43.04	43.06
9000	43.08	43.10	43.11	43.13	43.15	43.17	43.18	43.20	43.21	43.23

Table 11-13
Continued

VISC, CST	H	VISC, CST	H	VISC, CST	H	VISC, CST	H
10,000	43.25	100,000	46.49	1,000,000	49.14	10,000,000	51.38
20,000	44.30	200,000	47.34	2,000,000	49.85	20,000,000	51.99
30,000	44.88	300,000	47.81	3,000,000	50.25	30,000,000	52.34
40,000	45.28	400,000	48.14	4,000,000	50.53	40,000,000	52.58
50,000	45.59	500,000	48.39	5,000,000	50.74	50,000,000	52.76
60,000	45.83	600,000	48.59	6,000,000	50.91	60,000,000	52.91
70,000	46.03	700,000	48.76	7,000,000	51.06	70,000,000	53.04
80,000	46.21	800,000	48.90	8,000,000	51.18	80,000,000	53.14
90,000	46.36	900,000	49.03	9,000,000	51.29	90,000,000	53.24

NOTES:
VISCOSITIES OF PETROLEUM PRODUCTS DO NOT BLEND LINEARLY ON VOLUME OR WEIGHT BASIS. BLENDING INDICES ARE THEREFORE EMPLOYED. INDICES FOR A WEIGHT BASIS BLENDING (ALSO CALLED *REFUTAS FUNCTIONS*) ARE PRESENTED IN THE TABLE.
THE UNITS OF KINEMATIC VISCOSITY ARE IN CENTISTOKES.
BLENDING INDICES FOR VARIOUS VISCOSITY RANGES ARE PRESENTED. THESE ARE CALCULATED BY THE FOLLOWING RELATIONSHIPS:
$I = 23.097 + 33.468 * LOG\,LOG(V + 0.8)$
V = *KINEMATIC VISCOSITY IN CENTISTOKE UNITS*
IN CASE VISCOSITY BLENDING INDEX IS KNOWN, VISCOSITY IN CENTISTOKES IS CALCULATED AS FOLLOWS:
$V = \left(10^{(10(VI-23.097)/33.468)}\right) - 0.8$
I = *VISCOSITY INDEX (WT. BASIS BLENDING)*

COMPONENT	VISCOSITY, CST	BLEND INDEX, *H*	VOL%
VACUUM RESID	80,000	754	0.528
CUTTER STOCK	8.0	135	0.472
BLEND	175	460	1.000

BLENDING MARGIN

A blending margin of 4–5 *H* is normally allowed. Therefore, to meet a guaranteed specification of 464 *H*, corresponding to 180 cst, fuel oil must be blended to 460 *H* or 170 cst.

POUR POINT BLENDING

The pour point and freeze point of the distillate (kerosene, diesels, etc.) do not blend linearly, and blending indices are used for linear blending by volume. Tables 11-14 and 11-15 show the blending indices used to estimate the pour point and cloud point of distillate petroleum products.

A blending margin of 10 PI (pour index) is allowed between the guaranteed specification and the refinery blending. For example, to guarantee a pour spec of −6°C (21.2°F, pour index 336.3), the blending target would be 326.3 PI. In terms of the pour point, this corresponds to a blending margin of 1°F.

EXAMPLE 11-5

Determine the amount of kerosene that must be blended into diesel with a 43°F pour point to lower the pour point to 21°F. The properties of kerosene and diesel stream are as follows:

	KEROSENE	DIESEL
SPECIFIC GRAVITY	0.7891	0.8410
POUR POINT	−50°F	43°F

To determine the pour point of the blend, determine the pour indices, from the pour blend table, corresponding to the pour points of kerosene and diesel, then the target pour point and blend linearly as follows:

BLEND COMPONENT	POUR POINT, °F	BLEND INDEX	VOL%
DIESEL	43	588	53.6
KEROSENE	−50	46	46.4
BLEND	21	336	100.0

Therefore, to lower the pour point to 21°F, 46.4% kerosene by volume must be blended.

FLASH POINT BLENDING

The flash point of a blend can be estimated from the flash point of the blend components using flash point blend indices, which blend linearly

Table 11-14
Pour Point of Distillate Blends

POUR POINT, °R	INDEX	POUR POINT, °R	INDEX	POUR POINT, °R	INDEX
360	8.99	394	27.78	428	78.16
361	9.31	395	28.67	429	80.48
362	9.63	396	29.59	430	82.85
363	9.97	397	30.54	431	85.30
364	10.32	398	31.51	432	87.80
365	10.68	399	32.52	433	90.38
366	11.05	400	33.55	434	93.02
367	11.44	401	34.62	435	95.74
368	11.83	402	35.71	436	98.52
369	12.24	403	36.84	437	101.39
370	12.66	404	38.00	438	104.32
371	13.10	405	39.19	439	107.34
372	13.54	406	40.41	440	110.44
373	14.01	407	41.68	441	113.62
374	14.48	408	42.97	442	116.88
375	14.97	409	44.31	443	120.23
376	15.48	410	45.68	444	123.67
377	16.00	411	47.10	445	127.19
378	16.54	412	48.55	446	130.81
379	17.10	413	50.04	447	134.53
380	17.67	414	51.58	448	138.34
381	18.26	415	53.16	449	142.25
382	18.87	416	54.78	450	146.26
383	19.50	417	56.45	451	150.37
384	20.14	418	58.17	452	154.59
385	20.81	419	59.93	453	158.92
386	21.49	420	61.74	454	163.37
387	22.20	421	63.60	455	167.92
388	22.93	422	65.52	456	172.59
389	23.68	423	67.49	457	177.38
390	24.45	424	69.51	458	182.30
391	25.24	425	71.59	459	187.34
392	26.06	426	73.72		
393	26.91	427	75.91		

NOTES:
ALSO APPLICABLE TO FREEZE POINTS AND FLUIDITY BLENDING IS ON A VOLUME BASIS.
POUR POINT BLEND INDEX = 3262000* × (POUR POINT, °R/1000)$^{12.5}$
POUR POINT, °R = 1000 × (Index/316000)$^{0.08}$
POUR POINT, °F = POUR POINT (°R) − 460
*CORRELATION OF HU AND BURNS.

Table 11-14
Continued

POUR POINT, °R	INDEX	POUR POINT, °R	INDEX	POUR POINT, °R	INDEX
460	192.50	496	493.72	532	1185.38
461	197.80	497	506.31	533	1213.53
462	203.23	498	519.19	534	1242.30
463	208.80	499	532.38	535	1271.70
464	214.50	500	545.87	536	1301.73
465	220.36	501	559.67	537	1332.42
466	226.35	502	573.80	538	1363.77
467	232.50	503	588.25	539	1395.79
468	238.80	504	603.04	540	1428.51
469	245.26	505	618.16	541	1461.93
470	251.88	506	633.64	542	1496.07
471	258.66	507	649.47	543	1530.95
472	265.61	508	665.67	544	1566.56
473	272.73	509	682.24	545	1602.94
474	280.02	510	699.18	546	1640.10
475	287.50	511	716.51	547	1678.04
476	295.15	512	734.24	548	1716.80
477	303.00	513	752.37	549	1756.37
478	311.04	514	770.90	550	1796.78
479	319.27	515	789.86	551	1838.05
480	327.70	516	809.25	552	1880.18
481	336.34	517	829.07	553	1923.21
482	345.18	518	849.34	554	1967.13
483	354.24	519	870.07	555	2011.98
484	363.52	520	891.26	556	2057.77
485	373.02	521	912.92	557	2104.51
486	382.75	522	935.07	558	2152.23
487	392.71	523	957.71	559	2200.95
488	402.91	524	980.85	560	2250.67
489	413.35	525	1004.51		
490	424.05	526	1028.69		
491	434.99	527	1053.40		
492	446.20	528	1078.66		
493	457.67	529	1104.48		
494	469.41	530	1130.86		
495	481.42	531	1157.83		

Table 11-15
Cloud Point of Distillate Blends

CLOUD POINT, °R	INDEX	CLOUD POINT, °R	INDEX	CLOUD POINT, °R	INDEX
−100	0.259	**−64**	0.930	**−28**	3.333
−99	0.269	**−63**	0.963	**−27**	3.454
−98	0.278	**−62**	0.998	**−26**	3.578
−97	0.288	**−61**	1.034	**−25**	3.708
−96	0.299	**−60**	1.072	**−24**	3.842
−95	0.310	**−59**	1.110	**−23**	3.980
−94	0.321	**−58**	1.150	**−22**	4.124
−93	0.332	**−57**	1.192	**−21**	4.273
−92	0.344	**−56**	1.235	**−20**	4.427
−91	0.357	**−55**	1.279	**−19**	4.587
−90	0.370	**−54**	1.326	**−18**	4.752
−89	0.383	**−53**	1.373	**−17**	4.924
−88	0.397	**−52**	1.423	**−16**	5.102
−87	0.411	**−51**	1.474	**−15**	5.286
−86	0.426	**−50**	1.528	**−14**	5.477
−85	0.442	**−49**	1.583	**−13**	5.675
−84	0.457	**−48**	1.640	**−12**	5.879
−83	0.474	**−47**	1.699	**−11**	6.092
−82	0.491	**−46**	1.761	**−10**	6.312
−81	0.509	**−45**	1.824	**−9**	6.540
−80	0.527	**−44**	1.890	**−8**	6.776
−79	0.546	**−43**	1.958	**−7**	7.020
−78	0.566	**−42**	2.029	**−6**	7.274
−77	0.586	**−41**	2.102	**−5**	7.536
−76	0.608	**−40**	2.178	**−4**	7.808
−75	0.629	**−39**	2.257	**−3**	8.090
−74	0.652	**−38**	2.338	**−2**	8.382
−73	0.676	**−37**	2.423	**−1**	8.685
−72	0.700	**−36**	2.510		
−71	0.725	**−35**	2.601		
−70	0.752	**−34**	2.694		
−69	0.779	**−33**	2.792		
−68	0.807	**−32**	2.893		
−67	0.836	**−31**	2.997		
−66	0.866	**−30**	3.105		
−65	0.897	**−29**	3.217		

NOTES:
FOR CLOUD POINTS BELOW 0°F
THE INDEX SHOULD BE BLENDED ON WEIGHT BASIS.
THE BLENDING INDEX IS GIVEN BY THE FOLLOWING EQUATION:
$I = \exp[2.303(0.954 + 0.0154 \times T)]$.

Table 11-15
Continued

CLOUD POINT, °F	INDEX	CLOUD POINT, °F	INDEX	CLOUD POINT, °F	INDEX
0	8.999	**36**	32.261	**72**	115.657
1	9.323	**37**	33.425	**73**	119.832
2	9.660	**38**	34.632	**74**	124.159
3	10.009	**39**	35.882	**75**	128.641
4	10.370	**40**	37.178	**76**	133.285
5	10.744	**41**	38.520	**77**	138.097
6	11.132	**42**	39.911	**78**	143.083
7	11.534	**43**	41.351	**79**	148.249
8	11.951	**44**	42.844	**80**	153.601
9	12.382	**45**	44.391	**81**	159.146
10	12.829	**46**	45.994	**82**	164.892
11	13.292	**47**	47.654	**83**	170.845
12	13.772	**48**	49.375	**84**	177.013
13	14.269	**49**	51.157	**85**	183.404
14	14.785	**50**	53.004	**86**	190.025
15	15.318	**51**	54.918	**87**	196.885
16	15.871	**52**	56.901	**88**	203.993
17	16.444	**53**	58.955	**89**	211.358
18	17.038	**54**	61.083	**90**	218.989
19	17.653	**55**	63.288	**91**	226.895
20	18.291	**56**	65.573	**92**	235.086
21	18.951	**57**	67.941	**93**	243.573
22	19.635	**58**	70.394	**94**	252.367
23	20.344	**59**	72.935	**95**	261.478
24	21.078	**60**	75.568	**96**	270.918
25	21.839	**61**	78.296	**97**	280.699
26	22.628	**62**	81.123	**98**	290.833
27	23.445	**63**	84.052	**99**	301.333
28	24.291	**64**	87.086		
29	25.168	**65**	90.230		
30	26.077	**66**	93.488		
31	27.018	**67**	96.863		
32	27.994	**68**	100.360		
33	29.004	**69**	103.983		
34	30.052	**70**	107.737		
35	31.136	**71**	111.627		

on volume basis. The flash point (in °F) vs. flash blend indices are presented in Table 11-16.

EXAMPLE 11-6

Determine the flash point of a blend containing 30 vol% component A with a flash point of 100°F, 10% component B, with a flash of 90°F, and 60% component C with a flash point of 130°F.

From the flash blending tables, find the blend indices for the three components and blend linearly with the volume as follows:

COMPONENT	VOL%	FLASH POINT, °F	FLASH INDEX	VOLUME × FLASH INDEX
A	0.3	100	754	226.2
B	0.10	90	1169	116.9
C	0.60	130	225	135.0
BLEND	1.00	111		478.1

The flash index of the blend is computed at 478.1, which corresponds to a flash point of 111°F.

ALTERNATIVE METHOD FOR DETERMINING THE BLEND FLASH POINT

The flash index is first determined from Table 11-17. Two empirical indices are worked out, the 154 index and the 144 index. The 154 index is a criteria for meeting the 154°F flash point and 144 index is criteria for meeting the 144°F flash point.

If the value of the 154 index is positive for any component or blend, it will meet the 154°F flash criteria; that is, the flash will be equal to or higher than 154°F. Similarly, if the 144 flash index is positive, it will meet the 144°F flash criteria. If the 144 index is negative, the corresponding flash point will be lower than 144°F:

$$154 \text{ index} = (0.4240 - 0.0098 \times \text{FI}) \times \text{MB}$$

where FI is the flash index (Table 11-17) and MB is moles/barrel.

Table 11-16
Flash Point (Abel) vs. Flash Blending Index[4]

FLASH POINT, °F	0	1	2	3	4	5	6	7	8	9
80	1845.4	1761.4	1681.6	1605.7	1533.5	1464.9	1399.6	1337.5	1278.4	1222.2
90	1168.6	1117.6	1069.0	1022.8	978.7	936.6	896.6	858.4	822.0	787.3
100	754.2	722.6	692.4	663.7	636.2	610.0	584.9	561.0	538.1	516.3
110	495.4	475.5	456.4	438.2	420.8	404.1	388.2	372.9	358.3	344.3
120	331.0	318.1	305.9	294.1	282.9	272.1	261.8	251.9	242.4	233.3
130	224.6	216.2	208.2	200.5	193.1	186.0	179.2	172.7	166.5	160.4
140	154.7	149.1	143.8	138.7	133.8	129.0	124.5	120.1	115.9	111.9
150	108.0	104.3	100.7	97.3	93.9	90.7	87.7	84.7	81.8	79.1
160	76.5	73.9	71.5	69.1	66.8	64.6	62.5	60.5	58.5	56.6
170	54.8	53.0	51.3	49.7	48.1	46.6	45.1	43.7	42.3	41.0
180	39.7	38.5	37.3	36.2	35.1	34.0	32.9	31.9	31.0	30.0
190	29.1	28.3	27.4	26.6	25.8	25.1	24.3	23.6	22.9	22.2
200	21.6	21.0	20.4	19.8	19.2	18.7	18.1	17.6	17.1	16.6
210	16.2	15.7	15.3	14.9	14.4	14.0	13.7	13.3	12.9	12.6
220	12.2	11.9	11.6	11.3	11.0	10.7	10.4	10.1	9.8	9.6
230	9.3	9.1	8.8	8.6	8.4	8.2	8.0	7.8	7.6	7.4
240	7.2	7.0	6.8	6.6	6.5	6.3	6.2	6.0	5.9	5.7
250	5.6	5.4	5.3	5.2	5.0	4.9	4.8	4.7	4.6	4.5
260	4.4	4.3	4.1	4.1	4.0	3.9	3.8	3.7	3.6	3.5
270	3.4	3.4	3.3	3.2	3.1	3.1	3.0	2.9	2.9	2.8
280	2.7	2.7	2.6	2.5	2.5	2.4	2.4	2.3	2.3	2.2
290	2.2	2.1	2.1	2.0	2.0	2.0	1.9	1.9	1.8	1.8
300	1.8	1.7	1.7	1.6	1.6	1.6	1.5	1.5	1.5	1.4

NOTES:
FLASH INDEX $= 10^{(-6.1188+4345.2/(\text{FLASH POINT}+383)}$
FLASH POINT = 4345.2/(LOG (FLASH INDEX) + 6.1188) – 383.0.
WHERE
FLASH POINT (ABEL) IS IN °F.

$$144 \text{ index} = (0.6502 - 0.01107 \times \text{FI}) \times \text{MB}$$

This estimation requires data on molecular weight of the fraction. For routinely blended stocks, the values of the 144 and 154 indices are prepared, and these can be used to determine whether or not the given blend will meet the flash index. Each index blends linearly with volume and has zero as a reference point.

Table 11-17
Flash Point vs. Flash Index (for 154 and 144 Indices)

FLASH POINT, °C	0.0	0.5	1.0	1.5	2.0	2.5	3.0	3.5	4.0	4.5
10	784.80	762.07	740.07	718.78	698.17	678.23	658.92	640.22	622.12	604.59
15	587.61	571.16	555.23	539.80	524.85	510.36	496.31	482.70	469.51	456.72
20	444.33	432.31	420.65	409.35	398.38	387.75	377.43	367.42	357.71	348.29
25	339.15	330.27	321.66	313.30	305.18	297.30	289.65	282.22	275.00	267.99
30	261.18	254.57	248.14	241.90	235.83	229.94	224.21	218.64	213.23	207.97
35	202.85	197.88	193.04	188.34	183.77	179.32	174.99	170.88	166.69	162.70
40	158.83	155.05	151.38	147.81	144.33	140.95	137.65	134.44	131.32	128.28
45	125.31	122.43	119.62	116.88	114.22	111.62	109.09	106.63	104.22	101.88
50	99.66	97.38	95.21	93.10	91.04	89.03	87.08	85.17	83.31	81.49
55	79.72	77.99	76.31	74.67	73.07	71.50	69.98	68.49	67.04	65.62
60	64.24	62.89	61.57	60.28	59.03	57.80	56.60	55.43	54.29	53.18
65	52.09	51.03	49.99	48.98	47.98	47.02	46.07	45.15	44.24	43.36
70	42.50	41.66	40.83	40.03	39.24	38.47	37.72	36.99	36.27	35.57
75	34.88	34.21	33.55	32.91	32.28	31.66	31.06	30.47	29.90	29.34
80	28.79	28.25	27.72	27.20	26.70	26.20	25.72	25.25	24.78	24.33
85	23.88	23.45	23.02	22.61	22.20	21.80	21.41	21.07	20.65	20.28
90	19.92	19.57	19.22	18.88	18.55	18.22	17.91	17.59	17.29	16.99
95	16.69	16.41	16.12	15.85	15.58	15.31	15.05	14.79	14.54	14.30
100	14.06	13.82	13.59	13.36	13.14	12.92	12.71	12.50	12.29	12.09
105	11.89	11.70	11.51	11.32	11.14	10.96	10.78	10.61	10.44	10.27
110	10.10	9.94	9.78	9.63	9.48	9.33	9.18	9.04	8.90	8.76
115	8.62	8.49	8.36	8.23	8.10	7.98	7.85	7.73	7.61	7.50
120	7.38	7.27	7.16	7.06	6.95	6.85	6.74	6.64	6.54	6.45
125	6.35	6.26	6.17	6.07	5.99	5.90	5.81	5.73	5.64	5.56
130	5.48	5.40	5.33	5.25	5.17	5.10	5.03	4.96	4.89	4.82
135	4.75	4.68	4.62	4.55	4.49	4.43	4.37	4.30	4.25	4.19
140	4.13	4.07	4.02	3.96	3.91	3.85	3.80	3.75	3.70	3.65
145	3.60	3.55	3.51	3.46	3.41	3.37	3.32	3.28	3.24	3.19
150	3.15	3.11	3.07	3.03	2.99	2.95	2.91	2.88	2.84	2.80
155	2.77	2.73	2.70	2.66	2.63	2.59	2.56	2.53	2.50	2.47
160	2.44	2.41	2.38	2.35	2.32	2.29	2.26	2.23	2.20	2.18
165	2.15	2.12	2.10	2.07	2.05	2.02	2.00	1.97	1.95	1.93
170	1.90	1.88	1.86	1.84	1.82	1.79	1.77	1.75	1.73	1.71
175	1.69	1.67	1.65	1.63	1.61	1.60	1.58	1.56	1.54	1.52
180	1.51	1.49	1.47	1.45	1.44	1.42	1.41	1.39	1.37	1.36
185	1.34	1.33	1.31	1.30	1.28	1.27	1.26	1.24	1.23	1.22
190	1.20	1.19	1.18	1.16	1.15	1.14	1.13	1.11	1.10	1.09
195	1.08	1.07	1.06	1.04	1.03	1.02	1.01	1.00	0.99	0.98
200	0.97	0.96	0.95	0.94	0.93	0.92	0.91	0.90	0.89	0.88
205	0.87	0.86	0.86	0.85	0.84	0.83	0.82	0.81	0.80	0.80
210	0.79	0.78	0.77	0.76	0.76	0.75	0.74	0.73	0.73	0.72
215	0.71	0.71	0.70	0.69	0.69	0.68	0.67	0.67	0.66	0.65
220	0.65	0.64	0.63	0.63	0.62	0.62	0.61	0.60	0.60	0.59
225	0.59	0.58	0.58	0.57	0.57	0.56	0.55	0.55	0.54	0.54
230	0.53	0.53	0.52	0.52	0.52	0.51	0.51	0.50	0.50	0.49
235	0.49	0.48	0.48	0.47	0.47	0.47	0.46	0.46	0.45	0.45
240	0.45	0.44	0.44	0.43	0.43	0.43	0.42	0.42	0.41	0.41

$FLASH\ INDEX = 10^{((2050.86/(F+273.16)-4.348))}$

$F = FLASH\ POINT\ (^{\circ}C)$

EXAMPLE 11-7

Determine whether the following fuel oil blend will meet the 154° and 144° flash criteria:

STREAM	VOL%	SPECIFIC GRAVITY	API	MB	FLASH, °C	FI	144° INDEX	154° INDEX
VACUUM RESID	63.53	1.0185	7.43	0.361	250.0	0.40	0.2332	0.1518
FCC CUTTER	25.20	0.9348	19.87	0.772	89.0	20.6	0.3258	0.1829
KEROSENE	3.12	0.7901	47.59	1.687	41.0	20.6	0.7120	0.3996
LT. DIESEL	8.15	0.8428	36.39	1.317	90.0	20.0	0.5647	0.3192
BLEND	100.00						1.8357	1.0535

As the 144 and 154 indices are positive for this blend, the blend meets both 144 and 154 flash specifications.

REID VAPOR PRESSURE BLENDING FOR GASOLINES AND NAPHTHAS

Gasolines of different Reid vapor pressures (RVPs) do not blend linearly. For accurately estimating the RVP of the blends, RVP blend indices are used. These are presented in Table 11-18.

EXAMPLE 11-8

Calculate the RVP of a blend of *n*-butane, alkylate, and cat reformate with following properties:

COMPONENT	VOLUME FRACTION	VAPOR PRESSURE (VP), kPa	VP BLEND INDEX (VPBI)	VOL*VPBI
n-BUTANE	0.02	355.8	138	2.76
ALKYLATE	0.45	29.7	6.19	2.79
REFORMATE	0.53	35.2	7.66	4.06
BLEND	1.00	42.1		9.6

Table 11-18
Vapor Pressure vs. RVP Index of Gasolines

VAPOR PRESSURE, kPa	RVP INDEX									
	0	1	2	3	4	5	6	7	8	9
0	0.00	0.09	0.21	0.35	0.51	0.67	0.84	1.02	1.20	1.40
10	1.59	1.79	2.00	2.21	2.42	2.64	2.86	3.09	3.32	3.55
20	3.79	4.02	4.26	4.51	4.75	5.00	5.25	5.51	5.77	6.02
30	6.28	6.55	6.81	7.08	7.35	7.62	7.89	8.17	8.44	8.72
40	9.00	9.29	9.57	9.86	10.14	10.43	10.72	11.01	11.31	11.60
50	11.90	12.20	12.50	12.80	13.10	13.41	13.71	14.02	14.33	14.64
60	14.95	15.26	15.57	15.89	16.20	16.52	16.84	17.16	17.48	17.80
70	18.12	18.45	18.77	19.10	19.43	19.76	20.09	20.42	20.75	21.08
80	21.42	21.75	22.09	22.42	22.76	23.10	23.44	23.78	24.12	24.47
90	24.81	25.16	25.50	25.85	26.20	26.55	26.90	27.25	27.60	27.95
100	28.30	28.66	29.01	29.37	29.73	30.08	30.44	30.80	31.16	31.52
110	31.89	32.25	32.61	32.98	33.34	33.71	34.07	34.44	34.81	35.18
120	35.55	35.92	36.29	36.66	37.04	37.41	37.78	38.16	38.54	38.91
130	39.29	39.67	40.05	40.43	40.81	41.19	41.57	41.95	42.34	42.72
140	43.10	43.49	43.87	44.26	44.65	45.04	45.43	45.81	46.20	46.59
150	46.99	47.38	47.77	48.16	48.56	48.95	49.35	49.74	50.14	50.54
160	50.93	51.33	51.73	52.13	52.53	52.93	53.33	53.73	54.14	54.54
170	54.94	55.35	55.75	56.16	56.56	56.97	57.38	57.79	58.19	58.60
170	54.94	55.35	55.75	56.16	56.56	56.97	57.38	57.79	58.19	58.60
190	63.14	63.55	63.97	64.39	64.80	65.22	65.64	66.06	66.48	66.90
200	67.32	67.74	68.16	68.58	69.01	69.43	69.85	70.28	70.70	71.13

RVP INDEX OF LPG GASES FOR GASOLINE BLENDING:

COMPONENT	VAPOR PRESSURE, kPa	RVP INDEX*
PROPANE	1310	705.42
i-BUTANE	497.8	210.46
n-BUTANE	355.8	138.31

*RVP INDEX $(VP/6.8947)^{1.25}$

Given the RVP of the blend components the vapor pressure blend index for individual components is read from the RVP vs. RVP indices table.

The RVP index for the blend is next estimated by linear blending the component RVP indices. Thus, a blend index of 9.6 corresponds to a RVP of 42.1 kPa for this blend.

ANILINE POINT BLENDING

The aniline point is of a gas oil indicative of the aromatic content of the gas oil. The aromatic hydrocarbon exhibits the lowest and paraffins the highest values. Aniline point (AP) blending is not linear, and therefore blending indices are used.

The following function converts aniline point to the aniline index:

$$\text{ANLIND} = 1.25^{*}\text{AP} + 0.0025^{*}(\text{AP})^{2}$$

where ANLIND is the aniline index and AP is the aniline point.

The aniline index can be converted back into the aniline point by the following function:

$$\text{AP} = 200 \times \left(\text{SQRT}\left(1.5625 + \frac{\text{ANLIND}}{100}\right) - 1.25\right)$$

EXAMPLE 11-9

Determine the aniline point and diesel index of the following gas oils blend:

BLEND COMPONENTS	VOL%	SPECIFIC GRAVITY	ANILINE POINT, °F
LIGHT DIESEL	0.5000	0.844	159.8
KEROSENE	0.2000	0.787	141.3
L. CYCLE GAS OIL	0.3000	0.852	98.3
BLEND	1.0000	0.835	138.57

The aniline point of the blend is determined by first estimating the aniline blend index of each component, then blending them volumetrically. The diesel index is determined as function of aniline point and API gravity:

BLEND COMPONENTS	VOL%	SPECIFIC GRAVITY	ANILINE POINT, °F	A. P. INDEX	VOL*API
LIGHT DIESEL	0.5000	0.844	159.8	263.59	131.795
KEROSENE	0.2000	0.787	141.3	226.54	45.307
L. CYCLE GAS OIL	0.3000	0.852	98.3	147.03	44.109
BLEND	1.0000	0.835	138.57		221.215

$$\begin{aligned}\text{Diesel Index} &= (\text{API GRAVITY} \times \text{ANILINE POINT})/100 \\ &= 37.96 \times 138.57/100 \\ &= 52.60\end{aligned}$$

CRUDE OIL ASSAYS

A frequent problem encountered in process studies is the determination of the yield and properties of crude oil cuts. The usual crude assay data provide information on limited number of cuts, and in general, the desired information on specific cuts has to be obtained by some form of interpolation or approximation. Also, the property of a wide cut may not be the same as the combined properties of two adjacent narrow cuts spanning the same cut points as the wide cut. The same criticism applies to most nomographs (property vs. start and end of cut and also property vs. mid-volume percent or middle boiling point) presented in many crude assays; that is, they are not property conservative.

Typically, the crude TBP between 100°F and vacuum resid end point is cut into narrow cuts, each spanning 20°F. The entire boiling range of crude oil, excluding light ends, is therefore cut into 56 narrow cuts, which are defined by standard TBP boiling ranges. These are called *pseudocomponents*, since their boiling ranges are narrow, and for all intents and purposes, they act like pure components.

The properties of the narrow cuts are generated in such a way that, when reconstituted into the wide cuts, given as input data in the assay, the errors in the back-calculated properties are minimized in a least square sense. This procedure has the advantage of being property conservative; that is, the property of the wide cut is consistent with the properties of any combination of narrower cuts of the same composition. The whole crude is the widest possible cut and therefore any breakdown of it conserves mass, sulfur, and other extensive properties. Intensive properties such as viscosity and pour point, which in general do not blend linearly, are handled using linear blending indices, which are also conserved.

CRUDE CUTTING SCHEME

The first seven crude cuts are pure components. The crude assay light ends analysis provides the vol% of these constituent in the crude: (1) H_2S, (2) H_2, (3) C_1, (4) C_2, (5) C_3, (6) iC_4, (7) nC_4. The ($iC_5 + nC_5$) are, together, treated as the eighth cut, with a TBP end point of 100°F. The sum of light ends through nC_5 should be equal to the lights vol% distilled at 100°F. If the two numbers do not agree, the light ends yields are assumed to be correct and the TBP yields between 100 and 200°F are adjusted to provide a smooth transition between light ends and the reminder TBP curve.

The ninth cut has a TBP cut point of (100–120°F). Each successive cut has a TBP end point 20°F higher than the preceding component. The 63rd component has end point of 1200°F. The TBP end point of a crude is difficult to establish with any degree of certainty in the laboratory and therefore always is assumed to be 1292°F. The 64th (last) cut is assumed to have a TBP cut range of (1200–1292°F). Table 11-19 shows the TBP cuts of all pseudocomponents in the crude. This cutting scheme is not unique; any other convenient cutting scheme can be used, provided the cuts are narrow.

DETERMINATION OF NARROW CUT (PSEUDOCOMPONENT) PROPERTIES

The vol% of each narrow cut or pseudocomponent is known from TBP curve of the crude, once the cut points of pseudocomponents are defined. If their properties can be determined, these can be combined to determine the yield and properties of any other wide cut of that crude not available in the crude assay data. Also, narrow cut properties from different crudes can be combined to predict the wide cut yields and the properties from mixed crudes.

Narrow cut properties are determined from the wide cut properties data in the crude assay. Every crude assay provides data on the properties of different wide cuts (naphthas, kerosenes, diesels, vacuum gas oils, vacuum resides, etc.). The crude assay also provides data on the proper-

Table 11-19
Pseudocomponents Narrow Cut Points

CUT INDEX	CUT NAME	TBP TEMPERATURE END, °F
1	H_2S	−76.5
2	H_2	−422.9
3	C_1	−258.7
4	C_2	−127.5
5	C_3	−43.7
6	IC_4	10.9
7	NC_4	31.1
8	CUT NC4–100°F	100
9	CUT 100–120	120
10	CUT 120–140	140
11	CUT 140–160	160
12	CUT 160–180	180
13	CUT 180–200	200
14	CUT 200–220	220
15	CUT 220–240	240
16	CUT 240–260	260
17	CUT 260–280	280
18	CUT 280–300	300
19	CUT 300–320	320
20	CUT 320–340	340
21	CUT 340–360	360
22	CUT 360–380	380
23	CUT 380–400	400
24	CUT 400–420	420
25	CUT 420–440	440
26	CUT 440–460	460
27	CUT 460–480	480
28	CUT 480–500	500
29	CUT 500–520	520
30	CUT 520–540	540
31	CUT 540–560	560
32	CUT 560–580	580
33	CUT 580–600	600
34	CUT 600–620	620
35	CUT 620–640	640
36	CUT 640–660	660
37	CUT 660–680	680
38	CUT 680–700	700
39	CUT 700–720	720
40	CUT 720–740	740
41	CUT 740–760	760

Table 11-19
Continued

CUT INDEX	CUT NAME	TBP TEMPERATURE END, °F
42	CUT 760–780	780
43	CUT 780–800	800
44	CUT 800–820	820
45	CUT 820–840	840
46	CUT 840–860	860
47	CUT 860–880	880
48	CUT 880–900	900
49	CUT 900–920	920
50	CUT 920–940	940
51	CUT 940–960	960
52	CUT 960–980	980
53	CUT 980–1000	1000
54	CUT 1000–1020	1020
55	CUT 1020–1040	1040
56	CUT 1040–1060	1060
57	CUT 1060–1080	1080
58	CUT 1080–1100	1100
59	CUT 1100–1120	1120
60	CUT 1120–1140	1140
61	CUT 1140–1160	1160
62	CUT 1160–1180	1180
63	CUT 1180–1200	1200
64	CUT 1200–1292	1292

ties of whole crude (specific gravity, sulfur, RVP, salt content, etc.), and this is, in fact, the largest wide cut, spanning the 0–100% TBP distillation range. With the input of each wide cut from the crude assay, the properties are processed to make them blend volumetrically (i.e., sulfur is multiplied with specific gravity, viscosity is converted into viscosity blending index, and so forth). The composition of each wide cut is expressed in terms of volume of pseudocomponents such that the sum of volume fractions of the pseudocomponents in the wide cut add to 1.

Then, for each property, it is postulated that there is a set of property values for all pseudocomponents, so that adding the appropriate volume of pseudocomponent times its respective property value equals the property of each wide cut.

Thus, for each wide cut,

$$p_1 \times v_1 + p_2 \times v_2 + p_3 \times v_3 \ldots p_j \times v_j = P(k) + E(k) \qquad (11\text{-}5)$$

here

$v_1, v_2, v_3, \ldots, v_j$ = the volumetric fraction of the pseudocomponents 1, 2, 3, …, j contained in the wide cut. The sum of these, for each wide cut, is 1.

$p_1, p_2, p_3, \ldots, p_j$ = the property of pseudocomponents 1, 2, 3, …, j, respectively. For example, p_1, p_2, and p_3 could be the specific gravity, RON, or any other linearly blendable property. These are unknown and to be determined.

$P(k)$ = the property of wide cut k (known).

$E(k)$ = the error in calculation of wide cut property from assumed pseudocomponent property.

The property of both the wide cut and the pseudocomponents or narrow cuts must be in volumetric linear blending units. For example, the wide cut viscosity must be converted into corresponding viscosity blending index to determine the viscosity blending index of the pseudo-components spanning the wide cut.

If there are *m* wide cuts, for which the property in question is known, it is possible to write *m* simultaneous equations such as the preceding.

In general, *m* is less than *n*, the number of pseudocomponents, so that the system of equations cannot be solved. Some additional constraints are needed. These are provided by the following:

1. The $p_1, p_2, \ldots, p_j$ set of values (pseudocomponent properties) must be smooth and continuous as a function of volume.
2. The sum of square of errors should be minimum.

The solution is carried out with an iterative procedure. This will be demonstrated later with an example.

The *p*(*i*) values are initialized to an arbitrary starting point for the first iteration. A constant value, equivalent to the average of the properties of the wide cut data supplied, is used:

$$p(1) = p(2) = p(3) = \cdots = p(n) = \frac{(\text{sum from } k = 1 \text{ to } k = m \text{ of } P(k))}{m}$$

On subsequent iterations, of course, the latest values of $p(i)$ are used.

Equation (11-5) is solved for the error $E(k)$ in prediction of each wide cut property, using the assumed data of previous step:

$$E(k) = (\text{sum from } i = 1 \text{ to } j, p(i)^* v(i) - P(k))$$

The values of $v(i)$, the volume fraction of each pseudocomponent in the wide cut k has already been established by a TBP curve.

The set of $p(i)$ values are then adjusted by a composite set of correction terms in the direction opposite to error, in each equation, to bring the estimated wide cut property closer to the actual.

A volumetric average of all applicable error terms is used to correct the old property value of the pseudocomponent. In the averaging calculations, the volumetric weight of the pseudocomponent in a wide cut is calculated and multiplied with the wide cut error to arrive at the correction for that wide cut. Each wide cut error introduces a correction term to the pseudocomponent property.

After the first iteration, the set of $p(i)$ values (pseudocomponent properties) is discontinuous, with several groups of adjacent $p(i)$ having the same values and with discontinuities occurring at the boundaries of the wide cuts. The set of $p(i)$ values is passed through a smoothing procedure, and the smoothed set of $p(i)$ values is used to start the next iteration.

Each iteration introduces some irregularity in the set of $p(i)$ values while improving the overall fit. After each iteration, the smoothing procedure restores continuity with some loss of goodness of fit.

After each iteration, the standard deviation of the wide cut property error is computed:

$$\sigma = \sqrt{\frac{(E_{\text{WCUT1}})^2 + (E_{\text{WCUT2}})^2 + (E_{\text{WCUT3}})^2 + \cdots + (E_{\text{WCUT}m})^2}{m - 1}}$$

where

σ = standard deviation;

m = number of wide cuts in the input;

$E_{\text{WCUT}j}$ = error (input – calculated) of a property for the jth cut.

The standard deviation is generally seen to improve dramatically at first, approaching a final value asymptotically. The standard deviation is computed before smoothing. The later iterations, which apply very small corrections to the set of $p(i)$ values, have little effect on the smoothness of continuity of $p(i)$ values.

The whole procedure terminates when either the standard deviation increases by less than 1% between successive iterations or a preset number of iterations (usually 20) has been reached.

Smoothing Procedure

A problem with the method discussed so far is that, although the pseudocomponents blend back to the original wide cut input data better after the first iteration than with the first estimate values, many of the adjacent pseudocomponents still have identical values and there are sharp discontinuities where wide cut boundaries occur. The next step helps smooth out these sharp discontinuities.

For most pseudocomponents, four components are required to determine a new smoothed property for each pseudocomponent:

1. The pseudocomponent's unsmoothed property (already determined).
2. The linear interpolated property of the immediately adjacent pseudocomponents (to be determined).
3. The linear interpolated property of the adjacent but once removed pseudocomponents (to be determined).
4. The smoothed value is computed as equal to (40% of (1)+ 40% of (2) + 20% of (3)).

As an example consider the data in the following table after the first iteration and after the first smoothing.

PSEUDO-COMPONENT, °F	FIRST ITERATION	SMOOTHED VALUE
n	a	a
$n+1$	b	$0.5 \times b + 0.5 \times (a+c)/2$
$n+2$	c	$0.4 \times c + 0.4 \times (b+d)/2 + 0.2 \times (a+e)/2$
$n+3$	d	$0.4 \times d + 0.4 \times (c+e)/2 + 0.2 \times (b+f)/2$
$n+4$	e	$0.5 \times e + 0.5(d+f)/2$
$n+5$	f	f

We see that

1. The n and $n+5$ pseudocomponents properties are not smoothed, having adjacent components only on one side.
2. The $n+1$ and $n+4$ pseudocomponents are smoothed using 50% of (1) and 50% of (2). The unsmoothed specific gravities and middle vol% are used for interpolation calculation.
3. The $n+2$ and $n+3$ pseudocomponents properties are smoothed by adding 40% of the unsmoothed value, 40% of the adjacent value average, and 20% of adjacent but once removed value.

The smoothing procedure is also shown in Figure 11-2.

The process of cutting the crude oil into narrow cuts is best illustrated with an example. Suppose the following crude oil data were supplied. (For simplicity, only one property and a short range of TBP are used for this illustration. Actual crude oil cutting into narrow cuts requires the TBP distillation of whole crude.)

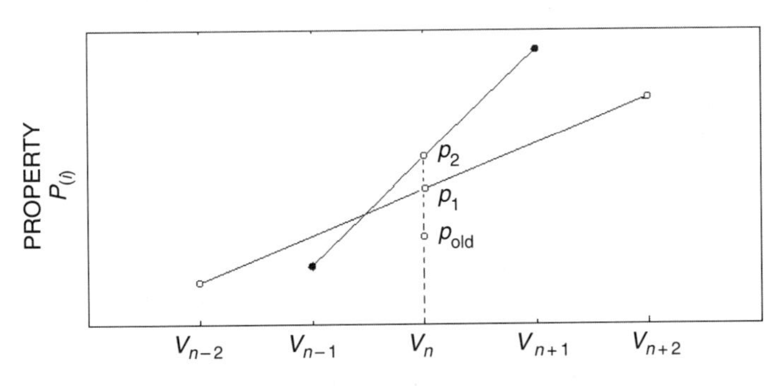

p_{old} = value of property nth cut (unsmoothed value)

p_1 = linear interpolation between V_{n-2} and V_{n+2}

p_2 = linear interpolation between V_{n-1} and V_{n+1}

p_{new} = value of property nth cut (smoothed value)

$$p_{new} = 0.4 \times p_{old} + 0.4 \times p_2 + 0.4 \times p_1$$

Figure 11-2. Smoothing procedure.

The given TBP distillation data are as follows:

TBP CUT, °F	VOL%
400–420	1.7400
420–440	1.8450
440–460	1.3050
460–480	2.6300
480–500	1.9150
500–520	1.9850
520–540	2.1200
540–560	2.0200
560–580	2.0250
580–600	2.0775

And, the given wide cut data are

TBP CUT POINTS, °F	SPECIFIC GRAVITY
400–500	0.8244
500–600	0.8562
400–600	0.8410

We need to determine the specific gravities of the pseudocomponents consistent with the wide cut specific gravities available from the assay data.

The first step is to assume an initial set of specific gravities for each of the pseudocomponents. These are the average specific gravities of the wide cuts input. In this case,

$$\text{Initial value} = (0.8244 + 0.8562 + 0.8410)/3 = 0.8405$$

Therefore, to begin with, all the pseudocomponents have a specific gravity of 0.8405.

The next step is to calculate the vol% of each pseudocomponent in the wide cut; for this example, these are

PSEUDO-COMPONENTS, °F	CRUDE	WIDE CUT, VOL.%		
		400–500°F	500–600°F	400–600°F
400–420	1.7400	18.44		8.85
420–440	1.8450	19.55		9.38
440–460	1.3050	13.83		6.64
460–480	2.6300	27.87		13.38
480–500	1.9150	20.30		9.74
500–520	1.9850		19.41	10.10
520–540	2.1200		20.73	10.78
540–560	2.0200		19.75	10.27
560–580	2.0250		19.80	10.30
580–600	2.0775		20.31	10.57
TOTAL	19.6625	100.00	100.00	100.00

Next the pseudocomponents are blended to calculate the specific gravities of each wide cut. This is easy for the first iteration. For the 400–500°F cut, this equals the initial assumption for each pseudocomponent's specific gravity (SG):

$$\begin{aligned} SG_{400-500} &= 18.44/100 \times 0.8405 + 19.55/100 \times 0.8405 \\ &\quad + 13.83/100 \times 0.8405 + 27.87/100 \times 0.8405 \\ &\quad + 20.30/100 \times 0.8405 \\ &= 0.8405 \end{aligned}$$

Since all the pseudocomponents have the same specific gravity in this initial iteration, the specific gravities of other wide cuts are calculated to be 0.8405 as well.

Next, an error term is calculated for each wide cut:

$$\text{Error} = \text{estimated SG} - \text{input SG}$$

Therefore,

$$\begin{aligned} \text{Error}_{400-500} &= 0.8405 - 0.8244 \\ &= 0.0161 \\ \text{Error}_{500-600} &= 0.8405 - 0.8562 \\ &= -0.0157 \\ \text{Error}_{400-600} &= 0.8405 - 0.8410 \\ &= 0.0004 \end{aligned}$$

Next, a volumetric average of all applicable error terms is used to correct the old specific gravity (0.8405) of all pseudocomponents.

For the 400–420°F pseudocomponent, only the 400–500°F and 400–600°F error terms apply, since the 400–420°F pseudocomponent is not a part of the wide cut 500–600°F. In the averaging calculations, the weight of the 400–500°F error terms is

$$\text{Weighted error}_{400-500} = 18.44/(18.44 + 8.85) \\ = 0.6757$$

Similarly, the weight of the 400–600°F error terms is

$$\text{Weighted error}_{400-600} = 8.85/(18.44 + 8.85) \\ = 0.3243$$

The new specific gravity value of the 400–420°F pseudocomponent is

$$\text{New SG} = \text{old SG} - \text{weighted error } 400 - 500°\text{F cut} \\ - \text{weighted error } 400 - 600°\text{F cut}$$

Therefore,

$$\text{New SG} = 0.8405 - 0.6757 \times (0.0161) - 0.3243 \times (-0.0004) \\ = 0.8298$$

The same procedure is carried out for all other pseudocomponents.

PSEUDOCOMPONENT, °F	INITIAL ESTIMATE	FIRST ITERATION
400–420	0.8405	0.8298
420–440	0.8405	0.8298
440–460	0.8405	0.8298
460–480	0.8405	0.8298
480–500	0.8405	0.8298
500–520	0.8405	0.8510
520–540	0.8405	0.8510
540–560	0.8405	0.8510
560–580	0.8405	0.8510
580–600	0.8405	0.8510

After each iteration, the property of the wide cut is recalculated and the standard deviation of the wide cut property error is computed:

WIDE CUT, °F	INPUT SG	CALCULATED SG	ERROR
WIDE CUT 400–500	0.8244	0.8298	−0.0054
WIDE CUT 500–600	0.8562	0.8510	0.0052
WIDE CUT 400–600	0.8410	0.8408	0.0001

The standard deviation is 0.0053.

The smoothing procedure is carried out next, on the pseudocomponent properties computed after each iteration by the procedure described earlier:

PSEUDOCOMPONENT, °F	FIRST ITERATION	FIRST SMOOTHING
400–420	0.8298	0.8298
420–440	0.8298	0.8298
440–460	0.8298	0.8298
460–480	0.8298	0.8319
480–500	0.8298	0.8361
500–520	0.8510	0.8446
520–540	0.8510	0.8489
540–560	0.8510	0.8510
560–580	0.8510	0.8510
580–600	0.8510	0.8510

The second iteration is now started with smoothed property (SG) data of the first iteration:

TBP CUT, °F	VOL%	SG (SMOOTHED VALUES, FIRST ITERATION)
400–420	1.7400	0.8298
420–440	1.8450	0.8298
440–460	1.3050	0.8298
460–480	2.6300	0.8319
480–500	1.9150	0.8361
500–520	1.9850	0.8446
520–540	2.1200	0.8489
540–560	2.0200	0.8510
560–580	2.0250	0.8510
580–600	2.0775	0.8510

Wide cut SGs are calculated on the basis of this data; and the error, for each wide cut, is calculated:

WIDE CUT	ESTIMATED SG	INPUT SG	ERROR (SG)
WIDE CUT 400–500	0.8316	0.8244	0.0072
WIDE CUT 500–600	0.8493	0.8562	−0.0069
WIDE CUT 400–600	0.8408	0.8410	−0.0002

Wide cut errors are again distributed to pseudocomponents by the same procedure as in the first iteration. After second iteration, the data are again smoothed by the procedure described earlier. The results of second iteration and second smoothing are as follows:

PSEUDOCOMPONENT, °F	SECOND ITERATION	SECOND SMOOTHING
400–420	0.8249	0.8249
420–440	0.8249	0.8249
440–460	0.8249	0.8260
460–480	0.8270	0.8295
480–500	0.8313	0.8356
500–520	0.8492	0.8449
520–540	0.8535	0.8510
540–560	0.8556	0.8545
560–580	0.8556	0.8556
580–600	0.8556	0.8556

This procedure is continued until the standard deviation of the error terms, calculated after each iteration, ceases to get significantly smaller. The process is repeated for all properties of the wide cuts in the crude assay data. The following properties are generally determined for each of the narrow cuts from wide cut data:

Specific gravity.
Sulfur, wt %.
Mercaptan sulfur, ppmw.
RON clear.

MON clear.
Aromatics.
Naphthenes.
Refractive index.
Cetane index.
Pour point .
Cloud point.
Freeze point.
Aniline point.
Conradson carbon.
Nitrogen content, ppm.
Vanadium, ppm.
Nickel, ppm.
Iron, ppm.
Viscosity (blending index), at 122°F, in cst.
Viscosity (blending index), at 210°F, in cst.

In the preceding explanation, it is assumed that all the pseudocomponents are computed. This is rarely the case. Either some of the lighter components do not exist or the property is not applicable over the whole boiling range. When the latter happens, the procedure is identical but only the $p(i)$ within the range are computed, with the others set to 0. This ensures that the property values are not extrapolated beyond the range but only interpolated.

Wide cut input data must be in linearly blendable units. For example,

- Sulfur, mercaptants, nitrogen, vanadium, nickel, iron, and the like (in weight units) must be multiplied by their specific gravities.
- All viscosities, pour points, and cloud points must be converted to corresponding blend indices.

The cutting of crude TBP into narrow cuts, as just shown, is best done with a computer program. Software from many vendors is available. There may be differences in the TBP cutting scheme or smoothing procedure, but the basic logic of these programs is broadly similar (Figure 11-3).

Crude assay data are cut into narrow cuts and the narrow cut data are stored in the form of a crude library. Such data can readily be used for diverse purposes, such as estimating the properties of user-specified wide cuts from single or mixed crudes, spiking crude with light ends, or blending crude with condensate.

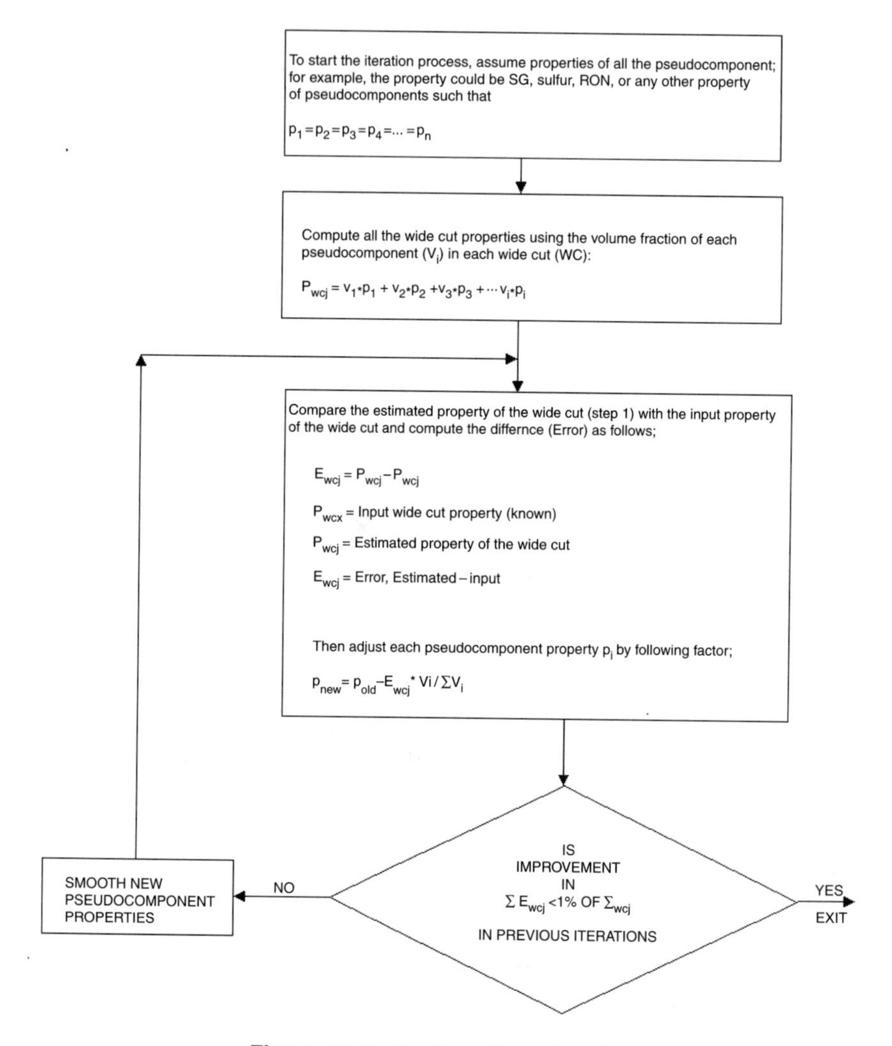

Figure 11-3. Crude cutting algorithm.

A successful run of the procedure calculates a good fit of narrow cut properties, which blend with little error back to original wide cuts properties entered. Generally, the more wide cuts entered with data, the better is the prediction of narrow cut properties. When only a small number of wide cut data points are available, the prediction of the original or different wide cut properties is much poorer.

Problems with procedure occur when

- The assay has too few wide cuts (usually less than six).
- A large stretch of the crude's TBP is described by only one wide cut (for example, an atmospheric residuum cut as opposed to one or two vacuum gas oil cuts and a vacuum residuum cut).
- A property is given for less than three adjacent or overlapping cuts. The smoothing algorithm may distort a property if too few data points are given, one or two data points are for very wide cuts, or there is a major change in the value of a property between adjacent cuts. This can occur, for example, for nitrogen content or metals (ppm), which may change by two orders of magnitude between adjacent cuts. Because of smoothing algorithm, the narrow cuts may show no orderly increase with respect to increasing TBP. Instead, they may oscillate and even go negative while maintaining average wide cut input data. In this case, pseudocomponent properties showing negative values could be manually changed to zeros at the risk of compromising the fit of pseudocomponent properties with respect to wide cuts input data.
- If the properties of different cuts with overlapping TBPs are given, there may be conflicting data, since the data values for each property would be average for respective wide cut and not the instantaneous values.

PROPERTIES OF THE WIDE CUTS

The yield and properties required of a wide cut is simply the volumetric linear blend of the appropriate narrow cuts. In other words, the yield on the crude of a wide cut is the sum of yields on the crude of the narrow cuts that constitute the wide cut. The properties of the wide cut are the volumetric weighted average properties of the same narrow cuts. If the TBPs at the start and end of the cut do not coincide with the narrow cuts, portions of narrow cuts at the start and end are used. These portions are calculated basis a simple ratio of the narrow cut TBP that is part of the wide cut to the whole TBP of the narrow cut.

EXAMPLE 11-10

Calculate the specific gravity and sulfur of diesel cut 500–650°F TBP from the following pseudocomponent data:

PSEUDOCOMPONENT, °F	CRUDE, vol%	SG	SULFUR, w%
500–520	1.9850	0.8448	1.3090
520–540	2.1200	0.8503	1.4540
540–560	2.0200	0.8561	1.5780
560–580	2.0250	0.8618	1.6820
580–600	2.0780	0.8679	1.7720
600–620	2.0680	0.8745	1.8510
620–640	1.9650	0.8806	1.9200
640–660	1.9650	0.8863	1.9780

As the required end point of diesel is 650°F, the properties of the 640–650°F pseudocomponent are estimated from the 640–660°F pseudocomponent as follows:

$$\text{Boiling range } 640-650°\text{F pseudocomponent} = 10°\text{F}$$

$$\text{Boiling range } 640-660°\text{F pseudocomponent} = 20°\text{F}$$

$$\text{Ratio of } (640-650) \text{ cut to } (640-660) \text{ cut} = 0.5$$

$$\text{Vol\% } (640-650) \text{ cut} = 0.5 \times 1.965 = 0.9825$$

By linear interpolation,

$$\text{Specific gravity of } (640-650) \text{ cut} = 0.8806 + 0.5 \times (0.8863 - 0.0.8806) = 0.8835$$

$$\text{Sulfur (wt \%)} = 1.9200 + 0.5 \times (1.9780 - 1.9200) = 1.949$$

PSEUDO-COMPONENT, °F	CRUDE, VOL%	SG	SULFUR, Wt %	VOL× SG	SG× SULFUR	(SG×SULFUR) × VOLUME
500–520	1.9850	0.8448	1.3090	1.6769	1.1058	2.1951
520–540	2.1200	0.8503	1.4540	1.8026	1.2363	2.6210
540–560	2.0200	0.8561	1.5780	1.7293	1.3509	2.7289
560–580	2.0250	0.8618	1.6820	1.7451	1.4495	2.9353
580–600	2.0780	0.8679	1.7720	1.8034	1.5379	3.1958
600–620	2.0680	0.8745	1.8510	1.8084	1.6187	3.3475
620–640	1.9650	0.8806	1.9200	1.7303	1.6908	3.3223
640–650	0.9825	0.8835	1.9490	0.8680	1.7218	1.6917
TOTAL	15.2435			13.1644		22.0376

$$
\begin{aligned}
\text{SG of wide cut} &= \Sigma(\text{SG} \times \text{VOL})/\text{VOL} \\
&= 13.1644/15.2435 \\
&= 0.8636 \\
\text{SG} \times \text{sulfur of wide cut} &= \Sigma(\text{SG} \times \text{SULFUR} \times \text{VOL})/\Sigma\text{VOL} \\
&= 22.0376/15.2435 \\
&= 1.4457 \\
\text{Sulfur (wt\%)} &= \text{SG} \times \text{sulfur}/\text{SG} \\
&= 1.4457/0.8636 \\
&= 1.6740
\end{aligned}
$$

Many other properties of the wide cuts are estimated from already determined properties, using correlations, as follows.

API gravity is thus determined:

$$
{}^{\circ}\text{API} = 141.5/\text{SG} - 131.5
$$

where SG is specific gravity.

ASTM DISTILLATION FROM TBP DISTILLATION

ASTM distillation of the wide cut is obtained by conversion of the TBP to ASTM (D-86) by the method used in the API data book.[5] As the TBP of the wide cut is known from input data, the ASTM distillation of the wide cut can be determined.

ASTM Specification D-86 Distillation 50 and 99% Temperatures from TBP Temperatures (°F)

Calculate factors A and B as follows:

$$
A = \text{TBP}_{50} - \text{TBP}_{10}
$$

$$
B = \text{TBP}_{90} - \text{TBP}_{50}
$$

Calculate the 50% ASTM boiling point as follows:

1. Calculate delta (D):

$$D = 0 \text{ (if A.LT. 50)}$$

$$D = 0.735 \times A - 3.765 \text{ (if 50 GE A BUT LT 220)}$$

$$D = 0.027 \times A + 6.555 \text{ (if A.GE. 220)}$$

2. Calculate the uncorrected D-86 50% boiling point:

$$\text{ASTM}_{50\%} = \text{TBP}_{50\%} - D$$

Calculate the 99% ASTM boiling point as follows:

1. Calculate delta (D):

$$D = 0.00398 \times B^2 - 0.556 \times B \text{ (if B.LT. 90)}$$

$$D = -74.914 + 12.672 \times \ln(B) \text{ (if B.GE. 90)}$$

2. Calculate the uncorrected D-86

$$\text{ASTM}_{99\%} = \text{TBP}_{99\%} - D$$

Now correct the D-86 points using stem correction, if temperature greater than 274°F. (Note the iterative calculation, since stem correction is based on corrected D-86 points.)

$$T_{50\%\text{corrected}} = T_{50\%\text{uncorrected}} - \text{stem correction}_{50\%}$$

$$T_{99\%\text{corrected}} = T_{99\%\text{uncorrected}} - \text{stem correction}_{99\%}$$

$$\text{Stem correction}_{50\%} = 5.04 + 0.00006751 \times T_{50\%\text{corrected}}$$

$$\text{Stem correction}_{99\%} = 5.04 + 0.00006751 \times T_{99\%\text{corrected}}$$

"K" WATSON AND MOLECULAR WEIGHT FROM THE ASTM DISTILLATION TEMPERATURE

1. From the ASTM distillation temperature, calculate the (uncorrected) average boiling point (V) of the petroleum fraction:

$$V = (T_{10\%} + T_{30\%} + T_{50\%} + T_{70\%} + T_{90\%})/5$$

2. Calculate the slope (S) of the 10–90% ASTM distillation curve:

$$S = (T_{90\%} - T_{10\%})/(\mathrm{VOL}_{90\%} - \mathrm{VOL}_{10\%})$$

3. Correct the average boiling point by the following correction factor (C):

$$C = -2.34813 - 5.52467 \times S + 0.01239 \times V - 0.87567 \times S \times S - 1.33817\mathrm{E}{-05} \times \mathrm{V} \times \mathrm{V} + 4.39032\mathrm{E}{-03} \times \mathrm{V} \times S - 3.29273\mathrm{E}{-02} \times S \times S \times S + 3.553\mathrm{E}{-07} \times S \times S \times V \times V$$

4. Add the correction factor to the boiling point in (1) and convert the temperatures in °F to temperature in °R (Rankine).

$$T = V + C + 460$$

5. Then calculate the Watson K as follows;

$$\mathrm{K} = T^{(\frac{1}{3})}/\mathrm{density}$$

6. Molecular weight (MW):

$$\mathrm{MW} = 100 \exp(0.001 \times A \times T + C)$$

where

$A = 2.386 - 0.710 \times D$
$C = 0.546 - 0.460 \times D$
D = density of the cut.
T = average boiling point in °F, calculated from the ASTM boiling point.

EXAMPLE 11-11

ASTM distillation of a heavy straight-run (HSR) naphtha with an API gravity of 61.33 follow. Calculate the Watson K and molecular weight of the naphtha.

VOL%	TEMPERATURE, °F
IBP	209
10	214
30	230
50	250
70	269
90	288
END POINT	297

Referring to this procedure,

$$V = 250.2$$

$$S = (288 - 214)/(80)$$

$$= 0.925$$

$$C = -4.93637$$

$$T = 245.26°\text{F}$$

$$= 705.26°\text{R}$$

$$\text{API} = 61.33$$

$$\text{Density} = 0.7338$$

$$\text{K} = (705.26)^{(\frac{1}{3})}/0.7338$$

$$= 12.13$$

Calculate the molecular weight:

$$A = 1.864997$$

$$C = 0.208449$$

$$\text{Molecular weight} = 114.6$$

CHARACTERIZATION FACTOR FROM API GRAVITY AND KINEMATIC VISCOSITY AT 210°F (FOR VISCOSITIES GREATER THAN 3.639 CST AT 210°F)

Calculate factor 1 (FCTR1), a function of viscosity:

$$\text{FCTR1} = B^2 - 4 \times \text{C} \times [\text{A} - \ln(\text{viscosity})]$$

where

$$A = 384.5815;$$

$$B = -74.2124;$$

$$C = 3.592245;$$

$$S = 0.082;$$

$$\text{FCTR2} = ((\text{FCTR1})^{0.5} - B)/2 \times C;$$

$$\text{Watson K} = (\text{FCTR2}) + S \times (\text{API} - 10).$$

EXAMPLE 11-12

Calculate the Watson K for an atmospheric resid with a viscosity of 26.44 cst at 210°F and an API gravity of 16.23.

Using the constants just described,

$$\text{FCTR1} = 26.49309$$

$$\text{FCTR1} = 11.0725$$

$$\text{Watson K} = 11.5833$$

MOLECULAR WEIGHT OF NARROW CUTS WITH TEMPERATURES GREATER THAN 500°F

Molecular weight is calculated as a function of API gravity and the Watson K by the following correlation:[6]

$$MW = 51.38 \times API - 1017^*WK + 0.2845 \times (API)^2 + 55.73 \times (WK)^2 - 0.001039 \times (API)^3 - 3357 \times \exp(-WK/15) - 6.606 \times (WK) \times (API) + 6633.855$$

where

MW = molecular weight;
API = API (specific gravity);
WK = Watson K.

EXAMPLE 11-13

Calculate the molecular weight of the atmospheric resid in the preceding example. The API gravity of the resid is 16.23.

The Watson K was computed earlier at 11.583. The molecular weight is estimated by the previous correlation at 442.69

DETERMINATION OF THE (LOWER) HEATING VALUE OF PETROLEUM FRACTIONS

1. If the API gravity of distillate is above 42°, the net heat (H) of combustion in Btu/lb can be estimated by following correlation:[7]

$$H = 8505.4 + 846.81 \times (K) + 114.92 \times (API) + 0.12186 \times (API)^2 - 9.951 \times (K) \times (API)$$

2. If the API gravity of distillate or residua is below 42°, the net heat (H) of combustion in Btu/lb can be estimated by following correlation:[8]

$$H = 17213 + 5.22 \times API - 0.5058 \times (API)^2$$

3. Correct the preceding heating values for sulfur content of the distillate or residua:

$$\text{Heat} = H \times (1 - 0.01 \times S) + 43.7 \times S$$

where

Heat = lower heating value (Btu/lb);
K = Watson K;
API = API gravity;
S = Sulfur content (wt%).

VAPOR PRESSURE OF NARROW HYDROCARBON CUTS

Vapor pressure of narrow hydrocarbon cuts can be estimated from their boiling point and temperature by correlation of Van Kranen and Van Ness, as follows.

Determine factor x:

$$I = 6.07918 - 3.19837 \times \frac{(232.0 + B)}{(232 + T)} \times \left[\frac{(1120 - T)}{(1120 - B)}\right]$$

$$VP_{mmHg} = 10^x$$

$$VP_{psi} = 10^x \times 0.01932$$

where

B = average boiling point, °C.
T = temperature, °C
VP_{mmHg} = vapor pressure, mmHg;
VP_{psi} = vapor pressure, psi.

DETERMINATION OF RAMSBOTTOM CARBON RESIDUE[9] FROM CONRADSON CARBON[10]

The carbon residue value of burner fuel oil serves as a rough approximation of the tendency of the fuel to form deposits in vaporizing pot-type and sleeve-type burners. Provided additives are absent, the carbon residue of diesel fuel correlates approximately with combustion chamber deposits. Ramsbottom carbon can be estimated from Conradson carbon[11] by the following correlation:

$$\text{RAMSBTM} = \text{EXP}\left(\left(\frac{(((0.1314{*}\text{LN(CONCARBON)} + 0.8757)^2) - 0.76685)}{0.2628}\right) - 0.2141\right)$$

PENSKEY MARTIN FLASH[12] AS A FUNCTION OF IBP (TBP) AND CUT LENGTH (NOT FOR COMMERCIALLY FRACTIONATED CUTS)

Estimate the cut length (CUTL) as follows:

$$\text{CUTL} = T_{\text{end}} - T_{\text{start}}$$

where

T_{start} = initial boiling point (IBP), °F;
T_{end} = final boiling point (FBP), °F.

Calculate a factor X as follows:

$$X = 0.7129 \times (\text{IBP}) - 122.227$$

$$\text{Penesky Martin Flash (PMFL)} = X + (2.571 \times \text{CUTL} - 225.825)^{0.5}$$

FLASH POINT, °F (ABEL), AS A FUNCTION OF AVERAGE BOILING POINT

The correlation is applicable up to 1050°F:

$$\text{Flash} = 0.537 \times T - 93 \text{ if } (T \leqslant 174°\text{F})$$
$$\text{Flash} = 0.67 \times T - 116 \text{ if } (174 \leqslant T < 303)$$
$$\text{Flash} = 0.7506 \times T - 140.3 \text{ if } (303 \leqslant T < 577)$$
$$\text{Flash} = 0.674 \times T - 94.5 \text{ if } (577 \leqslant T < 1050)$$

where T is the average boiling temperature, °F.

SMOKE POINT OF KEROSENES

The smoke point (S) of a kerosene can be estimated by the following correlation with the aromatic content of the stream.

$$S = \left(53.7/\text{arom}^{0.5}\right) + 0.03401 \times \text{API}^{1.5} + 1.0806$$

where

$$S = \text{Smoke point, mm;}$$
$$\text{arom} = \text{aromatics content, LV\%;}$$
$$\text{API} = \text{API gravity of the cut.}$$

LUMINOMETER NUMBER

The luminometer number is qualitatively related to the potential radiant heat transfer from the combustion products of aviation turbine fuels. It provides an indication of the relative radiation emitted by combustion products of gas turbine fuels.

The luminometer number (L_n) is related to the smoke point by following correlation (ASTM specification D-1740):

$$L_n = -12.03 + 3.009 \times \text{SP} - 0.0104 \times \text{SP}^2$$
$$\text{SP} = 4.16 + 0.331 \times L_n + 0.000648 \times {L_n}^2$$

where L_n is the luminometer number and SP is the smoke point.

LUMINOMETER NUMBER BLENDING

Luminometer numbers do not blend linearly. Therefore, blending indices are employed that blend linearly by weight. The procedure is to first determine component Luminometer blending indices, blend component indices on weight basis to obtain blended index, next calculate stream Luminometer number from following relations.

The blending index is

$$L_i = \exp(\exp[(139.5 - L_n)/128.448])$$

where

L_i = luminometer number blending index;

L_n = luminometer number;

$$L_n = 139.35 - 128.448 \times \log_e(\log_e L_i)$$

CETANE INDEX

The cetane index (CI) is estimated from the API gravity and 50% ASTM distillation temperature in °F of the diesel, by the following correlation.[13]

$$CI = -420.34 + 0.016 \times (API)^2 + 0.192 \times (API) \times \log(M) + 65.01(\log M)^2 - 0.0001809 \times M^2$$

or

$$CI = 454.74 - 1641.416 \times D + 774.74 \times D^2 - 0.554 \times B + 97.803 \times (\log B)^2$$

where

API = API gravity;
M = mid-boiling temperature, °F;
D = density at 15°C;
B = mid-boiling temperature, °C.

DIESEL INDEX

The diesel index (DI) is correlated with the aniline point of diesel by the following relationship:

$$DI = \text{aniline point} \times \text{API gravity}/100$$

The diesel index can be approximately related to cetane index of diesel by following relation:

$$DI = (CI - 21.3)/0.566$$

(U.S.) BUREAU OF MINES CORRELATION INDEX (BMCI)

$$BMCI = 87{,}552/(VAPB) + 473.7 \times (SG) - 456.8$$

where VABP is the volume average boiling point, °Rankine, and SG is the specific gravity. The BMCI has been correlated with many characteristics of a feed, such as crackability, in-steam cracking, and the paraffinic nature of a petroleum fraction. Paraffinic compounds have low BMCIs and aromatics high BMCIs.

AROMATICITY FACTOR

The aromaticity factor is related to the boiling point, specific gravity, aniline point, and sulfur as follows:

$$AF = 0.2514 + 0.00065 \times VABP + 0.0086 \times S + 0.00605 \times AP + 0.00257 \times (AP/SG)$$

where

AF = aromaticity factor;
$VAPB$ = volume average boiling point, °F;
S = sulfur, wt%;
AP = aniline point, °F;
SG = specific gravity.

FLUIDITY OF RESIDUAL FUEL OILS

The low temperature flow properties of a waxy fuel oil[14] depend on its handling and storage conditions. These properties may not be truly indicated by the pour point. The pour point of residual fuel oils are influenced by the previous thermal history of thc oil. A loosely knit wax structure built up on cooling of oil can normally be broken by the application of relatively little pressure. The usefulness of pour point test in relation to residual fuel oils is open to question, and the tendency to regard pour point as the limiting temperature at which a fuel oil will flow can be misleading.

In addition, the pour point test does not indicate what happens when an oil has a considerable head of pressure behind it, such as when gravitating from a storage tank or being pumped from a pipeline. Failure to flow at the pour point is normally attributed to a separation of wax from the fuel. However, it can also be due to viscosity in case of very viscous fuel oils.

Fluidity Test

The problem of accurately specifying the handling behavior of fuel oils is important. Because of technical limitation of the pour point test, various pumpability tests[15] have been devised. However, most of these test methods tend to be time consuming. A method that is relatively quick and easy to perform has found limited acceptance, as the "go/no go" method is based on an appendix to the former ASTM specification D-1659-65. This method covers the determination of fluidity of a residual fuel oil at a specified temperature in an "as-received" condition.

Summary of the Test Method

A sample of fuel oil in its as-received condition is cooled at the specified temperature for 30 min in a standard U-tube and tested for movement under prescribed pressure conditions.

The sample is considered fluid at the temperature of the test if it will flow 2 mm under a maximum pressure of 152 mmHg. The U-tube for the test is 12.5 mm in diameter.

This method may be used as a go/no go procedure for operational situations where it is necessary to ascertain the fluidity of a residual oil under prescribed conditions in an as-received condition.

The conditions of the method simulate those of a pumping situation, where oil is expected to flow through a 12 mm pipe under slight pressure at specified temperature.

Fluidity, like the pour point of specification D-97, is used to define cold flow properties. It differs from D-97 in that

1. It is restricted to fuel oils.
2. A prescribed pressure is applied to the sample. This represents an attempt to overcome the technical limitation of pour point method, where gravity-induced flow is the criterion.

The pumpability test (ASTM specification D-3245) represents another method for predicting field performance in cold-flow conditions. The pumpability test, however, has limitations and may not be suitable for very waxy fuel oils, which solidify so rapidly in the chilling bath that no reading can be obtained under conditions of the test. It is also time consuming and therefore not suitable for routine control testing.

Fluidity Blending

Fluidity, like pour points, do not blend linearly. Blending indices are therefore used for blending the fluidity of two fuel oils. The blending indices used are the same as those used for distillate fuels (Table 11-14).

EXAMPLE 11-14

Determine the fluidity of a blend of two fuel oil grades. The blend contains 30 vol% grade 964 and 70 vol% grade 965. As per fluidity tests, grade 964 is fluid at 32°F and grade 965 is fluid at 59°F.

The fluidity of the blend is determined by first determining the fluidity index of the two grades and blending linearly. The fluidity index is then converted into a fluidity temperature, as follows. The fluidity index can be read directly from Table 11-14 or determined from the following correlations:

$$\text{Fluidity index} = 3162000 \times [(\text{fluidity}, {}^\circ\text{R})/1000]^{12.5}$$

$$\text{Fluidity } ({}^\circ\text{R}) = 1000 \times [(\text{fluidity index})/3162000]^{0.08}$$

GRADE	VOL%	FLUIDITY TEMPERATURE, °R	FLUIDITY INDEX	BLEND
964	30.0	492	446.20	133.86
965	70.0	519	870.07	609.07
BLEND				742.91

The corresponding fluidity of the blend is 512.5°R or 52.5°F.

CONVERSION OF KINEMATIC VISCOSITY TO SAYBOLT UNIVERSAL VISCOSITY OR SAYBOLT FUROL VISCOSITY

Saybolt universal viscosity is the efflux time in seconds for a 60-cm^3 sample to flow through a standard orifice in the bottom of a tube. The orifice and tube geometry are specified in the ASTM standards. Saybolt furol viscosity is determined in the same manner as Saybolt universal viscosity except that a larger orifice is used.

Conversion to Saybolt Universal Viscosity

Kinematic viscosity in centistokes (mm^2/sec) can be converted to Saybolt universal viscosity in Saybolt universal seconds (SUS) units at the same temperature by API databook[16] procedure or using ASTM conversion tables:[17]

$$\text{SUS} = 4.6324 \times (\text{cst}) + \frac{[1.0 + 0.03264 \times (\text{cst})]}{[3930.2 + 262.7 \times (\text{cst}) + 23.97 \times (\text{cst})^2 + 1.646 \times (\text{cst})^3] \times 10^{-5}} \quad (11\text{-}6)$$

where

SUS = Saybolt universal viscosity in Saybolt universal seconds units at 37.8°C;

cst = kinematic viscosity, centistokes (mm^2/sec).

SUS viscosity at 37.8°C can be converted to SUS viscosity at another temperature by following relationship:

$$\text{SUS}_t = [1 + 0.000110 \times (t - 37.8)] \times \text{SUS} \quad (11\text{-}7)$$

where

SUS = SUS viscosity at 37.8°C;

SUS_t = SUS viscosity at the required temperature;

t = temperature at which SUS viscosity is required, °C.

EXAMPLE 11-15

An oil has a kinematic viscosity of 60 centistokes at 37.8°C. Calculate the corresponding Saybolt universal viscosity at 37.8 and 98.9°C.

Using equation (11-6), the Saybolt universal viscosity at 37.8°C is calculated at 278.59 sec.

The viscosity at 98.9°C is calculated, by equation (11-7), at 280.45 sec.

Conversion to Saybolt Furol Viscosity

Kinematic viscosity in centistokes (mm^2/sec) at 50 and 98.9°C can be converted to Saybolt furol viscosity in Saybolt furol seconds (SFS) units at the same temperature by API databook[18] procedure:

$$SFS_{50} = 0.4717 \times VIS_{50} + \frac{13.924}{(VIS_{50})^2 - 72.59 \times VIS_{50} + 6.816} \quad (11\text{-}8)$$

$$SFS_{98.9} = 0.4792 \times VIS_{98.9} + \frac{5.610}{(VIS_{98.9})^2 + 2.130} \quad (11\text{-}9)$$

where

$$\begin{aligned} SFS_{50} &= \text{Saybolt furol viscosity at } 50°\text{C, Saybolt furol seconds;} \\ VIS_{50} &= \text{kinematic viscosity at } 50°\text{C, centistokes } (mm^2/sec); \\ SFS_{98.9} &= \text{Saybolt furol viscosity at } 98.9°\text{C, Saybolt furol seconds;} \\ VIS_{98.9} &= \text{kinematic viscosity at } 98.9°\text{C, centistokes } (mm^2/sec). \end{aligned}$$

EXAMPLE 11-16

1. Estimate the Saybolt furol viscosity of an oil at 50°C, if the kinematic viscosity at 50°C is 3000 centistokes. Saybolt furol viscosity at 50°C is estimated from kinematic viscosity by equation (11-8) at 1415.1 sec.
2. Estimate the Saybolt furol viscosity of an oil at 98.9°C, if the kinematic viscosity at 98.9°C is 120 centistokes.

   Saybolt furol viscosity at 98.9°C is estimated from kinematic viscosity by equation (11-9) above at 57.50 sec.

REFRACTIVE INDEX OF PETROLEUM FRACTIONS

The refractive index of a petroleum fraction can be predicted from its mean average boiling point, molecular weight, and relative density using the API databook procedure.[19]

The method may be used to predict refractive index for a petroleum fraction with normal boiling point up to 1100 Kelvins.

$$n = \left(\frac{1 + 2 \times I}{1 - I}\right)^{\frac{1}{2}} \tag{11-10}$$

$$I = 3.587 \times 10^{-3} \times T_b^{1.0848} \times \left(\frac{M}{d}\right)^{-0.4439} \tag{11-11}$$

where

n = refractive index at 20°C;

I = characterization factor of Huang[20] at 20°C;

T_b = mean average boiling point, Kelvins;

M = molecular weight of petroleum fraction;

d = liquid density at 20°C and 101.3 kPa, in kilogram per cubic decimeter (kg/dm^3).

EXAMPLE 11-17

Calculate the refractive index of a petroleum fraction with a liquid density of 0.7893 kg/dm^3, molecular weight of 163.47, and mean average boiling point of 471.2 K.

Here,

$d = 0.7893$;

$M = 163.47$;

$T_b = 471.2\,K$.

First calculate Huang characterization factor by equation (11-11):

$$I = 3.587 \times 10^{-3} \times (471.2)^{1.0848} \times (163.47/0.7893)^{-0.4439}$$
$$= 0.2670$$

Next calculate the refractive index by equation (11-10):

$$n = [(1 + 2 \times 0.2670)/(1 - 0.2670)]^{0.5}$$
$$= 1.4466$$

DETERMINATION OF MOLECULAR-TYPE COMPOSITION

An estimate of fractional composition of paraffins, naphthenes, and aromatics contained in light and heavy petroleum fractions can be done if the data on viscosity, relative density, and refractive index of the desired fraction are available. The algorithm is based on a procedure of API databook.[21]

$$X_p = a + b \times (R_i) + c \times (\mathrm{VG}) \tag{11-12}$$

$$X_n = d + e \times (R_i) + f \times (\mathrm{VG}) \tag{11-13}$$

$$X_a = g + h \times (R_i) + I \times (\mathrm{VG}) \tag{11-14}$$

where

X_p = mole fraction of paraffins;

X_n = mole fraction of naphthenes;

X_a = mole fraction of aromatics;

R_i = refractivity intercept as given by equation (11-15);

a, b, c, d, e, f, g, h, and i = constants varying with molecular weight range as given later;

VG = viscosity gravity function VGC, as given by equations (11-16) and (11-17) for heavy fractions, or viscosity gravity function VGF, as given by equations (11-18) and (11-19) for light fractions.

$$R_i = n - d/2 \tag{11-15}$$

where n is the refractive index at 20°C and 101.3 kPa and d is the liquid density at 20°C and 101.3 kPa in kg/dm^3.

The constants used in equations (11-12) to (11-14) are as follows:

CONSTANT	LIGHT FRACTION (MW = 80–200)	HEAVY FRACTION (MW = 200–500)
a	−23.940	−9.000
b	24.210	12.530
c	−1.092	−4.228
d	41.140	18.660
e	−39.430	−19.900
f	0.627	2.973
g	−16.200	8.660
h	15.220	7.370
i	0.465	1.255

For heavy fractions (molecular weight 200–500), the viscosity gravity constant is determined by following equations:

$$\mathrm{VGC} = \frac{10 \times d - 1.0752 \times \log(V_{311} - 38)}{10 - \log(V_{311} - 38)} \tag{11-16}$$

or

$$\mathrm{VCG} = \frac{d - 0.24 - 0.022 \times \log(V_{372} - 35.5)}{0.755} \tag{11-17}$$

where d is the relative density at 15°C and 101.3 kPa and V is the Saybolt universal viscosity at 311 or 372 K in Saybolt universal seconds.

For light fractions (molecular weight 80–200), the viscosity gravity constant is determined by following equations:

$$\mathrm{VGF} = -1.816 + 3.484 \times d - 0.1156 \times \ln(v_{311}) \tag{11-18}$$

or

$$\mathrm{VGF} = -1.948 + 3.535 \times d - 0.1613 \times \ln(v_{372}) \tag{11-19}$$

where v is the kinematic viscosity at 311 or 372 K, in mm^2/sec.

The viscosity-gravity (VG) constant[22] is a useful function for approximate characterization of viscous fractions of petroleum. It is relatively insensitive to molecular weight and related to the composition of the fraction. Values of VG near 0.8 indicate samples of paraffinic character, while those close to 1.0 indicate a preponderance of aromatic molecules. The VG, however, should not be applied to residual oils or asphaltic materials.

EXAMPLE 11-18

Calculate the molecular type distribution of a petroleum fraction of relative density 0.9433, refractive index 1.5231, normal boiling point 748 K, and a viscosity of 695 Saybolt universal seconds at 311 K.

First, determine the molecular weight of the fraction as a function of the mean average boiling point and relative density at 15°C by API databook procedure.[23]

$$\mathrm{MW} = 2.1905 \times 10^2 \times \exp(0.003924 \times T) \times \exp(-3.07 \times d) \times (T)^{0.118} \times (d)^{1.88}$$

where MW is the molecular weight, d is the relative density at 15°C, and T is the mean average boiling point in K. Therefore,

$$\begin{aligned}\mathrm{MW} &= 2.1905 \times 10^2 \times \exp(0.003924 \times 748) \\ &\quad \times \exp(-3.07 \times 0.9433) \times (748)^{0.118} \times (0.9433)^{1.88} \\ &= 446\end{aligned}$$

As the molecular weight is greater than 200, the fraction is termed *heavy*. Viscosity gravity constant VGC is determined as per equation (11-16):

$$\begin{aligned}\mathrm{VGC} &= \frac{10 \times 0.9433 - 1.0752\ \log(695 - 38)}{[10 - \log(695 - 38)]} \\ &= 0.8916\end{aligned}$$

Next, calculate the refractivity index as per equation (11-15):

$$\begin{aligned}R_i &= 1.5231 - 0.9433/2 \\ &= 1.0514\end{aligned}$$

The mole fraction of paraffins, naphthenes, and aromatics are next determined using equations (11-12) to (11-14) and constants for heavy fractions:

$$X_p = -9.00 + 12.53 \times 1.0514 - 4.228 \times 0.8916$$
$$= 0.4051$$
$$X_n = 18.66 - 19.90 \times 1.0514 + 2.973 \times 0.8916$$
$$= 0.3869$$
$$X_a = -8.66 + 7.37 \times 1.0514 + 1.225 \times 0.8916$$
$$= 0.2080$$

EXAMPLE 11-19

Calculate the molecular-type distribution of a petroleum fraction of relative density 0.8055, refractive index 1.4550, normal boiling point 476 K, and a viscosity of 1.291 mm^2/sec at 311 K.

First, determine the molecular weight of the fraction as a function of the mean average boiling point and relative density at 15°C:

$$\text{MW} = 2.1905 \times 10^2 \times \exp(0.003924 \times 476)$$
$$\times \exp(-3.07 \times 0.8055) \times (476)^{0.118} \times (0.8055)^{1.88}$$
$$= 164.73$$

As the molecular weight is less than 200, the fraction is termed *light*. Viscosity gravity constant VGF is determined as per equation (11-18) or (11-19):

$$\text{VGF} = -1.816 + 3.484 \times 0.8055 - 0.1156 \times \ln(1.291)$$
$$= 0.9608$$

Next, calculate the refractivity index as per equation (11-15):

$$R_i = 1.4550 - 0.8055/2$$
$$= 1.0522$$

The mole fraction of paraffins, naphthenes, and aromatics are next determined using equations (11-12) to (11-14) and constants for light fractions:

$$X_p = -23.94 + 24.21 \times 1.0522 - 1.092 \times 0.9608$$
$$= 0.4858$$
$$X_n = 41.14 - 39.43 \times 1.0522 + 0.627 \times 0.9608$$
$$= 0.2521$$
$$X_a = -16.2 + 15.22 \times 1.0522 + 0.465 \times 0.9608$$
$$= 0.2621$$

DETERMINATION OF VISCOSITY FROM VISCOSITY/ TEMPERATURE DATA AT TWO POINTS

Kinematic viscosity/temperature charts[24] are a convenient way to determine the viscosity of a petroleum oil or liquid hydrocarbon at any temperature within a limited range, provided viscosities at two temperatures are known. The procedure is to plot two known kinematic viscosity/ temperature points on the chart and draw a straight line through them. A point on this line within the range defined shows the kinematic viscosity at the corresponding desired temperature and vice versa.

The kinematic viscosity of a petroleum fraction can also be expressed as a linear function of temperature by following equations. These equations agree closely with the chart scales. They are necessary when calculations involve viscosities smaller than 2 cst:

$$\log \log Z = A - B \log T \tag{11-20}$$

$$Z = v + 0.7 + \exp(-1.47 - 1.84 \times v - 0.51 \times v^2) \tag{11-21}$$

$$v = (Z - 0.7) - \exp[-0.7487 - 3.295 \times (Z - 0.7) + 0.6119 \times (Z - 0.7)^2 - 0.3193 \times (Z - 0.7)^3] \tag{11-22}$$

where

v = kinematic viscosity, cst $(\mathrm{mm}^2/\mathrm{sec})$;
T = Temperature, K or °R;
A and B = constants.

Constants A and B for a petroleum fraction can be calculated when the temperature and corresponding viscosity are available for two points, as follows. Writing equation (11-20) for the two temperature/viscosity pairs gives the following:

$$\log \log Z_1 = A - B \log T_1 \tag{11-23}$$

$$\log \log Z_2 = A - B \log T_2 \tag{11-24}$$

where

$$Z_1 = v_1 + 0.7 + \exp(-1.47 - 1.84 \times v_1 - 0.51 \times v_1^2) \tag{11-25}$$

$$Z_2 = v_2 + 0.7 + \exp(-1.47 - 1.84 \times v_2 - 0.51 \times v_2^2) \tag{11-26}$$

v_1 and v_2 are the kinematic viscosities at temperatures t_1 and t_2, in K or °R, respectively.

Let

$$k_1 = \exp(-1.47 - 1.84 \times v_1 - 0.51 v_1^2) \tag{11-27}$$

$$k_2 = \exp(-1.47 - 1.84 \times v_2 - 0.51 v_2^2) \tag{11-28}$$

$$y_1 = \log Z_1 = \log(v_1 + 0.7 + k_1) \tag{11-29}$$

$$y_2 = \log Z_2 = \log(v_2 + 0.7 + k_2) \tag{11-30}$$

Thus equations (11-23) and (11-24) can be rewritten as follows:

$$\log y_1 = A - B \log T_1 \tag{11-31}$$

$$\log y_2 = A - B \log T_2 \tag{11-32}$$

Substituting the value of A from (11-31) in equation (11-32) yields

$$B = \frac{\log\left(\dfrac{Z_1}{Z_2}\right)}{\log\left(\dfrac{T_2}{T_1}\right)} \tag{11-33}$$

$$A = \log(Z_1) + B + \log(T_1) \tag{11-34}$$

With the values of A and B known, viscosity at any other temperature in K or °R can be determined from equations (11-20), (11-21), and (11-22).

EXAMPLE 11-20

A kerosene stream of Dubai crude has a kinematic viscosity of 1.12 cst at 323 K and a viscosity of 0.70 cst at 371.9 K. Determine the kinematic viscosity of the stream at 311 K. Here,

$$\nu_1 = 1.12\,\text{cst}$$
$$\nu_2 = 0.70\,\text{cst}$$

First, calculate factors k_1 and k_2:

$$k_1 = \exp[-1.47 - 1.84 \times (1.12) - 0.51 \times (1.12)^2] = 0.0154$$

$$k_2 = \exp[-1.47 - 1.84 \times (0.70) - 0.51 \times (0.70)^2] = 0.0494$$

Next, calculate Z_1 and Z_2:

$$Z_1 = \log(1.17 + 0.7 + 0.0154) = 0.2637$$

$$Z_2 = \log(0.7 + 0.7 + 0.0494) = 0.1612$$

Calculate constants A and B for the petroleum fraction:

$$B = \frac{\log\left(\dfrac{0.2637}{0.1612}\right)}{\log\left(\dfrac{371.9}{323}\right)} = 3.4942$$

$$A = \log(0.2637) + 3.4942 \times \log(323) = 8.1889$$

The viscosity at 311 K can now be determined as follows:

$$\log\log Z_{311} = 8.1889 - 3.4942 \times \log(311) = -0.5214$$

$$Z_{311} = 2.0001$$

The corresponding viscosity at 311 K can be determined by equation (11-22):

$$\begin{aligned} \nu_{311} &= (2.001 - 0.7) - \exp[-0.7478 - 3.295 \times (2.001 - 0.7) \\ &\quad + 0.6119 \times (2.001 - 0.7)^2 - 0.3193 \times (2.001 - 0.7)^3] \\ &= 1.2910\,\text{cst} \end{aligned}$$

NOTES

1. Healy, Maason, and Peterson. "A New Approach to Blending Octane." API midyear meeting, May 1959.
2. William E. Morris. "Interaction Approach to Gasoline Blending." National Petroleum Refiners Association 73rd Meeting, March 23–25, 1975.
3. G. I. Jenkin. *Institute of Petroleum Journal* 54 (1968), pp. 80–85.
4. R. O. Wickey and D. M. Chittenden. "Flash Point of Blends Correlated." *Petroleum Refiner* 42, no. 6 (June 1953), pp. 157–158.
5. Ronald P. Danner and Thomas E. Danbert. "Technical Data Book Petroleum Refining." American Petroleum Institute, API technical databook (Petroleum Refining), 1977, Washington, DC.
6. Ibid.
7. Ibid.
8. ASTM equation.
9. ASTM specification D-524.
10. ASTM specification D-189.
11. ASTM specification D-524, under nonmandatory information, presents a graph of Ramsbottom carbon as a function of Conradson carbon.
12. ASTM specification D-93.
13. ASTM test method specification D-976–80.
14. ASTM specification D-97 (supplement; nonmandatory information).
15. ASTM specification D-3245.
16. API databook procedure 11A1.1.
17. ASTM specification D-2161.
18. API databook procedure 11A1.4.
19. API databook procedure 2B5.1.
20. P. K. Huang. "Characterization and Thermodynamic Correlations for Undefined Hydrocarbon Mixtures." Ph.D thesis, Department of Chemical Engineering, Pennsylvania State University, 1977.

21. API databook procedure 2B4.1. See also, M. R. Raizi and T. E. Daubert. "Prediction of Composition of Petroleum Fractions." *Industrial Engineering and Chemistry, Process Design and Development* 16 (1980), p. 289.
22. ASTM specification D-2501.
23. API databook procedure 2B2.1.
24. ASTM specification D-341.

CHAPTER TWELVE

Refinery Stock Balancing

Before the advent of linear programming (LP) models, process-planning studies were done by hand with desktop calculators and usually large printed or duplicated worksheets. Little optimization was possible via trial and error, as this would involve calculating a stock balance over and over again until a satisfactory answer was arrived at.

Refinery LP models now do stock balancing. Many LP packages are available that facilitate plant yield calculations and optimize product blending.

However, stock balancing must be done by hand at times. A refinery operation planner may take an LP-optimized stock balance and redo it by hand, taking into account the conditions in the refinery that cannot be conveniently incorporated into the LP model,[1] for example, a critical pump-out of service, partially coked-up furnace, catalyst bed with high pressure drop or low activity, a delayed ship causing severe ullage constraints, or a change of specifications can upset the best-laid plans.

LP models are price driven and cannot handle nonlinear blending. LP models sometimes give complicated solutions to simple problems, which often need to be compromised for practical reasons. Also, LP solutions may require large number of changes to the model to realize small real benefits and tend to overoptimize, unless they are very sophisticated. For these reasons, they are not considered a good tool for producing a practical plan for the refinery operations.

Long-term process planning studies may also be done by hand when no LP model of the refinery in question is available; and putting together an LP model and testing it takes more time than a simple hand balance. Hand balancing is done on a personal computer (PC) with a spreadsheet program. The spreadsheet simulates a typical refinery flow diagram. Each box on the spreadsheet corresponds to a refinery unit. Each unit is represented by a performance equation that relates the output of the unit to change in the input or its operating conditions. The equations need not be linear.

DATA FOR MODEL BUILDING

Much of the data required for building a spreadsheet program are the same as required for building an LP model. As a matter of fact, a refinery's spreadsheet program and the matrix of an LP model have much in common. Both the models require data on the unit's possible operating modes, minimum and maximum capacities, operating factor, yields, stream qualities, and product specifications. Possible sources of these data are discussed next.

OPERATING MODES AND YIELDS

This information is available from refinery's stock balancing manual. This information is developed from crude oil assay and refinery test runs on the units. If no information is available, distillation yields can be estimated from crude assay and ASTM distillation of the cuts.

The process yields of secondary units such as cat reformers, FCCU, visbreakers, and hydrocracker units are available from the latest refinery test runs or the process licensor data. From whatever source the yield data is obtained, the feed composition and operating severity of the unit has to be decided on before a good estimate can be made. Therefore, for example, for a cat reformer, the feed PONA (paraffin, olefin, napthlene, and aromatic content of a feed) must be known and the severity has to be decided on before the unit yield can be estimated.

STREAM QUALITIES

Stream qualities, such as density, sulfur, octane number, smoke point, and pour point, can be obtained from the same source, such as crude assay data or results of the latest test runs on different units. To minimize the stock balancing calculations, experience and engineering judgment are required to decide which qualities would be most restrictive and control the stock balance. For example, if the diesel end point from a given crude is determined to meet the pour point specifications, the sulfur specification may not be a problem and need not be calculated. Often, a stock balance has to be calculated several times. The effort of laying out the calculations and including all necessary yields and stream qualities in a spreadsheet can save considerable time.

PRODUCT SPECIFICATIONS

All streams from different processing units are blended to produce saleable finished products at certain specifications. The major product groups are naphtha, gasoline, kerosene, diesel, and fuel oil. However, each product group may have a large number of product grades to meet the requirements of the product in different regions of the world. For example, a refinery may produce 10 or more grades of diesel with different pour points, sulfur, cetane indices, and the like to meet its client requirements, with different climatic conditions or different environmental regulations in force. The quality of the crude and processing unit capability decide the specifications a refinery can economically produce for each product group, to meet market demand. Information on the product grades a refinery can produce and sell are published in the form of product specification book, which is constantly updated.

UNIT CAPACITIES AND OPERATING FACTOR

All refinery units have a maximum and minimum operating capacity in terms of throughput in barrels per stream day. These data are available from the previous test run reports of the unit. However, the unit may not be available for a given period because of scheduled and unscheduled maintenance work. All refineries maintain a maintenance schedule for at least 1 year in advance. This schedule is constantly updated. Therefore, a unit operating factor can be worked out for every processing unit to estimate the available unit capacity in a given time period.

CALCULATION PROCEDURE

The objective of the calculations, otherwise known as problem statements, may have some control over the sequence of the steps. Typically, either the crude feed rate is known or the product requirements are given. For the latter case, the crude rate is estimated by totaling the product volume requirements. Next, to process the given crude rate, the various units capacity utilization are determined. Product blending calculations can be made once the blending volumes from various units are available. A good run ensures that the available unit capacities of all important units, such as distillation and key conversion units, are fully or nearly fully utilized. In the product blending part, there should be no unnecessary quality giveway. For example, if a fuel oil specification demands a product with 400 centistoke viscosity, any blend viscosity less than say 390 centistoke would constitute giveaway on viscosity and unnecessary loss of cutter stock, which could have been utilized for blending a higher-valued product.

If many different product grades are to be made, there are many ways to simplify the calculations. The different grades of same group (for example, all grades of fuel oils) can be pooled and pool specifications calculated, if product requirements are given. Stock balancing calculations may be carried out to determine what crude rate and downstream secondary unit feed rate will do the job. Conversely, if crude feed rate is known, only the balancing grade fuel oil production must be estimated.

The blending components of the pool must have diverse enough qualities to meet the demand for grades with extreme specifications. For example, if there is demand for equal volumes of two grades of gasoline at RON 90 and 100, a blend stock of RON 95 may satisfy the pool requirement for the two grades, assuming linear blending, but would be unsatisfactory for blending each of the individual grades. Although it could be used to blend RON 90 gasoline, there could be a lot of octane giveaway, and it could not used to blend RON 100 gasoline, without using another, much higher-octane blend stock.

As long as blending stocks are sufficiently diverse, blending individual grades may not even be required, depending on the problem statement; but if blends of individual product grades are required, these calculations should be done after pool specifications have been met.

Fixed blend "recipes" can be used for low-volume product grades. Ideally, this will decrease the unknowns down to one or two balancing grades for each product group.

Usually, one balancing grade is sufficient. Balancing grades tend to be those products that have the largest volume and are sold in the spot market. Any change that occurs in stock balance is absorbed on recalculation in the production of balancing grades only.

For example, if the fuel oil group has several grades with different viscosities and sulfur levels, blends of most of these grades can be fixed during the first calculation. The balancing grade may require one high-volume grade of cutter stock to meet viscosity plus another high-volume grade cutter to meet sulfur specs. Usually, one of these qualities controls the cutter requirement of each grade. Any changes to the volume of blend stocks available is reflected in these two grades. Each recalculation must include a recalculation of the volume of cutter stock required to meet controlling specification.

BLENDING MARGINS

Blending methods have always some level of uncertainty. It is necessary to incorporate a margin for error in critical specifications. The magnitude of this margin is decided on the basis of past experience. Some suggested blending margins used in actual practice follow. However, we emphasize that margins are, in fact, giveaways on quality and thus an economic penalty to refinery and it should be minimized. The magnitude of blending margins should be weighed against any economic penalty resulting from failure to meet a guaranteed specification.

QUALITY	BLENDING MARGIN
SPECIFIC GRAVITY	0.01
OCTANE NUMBER, RON/MON	1.0
VISCOSITY BLENDING INDEX	5.0 vol
SULFUR	0.05 Wt%
CETANE INDEX	2.0
POUR POINT INDEX	3.0
SMOKE POINT	2.0 mm
AROMATICS	0.50 vol%
REID VAPOR PRESSURE	3.5 kPa

REFINERY MATERIAL BALANCE SPREADSHEET PROGRAM

To run the program the following data in the spreadsheet are updated.

CRUDE AND VACUUM DISTILLATION UNITS

1. Time period or number of days in the month.
2. Crudes to be processed.
3. Total crude rate to each crude distillation unit, in thousands of barrels per day.
4. Operation mode of each crude and vacuum column.

5. Unit capacities available for each crude and vacuum column.
6. Disposition of atmospheric resids to various vacuum distillation columns.

The distribution of various crudes to crude distillation units (CDUs) and their operation mode is decided by the user; the spreadsheet program computes the flow rates and properties of various crude cuts on the basis of crude assays data and the unit test runs.

Disposition of atmospheric resids from CDUs to various vacuum distillation units (VDUs) is decided by the capacity of the VDU, its mode of operation, and sometimes the need to segregate certain feedstocks. For example, one VDU may be reserved to produce asphalt from certain heavy crude and another VDU may choose feedstocks to produce lubricating oil distillate only.

VACUUM RESID DISPOSITION

The disposition of vacuum resids is decided next. Vacuum resids from a VDU may have the following possible dispositions: to a visbreaker or other conversion unit, such as delayed coker, resid hydrocracker (H-oil etc.) or the asphalt converter; to fuel oil blending; or to inventory buildup for later processing or export.

Conversion units, such as resid hydrocracking, visbreaking, or asphalt converter, are filled up first, and the remaining stock goes to fuel oil blending or inventory buildup.

HEAVY DIESEL/HVGO DISPOSITION TO CONVERSION UNITS

Heavy-vacuum gas oils from vacuum distillation units and heavy diesels are pooled. Heavy-vacuum gas oil (HVGO) have the following possible dispositions: feed to the hydrocracker, feed to the fluid cat cracker (FCCU), use for fuel oil blending, or to inventory for later processing or export.

Conversion units are filled to capacity first. The operation mode of the processing unit is chosen by the user. The program computes the unit material balance and product streams qualities from the built in yield and quality data.

DISPOSITION OF STRAIGHT-RUN DIESELS AND LIGHT-CYCLE GAS OIL TO THE DIESEL DESULFURIZER

Material balance for the diesel desulfurizer is taken up next. The spreadsheet displays the volume and properties of various diesel streams from the CDU (light diesels), VDU (light-vacuum gas oil, LVGO), and FCCU (light-cycle gas oil, LCGO). Light cycle gas oil must be hydrotreated to send it to diesel pool because of product stability considerations.

The volume of the feedstream to the diesel desulfurizer is manually adjusted to fill the unit. The objective is to give priority to high-sulfur streams. A part of the LCGO from the FCCU is sent to diesel desulfurizer unit. The only other disposition for LCGO in fuel oil is as cutter, so there is every incentive to blend as much LCGO into diesel as possible. The primary purpose is to improve the stability of the LCGO rather than desulfurize it. The remaining capacity is utilized for desulfurizing straight-run diesel streams, starting with the highest-sulfur streams, until the unit is full.

DISPOSITION OF MEDIUM NAPHTHA TO THE PRETREATER/ CATALYTIC REFORMER UNIT

A cat reformer can have a number of medium naphtha feeds. Also, a unit may run on a number of different severities. The disposition of feed to different severites or modes must be decided before the unit material balance can be worked out.

FUEL OIL BLENDING

All available vacuum resids, visbroken resids, and atmospheric long resids are pooled to compute the available volumes and their properties. To these are added the available cutter stocks, such as light and heavy cycle oils and heavy cat naphtha from the FCCU. The resid and the cutter stock constitute the fuel oil pool. The program calculates the fuel pool volume and properties (viscosity, sulfur, Con carbon, etc.).

The volume and properties (specifications) of fixed fuel grades are known from the operating plan of the refinery for that month. These volumes and properties are pooled and deducted from the total fuel pool to arrive at the balancing grade fuel production and its qualities. The properties of the balancing grade (viscosity, sulfur, Con carbon, gravity) are adjusted by the addition of diesel oil to meet the specifications of the balancing grade fuel oil. The amount of diesel cutter is adjusted by trial and error until the properties of the balancing grade fuel oil are within its specification limits.

DIESEL BLENDING

All the remaining diesel blend streams, after feeding the diesel desulfurizer unit, and the desulfurized diesel stream from that unit are blended together to estimate the diesel pool volume and its properties. Next, fixed-grade diesel volumes and their properties are deducted from the pool to arrive at the balancing-grade diesel volume and its properties. The balancing-grade diesel pool properties are adjusted by the addition of kerosene until all the balancing-grade diesel properties (pour point, sulfur, diesel index, etc.) are within the limits required by the specifications of the balancing-grade diesel.

GASOLINE BLENDING

Gasoline blending is taken up next. Feed to the catalytic reformer is specified and so are the operation severities. The cat reformer material balance is computed by the program, on the basis of built-in yields of the cat reformer.

All the gasoline streams are pooled, and the average pool properties (RON, MON, Reid vapor pressure, specific gravity, etc.) determined. Next, fixed grades gasoline requirements are pooled and deducted from the gasoline pool to arrive at the balancing-grade gasoline production. If any property such as RON, MON, or Reid vapor pressure (RVP) of the balancing-grade gasoline fails to meet the specs, gasoline pool composition could be varied by changing the reformer severity or adjusting the butane or more volatile components of the blend.

NAPHTHA BLENDING

The only significant properties of naphtha blending are RVP and specific gravity (SG). Blending is done by adjusting the light straight-run, whole straight-run (WSR), and butane content of each grade to meet SG and RVP specs.

EXAMPLE 12-1

A refinery (Figure 12-1) has the following process units. The capacity of the major processing units indicated is nominal capacity, in barrels per stream day (bpsd):

PROCESS	NOMINAL CAPACITY
CRUDE DISTILLATION	260,000 bpsd
VACUUM DISTILLATION	115,000 bpsd
UNIFINER/CAT REFORMER	15,000 bpsd
DIESEL DESULFURIZER	20,000 bpsd
PARTIAL HYDROCRACKER	50,000 bpsd
FLUID CAT CRACKER	36,000 bpsd
POLYMER GASOLINE PLANT	2,400 bpsd
VISBREAKER	20,000 bpsd
KEROSENE TREATING	42,000 bpsd
HYDROGEN PLANT	27 mmscfd
SULFUR PLANT	150 (tons/day)

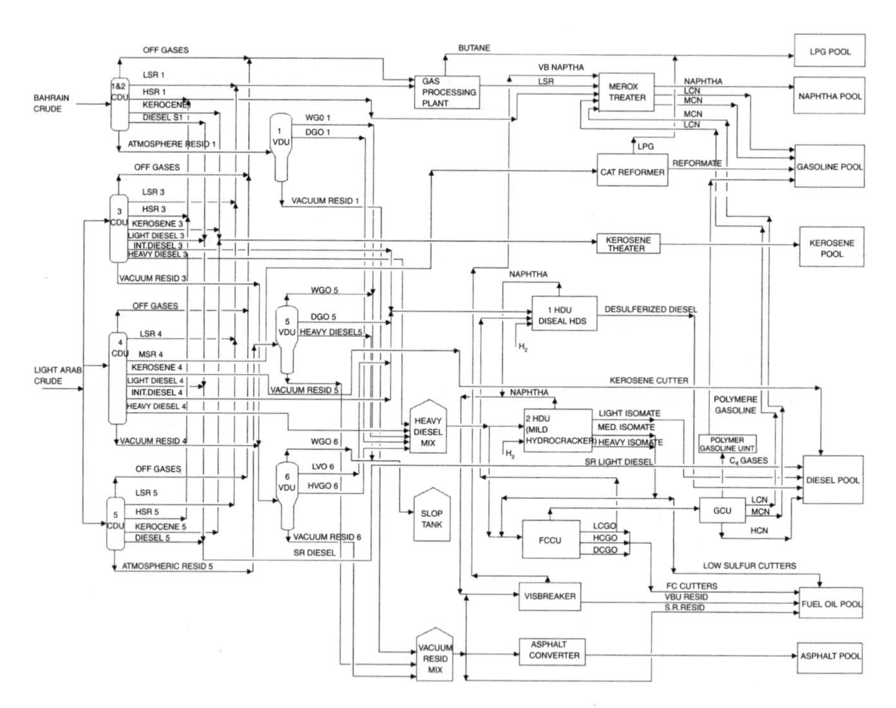

Figure 12-1. Refinery configuration for Example 12-1. LSR = light straight run; MSR = medium straight run; HSR = heavy straight run; INT. = intermediate; DGO = diesel gas oil; LVGO = light vacuum gas oil; HVGO = heavy vacuum gas oil; CDU = crude distillation unit; VDU = vacuum distillation unit; HDU = heavy diesel unit; LPG = liquefied petroleum gas; WGO = wet gas oil; VBU = visbreaker unit; LCGO = light cycle gas oil; HCGO = heavy cycle gas oil; GCU = gas concentration unit; LCN = light cat naphtha; MCN = medium cat naphtha; FC = FCCU cutters; DGCO = decant gas oil; HDS = hydrodesulfurization unit.

We want to process the following crudes during the month:

Light Arabian, 201 thousands bpsd
Bahrain, 42 thousands bpsd

The processing scheme of the refinery is shown in Figure 12-1. The maximum available unit capacities and estimated operating factor for the crude and other processing units, per month (30 days), are shown in Tables 12-1 to 12-3.

Bahrain crude is processed on crude units 1 and 2, and light Arabian crude is processed on crude units 3–5. Atmospheric resid is further distilled in vacuum distillation units 1, 5, and 6. A part of the vacuum resid is visbroken in the visbreaker unit. Both visbroken and straight-run vacuum resid are blended with FCCU cutters to fuel oil grades. Vacuum gas oils from vacuum distillation units are pooled and sent to the mild hydrocracker unit (with approximately 30% conversion) and FCCU. Unconverted, desulfurized vacuum gas oil (medium and heavy isomate) is used as feed to the FCCU or low-sulfur cutter stock for fuel oil.

We want to make an estimate of the product slate in barrels per month, assuming 30 days operation, unit capacity utilization, and the inventory changes required to sustain this operation.

The format of the spreadsheets is shown in Tables 12-1 to 12-35. Most of the data on unit yield and stream qualities for blending are built into the spreadsheet model and need not be revised for most routine estimates.

Table 12-1 lists data on the number of processing days and individual crudes processed. Tables 12-2 and 12-3 list the maximum unit capacities, operating factor, and available unit capacities. Tables 12-4 and 12-5 compute the overall yield of various products from crude units. Tables

Table 12-1
Crude Processed

CRUDE	mbpcd*	TOTAL mb**
ARABIAN	201.00	6030.00
BAHRAIN	42.00	1260.00
MURBAN	0.00	0.00
DUBAI	0.00	0.00
TOTAL	243.00	7290.00

*mbpcd = 1000 barrels per calender day.
**mb = 1000 of barrels.

Table 12-2
Crude Distillation Unit (CDU) Capacities

UNIT	NAME	CAPX, mbpcd	OPFACT	CAPACITY, mbl
CRUDE UNIT 1	CDU1	20.00	1.000	600.00
CRUDE UNIT 2	CDU2	20.00	1.000	600.00
CRUDE UNIT 3	CDU3	64.00	1.000	1920.00
CRUDE UNIT 4	CDU4	93.00	1.000	2790.00
CRUDE UNIT 5	CDU5	46.00	1.000	1380.00
TOTAL CDU		243.00		7290.00

CAPX = MAXIMUM CAPACITY
CAPACITY = AVAILABE CAPACITY
OPFACT = UNIT OPERATING FACTOR

Table 12-3
Other Processing Unit Capacities

UNIT	NAME	CAPX, mbpcd	OPFACT	CAPACITY, mb
VACUUM DISTILLATION UNIT 1	VC1	9.60	0.843	242.78
VACUUM DISTILLATION UNIT 5	VC5	33.00	0.843	834.57
VACUUM DISTILLATION UNIT 6	VC6	70.00	0.843	1770.30
KEROSENE TREATING UNIT	KTU	45.00	0.843	1138.05
VISBREAKER	VB	20.00	0.843	505.80
FLUID CAT CRACKER	FCCU	44.00	0.818	1079.38
DIESEL HYDRODESULFURIZER	HD1	22.00	0.843	556.38
PARTIAL HYDROCRACKER	HD2	52.00	0.843	1315.08
CAT REFORMER	CR	18.00	0.750	405.00

12-6 to 12-12 calculate the material balance of vacuum distillation units no 1, 5, and 6. Table 12-13 shows pooling of vacuum resids from various vacuum distillation unit and its disposition to visbreaker, asphalt converter and fuel oil blending. Tables 12-14 and 12-15 show material balance and stream qualities for asphalt converter and visbreaker unit. Table 12-16 shows the composite volumes of vacuum resid and their estimated properties for blending into fuel oil. Table 12-17 show all the blend components

Table 12-4
Crude Unit Yields

CDU NO.	FEED RATE, mbpcd	CRUDE	VOLUME, mb	BUTANE YIELD %VOL	mb	LSR YIELD	mb	MSR YIELD	mb	sg	KEROSENE YIELD	mb	LIGHT DIESEL YIELD	mb	MEDIUM DIESEL YIELD	mb	HEAVY DIESEL YIELD	mb	RESID YIELD	mb	LOSSES, % CRUDE	mb	TOTAL, %
1	0	A	0	1.7	0	7.9	0	10.50	0.00	0.7380	16.60	0.00	11.8	0.00	0.00	0.00	0.00	0.00	51.20	0.00	0.30	0.00	100.00
2	0	A	0	1.8	0	7.4	0	9.70	0.00	0.7346	20.00	0.00	10.3	0.00	0.00	0.00	0.00	0.00	50.50	0.00	0.30	0.00	100.00
3	62	A	1860	1.8	33.48	6.4	119.04	11.70	217.62	0.7331	16.90	314.34	21.9	407.34	0.00	0.00	1.30	24.18	39.70	738.42	0.30	5.58	100.00
4	93	A	2790	1.3	36.27	11.1	309.69	9.50	265.05	0.7458	17.50	488.25	7.4	206.46	12.70	354.33	12.90	359.91	27.30	761.67	0.30	8.37	100.00
5	46	A	1380	1.7	23.46	7.6	104.88	10.40	143.52	0.7356	16.00	220.80	13.5	186.30	0.00	0.00	0.00	0.00	50.50	696.90	0.30	4.14	100.00
1	20	B	600	0.6	3.6	5.1	30.6	10.50	63.00	0.7338	16.60	99.60	13.3	79.80	0.0	0.00	0.0	0.00	53.60	321.60	0.3	1.80	100.00
2	20	B	600	0.7	4.2	4.6	27.6	9.70	58.20	0.7310	20.00	120.00	11.8	70.80	0.0	0.00	0.0	0.00	52.90	317.40	0.3	1.80	100.00
3	2	B	60	0.7	0.42	3.9	2.34	11.40	6.84	0.7284	17.40	10.44	23.80	14.28	0.0	0.00	1.4	0.84	41.10	24.66	0.3	0.18	100.00
4	0	B	0	0.3	0	8.1	0	9.00	0.00	0.7460	18.60	0.00	8.00	0.00	14.1	0.00	13.9	0.00	27.70	0.00	0.3	0.00	100.00
5	0	B	0	0.7	0	4.8	0	10.10	0.00	0.7318	16.80	0.00	14.40	0.00	0.0	0.00	0.0	0.00	52.90	0.00	0.3	0.00	100.00
1	0	C	0	1.7	0	7.9	0	9.40	0.00	0.7380	17.70	0.00	11.8	0.00	0.0	0.00	0.0	0.00	51.20	0.00	0.3	0.00	100.00
2	0	C	0	1.8	0	7.4	0	9.60	0.00	0.7346	20.10	0.00	10.3	0.00	0.0	0.00	0.0	0.00	50.50	0.00	0.3	0.00	100.00
3	0	C	0	1.8	0	6.4	0	10.30	0.00	0.7331	18.30	0.00	21.90	0.00	0.0	0.00	1.3	0.00	39.70	0.00	0.3	0.00	100.00
4	0	C	0	1.3	0	11.1	0	7.50	0.00	0.7458	19.50	0.00	7.40	0.00	12.7	0.00	12.9	0.00	27.30	0.00	0.3	0.00	100.00
5	0	C	0	1.7	0	7.6	0	9.60	0.00	0.7356	16.80	0.00	13.50	0.00	0.0	0.00	0.0	0.00	50.50	0.00	0.3	0.00	100.00
1	0	D	0	0.6	0	5.1	0	9.40	0.00	0.7338	17.70	0.00	13.3	0.00	0.0	0.00	0.0	0.00	53.60	0.00	0.3	0.00	100.00
2	0	D	0	0.7	0	4.6	0	9.60	0.00	0.7310	20.10	0.00	11.8	0.00	0.0	0.00	0.0	0.00	52.90	0.00	0.3	0.00	100.00
3	0	D	0	0.7	0	3.9	0	10.00	0.00	0.7284	18.80	0.00	23.80	0.00	0.0	0.00	1.4	0.00	41.10	0.00	0.3	0.00	100.00
4	0	D	0	0.3	0	8.1	0	8.00	0.00	0.7460	19.60	0.00	8.00	0.00	14.1	0.00	13.9	0.00	27.70	0.00	0.3	0.00	100.00
5	0	D	0	0.7	0	4.8	0	9.50	0.00	0.7318	17.40	0.00	14.40	0.00	0.0	0.00	0.0	0.00	52.90	0.00	0.3	0.00	100.00
TOTAL	243		7290		101.43		594.15		754.23			1253.43		964.98		354.33		384.93		2860.65		21.87	

A = ARABIAN CRUDE.
B = BAHRAIN CRUDE.
C = MURBAN CRUDE.
D = DUBAI CRUDE.

Table 12-5
Crude Unit Overall Material Balance

	mbpcd	mb	VOL %
INPUT			
ARABIAN	201	6030	82.72%
BAHRAIN	42	1260	17.28%
MURBAN	0	0	0.00%
DUBAI	0	0	0.00%
TOTAL	243	7290	100.00%
OUTPUT			
BUTANE	3.4	101.4	1.39%
LSR	19.8	594.2	8.15%
MSR	25.1	754.2	10.35%
KEROSENE	41.8	1253.4	17.19%
LIGHT DIESEL	32.2	965.0	13.24%
MEDIUM/INTER. DIESEL	11.8	354.3	4.86%
HEAVY DIESEL	12.8	384.9	5.28%
ATM RESID	95.4	2860.7	39.24%
LOSS	0.7	21.9	0.30%
TOTAL	243.0	7290.0	100.00%

Table 12-6
No. 1 Vacuum Distillation Unit (VDU), ASPHALT MODE

	YIELD, LV	mb	SG	SULFUR, wt%	SG*S	VBI, H
FEED	1.0000	267.30				
WGO	0.0740	19.78	0.8585	1.171	1.0050	−29
DGO	0.6230	166.53	0.9174	2.691	2.4690	280
BSGO	0.0460	12.30	0.9918	4.260	4.2250	638
VACUUM RESID	0.2570	68.70	1.0350	4.925	5.0970	827
TOTAL	1.0000	267.30				

VBI = VISCOSITY BLENDING INDEX (VOLUME BASIS).
WGO = WET GAS OIL.
DGO = DISTILLATE GAS OIL.
BSGO = HVGO.

of fuel oil pool, their volumes, properties and also overall pool volume and properties. The production of fixed-grade volumes is known or given (Table 12-18) and the production of balancing-grade fuel oil (I-961) is computed by

Table 12-7
Atmospheric Resid Distribution to Vacuum Units

RESID	AVAILABLE, mb/mol	TO 1 VDU FUEL	TO 1 VDU ASPHALT*	TO 5VDU (33 mb/day)**	TO 5 VDU ASPHALT	TO 6VDU (65 mb/day)***	TO FCCU	TO FUEL	TO INV.
1A RESID	0.00	0.00	0.00	0.00	0.00	0.00	0.00	0.00	0.00
2A RESID	0.00	0.00	0.00	0.00	0.00	0.00	0.00	0.00	0.00
3A RESID	738.42	0.00	0.00	287.40	0.00	451.02	0.00	0.00	0.00
4A RESID	761.67	0.00	0.00	287.00	0.00	474.67	0.00	0.00	0.00
5A RESID	696.90	0.00	0.00	250.60	0.00	446.30	0.00	0.00	0.00
1B RESID	321.60	0.00	267.30	18.40	0.00	35.90	0.00	0.00	0.00
2B RESID	317.40	0.00	0.00	116.70	0.00	200.70	0.00	0.00	0.00
3B RESID	24.66	0.00	0.00	0.00	0.00	24.66	0.00	0.00	0.00
4B RESID	0.00	0.00	0.00	0.00	0.00	0.00	0.00	0.00	0.00
5B RESID	0.00	0.00	0.00	0.00	0.00	0.00	0.00	0.00	0.00
1VDU WGO	19.78	0.00	0.00	0.00	0.00	19.78	0.00	0.00	0.00
TOTAL, mb	2860.65	0.00	267.30	960.10	0.00	1633.25	0.00	0.00	0.00
mbpcd	95.36	0.00	8.91	32.00	0.00	54.44	0.00	0.00	0.00

UNIT CAPACITY:
**270*
***990*
****1950*

Table 12-8
No. 1 VDU, Fuel Oil Mode

	YIELD, LV	mb	SG	SULFUR, wt%	SG*S	VBI, H
FEED	1.0000	0.00				
WGO	0.1390	0.00	0.8713	1.795	1.5640	62
DGO	0.4790	0.00	0.9169	2.634	2.4151	281
BSGO	0.0000	0.00				
VACUUM RESID	0.3820	0.00	1.0247	4.722	4.8386	754
TOTAL	1.0000	0.00				

deducting from composite fuel oil pool (Table 12-17) the volumes and properties fixed-grade pool (Table 12-18). We see, however, that fuel oil thus produced does not meet the viscosity specification (180 cst, viscosity blend index = 480), so further cutting with diesel is done to reduce the VBI from 586 to 480, thus adding to fuel oil volume (Table 12-19).

Table 12-20 shows the pooling of all heavy diesels produced by crude or vacuum distillation units. Table 12-21 shows the disposition of these HVGO streams to processing units. Hydrocracker and cat cracker units are filled first, and anything left is either blended to fuel oil or sent to inventory for export or later use. Table 12-22 shows the material balance and product properties of a mild hydrocracker unit (2 HDU). Unconverted but desulfurized HVGO from mild hydrocracker, called *isomate*, is used as feed to the FCCU (Table 12-23), and any surplus isomate may be used as cutter to fuel oil. Light isomate, which is in fact desulfurized diesel, is sent to the diesel pool.

Tables 12-24 to 12-26 show yield from the FCCU and product properties. Light and medium cat naphtha are blended to gasoline, while heavy cat naphtha is routed to diesel. Light cycle gas oil is partly routed to diesel pool after hydrotreating in the diesel hydrotreating unit. All remaining LCO (light cycle oil), HCGO, and decant oil are used as cutter in fuel oil blending. Table 12-27 shows feed to the diesel desulfurizer unit. Knowing the available capacity of the unit enables computing the total feed to the unit. Light cycle gas oil from the FCCU is a feed that must be hydrotreated before it can be blended into diesel. A certain fraction of the unit capacity is used up for this stream. The rest of the unit capacity is used to desulfurize untreated diesel, starting with the stream of highest sulfur content.

Table 12-9
Yield from Vacuum Unit No. 5

	YIELD 1A	mb	YIELD 2&5A	mb	YIELD 3A	mb	YIELD 4A	mb	YIELD 1B	mb	YIELD 2&5B	mb	YIELD 3B	mb	YIELD 4B	mb	TOTAL
FEED	1.0000	0.00	1.0000	250.60	1.0000	287.40	1.0000	287.00	1.0000	18.40	1.0000	116.70	1.0000	0.00	1.0000	0.00	960.10
WGO	0.0060	0.00	0.0060	1.50	0.0080	2.30	0.0000	0.00	0.0060	0.11	0.0060	0.70	0.0000	0.00	0.0000	0.00	4.61
DGO	0.2680	0.00	0.2600	65.16	0.0760	21.84	0.0000	0.00	0.2800	5.15	0.2730	31.86	0.1010	0.00	0.0000	0.00	124.01
HVGO	0.4410	0.00	0.4430	111.02	0.5330	153.18	0.3960	113.65	0.4400	8.10	0.4420	51.58	0.5240	0.00	0.3940	0.00	437.53
VACUUM RESID	0.2850	0.00	0.2910	72.92	0.3830	110.07	0.6040	173.35	0.2740	5.04	0.2790	32.56	0.3750	0.00	0.6060	0.00	393.95
TOTAL	1.0000	0.00	1.0000	250.6000	1.0000	287.4000	1.0000	287.0000	1.0000	18.4000	1.0000	116.7000	1.0000	0.0000	1.0000	0.0000	960.1000

NOTES:
FEED 1A = REDUCED CRUDE FROM CDU 1 PROCESSING ARABIAN CRUDE.
FEED 2A = REDUCED CRUDE FROM CDU 2 PROCESSING ARABIAN CRUDE.
FEED 3A = REDUCED CRUDE FROM CDU 1 PROCESSING ARABIAN CRUDE.
FEED 1B = REDUCED CRUDE FROM CDU 1 PROCESSING BAHRAIN CRUDE.

Table 12-10
Properties of Vacuum Distillates from VDU 5

	mb	SG	SULFUR	PI	VBI	SG*S
FEED	960.10					
WGO	4.61	0.8096	0.37	76	−197	0.30
DGO	124.01	0.8740	1.96	588	82	1.71
HVGO	437.53	0.9437	2.97	2919	413	2.80
VACUUM RESID	393.95	1.0177	4.26		765	4.34
TOTAL	960.1	0.9644	3.36	1406	512	3.236

NOTES:
VBI = VISCOSITY BLENDING INDEX (VOLUME BASIS).
PI = POUR POINT BLENDING INDEX.

Table 12-28 shows certain special blends, such as marine diesel. These are generally blended to specific formulas based on previous shipments. Table 12-29 show fixed grades diesel blending. Table 12-30 shows the total blend components, their volumes and blending properties, and the average pool properties. After deducting the properties of the fixed and special grades, the remaining volume of the pool and its blending properties are estimated. Kerosene is blended into it to meet the sulfur or pour properties of the balancing-grade diesel, whichever is limiting. Tables 12-31 to 12-33 show yields from the cat reformer unit and gasoline blending from LCN, cat reformate, light straight-run naphtha, and so forth. Tables 12-35 and 12-36 show the production estimates for kerosene. Some kerosene may be used up in special military blends such as JP-4 (a blend of kerosene, naphtha, and butane). The remaining kerosene pool is used first to meet fixed-grade requirements and next for balancing-grade production (Tables 12-35 and 12-36).

Blending naphthas, LSR and WSR, is taken up next. Most of the light and whole straight-run naphtha streams emanate from crude units. These are shown in Tables 12-37 to 12-39. The critical properties are the naphtha density and Ried vapor pressure. The RVP can be increased by blending butane, as there is generally economic incentive to blend the naphtha RVP close to specification.

If the refinery has facilities for liquefied petroleum gas recovery, it is recovered from crude, FCCU, and cat reformer units (Table 12-40). LPG is disposed of in gasoline, naphtha blending, and as LPG sale. The remaining LPG, if any, is spent as refinery fuel.

Table 12-11
Overall Yield from VDU 6

	YIELD 1A	mb	YIELD 2&5A	mb	YIELD 3A	mb	YIELD 4A	mb	YIELD 1B	mb	YIELD 2&5B	mb	YIELD 3B	mb	YIELD 4B	mb	YIELD 1VDU WGO	mb	TOTAL
FEED	1.0000	0.00	1.0000	446.30	1.0000	451.02	1.0000	474.67	1.0000	35.90	1.0000	200.70	1.0000	24.66	1.0000	0.00	1.0000	19.78	1653.03
WGO	0.0040	0.00	0.0050	2.23	0.0000	0.00	0.0000	0.00	0.0040	0.14	0.0050	1.00	0.0000	0.00	0.0000	0.00	0.0250	0.49	3.87
DGO	0.2590	0.00	0.2490	111.13	0.0650	29.32	0.0000	0.00	0.2710	9.73	0.2600	52.18	0.0700	1.73	0.0000	0.00	0.9530	18.85	222.93
HVGO	0.4400	0.00	0.4430	197.71	0.5090	229.57	0.3210	152.37	0.4420	15.87	0.4480	89.91	0.5220	12.87	0.3310	0.00	0.0220	0.44	698.74
VACUUM RESID	0.2970	0.00	0.3030	135.23	0.4260	192.13	0.6790	322.30	0.2830	10.16	0.2870	57.60	0.4080	10.06	0.6690	0.00	0.0000	0.00	727.49
TOTAL	1.0000	0.00	1.0000	446.30	1.0000	451.02	1.0000	474.6700	1.0000	35.9000	1.0000	200.70	1.0000	24.6600	1.0000	0.0000	1.0000	19.78	1653.03

Table 12-12
VDU 6 Stream Properties

	mb	SG	SULFUR	PI	VBI	SG*S
FEED	1653.03					
WGO	3.87	0.8127	0.51	71	−185	0.41
DGO	222.93	0.8439	1.87	537	67	1.58
HVGO	698.74	0.9375	2.89	2703	393	2.71
VACUUM RESID	727.49	1.0164	4.22		750	4.28
TOTAL	1653.03	0.95932	3.33		505	3.195

Table 12-13
Vacuum Resid Production and Disposition

UNIT	OPERATION MODE	PRODUCTION, mb	TO VISBREAKER	TO ASPHALT CONVERTER	ASPHALT VDU 1	TO FUEL OIL BLENDING
VDU 1	ASPHALT	68.70			68.70	0.00
VDU 1	FUEL OIL	0.00				0.00
VDU 5	FUEL OIL	393.95	393.95			0.00
VDU 6	FUEL OIL	727.49	200.05	23.26		504.18
TOTAL		1190.13	594.00	23.26	68.70	504.17

Table 12-14
Asphalt Converter Yield

STREAM	VOL% YIELD	mb
ASPHALT REQUIREMENTS		90
ASPHALT PRODUCTION FROM VDU 1		68.70
ASPHALT REQUIRED FROM CONVERTER		21.30
ASPHALT CONVERTER FEED		23.26
FEED	100.00	23.26
LOSS	1.00	0.23
FUEL OIL DURING REGULATION	7.40	1.72
ASPHALT	91.60	21.30
TOTAL	100.00	23.26

Table 12-15
Visbreaker Unit

	AVAILABLE mb	USED mb	SG	SG*S	H	YIELD LV %
FEED						
5VR	393.95	393.95	1.0177	4.337	764.71	
6VR	727.49	200.05	1.0164	4.285	749.73	
TOTAL	1121.43	594	1.0173	4.319	759.67	
PRODUCT						
LOSS		2.38				0.40
NAPHTHA		16.04				2.70
VISBREAKER RESID		575.59	1.03	4.653	669.67	96.90
TOTAL		594.00				100.00

NOTES:
H = VBI, VISCOSITY BLENDING INDEX.
VISBREAKER RESID = VISBROKEN RESID FROM VISBREAKER.
SG*S = PRODUCT OF SPECIFIC GRAVITY AND SULFUR WT%.
FEED RATE = 19.80 mbpcd.
VR = VACUUM RESID
LV = LIQUID VOLUME

Table 12-16
Resid Pool

RESIDS	VOL	H	SG	SG*S	SULFUR WT %	CON CARBON
4A VR	0.00	646	0.9908	3.9400	3.98	13.7
5VDU VR	0.00	765	1.0177	4.3365	4.26	21.5
6VDU VR	504.18	750	1.0164	4.2846	4.22	21.1
VB RESID	575.59	760	1.0173	4.3191	4.25	23.1
ASPH 1/5	0.00	791	1.0350	5.0800	4.91	21.6
ASPHALT CONVERTER	1.72	832	1.0220	4.3500	4.26	26.9
3A VR	0.00	486	0.9643	3.2440	3.36	26.9
TOTAL RESID INCLUDING VB RESID	1081.48	755	1.0169	4.3031	4.23	22.2
TOTAL STRAIGHT-RUN RESID	505.89	750	1.0165	4.2849	4.22	21.1

Table 12-17
Fuel Oil Blending

FO BLEND STREAM	VOL	H	SG	SG*S	SULFUR	CON CARBON
TOTAL V.RESID	1081.48	755	1.0169	4.3031	4.23	22.17
TOTAL V.R w/O v.B	505.89	750	1.0165	4.2849	4.22	21.12
VBU RESID	575.59	760	1.0173	4.3191	4.25	23.10
FCC CUTTERS	316.92	173	0.9316	1.0736	1.15	0.80
MED ISOMATE	6.79	241	0.8844	0.0000	0.00	0.00
HEAVY ISOMATE	0.00	378	0.9018	0.0000	0.00	0.00
HEAVY CAT NAPHTHA	110.31	−250	0.7800	0.0780	0.10	0.00
4A M/I DIESEL	0.00	61	0.8709	1.6240	1.86	0.00
HVGO	74.49	364	0.9322	2.6200	2.81	0.00
TANKAGE	0.00	0	0.9602	2.5280	2.63	0.00
TOTAL	1590.00	549	0.9789	3.2690	3.25	15.24

Table 12-18
Fixed-Grade Fuel Oil Pool

FIXED FUEL GRADES	PROPERTIES					
	VOLUME	H	SG	SG*S	SUL	CON CARBON
I-925	0.00	458	0.9490	2.2300	2.35	15.00
I-928	360.00	458	0.9550	2.5800	2.70	15.00
I-934	0.00	427	0.9520	2.7100	2.85	15.00
I-933	0.00	394	0.9480	2.7200	2.87	15.00
I-957	0.00	349	0.9480	2.5900	2.73	15.00
I-957LS	0.00	338	0.9310	1.7800	1.91	15.00
I-960	0.00	430	0.9600	3.2100	3.35	15.00
I-962	0.00	484	0.9650	3.2300	3.35	15.00
I-964	0.00	439	0.9630	3.0700	3.19	15.00
I-971	135.00	488	0.9710	3.2900	3.39	15.00
I-961 (80 cst)	0.00	396	0.9480	2.7700	2.92	15.00
TOTAL FIXED GRADES	495.00	466.18	0.96	2.77	2.89	15.00

Table 12-19
Balancing-Grade Fuel Oil Blending

STREAM	AVAIL	H	SG	SG *S	SULFUR	CON CARBON
FUEL OIL POOL	1590.00	549	0.9789	3.2690	3.25	15.24
FIXED GRADES FUEL OIL	495.00	466	0.9594	2.7736	2.89	15.00
BALANCING GRADE FUEL OIL	1095.00	586	0.9878	3.4929	3.41	15.35
I-888 CUTTER (DIESEL)	210.00	−30	0.8530	0.8530	1.00	1.00
I-961 POOL (BALANCING GRADE)	1325.00	480	0.9515	3.0218	3.18	12.84
TO TANKS	0.00	480	0.9515	3.0218	3.18	12.84
I-961 POOL	1325.00	480	0.9515	3.0218	3.18	12.84

Table 12-20
Heavy Diesels Yield Summary

UNITS	STREAM	VOL	H	SG	SG*S
CDU 3	3A HDO	25.02	215	0.9020	2.170
CDU 4	4A HDO	359.91	290	0.9160	2.330
VDU 1	DGO (ASPHALT MODE)	178.82	299	0.9200	2.470
VDU 5	VDU 5 VHD	437.53	413	0.9437	2.802
VDU 6	VDU 6 VHD	698.74	393	0.9375	2.710
	TOTAL	1700.02	364	0.9322	2.620
	TO INVENTORY, +/−	−94.57			
	TOTAL, mb	1605.45			

Table 12-21
Heavy Diesel Disposition

STREAM	mb
TO I-725 (HVGO)	0.00
TO HYDROCRACKER, HDU 2	1315.08
TO FCCU	215.88
TO FUEL BLENDING	74.49
TOTAL	1605.45

Table 12-22
HDU 2 (Mild Hydrocracker) Unit Yield Summary

	VOLUME mb	DIESEL INDEX, DI	PI	SUL	H	SG
FEED	1315.08					
LIGHT DIESEL	18.72	64.0	46	0.030	−250	0.7950
WSR NAPHTHA	22.74					0.7034
LIGHT ISOMATE	393.21	38.0	450	0.120	25	0.8839
MEDIUM ISOMATE	316.30			0.240	241	0.8844
HEAVY ISOMATE	603.18			0.280	378	0.9018
TOTAL	1354.15					
VOLUME GAIN	39.07					

Table 12-23
Distribution of Isomates from HDU 2 Unit

	LIGHT ISOMATE	MEDIUM ISOMATE	HEAVY ISOMATE
PRODUCED	393.21	316.30	603.18
INVENTORY, +/−	0.00	−49.19	0.00
TOTAL	393.21	267.11	603.18
DISPOSITION			
TO DIESEL	393.21	0.00	0.00
TO FUEL	0.00	6.79	0.00
TO FCCU	0.00	260.32	603.18
TOTAL	393.21	267.11	603.18

Table 12-24
FCCU Feed Summary

	MODE 1, mb	MODE 2, mb
ISOMATE FEED	863.51	0.00
HVGO FEED	0.00	215.88
TOTAL	863.51	215.88
ISOMATE %	80.00	20.00
RUN DAYS	30.00	30.00
FEED RATE, mbpcd	28.78	7.20

Table 12-25
FCCU Yield Summary

	YIELDS LV%						
PRODUCT	MODE 1	MODE 2	VOLUME, mb	SG	H	SG *S	DI
LIGHT CAT NAPHTHA	0.2950	0.2320	304.82				
MEDIUM CAT NAPHTHA	0.0613	0.1180	78.41				
POLYMER GASOLINE	0.0337	0.0336	36.35				
HEAVY CAT NAPHTHA	0.1060	0.0870	110.31	0.78	−250.00	0.08	
BUTANE	0.0295	0.0258	31.04				
LIGHT CYCLE GAS OIL	0.2810	0.2590	298.56	0.89	−82.80	0.65	33.00
HEAVY CYCLE + DECANT OIL	0.2140	0.2600	240.92	0.95	253.20	1.21	
TOTAL	1.0205	1.0154	1100.41				
GAIN			21.03				
CUTTERS			539.48	0.91	67.25	0.90	

Table 12-26
FCCU Cutter Quality

CUTTER BLEND	mb	H	SG	SG*S
STREAM				
TOTAL LCGO	298.56			
LCGO TO HDU 1	222.55			
LCGO TO FUEL OIL AS CUTTER	76.01			
LCGO	76.00	−82.80	0.89	0.65
HCGO + DECANT OIL	240.92	253.20	0.95	1.21
CUTTER QUALITY	316.92	172.62	0.9316	1.0736

1 HDU CAPACITY, 18.546 mbpcd, 30 DAYS

Table 12-27
Gas Oil (Diesel) Blending from HDU 1 Diesel Hydrodesulfurizer

STREAM	AVAILABLE	TO HDU 1	DI	PI	H	SULFUR	BALANCE
LCGO 1	222.55	222.55	33.1	107	−83	0.296	0.00
LCGO 2	0.00	0.00	33.1	107	−83	2.813	0.00
1A LD	0.00	0.00	63.0	197	−68	1.111	0.00
2A LD	0.00	0.00	61.3	235	−37	1.250	0.00
3A LD	407.34	0.00	61.2	308	−19	1.319	407.34
4A LD	206.46	0.00	57.4	190	−76	0.840	206.46
5A LD	186.30	0.00	62.5	240	−50	1.169	186.30
1B LD	79.80	0.00	55.0	197	−72	1.029	79.80
2B LD	70.80	0.00	56.3	220	−45	1.169	70.80
3B LD	14.28	0.00	54.2	300	−24	1.270	14.28
4B LD	0.00	0.00	0.0	127	−82	0.804	0.00
5B LD	0.00	0.00	61.4	240	−50	1.148	0.00
4A M/I D	354.33	0.00	57.2	524	65	1.677	354.33
5VDU HDO	437.53	334.03	56.5	588	0	1.955	103.50
6VDU HDO	222.93	0.00	56.5	537	0	1.873	222.93
TOTAL	2202.32	556.58	47.1	396	−33	1.292	1645.74
HDU 1 FEED		556.58	47.1	396	−33	1.292	

UNIT CAPACITY, 18.546 bpcd
TOTAL FEED, 556.38 mb
DAYS, 30

Table 12-28
Special Blends

STOCK	VOLUME	SG	H	CON CARBON	SG* SULFUR	SULFUR
VACUUM RESID 4A	0.00	0.9908	640	13.4	3.772	3.807
LD 3	0.00	0.8511	−19	0.0	1.123	1.319
HDU 1	6.00	0.8500	20	0.0	0.196	0.231
M/I 4A	4.00	0.8681	65	0.0	0.701	0.808
I-961	1.40	0.9795	461	15.0	3.411	3.482
TOTAL	11.40	0.87	89.95	1.84	0.77	0.832

I-892 MARINE DIESEL REQUIREMENTS, 11.15 mb
H = 90 MAX, CON CARB = 2.0 MAX, SULFUR = 1.6% MAX)

Table 12-29
Diesel Fixed Grades

GRADE	VOLUME	DI	PI	SG*SUL	SULFUR	FLASH INDEX
I-800	300	47.0	585	0.8300	0.9700	0.001
I-875	0	55.0	294	0.8200	0.9700	0.001
I-876	510	53.2	338	0.8230	0.9700	0.001
I-876ZP	0	53.2	190	0.3290	0.4000	0.001
I-885	584	51.9	365	0.4090	0.4800	0.001
I-88801	0	51.2	389	0.8280	0.9700	0.001
I-88802	0	45.6	389	0.8280	0.9700	0.001
I-88805	0	53.5	389	0.8280	0.9700	0.001
I-88803	0	51.2	446	0.8340	0.9700	0.001
I-88807	0	53.5	389	0.8280	0.9700	0.001
TOTAL	1394	51.3	402	0.6511	0.7647	0.0010

Table 12-30
Diesel Blend Pool

STREAM	AVAILABLE, mb	VOL, BLENDED, mb	DI	PI	SG* SUL	SUL	FI	H
LCGO 1	0.00	0.00	31.5	76	0.210	0.240	0.380	−83
LCGO 2	0.00	0.00	26.7	255	2.750	2.990	0.330	−50
1A LD	0.00	0.00	57.9	197	0.932	1.100	0.380	−68
2A LD	0.00	0.00	56.6	235	1.058	1.251	0.380	−37
3A LD	407.34	407.34	57.1	308	1.123	1.320	0.240	−19
4A LD	206.46	206.46	57.4	190	0.701	0.840	0.450	−76
5A LD	186.30	186.30	57.1	240	0.985	1.160	0.380	−50
1B LD	79.80	79.8	56.5	197	0.867	1.029	0.380	−72
2B LD	70.80	70.8	53.7	220	0.994	1.169	0.380	−45
3B LD	14.28	14.28	52.5	300	1.087	1.271	0.220	−24
4B LD	0.00	0	0.0	0	0.676	0.804	0.000	0
5B LD	0.00	0	61.4	240	0.974	1.148	0.380	−50
4A M/I D	354.33	354.33	51.9	524	1.456	1.677	0.000	57
VDU 5 DGO	103.50	103.5	48.7	569	1.618	1.856	0.350	0
VDU 6 DGO	222.93	222.93	50.1	531	1.579	1.817	0.350	0
LIGHT ISOMATE	393.21	393.21	38.8	450.0	0.062	0.070	0.170	25
HY CAT NAPHTHA	0.00	0	30.8	46	0.330	0.400	−1.910	−250
HDU 1 DIESEL	550.58	550.58	50.3	375	0.194	0.228	0.300	20
HDU 2 DIESEL	18.72	18.72	64	46	0.024	0.030	−1.910	−250
TOTAL	2608.25	2608.252	51.2	380	0.635	0.941	0.242	−2
FIXED GRADES		1394.00	51.32	402.47	0.65	0.76	0.00	0
I-888 POOL		1214.25	51.1	355.1	0.6	1.1	0.52	−4
KERO CUTTER		270.00	64.0	46	0.158	0.170	−1.91	−300
TOTAL I-888 POOL		1484.25	53.4	298.9	0.53	0.97	0.08	−58
SPECIFICATIONS, I-888			51.2	389	0.828	0.970	0.001	

NOTE:
FI = 144 °F FLASH INDEX

Table 12-31
Catalytic Reformer Feed

FEED	AVAILABLE, mb	VOLUME, mb	SG	BALANCE, mb
MEDIUM STRAIGHT RUN FROM CDU 4A	265.05	100.00	0.7400	165.05
HSR FROM CDU 3A	217.62	108.00	0.7331	109.62
MEDIUM CAT NAPHTHA	78.41	0.00	0.7750	78.41
OTHERS	0.00	0.00	0.0000	0.00
TOTAL CAT REFORMER FEED	561.08	208.00	0.7364	353.08

Table 12-32
Product from Cat Reformer

STREAM	FEED, mb	DAYS	TOTAL FEED, mb	YIELD, %	PRODUCT, mb	C4 VOLUME, mb	LOSS, mb
90R REFORMATE	8.00	0.00	0.00	86.10	0.00	0.00	0.00
95R REFORMATE	8.00	0.00	0.00	80.50	0.00	0.00	0.00
97R REFORMATE	8.00	26.00	208.00	76.50	159.12	0.00	48.88
TOTAL			208.00		159.12	0.00	48.88

Table 12-33
Gasoline Streams for Blending

STREAMS	AVAILABLE, mb	VOLUME BLENDED	RON	RVP	SENSITIVITY RON-MON	BALANCE
97R REFORMATE	159.12	159.12	97.20	9.00	10.10	0.00
95R REFORMATE	0.00	0.00	95.20	9.00	9.80	0.00
90R REFORMATE	0.00	0.00	90.20	9.00	8.30	0.00
LIGHT CAT NAPHTHA	304.82	304.82	89.60	9.10	11.50	0.00
MEDIUM CAT NAPHTHA	78.41	78.41	84.00	1.70	10.00	0.00
POLYMER GASOLINE	36.35	36.35	97.50	9.40	16.70	0.00
VBU NAPHTHA	16.04	16.04	63.40	8.20	7.00	0.00
LSR NAPHTHA	594.15	0.00	55.20	9.70	1.00	594.15
BUTANE	101.43	0.00	96.20	60.00	4.50	101.43
POOL	1131.20	594.74	90.67	9.71	11.12	695.58

Table 12-34
Gasoline Grade Production

GRADE	VOLUME, mb	RON	RVP, psia	SENSIBILITY RON-MON
I-383	120.00	83.20	9.20	4.50
I-385	50.00	85.20	8.60	6.20
I-390	100.00	90.20	8.90	7.30
I-393	140.00	93.20	9.40	10.00
I-397	20.00	97.20	9.00	11.50
SUBTOTAL	430.00	88.97	9.12	7.47
TOTAL GASOLINE POOL	594.74	90.67	9.71	11.12
FIXED GRADES	430.00	88.97	9.12	7.47
BALANCING GRADE I-395	164.74	95.12		

Table 12-35
JP-4 (Jet Fuel) Blending

STREAM	AVAILABLE, mb	VOLUME, mb	RVP, psia	SG	FREEZE POINT, °F	FREEZE INDEX	FR.I* VOL
LSR NAPHTHA	594.15	0.00	10.3	0.6724	−105.0	7.548	0.00
MSR NAPHTHA	754.23	52.13	1.5	0.7313	−105.0	7.548	393.46
BUTANE	101.43	2.73	60.0	0.5692	−105.0	7.548	20.61
KEROSENE	1253.43	42.13	0.0	0.7911	−40.0	61.742	2601.17
MEDIUM CAT NAPHTHA TANKAGE	0.00	0.00	1.7	0.7750	−105.0	7.548	0.00
TOTAL	2703.24	96.99	2.5	0.7527	−62.4		31.09

SPECIFICATIONS I-434 (JP-4 FUEL) JET FUEL BLENDING:
SG MIN = 0.7525
FREEZE = −50°F

Table 12-36
Kerosene Grades Production

PRODUCTION FROM CRUDE UNITS, mb	1253.43
TO DIESEL BLENDING	270.00
TO FUEL OIL BLENDING	0.00
TO INVENTORY	0.00
NET AVAILABLE FOR BLENDING	983.43
TO KEROSENE TREATERS	983.43
KEROSENE TREATER FEED RATE, mbpcd	32.78
UNTREATED KEROSENE	0.00
KERO FIXED GRADES, mb	
I-400	200.10
I-411	8.10
I-434	42.13
I-419	238.00
TOTAL, FIXED GRADES	488.33
BALANCING-GRADE KEROSENE PRODUCTION, mb	495.10

Table 12-37
Naphtha Production

STREAM	AVAILABLE, mb	VOL BLENDED, mb	RVP, psia	SG
LSR	594.15	594.15	10.3	0.6732
MSR	494.10	494.10	1.5	0.7354
HDU 2 WSR	22.74	22.74	5.4	0.7034
TOTAL POOL	1110.99	1110.99	6.3	0.7015

Table 12-38
Light Naphtha Blending

	AVAILABLE, mb	VOL BLENDED, mb	RVP, psia	SG
LSR NAPHTHA	294.00	294.00	10.3	0.6732
BUTANE	0.00	0.00	60.0	0.5692
BLEND	294.00	294.00	10.3	0.6732

Table 12-39
Whole Straight-Run (WSR) Naphtha Blending

STREAM	AVAILABLE, mb	VOL. BLENDED, mb	RVP, psia	SG
LSR NAPHTHA	300.15	300.15	10.3	0.6689
MSR NAPHTHA	494.10	494.10	1.5	0.7250
HDU 2 WSR	22.74	22.74	5.4	0.7034
INVENTORY, +/−		0.00		
SUBTOTAL	816.99	816.99	4.8	0.7038
BUTANE		25.00	60.0	0.5692
TOTAL BLEND		841.99	6.5	0.6998

Table 12-40
LPG Production and Disposition

PRODUCTION	mb
LPG PRODUCTION FROM CRUDE UNITS	101.43
FROM FCCU	31.04
FROM CAT REFORMER	0.00
TOTAL PRODUCTION	132.47
DISPOSITION	
LPG PRODUCT	25.00
TO GASOLINE BLENDING	2.73
TO NAPHTHA BLENDING	25.00
TO SPECIAL JET FUEL (JP-4)	2.73
REMAINING LPG TO REFINERY FUEL	77.01
TOTAL DISPOSITION	132.47

Table 12-41
Unit Volume Losses and Gains

UNIT	mb
LOSSES	
CRUDE UNITS	21.87
ASPHALT CONVERTER	0.23
CAT REFORMER	48.88
VISBREAKER UNIT	2.38
TOTAL	73.36
UNIT GAINS	
FCCU	21.03
MILD HYDROCRACKER	39.07
TOTAL VOLUME GAIN	60.10
NET LOSSES	13.26

Table 12-42
Estimated Overall Material Balance

	mb	DOP* REQUIREMENT
FEED		
LIGHT ARABIAN CRUDE	6030.00	6030.00
BAHRAIN CRUDE	1260.00	1260.00
MURBAN CRUDE	0.00	0.00
DUBAI CRUDE	0.00	0.00
TOTAL CRUDE	7290.00	7290.00
PRODUCTS		
LPG	25.00	25.00
LIGHT NAPHTHA	294.00	
WSR NAPHTHA	841.99	
GASOLINES		
I-383	120.00	120.00
I-385	50.00	50.00
I-390	100.00	100.00
I-393	140.00	140.00
I-397	20.00	20.00
I-395 (BALANCING GRADE)	164.74	
TOTAL GASOLINES	594.74	430.00
KEROSENES		
I-400	200.10	200.10
I-411	8.10	8.10
I-434	96.99	96.99
I-419	238.00	238.00
I-440 (BALANCING GRADE)	495.10	
GROSS KEROSENE PRODUCTION	1038.29	
KEROSENE TO DIESEL BLENDING	270.00	
KEROSENE PRODUCTION	768.29	543.19
DIESELS		
I-800	300.00	300.00
I-875	0.00	
I-876	510.00	510.00
I-876ZP	0.00	
I-885	584.00	584.00
I-888 (BALANCING GRADE)	1484.25	
I-892	11.40	11.40
GROSS DIESEL	2889.65	
DIESEL TO FUEL OIL BLENDING	210.00	
DIESEL PRODUCTION	2679.65	1405.40
FUEL OILS		
I-925	0.00	
I-928	360.00	360.00
I-934	0.00	
I-933	0.00	
I-957	0.00	
I-957LS	0.00	
I-960	0.00	
I-962	0.00	
I-964	0.00	
I-971	135.00	135.00
I-961 (80 cst)	0.00	
I-961 (BALANCING GRADE)	1325.00	
TOTAL FUEL OIL	1820.00	495.00
ASPHALT	90.00	90.00
TOTAL PRODUCTS	7113.67	
INTERMEDIATE STOCKS, INVENTORY CHANGES**		
90R REFORMATE	0.00	
95R REFORMATE	0.00	
97R REFORMATE	0.00	
LIGHT CAT NAPHTHA	0.00	
MEDIUM CAT NAPHTHA	0.00	
HEAVY CAT NAPHTHA	0.00	
POLYMER GASOLINE	0.00	
HSR NAPHTHA	0.00	
KEROSENE BASE STOCK	0.00	
DIESEL	0.00	
LVGO (4 M/I DIESEL)	0.00	
HDU LIGHT DIESEL	0.00	
LIGHT ISOMATE	0.00	
MED ISOMATE	49.19	
HEAVY ISOMATE	0.00	
FCC CUTTER	0.00	
6VDU FEED/ATM RESID	0.00	
HVGO	94.57	
VACUUM RESID	0.00	
TOTAL	143.76	
TOTAL OUTPUT	7257.43	
LIQUID RECOVERY	99.55%	

*DOP REQUIREMENTS REFER TO CRUDE RUN AND FIXED GRADES ONLY.
*POSITIVE INVENTORY CHANGES INDICATE BUILDUP OF INVENTORY AND NEGATIVE INVENTORY CHANGES INDICATE DRAWDOWN FROM INVENTORY.

NOTES

1. J. R. White. "Use Spreadsheets for Better Refinery Operation." *Hydrocarbon Processing* (October 1986), p. 49. "Linear Programming Optimisation of Refinery Spreadsheets" *Hydrocarbon Processing* (November 1987), p. 90.

CHAPTER THIRTEEN

Refinery Linear Programming Modeling

OVERVIEW

The basic problem of linear programming (LP) is to maximize or minimize a function of several variables subject to a number of constraints. The functions being optimized and the constraints are linear. General linear programming deals with allocation of resources, seeking their optimization. In the context of an oil refinery, an LP model is a mathematical model of the refinery, simulating all refinery unit yields, unit capacities, utility consumption, and the like as well as product blending operations of the refinery by means of linear equations, each equation subject to a number of constraints. These equations are compiled in a matrix of rows and columns, the columns representing the unknowns or variables and the rows or equations representing the relations between variables. The values in the matrix are simply the coefficients that apply to unknowns in each equation. As the number of unknowns are more than the number of constraints relating them, a large number of solutions might satisfy all the problem parameters.

The optimal solution must be chosen from the set of only those solutions that satisfy all the problem parameters and, at the same time, maximize refinery profit or minimize operating cost. To aid the search for an optimum solution, LP is driven by a row in the matrix containing cost and revenue (the objective function row).

DEVELOPMENT OF THE REFINERY LP MODEL

In the oil industry, prior to the advent of LP techniques, all optimization studies were done by calculating several hand balances, moving toward an optimal solution by trial and error. Carrying out simplex procedure by hand was very tedious and time consuming. The typical refinery LP model used for planning has approximately 300–500 equations and 800–1500 activities to optimize. With a simplex algorithm available as a computer program, interest quickly developed in optimizing via a linear programming. In the 1950s, a standard input format to describe a matrix was agreed on, opening the market to LP software from different vendors. These software are generally of two types:

1. Programs in which the user enters all refinery data, such as unit yields, product properties, and unit capacities, in the form of spreadsheets that can easily be updated. These programs convert the data tables into matrix form by using special programming languages (Omni, Magan, etc.), thus saving many hours in producing matrix input correctly. These programs are called *matrix generators*. These programming languages can also be used to create a "report writer" program to print out the optimized results.
2. Optimizer programs[1] that read the matrix description in the standardized input format, optimize the problem, and report the results in an "unscrambled report," which is simply a list of rows and columns and the associated optimized values.

A refinery LP model is designed to model a wide variety of activities, including, among others, the following:

Distillation of crudes.
Downstream processing units, such as cat reformers, hydrocrackers, desulfurizers, and visbreakers in various processing modes.
Pooling of streams.
Recursing on the assumed qualities of a rundown tank's content.
Finished product blending.
Refinery fuel blending.
Importing feedstocks to meet product demand.
Exporting surplus refinery streams to other refineries.

The refinery LP model, in fact, is simply a set of data tables in the form of spreadsheets that are converted into a matrix using special programming languages. As many solutions to the problem are possible, the criteria for choosing an optimum solution is that which, apart from satisfying all equations, gives maximum profit to the refinery.

The optimum solution of a refinery LP model yields the following:

A complete, unitwise, material balance of all refinery units. The material balance could be on a volume or weight basis.

The unit capacities available and utilized.

Feedstocks available and used for processing or blending.

Utilities (fuel, electricity, steam, cooling water), chemical, and catalyst consumption for all processing units and the overall refinery.

Blend composition of all products and the properties of blended products.

An economic summary, which may include the cost of crude, other feedstocks, utilities, chemicals, and catalyst consumed and the prices of blended finished product.

THE STRUCTURE OF A REFINERY LP MODEL

ROW AND COLUMNS NAMES AND TYPES

Row and column naming conventions are followed for easy identification and manipulation of data by different vendors of LP software. (The row and column naming and also data tables naming convention followed here for rows, columns, and data tables is from the popular Process Industries Modeling System (PIMS), a PC-based refinery LP package.[2] Here, row and column names are seven characters long: The first four characters generally identify the type of the row or column, while the last three characters identify the stream.

Row Names

ROW	CODE	EXAMPLE
OBJECTIVE FUNCTION ROW	OBJFN	
MATERIAL BALANCE (VOLUME BASIS)	VBALxxx	VBALNAP,VBALKER
MATERIAL BALANCE (MASS BASIS)	WBALxxx	WBALRES
UTILITY BALANCE	UBALxxx	UBALKWH,UBALFUL
PROCESS UNIT CAPACITY	CCAPxxx	CCAPFCC
VOLUME BLEND BALANCE	EVBLxxx	EVBL888
MASS BLEND BALANCE	EWBLxxx	EWBL440
MINIMUM BLEND SPECIFICATION	Nsssxxx	NSPG395
MAXIMUM BLEND SPECIFICATION	Xsssxxx	XRON395
USER-DEFINED EQUALITY ROW	Eabcxyz	ECHGFCU
USER-DEFINED “LESS THAN OR EQUAL TO ROW”	Labcxyz	
USER-DEFINED “GREATER THAN OR EQUAL TO ROW”	Gabcxyz	
RECURSION MATERIAL BALANCE ROW	RBALxxx	RBAL397
RECURSION PROPERTY BALANCE	Rprpxxx	RRON397

Column Names

COLUMN	CODE	EXAMPLE
PURCHASE 1000 UNITS OF MATERIAL OR UTILITY	PURCxxx	PURCABP
SELL 1000 UNITS OF MATERIAL OR PRODUCT	SELLxxx	SELLNAP
SPECIFICATION VOLUME BLEND 1000 UNITS OF PRODUCT, prd	BVBLprd	BVBL888
SPECIFICATION MASS BLEND 1000 UNITS OF PRODUCT, prd	BWBLprd	BWBL961
OPERATE 1000 UNITS OF PROCESS UNIT "PRS" IN MODE mmm	Sprsmmm	SFCUFC1

UNITS OF A ROW

The units of a column multiplied by the units of a coefficient in a row equal the units of a row. Thus, if the column activity represents thousands of tons per day and a coefficient in the utility balance row has unit of 000' Btu/ton, the units of row are

$$= 000'\,\text{Tons/Day} * 000'\ \text{Btu/Ton or million Btu/Day}$$

The types of restrictions encountered in a refinery LP model are as follows: feed availability; product demand; process yield; utility, catalyst, and chemical consumption; and product blending.

Feed Availability

These equations reflect the purchase of feedstock, such as crude and imported intermediate streams or blend stocks. The disposition of crudes could be to various crude units in different operation modes. The disposition of intermediate feedstock could be to a secondary processing unit or to product blending.

Table 13-1 (BUY Table) shows a typical data entry and Table 13-2 shows the matrix generated by the feed availability constraints. The row names are derived from the row names of the BUY Table. For example, row ABP in Table 13-1 generates a row name VBALABP in the matrix.

Table 13-1
BUY Table

TABLE BUY	TEXT	MIN	MAX	FIX	COST
	Crude Oil Imports				
ABP	AL CRUDE			165.1842	16.809
BAH	BAHRAIN CRUDE			42.000	15.109
	TOTAL FEED			207.1842	

In row VBALABP, column PURCABP has a negative coefficient, reflecting that the Arab crude purchased is made equal to its disposition to various crude units.

Product Demand

These equations relate the product demand to its blending. The product demand is inserted in the SELL Table (Table 13-3). The row names in this table are product codes. These rows generate a matrix (Table 13-4) by adding the prefix VBAL to the table row name, denoting that each row is, in fact, a material balance row relating the demand to its blending. For example, row 150 in the SELL Table generates the row VBAL150 (the material balance row for LPG).

Row VBAL150 has coefficients 1 and –1, respectively, for its two columns, SELL150 and BVBL150, indicating that variable SELL150 is made equal to BVBL150 or volume of LPG blended in product LPG.

If the demand for a product is fixed, this is called a *fixed-grade product*, and the LP solution meets this demand by fixing the value of the variable SELL150. If demand for a product to be produced is not fixed, it is called a *balancing or free grade*, and its production is optimized, based on the price of the product, within maximum or minimum demand constraints, if these exist. It is, therefore, necessary to insert the prices of balancing-grade products whose production is to be optimized.

Column names in Table SELL are generated from the table name and row names. For example, in Table 13-3 (Table SELL), row 150 generates a column or variable SELL150, whose production is optimized on the basis of its unit price, indicated in column price. Variable SELL150 has a value within the minimum and maximum constraints shown in this table.

Table 13-2
Matrix for Table 13-1

ROW	**VBALABP**	**TYPE LE**
	PURCABP	−1
	SCRDA1N	1
	SCRDA1K	1
	SCRDA2N	1
	SCRDA2K	1
	SCRDA3N	1
	SCRDA3K	1
	SCRDA4N	1
	SCRDA4K	1
	SCRDA5N	1
	SCRDA5K	1
ROW	**VBALBAH**	**TYPE LE**
	PURCBAH	−1
	SCRDB1N	1
	SCRDB1K	1
	SCRDB2N	1
	SCRDB2K	1
	SCRDB3N	1
	SCRDB3K	1
	SCRDB4N	1
	SCRDB4K	1
	SCRDB5N	1
	SCRDB5K	1
ROW	**MINOBJ**	**TYPE FR**
	PURCABP	−16.809
	PURBAH	−15.109
COLUMNS		**TYPE FIX**
	PURCABP	165.184
	PURCBAH	42.000

NOTE: SCRDA1N, SCRDA1K, ETC. ARE VECTORS INDICATING DISPOSITION OF ARAB CRUDE TO VARIOUS CRUDE UNITS AND MODES.

Process Yields

A refinery has a large number of process units. The most important units are the crude distillation unit (CDU), consisting of atmospheric distillation of crude and vacuum distillation of atmospheric resid. Downstream of the CDU are a number of other processing units, such as the

Table 13-3
SELL Table

TABLE SELL	TEXT		MIN	MAX	FIX	PRICE
	Finished Products					
150	I-150 LPG	BAL	0.000			8.495
220	I-220 LSR.	BAL	0.000			12.059
210	I-210 WSR	BAL	0.000			12.454
97L	I-397L .4g/l		0.000	0.000		17.657
440	I-440 JET A1 FUEL BAL		0.000			18.168
76Z	I-876ZP DIESEL G.OIL		0.000	0.000	7.737	17.003
888	I-888 DIESEL G.O BAL		0.000			15.489
928	I-928 FUEL OIL		0.000	0.000	13.455	9.431
961	I-961	BAL	0.000			8.948
ASP	ASPHALT		0.000	0.000	2.128	30.000
SUL	SULFUR		0.000			25.000
	TOTAL FINISHED PRODUCT		0.000	0.00	23.320	

fluid cat cracker (FCCU), distillate hydrocracker, diesel desulfurizer, and cat reformer unit.

The basic structure of all process submodels are as follows. A process unit may operate in a number of modes. Each operating mode becomes a column in the process submodel. The column names may be feed names or processing modes of a unit. The rows are, in fact, material balance rows for each product produced as a result of processing. The coefficients are the yield of the product in the mode represented by column heading. Columns can be compared to pipes in a plant through which material flows and the various rows, to taps from which products are drawn off.

Tables 13-5 and 13-7 show the typical data input for two process units, FCCU and VBU. Tables 13-6 and 13-9 present the matrix generated by these tables. For example, referring to Table 13-5, the FCCU submodel, the column headings FC1 and FC2 indicate operating modes of the FCCU.

Rows VBALC4U, VBALPOR, VBALLCN, VBALMCN, VBALLCO, VBALHCO, and so forth indicate material balances for the yields of cracked LPG, light cat naphtha, medium cat naphtha, light cycle gas oil, heavy cycle gas oil, and the like, which are produced in the FCCU and disposed of somewhere else.

Table 13-4
Matrix for Table 13-3

ROW	**VBAL150**	**TYPE LE**
	SELL150	1
	BVBL150	−1
ROW	**VBAL220**	**TYPE LE**
	SELL210	1
	BVBL210	−1
ROW	**VBAL210**	**TYPE LE**
	SELL210	1
	BVBL210	−1
ROW	**VBAL97L**	**TYPE LE**
	SELL97L	1
	BVBL150	−1
ROW	**VBAL440**	**TYPE LE**
	SELL440	1
	BVBL440	−1
ROW	**VBAL76Z**	**TYPE LE**
	SELL76Z	1
	BVBL76Z	−1
ROW	**VBAL888**	**TYPE LE**
	SELL888	1
	BVBL888	−1
ROW	**VBAL928**	**TYPE LE**
	SELL928	1
	BVBL928	−1
ROW	**VBAL961**	**TYPE LE**
	SELL961	1
	BVBL961	−1
ROW	**VBALASP**	**TYPE LE**
	SELLASP	1
	BVBLASP	−1
ROW	**VBALSUL**	**TYPE LE**
	SELLSUL	1
	BVBLSUL	−1
ROW	**MINOBJ**	**TYPE FR**
	SELL150	8.495
	SELL220	12.059
	SELL210	12.454
	SELL97L	17.657
	SELL440	18.168
	SELL76Z	17.003
	SELL888	15.489
	SELL928	9.431
	SELL961	8.948
	SELLASP	30.000
	SELLSUL	25.000

Table 13-5
FCCU Submodel

TABLE SFCU	TEXT	FC1 HMI	FC2 HVGO
	Feed streams to FCCU		
EBALFC1	Isomate feed	1	
EBALFC2	HVGO feed		1
VBALC4U	UNSAT BUTANES	−0.0322	−0.0196
VBALPOR	RERUN POLYMER	−0.0381	−0.0337
VBALLCN	LIGHT CAT NAPHTHA	−0.2990	−0.2160
VBALMCN	MEDIUM CAT NAPHTHA	−0.0670	−0.0850
VBALHCN	HEAVY CAT NAPHTHA	−0.1150	−0.1250
VBALLCO	LIGHT CYCLE GAS OIL	−0.2600	−0.2080
VBALHDO	HEAVY/DECANT OIL	−0.2170	−0.3140
	BALANCE CHECK	−1.0283	−1.0013
UBALFUL	REBOILER FUEL	0.2524	0.2524
UBALCCC	CAT $ CHEM. $/BBL	0.0534	0.0534
CCAPFCU	FEED LIMITATION	1	1

The coefficient at a column/row intersection is the yield of a product represented by row name from operation in mode indicated by column name.

Utility, Catalyst, and Chemical Consumption

Similar to VBAL rows, every process submodel contains a number of UBAL or utility balance rows. The coefficients of these rows in any column indicate the utility, chemical, and catalyst consumption per unit feed processed in the mode indicated by column heading.

For material and utility balance rows, by convention, all feeds are shown with positive sign and all products with negative sign.

User-Defined Rows

In Table 13-5, EBALFC1 and EBALFC2 are user-defined equality rows, which make the disposition of the two FCCU feeds equal to their production.

Table 13-6
Matrix for Table 13-5

ROW	**EBALFC1**	**TYPE EQ**		
	SFCUHM1	−1		
	SFCUFC1	1		
ROW	**EBALFC2**	**TYPE EQ**		
	SFCUHV2	−1		
	SFCUFC2	1		
ROW	**VBALC4U**	**TYPE EQ**		
	SFCUFC1	−0.0322		
	SFCUFC2	−0.0196		
ROW	**VBALPOR**	**TYPE EQ**		
	SFCUFC1	−0.0381		
	SFCUFC2	−0.0337		
ROW	**VBALLCN**	**TYPE EQ**		
	SFCUFC1	−0.299		
	SFCUFC2	−0.216		
ROW	**VBALMCN**	**TYPE EQ**		
	SFCUFC1	−0.067		
	SFCUFC2	−0.085		
ROW	**VBALHCN**	**TYPE EQ**		
	SFCUFC1	−0.115		
	SFCUFC2	−0.125		
ROW	**VBALLCO**	**TYPE EQ**		
	SFCUFC1	−0.260		
	SFCUFC2	−0.208		
ROW	**VBALHDO**	**TYPE EQ**		
	SFCUFC1	−0.217		
	SFCUFC2	−0.314		
ROW	**UBALFUL**	**TYPE EQ**		
	SCFUFC1	0.2524		
	SCFUFC2	0.2524		
ROW	**UBALCCC**	**TYPE EQ**		
	SCFUFC1	0.0534		
	SCFUFC2	0.0534		
ROW	**CCAPFCU**	**TYPE LE**	RHS	38.556
	SFCUFC1	1		
	SFCUFC2	1		

Process Unit Capacities

We see that every process unit submodel has a capacity row. The various operation modes of the process have a coefficient in this row, showing capacity consumed for processing one unit of feed. Thus, refer-

Table 13-7
VBU Submodel

TABLE SVBU	TEXT	BAS
VBALVBP	VBU FEED POOL	1
VBALGA1	OFF GASES	−0.036
VBALNVB	VBU NAPHTHA	−0.049
VBALVRS	VISBROKEN RESID	−0.951
	BALANCE CHECK	−0.036
UBALFUL	FUEL REQ. MSCF	0.3015
CCAPVBU	VBU CAPACITY	1
CCAPVBX	MAX VISBREAKING	1

Table 13-8
Matrix for Table 13-7

ROW	**VBALVBP**	**TYPE EQ**
	SVBUV31	−1
	SVBUV11	−1
	SVBUVA4	−1
	SVBUV15	−1
	SVBUV16	−1
	SVBUV26	−1
	SVBUBAS	1
ROW	**VBALGA1**	**TYPE EQ**
	SVBUBAS	−0.036
ROW	**VBALNVB**	**TYPE EQ**
	SVBUBAS	−0.049
ROW	**VBALVRS**	**TYPE EQ**
	SVBUBAS	0.951

ring to Table 13-5 showing the FCCU submodel, row CCAPFCU (FCCU capacity), the columns named FC1 and FC2 are the two operation modes of the FCCU. The sum of the column activity multiplied by the respective row coefficient of these two vectors must be equal to or less than the RHS of the CCAPFCU row.

The right-hand side of this equation is provided by the CAP Table, which provides the maximum and minimum FCCU capacity available for processing the various feeds to the unit. Table 13-9 shows the format of Table "CAP." The first column is process unit name codes. Columns

Table 13-9
CAP Table

TABLE CAPS	TEXT	MAX	MIN
CAT1	CRUDE UNIT NO 1	18.468	0.000
CAT2	CRUDE UNIT NO 2	18.351	0.000
CAT3	CRUDE/VAC UNIT NO 3	58.238	0.000
CAT4	CRUDE/VAC UNIT NO 4	87.811	0.000
CAT5	CRUDE UNIT NO 5	42.215	0.000
CVD6	6 VDU	41.909	0.000
CVBU	VISBREAKER	7.534	0.000
CACR	ASPHALT CONVERTER	4.191	0.000
CHDU	1 HDU	17.701	0.000
CIXM	2 HDU – MED SEVERITY	44.708	0.000
CPTF	PLAT FEED CAPACITY	10.822	0.000
CFCU	FCCU	38.556	0.000
CH2P	H2 PRODUCTION UNIT	7.827	0.000
CSFR	SULF. RECOVERY UNIT	0.149	0.000

"MAX" and "MIN" are maximum and minimum values of capacity of the units. These become the lower and upper bound of RHS of the matrix (Table 13-10) generated by capacity rows.

Product Blending

These equations ensure that the quantity of streams produced by process units are equal to the quantity of blend stock available for blending plus any loss and that going to refinery fuel. Also, the quantity of blend stock used in each product is made equal to the quantity of finished product. By convention, material or utility consumed by a process is shown as positive and material produced by a process is shown as negative. For example, if a unit consumes 100 units of a crude, that is entered in the matrix as +100, whereas if this crude unit produces 20 units of naphtha, 25 units of kerosene, 30 units of diesel, and 25 units of topped crude, all these outputs are entered in the matrix with a negative sign.

Product blending data are entered in the model as a table blend mix or as the properties of blend streams.

The table BLNMIX is, in fact, a blending map. The rows are the stream names and the column headings are the names of various product grades. An

Table 13-10
Matrix for Table 13-9

ROW	**CCAPAT1**	**TYPE LE**	**RHS**	**18.468**
	SLC1A1N	1		
	SLC1A1K	1		
	SLC1B1N	1		
	SLC1B1K	1		
ROW	**CCAPAT2**	**TYPE LE**	**RHS**	**18.351**
	SLC2A2N	1		
	SLC2A2K	1		
	SLC2B2N	1		
	SLC2B2K	1		
ROW	**CCAPAT3**	**TYPE LE**	**RHS**	**58.2382**
	SLC3A3N	1		
	SLC3A3K	1		
	SLC3B3N	1		
	SLC3B3K	1		
ROW	**CCAPAT4**	**TYPE LE**	**RHS**	**87.8108**
	SLCA4N	1		
	SLCA4K	1		
	SLCB4N	1		
	SLCB4K	1		
ROW	**CCAPAT5**	**TYPE LE**	**RHS**	**42.2148**
	SLCA5N	1		
	SLCA5K	1		
	SLCB5N	1		
	SLCB5K	1		
ROW	**CCAPVD6**	**TYPE LE**	**RHS**	**41.9089**
	SVD6RA1	1		
	SVD6RB1	1		
	SVD6RA2	1		
	SVD6RB2	1		
	SVD6RA5	1		
	SVD6RB5	1		
	SVD6VA3	1		
	SVD6VA4	1		
	SVD6W11	1		
	SVD6W21	1		
	SVD6W31	1		
	SVD6ARD	1		
	SVD6BS1	1		
	SVD6BS3	1		
ROW	**CCAPVBU**	**TYPE LE**	**RHS**	**7.543**
	SVBUBAS	1		

Table 13-10
Continued

ROW	**CCAPACR**	**TYPE LE**	**RHS**	**4.191**
	SACRV15	1		
	SACRV16	1		
	SACRV26	1		
ROW	**CCAPHDU**	**TYPE LE**	**RHS**	**17.006**
	SHDUMH1	1		
	SHDUMH2	1		
ROW	**CCAPIXM**	**TYPE LE**	**RHS**	**44.705**
	SIXMIX1	1		
	SIXMIX2	1		
ROW	**CCAPPLT**	**TYPE LE**	**RHS**	**10.8217**
	SREFNM0	1		
	SREFMM0	1		
	SREFNM3	1.059		
	SREFMM3	1.059		
	SREFNM5	1.161		
	SREFMM5	1.161		
	SREFNM6	1.161		
	SREFMM6	1.161		
	SREFNM7	1.240		
	SREFMM7	1.240		
ROW	**CCAPFCU**	**TYPE LE**	**RHS**	**38.5562**
	SFCUFC1	1		
	SFCUFC2	1		
	SFCUFC3	1		
ROW	**CCAPH2P**	**TYPE LE**	**RHS**	**7.8268**
	SH2PGAS	1		
ROW	**CCAPSFR**	**TYPE LE**	**RHS**	**0.1492**
	SSFRH2S	0.39714		

entry of 1 on the intersection of a row and column indicates that the stream indicated by row name is allowed to be blended in the grade indicated by the column name. The lack of an entry at this row/column intersection implies that this blend stock is not allowed to be blended in that grade.

Table BLNMIX, shows the process streams allowed into fuel oil grade I-961. The row names are blend stocks stream code. For example, an entry of 1 under column 961, in row UKE, shows that blend stock UKE is allowed in I-961 blending. Table 13-12 shows the matrix generated by

Table 13-11
BLNMIX TABLE

TABLE BLNMIX	TEXT	961
UKE	UNTREATED KEROSENE	1
TLD	SR.DIESEL POOL	1
WGO	DIESEL CUTTERS POOL	1
LVO	LT.VAC. GASOIL POOL	1
HVO	HVY.VAC. GASOIL POOL	
TRS	RESID POOL	1
AS1	1VDU ASPHALT	1
AS5	5VDU ASPHALT	1
HCN	HVY CAT NAPHTHAS	1
MCD	MEDIUM CAT NAPHTHA D/D	
TFC	HVY/LT CYCLE POOL	1

Table 13-12
Matrix for Table 13-11

ROW	EVBL961	TYPE EQ
	BVBL961	1
	BUKE961	−1
	BTLD961	−1
	BWGO961	−1
	BLVO961	−1
	BTRS961	−1
	BAS1961	−1
	BHCN961	−1
	BTFC961	−1

this table. The matrix shows that, in row EVBL961, the following columns have an entry:

BVBL961	Volume of fuel oil grade 961 blended.
BUKE961	Volume of untreated kerosene blended in grade 961.
BLTD961	Volume of light diesel blended in grade 961.
BWGO961	Volume of wet gas oil blended in grade 961.
BLVO961	Volume of LVGO blended in grade 961.
BTRS961	Volume of vacuum resid blended in grade 961.
BAS1961	Volume of asphalt blended in grade 961
BHCN961	Volume of heavy catalytic naphtha from FCCU blended in fuel oil.
BTFC961	Volume of cutter stocks blended in fuel oil 961.

The coefficient in the column BVBL961 is +1, while the coefficients of all other columns are −1, showing that the volume of the blended 961 is equal to the sum of the volumes of the individual blend stocks, as EVBL961 is an equality row.

The properties of blend streams coming from the crude and vacuum distillation units are entered in an ASSAYS table. The properties of process unit streams are entered in a number of tables called *blend properties*, or BLNPROP.

If a stream is formed by pooling a number of streams, its properties are unknown. However, a first guess of its properties is required to start the optimization process. Such guessed data are also provided in a separate table.

PROPERTY PROPAGATION TO OTHER TABLES

Many LP packages have a system of property propagation from property tables to process submodels and other tables. Properties of straight-run streams are entered in either the ASSAYS or BLNPROP tables. These property data, if needed in any other table, need not be reentered, placeholders (999 or some other symbol) are put in, which are replaced with data from the relevant data table.

Let us see how place holders are resolved. Suppose the ASSAYS table contains an entry for quality QQQ in stream SSS. This can be used to resolve placeholders in column SSS with row name xQQQabc in any submodel table. The row and column must be suitably named for placeholder resolution to occur.

EXAMPLE 13-1

In an ASSAY table, the data SG of kerosene stream from A1N mode of CDU1 is entered in row ISPGKN1 as follows:

A1N	
ISPGKN1	0.7879

These data are retrieved in a submodel table by the following entry:

	KN1
ESPGFDP	−999

Here, placeholder −999 is replaced with 0.7879, the specific gravity of KN1 as entered in the ASSAYS table. By convention, a negative sign is used for streams entering a recursed pool and their properties and a positive sign for the pool produced. Placeholder resolution can occur in user-defined E-, L-, or G-type rows as well as in recursion rows. Placeholders for recursed properties are replaced by the latest value of the recursed property of a stream every time the matrix is updated.

BLENDING SPECIFICATIONS

These equations ensure that each optimal product blend meets the specifications set for it. For example, Table 13-13 shows a part of a product specification table, listing properties of a fuel oil grade 961. Table 13-14 shows the matrix generated from this table. Row XVB1961 has as columns the blend components of fuel grade 961. The column coefficients are the VBI of individual blend components (see Table 13-15).

SPECIALIZED RESTRICTIONS

Several equations may be included in the matrix to reflect special situations in a refinery. For example, these may be the ratio of crude processed, restricting a processing unit to a particular mode of operation, or the ratio of the two products to be produced.

STREAM POOLING (RECURSION PROCESS)

Another class of refinery operation is important and must be modeled. This is the pooling of streams. A number of streams may be pooled to

Table 13-13
Specifications for Fuel Oil Grade 961

TABLE BLNSPEC	TEXT	961
XSUL	MAX. SUL	3.4
XVBI	MAX. VISCOSITY INDEX	461
N144	MIN. F.I. AT 144F	0.0001
XCON	MAX. CON. CARBON	14.3
XKLP	MAX. KERO INCLUSION	5

Table 13-14
Matrix for Table 13-13

ROW	**XSUL961**	**TYPE GE**
	(MAX SULFUR IN 961 FUEL OIL)	
	BWBL961	3.4000
	BUKE961	0.1078
	BTLD961	1.0182
	BWGO961	0.4842
	BLVO961	1.5451
	BTRS961	3.9652
	BAS1961	5.0974
	BAS5961	5.0829
	BHCN961	0.0137
	BTFC961	0.2619
ROW	**XVBI961**	**TYPE GE**
	(MAX VISCOSITY INDEX OF 961 F.O.)	
	BVBL961	4.6100
	BUKE961	−281.5921
	BTLD961	−37.7430
	BWGO961	−158.8042
	BLVO961	64.9132
	BTRS961	631.2328
	BAS1961	827.0000
	BAS5961	827.0000
	BHCN961	−365.0000
	BTFC961	108.4017
ROW	**XCON961**	**TYPE GE**
	(MAX CON CARBON OF 961 F.O.)	
	BWBL961	1.4300
	BUKE961	0.0000
	BTLD961	0.0000
	BWGO961	−0.0082
	BLVO961	−0.0174
	BTRS961	−15.2657
	BAS1961	−27.8173
	BAS5961	−27.8173
	BHCN961	0.0000
	BTFC961	−0.2573
ROW	**N144961**	**TYPE LE**
	(MIN FLASH INDEX AT 144 DEG F FOR 961 F.O.)	
	BKN4961	−0.1544
	BUKE961	1.6442
	BTLD961	−0.4645
	BWGO961	−3.3418

Table 13-14
Continued

	BLVO961	−0.6529
	BTRS961	−0.1598
	BAS1961	−0.0019
	BAS5961	−0.0019
	BHCN961	1.1520
	BTFC961	−0.3623
		−0.7700
ROW	**XKLP961**	**TYPE GE**
	(MAX KERO INCLUSION IN 961 F.O.)	
	BVBL961	0.0500
	BUKE961	−1.0000
	BHCN961	−1.0000

Table 13-15
Fuel Oil Blend Components

TABLE	BLNPROP	SG	SUL	VBI	CON	144
UKE	KEROSENE POOL	0.7891	0.1366	−281.592	0.000	−164.424
TLD	DIESEL POOL	0.8465	1.2022	−37.743	0.000	46.451
WGO	WET GAS OIL	0.8165	0.6018	−158.804	0.010	−334.182
LVO	LVGO	0.8683	1.7674	64.913	0.020	65.289
HVO	HVGO	0.9388	2.9855	396.298	0.000	
TRS	RESID POOL	0.9940	3.9986	631.233	15.358	15.918
AS1	ASPHALT 1 VDU	1.0350	4.9250	827.000	26.900	0.193
AS5	ASPHALT 5 VDU	1.0350	4.9110	827.000	26.900	0.193
HCN	HVY CAT NAPHTHA	0.8070	0.0170	−365.000	0.000	−115.200
TFC	FCC CUTTER POOL	0.9221	0.3256	108.402	0.279	36.232

form a single stream, which may become feed to a process unit or used for blending one or more products.[3]

The user must supply data on the properties of the pooled stream for optimizer to reach a solution. The user, however, has a problem, because the composition of the pooled stream is not known until after the optimum solution is reached. An iterative approach (recursion) is employed. The user provides a first guess on the properties of pooled stream. The optimizer then solves the model with the estimated data in it. After solving the model, an external program recalculates the physical properties of the pooled stream, which was earlier guessed. The revised physical property data are inserted in the model, and the model is run again. The cycle continues until the delta between the input and output properties of the pooled stream are within specified tolerance limits. Recursion is, therefore, a process of solving a model, examining the optimal solution, using an external program, calculating the physical property data, updating the model using the calculated data, and solving the model again. This process is repeated until the changes in the calculated data are within the specified tolerance.

The structure for pooling a number of LVGO streams into a LVGO pool is shown in Table 13-16. The first recursion row is RBALLVO, which pools all streams into a LVO pool. By convention, the streams entering the pool have a negative sign and the stream produced by pooling has a positive sign. The name of this row is always RBALxxx, where xxx is the pool tag.

The next few rows have names starting with Rxxx; for example, RSPGxxx (specific gravity of the recursed stream), RSULxxx (the sulfur of the recursed stream), and RVBIxxx (the viscosity index of the recursed stream).

The properties (specific gravity, sulfur, VBI index, etc.) of the individual streams are known and provided in the model. The properties of the pooled stream are not known; however, a guess is provided in a separate PGUESS table, whose format follows:

	SG	SUL	VBI
LVO	0.8728	1.855	70.38

These PGUESS entries replace the placeholders (999) under column LVO and the first cycle of solving the matrix begins.

Suppose, after the model is solved based on these properties of the LVO pool, the activities of the vectors in the pooling model are as follows:

W11 = 0
W21 = 0
W31 = 0
L15 = 0
L25 = 0
L26 = 6.1380
MI4 = 10.9150

Table 13-16
LVGO Pooling

TABLE SLVO	LVGO POOLING TEXT	W11	W21	W31	L15	L25	L26	MI4	LVO
VBALW11	1 VDU WGO (FUEL MODE)	1							
VBALW21	1 VDU WGO (ASP MODE)		1						
VBALW31	1 VDU WGO (BSGO MODE)			1					
VBALL15	5VDU LVGO Normal				1				
VBALL25	5 VDU LVGO Asphalt					1			
VBALL26	6VDU LVGO (65 MBPD)						1		
VBALMI4	4A M/I DSL							1	
VBALLVO	LVGO POOL	−1	−1	−1	−1	−1	−1	−1	
	LVGO POOL								
RBALLVO		−1	−1	−1	−1	−1	−1	−1	1
RSPGLVO		−0.8672	−0.8498	0.8672	−0.8739	−0.8742	−0.8726	−0.8692	999
RSULLVO		−1.4977	−1.4676	1.4977	−1.5092	−1.5097	−1.5070	−1.5011	999
RVBILVO		−54.1025	29.0000	−54.1025	−82.4343	−63.0000	−78.1173	−54.3376	999

NOTE: THE COLUMN COEFFICIENTS IN RSULLVO ROW ARE THE PRODUCT OF THE SPECIFIC GRAVITY AND SULFUR (SG* SULFUR) OF THE BLEND COMPONENTS.

Substituting these values in row RSPGLVO gives

$$\text{S. gravity of LVO pool} = \frac{6.1380^{*}(0.8726) + 10.9150^{*}(0.8692)}{(6.1380 + 10.9150)}$$

$$= 0.8704$$

$$\text{Sulfur of LVGO pool} = \frac{6.1380^{*}1.5070 + 10.9150^{*}(1.5011)}{(6.1380 + 10.9150)}$$

$$= 1.5032$$

$$\text{Similarly VBI} = 62.896$$

The model is next run with these pool properties and the pool properties recalculated, based on new activities of the vectors. If the recalculated pool properties are unchanged or within the tolerance limits of the earlier values, the recursion process is stopped and the solution is said to have converged.

DISTRIBUTIVE RECURSION

In the simple recursion process, the difference between the user's guess and the optimum solved value is calculated in an external program, updated, and resolved. The distributive recursion model moves the error calculation procedure from outside the linear program to the LP matrix itself. With this arrangement, the optimum that is reached has the physical property data for all recursed streams exactly matching the composition of the pool used to create those properties.

After the current matrix is solved, using the initial physical property estimates or guesses, the new values are computed and inserted into the matrix for another LP solution. The major difference between distributive recursion and normal recursion is the handling of the difference between the actual solution and the guess. This difference is referred to as an *error*. When a user guesses at the recursed property of a pooled stream, the "error" is created because the user guess is always incorrect. The distributive part of the distributive recursion is that this error is distributed to where the quality is being used.

A pooled stream can go to a number of product grades or become feed to a process unit. The error vector is distributed wherever the pooled stream property is used. In other words, it can be said that the pooled stream properties are represented by two vectors, one is the initial guess of the property and the second is the error or correction vector that seeks to bring the property in line with property computed from the composition of the blend.

EXAMPLE 13-2

Consider three catalytic reformate streams, R90, R95, and R98, from a catalytic reformer. The reformate streams are pooled into a single stream, SPL. The pooled reformate stream is used for blending three gasoline grades: I-390, I-395, and I-397 (see Figure 13-1).

All three gasoline grades are blended from following blend components:

BUT	Butane
LSR	Light Naphtha
LCN	Light Cat Naphtha
SPL	Pooled reformate stream

And, the pooled reformate stream is

ROW	R90	R95	R98	SPL	RRONSPL
*					
VBALR90	1				
VBALR95		1			
VBALR98			1		
*					
VBALSPL	−1	−1	−1		
RBALSPL	−1	−1	−1	1	
RRONSPL	90	95	98	94	1

To start with, the recurse process, an initial guess, is made as to the RON of the pooled stream. Let us assume it is 94 as shown in the matrix, in row RRONSPL.

Now, an error vector, RRONSPL, is introduced in the matrix to absorb any error made in estimating the RON of the pooled octane (i.e., 94).

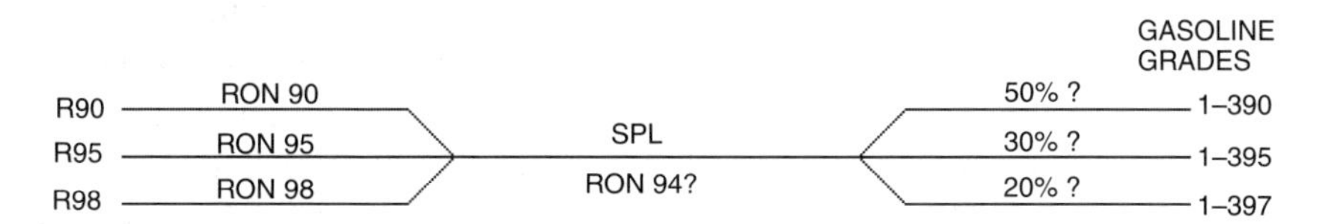

Figure 13-1. Principle of distributive recursion.

Let us assume that activity of columns R90, R95, and R98 are 5, 3, and 2, respectively, and calculate the activity of the error vector RRONSPL.

ROW	R90	R95	R98	SPL	RRONSPL
COL ACTIVITY	5	3	2		
RBALSPL	−1	−1	−1	1	
RRONSPL	−90	−95	−98	94	1
ROW ACTIVITY	−450	−285	−196	940	−9

By column arithmetic, the activity of the error vector RRONSPL is computed at −9.

This error is distributed in all the grades where the pooled stream is used. To start the distributive recursion process, a guess is made as to the distribution of the error in the three gasoline grades as follows:

I-390	50%
I-395	30%
I-397	20%

We consider the blending and octane balance of all these three gasoline grades. For I-390,

	BVBL390	BBUT390	BLSR390	BLCN390	BSPL390
EVBL390	1	−1	−1	−1	−1

VBL390 is volume blend balance. E shows that it is an equality row. For the RON of I-390,

	BVBL390	BBUT390	BLSR390	BLCN390	BSPL390	RRONSPL
NRON390	90.2	−93	−61.8	−91.2	94	(= −9*0.50)

Since I-390 gasoline happens to be one of the gasoline grades where pooled reformate stream SPL is being disposed of, we use the guessed value of its octane number for computing the RON of the I-390 blend. Also, a part of the error vector, RRONSPL (50%), is also included in this row. This error vector is designed to correct the error in guessing the octane number of SPL, and this aids in faster convergence of the solution to the optimum solution.

The matrix structures for I-395 and I-397 blending are similar, except for proportion of error vector included, which would be equal to the assumed distribution of the error vector.

For the RON of I-395,

	BVBL395	BBUT395	BLSR395	BLCN395	BSPL395	RRONSPL
NRON395	90.2	−93	−61.8	−91.2	94	= (− 9)*0.30

For the RON of I-397,

	BVBL397	BBUT397	BLSR397	BLCN397	BSPL397	RRONSPL
NRON397	90.2	−93	−61.8	−91.2	94	= (− 9)*0.20

The matrix representation of the pooled stream, if it becomes feed to another process unit, is discussed under delta-based modeling.

OBJECTIVE FUNCTION

The matrix picture discussed so far is not a complete LP model. To drive the optimization process, an extra row, called the *objective function*, is needed. The optimization process either minimizes or maximizes this function, depending on whether the function represents the refinery operating cost or profit. All cost vectors must have an entry in the objective function row. The coefficients in this row are the costs per unit of product produced. The cost of crude and other feeds, such as natural gas, or utilities are entered with a negative coefficient if the objective function represent overall profit to the refinery.

The unit cost data, which become the coefficients of the variables in the objective function row, are retrieved from following data tables of the model: the BUY table for crude and other feedstock prices, the SELL table for all product prices, or the UTILBUY (utility buy) table for all utilities, such as refinery fuel, electricity, cooling water, catalyst, and chemical costs.

OPTIMIZATION STEP

The simplex procedure for optimizing a set of linear equations was originally introduced in 1946 by Danzig[4] of the Rand Corporation (USA).

It did not become really popular until it was computerized in 1950s. In essence, the procedure is to first find a solution, any solution, that satisfies all the simultaneous equations. This may be as simple as assigning 0 to all unknowns, although this, by no means, assures a valid solution. Usually, it is best to start with a previous solution to a similar problem. Then, the activity of each matrix column or unknown is examined and the one is selected that yields the largest profit or, if no revenues are shown, the minimum cost per unit of use. Next, each row is examined to determine which equation restricts the use of this activity to the smallest value before the other activities in the equation are forced to go negative.

For example, in the equation $x + 2y = 10, x$ restricts the activity of y to equal to or less than 5. Otherwise, x has to take a negative value to satisfy the equation.

The activity for the most restrictive equation is solved ($y = 5 - x/2$), and the solution substituted in all other equations containing this activity, including the objective function. In fact, this activity has been made part of the solution, or in the LP jargon, "brought into the basis." The matrix element or coefficients are modified, and the procedure is repeated. When selection of any activity reduces profit or increases cost, as indicated by negative coefficient in the modified objective function, the procedure is concluded.

SOLUTION CONVERGENCE

Remember that properties of all pooled streams are assumed or guessed to start the optimization process. After the first cycle, the optimum activities of all columns or variables are known, the properties of all blend streams are again computed, and the matrix reoptimized. This process is repeated until there are no changes in the input and output properties of the streams. The optimized matrix is next sent to the "report writer," which prints a report in a preset format using matrix data (see Figure 13-2).

INTERPRETING THE SOLUTION

The output from the optimizer is in the form of an unscrambled report, which is a listing by rows, then by columns, in the same order as the input

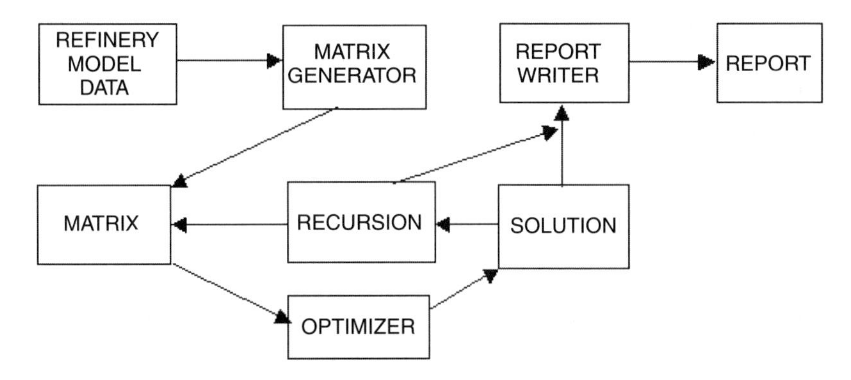

Figure 13-2. Refinery LP system.

in the original matrix. The basis is the collection of activities in the matrix, which are a part of the solution.

ROWS

The output for each row gives the row number and the row name supplied by the user.

The abbreviations used in the rows are:

BS. "Basis" is an indication that, in the final solution, this row is not limited to its upper or lower limit.

LL. "Lower limit" is an indication that the final solution is limited at a lower level by a lower bound established by this row.

UL. "Upper limit" is an indication that the final solution is limited at an upper level by an upper bound established by this row.

EQ. "Equal" means that the final solution is restrained to a fixed right-hand side. Such rows show upper and lower bounds that are identical.

Activity

This is the value of the left-hand side of the equation that the row represents, with values of the equation unknown supplied from final solution. For the objective function, activity is the maximum profit or minimum cost.

Slack Activity

This is the difference between upper (or lower) limits and the row activity just described. Values for slack activities are given only for rows that have “BS” status.

Lower Limit

This is the value given to the right-hand side in the original matrix that left-hand side must be greater than or equal to.

Upper Limit

This is the value given to the right-hand side in the original matrix that left-hand side must be less than or equal to.

Dual Activity

This is also known as *marginal cost*, *shadow price*, or *pi value*.

The pi value represents the rate of change of the objective function as the right-hand side of the row is increased. It is the total change in maximum profit or total minimum cost due to relaxing the upper or lower limit of a row by one unit. For example, if it is a crude availability row, it reflects the value of an incremental barrel of crude refined. It has nonzero value only if the limit is constraining. Runs that have the status of BS in the optimum solution show no dual activity. Marginal values are valid only over a small range around the optimal solution and should be used with caution.

COLUMNS

The output for each column gives

Column number. The column number starting from wherever the row numbers leave off.

Column name. Column names may be supplied by user or generated by matrix generation programs from input data tables.

Activity. This is the value of a variable in the final solution. Some variables in the basis may have no activity at all. Variables with zero activity may be in the basis because a certain number of variables (equal to the number of rows) are required to be in the basis by the nature of LP method.

Input costs. These are the coefficients found in the objective function row selected for optimization, and they allow total cost and revenues to be accounted for in the objective function row.

Lower limit. This is the lower limit (called *lower bound*) that value of an activity can have.

Upper limit. This is the upper limit (called *upper bound*) that value of an activity can have.

Reduced cost. This is also known as the *D-J* or *Delta-J* value. It is a reduction in net profit or increase in minimum cost, if an activity, not in the basis, is brought into the basis and given the value of 1. This is cost of using an activity that is not part of optimal solution. All reduced costs are zero or positive in an optimal solution.

REPORT WRITER PROGRAMS

The Report Writer Program retrieves data from the unscrambled solution and rearranges them in a more user-friendly format using a special programming language. A typical refinery LP solution may present the matrix solution report in the following format:

1. A summary report, showing an overall refinery material balance, starting with crude and other inputs; the overall production slate; and the refinery profit from operation or an economic summary.
2. The material balance of all refinery units.
3. A product blending report, showing the blend composition and properties of all blended product grades.

DELTA-BASED MODELING

Delta-based modeling (DBM) is a technique used to predict the yields and properties of the process units, when the yields and the properties are a function of the feed quality. In many situations, the feed is a pool of streams whose composition must be determined by the optimization process itself. Delta-based modeling is especially useful in these situations when combined with distributive recursion techniques. To implement the DBM, a feed quality parameter is defined that can be measured easily and related to the yield of the unit. For example, the yield of reformate is known to be a function of naphthene plus aromatic (N + A) content of the feed. Suppose the base case reformer yields are defined for a feed with an (N + A) content of 30.

In most LP applications, the feed is a pool of streams whose composition must be determined by the optimization process. Therefore, the properties of the feed to the unit are unknown. After pooling various cat reformer feed streams, if the model computes the reformer feed at (N + A) content at 45, a yield correction vector corrects the base yields to correspond to a feed with an (N + A) content of 45.

EXAMPLE 13-3

Consider the following cat cracker model (FCCU Table):

	SFCUBAS	SFCUSUF	SFCUCFP
VBALC4U	−0.0322	0.0047	
VBALPOR	−0.0381	0.0016	
VBALLCN	−0.2990	0.0307	
VBALMCN	−0.0670	−0.0067	
VBALHCN	−0.1150	−0.0037	
VBALLCO	−0.2600	0.0192	
VBALHDO	−0.2170	−0.0359	
ECHGFCU	1		−1
ESUFFCU	0.284	1.0000	2.986

The activity of the column SFCUCFP is made equal to the total feed to the cat cracker unit by user-defined equality row ECGHFCU. The cat cracker feed pool is formed in another submodel and has only one disposition, which is the cat cracker unit. Thus, if the activity of the column SFCUCFP is 10 mbpcd, the activity of the column SFCUBAS is also made equal to 10 mbpcd, as they are driven by the row ECHGFCU as follows:

	SFCUBAS	SFCUCFP	
ECHGFCU	1	−1	= 0

The base yield of the FCCU is determined for a sulfur content of 0.284% by weight. The FCCU feed pool is formed by combining a number of streams. Suppose, by the recursion process, the sulfur content of FCCU feed pool (SFCUCFP) is found to be 2.986%. In this case, the base case yields (FCUBAS) are corrected by a correction vector SFCUSUF. The numbers in this column are adjusted to the yield of FCCU for a 1% change in the sulfur content of the feed.

The activity of this column is determined by the row EFCUSUF, as follows:

	SFCUBAS	SFCUSUF	SFCUCFP	
ACTIVITY	10	X	10	
ECHGFCU	1		−1	=0
EFCUSUF	0.284	1.0	−2.986	

By matrix arithmetic,

$$EFCUSUF = 10^{*}0.284 + X + (-10^{*}2.986) = 0$$

$$X = 27.02$$

Let us next see how it effects the yields from unit SFU.

	SFCBAS	SFCSUF	TOTAL, mb	YIELD %
ACTIVITY	10	27.02		
VBALC4U	10*(−0.0322) = −0.322	27.02*0.0047 = 0.1260	−0.1960	−0.0196
VBALPOR	10*(−0.0381) = −0.0381	27.02*0.0016 = 0.0440	−0.3370	−0.0337
VBALLCN	10*(−0.2990) = −2.990	27.02*0.0307 = 0.83	−2.1600	−0.2160
VBALMCN	10*(−0.0670) = −0.6700	27.02*(−0.0067) = −0.1800	−0.8500	−0.0850
VBALHCN	10*(−0.1150) = −1.1500	27.02*(−0.0037) = −0.1000	−1.2500	−0.1250
VBALLCO	10*(−0.2600) = −2.6000	27.02*(0.0192) = 0.5200	−2.0800	−0.2080
VBALHDO	10*(−0.2170) = −2.170	27.02*(−0.0459) = −1.2402	−3.4102	−0.3410

The base yields were derived for a sulfur content of 0.284 wt %. As the sulfur content of the FCCU feed was 2.986 wt %, the shift vector SFCUSUF adjusted the base yield to that for a sulfur content of 2.986%.

Therefore, we see that the yields of all products from the FCCU are changed as follows:

FEED	0.284% SULFUR	2.986% SULFUR
C4U	−0.0322	−0.0196
POR	−0.0381	−0.0337
LCN	−0.2990	−0.2160
MCN	−0.0670	−0.0850
HCN	−0.1150	−0.1250
LCO	−0.2600	0.2080
HDO	−0.2170	−0.3410

It is possible for the model to have more than one shift vector. This model structure is similar to the one discussed earlier, as shown in the following example.

EXAMPLE 13-4

Let us consider a delta-based model segment with two shift vectors instead of one:

	SFCUBAS	SFCUSUF	SFCURES	SFCUCFP
VBALC4U	−0.0300	0.0047	−0.0012	
VBALPOR	−0.0370	0.0016	−0.0011	
VBALLCN	−0.3290	0.0307	−0.0114	
VBALMCN	−0.0980	−0.0067	−0.0020	
VBALHCN	−0.0790	−0.0037	0.0000	
VBALLCO	−0.2380	0.0192	−0.0007	
VBALHCO	−0.0800	0.0000	0.0186	
VBALHDO	−0.1300	−0.0359	−0.0020	
ECHGFCU	1			−1
ESUFFCU	0.284	1.0000		2.986
ERESFCU	0		1.0	5.000

Here, the first shift vector corrects the base case yields for different sulfur content of the feed, while the second shift vector modifies the base case yields for inclusion of 5.0% residuum in the FCCU feed. It is implied that the feed pool resid content is computed elsewhere by recursion of pooling streams and treats resid content as a property of the feed.

DATA FOR DELTA-BASED MODELS

The data used in the delta-based model must be developed by the user to reflect the parameters and their effect on the process yield. In the context of the refinery streams, only streams properties that are generally measured can be considered. For example, the cat reformer yield can be related to the naphthene plus aromatic content of the feed if the refinery has reference data to correlate the cat reformer yield at a given severity vs. (naphthene plus aromatic) content of the feed.

Care has to be taken that the shift vectors do not extrapolate data beyond the range for which these were developed. Also, as the shift vectors can take either positive or negative values, these should be declared free in the BOUND section of the LP model.

ATMOSPHERIC CRUDE DISTILLATION AND VDU MODELING

LP models for crude and vacuum distillation can be constructed exactly in the same manner as done in the case of other process unit submodels. Most refineries may have more than one crude unit and may be processing more than one crude oil. Also, a crude unit may operate in more than one mode; for example, one crude unit mode may maximize the production of naphtha and another may maximize the production of kerosene. It is, therefore, convenient to enter all data on yields and properties of the various cuts from different crude oils and crude and vacuum distillation units present and their utility consumption, pooling of various cuts from crude and vacuum units in three tables for easy updating.

ASSAYS TABLE

Data in the form of yields of various crude on different CDUs and properties of cut produced is entered in a single table, the Assay Table (Table 13-17). This table does not directly generate any matrix but only provides data for crude and vacuum units modeling. The yields in the Assay Table could be either in volume or weight units. The column names in the table are three character operation modes[5] of various crude columns. There are four type of rows in this table.

The first set of rows is used to provide cut yields. The row names are VBALxxx, WBALxxx, or DBAL.xxx. VBAL and WBAL indicate a volume or weight basis yield. The tag xxx is the name of the cut produced from the distillation. The DBAL row defines a cut being produced not on the crude distillation unit but from a downstream unit. Examples of such cuts in the refinery are numerous. For example, a crude distillation column may produce a broad cut containing naphtha and some kerosene. This cut is further fractionated into naphtha and kerosene in a downstream naphtha fractionation column. In this case, as naphtha is not being actually produced on the crude column but somewhere else, it will be represented in the Assay Table by the DBAL row instead of the VBAL row.

The second set of rows may be used to specify atmospheric or vacuum unit utility consumption that is crude specific. The row name for this set is in the form of UATMxxx or UVACxxx, where xxx is the utility consumed.

A third set of rows may be used to specify atmospheric or vacuum unit capacity consumption values that are crude specific. The row name for this set is CCAPATM or CCAPVAC, and the entries for each crude are the units of capacity consumed per unit charged to the atmospheric and vacuum towers.

The fourth set of rows is used to provide cut properties. The row names for this set are Ipppxxx where ppp is the property (SPG, SUL, API, etc.) and xxx is the cut name. The first letter, I, indicates that this is property data. The second through fourth characters of the row name identify the property, while the final three characters identify the process stream to which this property applies:

ISPGLN1	Specific gravity of LN1.
ISULW26	Sulfur content of cut W26.
IVBIL26	Viscosity blend index of cut L26.

The coefficients in the I rows are the values of the specific property of a specified stream. I rows do not themselves appear in the matrix. The LP model uses this data by multiplying, for example, ISPGLN1, with volumes in row VBALLN1 and placing the result in recursion row RSPGLN1 for computing the specific gravity of a blend of streams.

Table 13-17
Crude Assays

*	MODE	1CDU A1N	1CDU A1K	2CDU A2N	2CDU A2K	3CDU A3N	3CDU A3K	4CDU A4N	4CDU A4K	5CDU A5N	5CDU A5K	1CDU B1N	1CDU B1K	2CDU B2N	2CDU B2K	3CDU B3N	3CDU B3K	4CDU B4N	4CDU B4K	5CDU B5N	5CDU B5K
*	Atmos Unit Yields																				
VBALGA1	CDU Off Gas (C3 -)	0.0051	0.0051	0.0049	0.0049	0.0049	0.0049	0.0056	0.0056	0.0051	0.0051	0.0071	0.0071	0.0060	0.0060	0.0060	0.0060	0.0106	0.0106	0.0060	0.0060
VBALC41	CDU C3/C4	0.0194	0.0194	0.0205	0.0205	0.0205	0.0205	0.0148	0.0148	0.0194	0.0194	0.0129	0.0129	0.0151	0.0151	0.0151	0.0151	0.0065	0.0065	0.0151	0.0151
VBALLS1	CDU LSRs	0.0744	0.0744	0.0694	0.0694	0.0616	0.0616	0.1063	0.1063	0.0714	0.0714	0.0494	0.0494	0.0445	0.0445	0.0381	0.0381	0.0791	0.0791	0.0465	0.0465
VBALMN4	4A MSR							0.0874	0.0585									0.0843	0.0625		
VBALHN1	CDU HSR	0.0984	0.0765	0.0894	0.0565	0.1032	0.0873	0.1114		0.0964	0.0745	0.1001	0.0774	0.0912	0.0586	0.1026	0.0878	0.1192		0.0952	0.0734
VBALHA4	CDU Arab HSR for KRU								0.1402												
VBALHB4	CDU Bah HSR for KRU																		0.1409		
VBALKN1	CDU Kerosene	0.1651		0.1599		0.1730		0.0626	0.0626	0.1601		0.1664		0.2010		0.1802		0.0677	0.0677	0.1693	
VBALKA1	CDU Arab Ker for KRU		0.1870		0.2327		0.1889				0.1820										
VBALKB1	CDU Bah Ker for KRU												0.1891		0.2336		0.1950				0.1911
VBALDN1	CDU Light Dsl	0.1153	0.1153	0.1004	0.1004	0.2138	0.2138	0.0726	0.0726	0.1323	0.1323	0.1310	0.1310	0.1162	0.1162	0.2343	0.2343	0.0789	0.0789	0.1419	0.1419
VBALMI4	4A CDU Med/Int Dsl							0.1243	0.1243									0.1389	0.1389		
VBALRA1	Atmos Resid	0.5224	0.5224	0.5154	0.5154					0.5154	0.5154	0.5330	0.5330	0.5261	0.5261					0.5261	0.5261
VBALHD3	Heavy Diesels					0.0129	0.0129	0.1263	0.1263							0.0138	0.0138	0.1370	0.1370		
VBALVA3	CDU Vac Resids					0.4100	0.4100	0.2887	0.2887							0.4099	0.4099	0.2779	0.2779		
*		1	1	1	1	1	1	1	1	1	1	1	1	1	1	1	1	1	1	1	1
*	Total Yield																				
CCAPATM	CDU CAPACITY	1	1	1	1	1	1	1	1	1	1	1	1	1	1	1	1	1	1	1	1
CCAPVAC	VDU CAPACITY					1	1	1	1							1	1	1	1		
UATMFUL	GAS MSCF/BBL CRUDE	0.1613	0.1613	0.1613	0.1613	0.1963	0.1963	0.1753	0.1753	0.1613	0.1613	0.1613	0.1613	0.1613	0.1613	0.1963	0.1963	0.1753	0.1753	0.1613	0.1613
*	DEFERRED CUT YIELDS (Vacuum units)																				
*	1 VDU Normal Mode																				
DBALW11	1VDU WGO ON CRUDE	0.0462	0.0462	0.0446	0.0446							0.0545	0.0545	0.0522	0.0522						
DBALL11	1VDU DGO ON CRUDE	0.2544	0.2544	0.2499	0.2499							0.2709	0.2709	0.2663	0.2663						
DBALV11	1VDU VR ON CRUDE	0.2218	0.2218	0.2208	0.2208							0.2076	0.2076	0.2075	0.2075						
*	1 VDU Asphalt Operation																				
DBALLW1	1VDU LWGO ON CRUDE											0.0011	0.0011	0.0011	0.0011						
DBALW21	1VDU WGO ON CRUDE											0.0374	0.0374	0.0370	0.0370						
DBALL21	1VDU DGO ON CRUDE											0.3295	0.3295	0.3232	0.3232						
DBALBS1	1VDU BSGO ON CRUDE											0.0245	0.0245	0.0242	0.0242						
DBALAS1	1VDU ASP ON CRUDE											0.1404	0.1404	0.1407	0.1407						
*	1 VDU BSGO PRODUCTION																				
DBALW31	1VDU WGO ON CRUDE	0.0322	0.0322	0.0312	0.0312																
DBALL31	1VDU DGO ON CRUDE	0.3113	0.3113	0.3051	0.3051																
DBALBS3	1VDU BSGO ON CRUDE	0.0199	0.0199	0.0191	0.0191																
DBALV31	1VDU VACRES ON CRUDE	0.1591	0.1591	0.1601	0.1601																

Table 13-17
Continued

*	MODE	1CDU A1N	1CDU A1K	2CDU A2N	2CDU A2K	3CDU A3N	3CDU A3K	4CDU A4N	4CDU A4K	5CDU A5N	5CDU A5K	1CDU B1N	1CDU B1K	2CDU B2N	2CDU B2K	3CDU B3N	3CDU B3K	4CDU B4N	4CDU B4K	5CDU B5N	5CDU B5K
*	check yield = AR yld	0.5224	0.5224	0.5154	0.5154																
DBALW15	5VDU WGO ON CRUDE	0.0031	0.0031	0.0031	0.0031	0.0033	0.0033			0.0031	0.0031	0.0032	0.0032	0.0032	0.0032	0.0000	0.0000			0.0032	0.0032
DBALL15	5VDU LVGO ON CRUDE	0.1328	0.1328	0.1269	0.1269	0.0277	0.0277			0.1269	0.1269	0.1476	0.1476	0.1420	0.1420	0.0407	0.0407			0.1420	0.1420
DBALH15	5VDU HVGO ON CRUDE	0.2226	0.2226	0.2207	0.2207	0.2085	0.2085	0.1055	0.1055	0.2207	0.2207	0.2327	0.2327	0.2307	0.2307	0.2126	0.2126	0.1078	0.1078	0.2307	0.2307
DBALV15	5VDU VR ON CRUDE	0.1638	0.1638	0.1647	0.1647	0.1705	0.1705	0.1833	0.1833	0.1647	0.1647	0.1495	0.1495	0.1502	0.1502	0.1566	0.1566	0.1701	0.1701	0.1502	0.1502
*	5 VDU Asphalt Operation																				
DBALW25	5VDU WGO ON CRUDE																			0.0032	0.0032
DBALL25	5VDU LVGO ON CRUDE																			0.1422	0.1422
DBALH25	5VDU HVGO ON CRUDE																			0.2416	0.2416
DBALAS5	5VDU ASP ON CRUDE																			0.1391	0.1391
*	6VDU AT 65 MBPD (SVD7)																				
DBALW26	6VDU WGO ON CRUDE	0.0021	0.0021	0.0026	0.0026					0.0026	0.0026	0.0021	0.0021	0.0026	0.0026					0.0026	0.0026
DBALL26	6VDU LVGO ON CRUDE	0.1281	0.1281	0.1212	0.1212	0.0232	0.0232			0.1212	0.1212	0.1427	0.1427	0.1351	0.1351	0.0279	0.0279			0.1351	0.1351
DBALH26	6VDU HVGO ON CRUDE	0.2221	0.2221	0.2207	0.2207	0.1987	0.1987	0.0838	0.0838	0.2207	0.2207	0.2337	0.2337	0.2338	0.2338	0.2116	0.2116	0.0900	0.0900	0.2338	0.2338
DBALV26	6VDU VR ON CRUDE	0.1701	0.1701	0.1709	0.1709	0.1881	0.1881	0.2049	0.2049	0.1709	0.1709	0.1544	0.1544	0.1545	0.1545	0.1704	0.1704	0.1879	0.1879	0.1545	0.1545
*	VAC YIELDS (VOLUME) AS PROPERTY OF ATM RESID (1, 2, 5 CDU)																				
*	1 VDU Normal Mode																				
IW11RA1	1VDU on Atm Res-WGO	0.0885	0.0885	0.0866	0.0866							0.1023	0.1023	0.0993	0.0993						
IL11RA1	1VDU on Atm Res-DGO	0.4869	0.4869	0.4849	0.4849							0.5083	0.5083	0.5063	0.5063						
IV11RA1	1VDU on Atm Res-VRes	0.4245	0.4245	0.4285	0.4285							0.3895	0.3895	0.3945	0.3945						
*	CHECK	1	1	1	1							1	1	1	1						
*	1 VDU Asphalt mode																				
ILW1RB1	1VDU LWGO ON RESID											0.0020	0.0020	0.0020	0.0020						
IW21RB1	1VDU WGO ON AT RESID											0.0703	0.0703	0.0703	0.0703						
IL21RB1	1VDU DGO ON ATRESID											0.6183	0.6183	0.6143	0.6143						
IBS1RB1	1VDU BSGO ON AT RES.											0.0460	0.0460	0.0460	0.0460						
IAS1RB1	1VDU ASP ON AT RESID											0.2635	0.2635	0.2675	0.2675						
*	CHECK											1	1	1	1						
*	1 VDU WITH BSGO PRODUCTION																				
IW31RA1	1VDU WGO ON AT RESID	0.0616	0.0616	0.0606	0.0606																
IL31RA1	1VDU LVGO ON ATRESID	0.5959	0.5959	0.5919	0.5919																
IBS3RA1	1VDU BSGO ON AT RES.	0.0380	0.0380	0.0370	0.0370																
IV31RA1	1VDU VRES ON AT RESID	0.3045	0.3045	0.3105	0.3105																
*	CHECK	1	1	1	1																
*	5VDU AT 33 MBPD																				
IW15RA1	5VDU WGO ON AT RESID	0.0060	0.0060	0.0060	0.0060					0.0060	0.0060	0.0060	0.0060	0.0060	0.0060						
IL15RA1	5VDU LVGO ON ATRESID	0.2542	0.2542	0.2462	0.2462					0.2462	0.2462	0.2770	0.2770	0.2700	0.2700						

IH15RA1	5VDU HVGO ON ATRESID	0.4262	0.4262	0.4282	0.4282					0.4282	0.4282	0.4365	0.4365	0.4385	0.4385						
IV15RA1	5VDU VR ON AT RESID	0.3136	0.3136	0.3196	0.3196					0.3196	0.3196	0.2805	0.2805	0.2855	0.2855						
*	CHECK	1	1	1	1					1	1	1	1	1	1						
*	5 VDU Asphalt Mode																				
IW25RB1	5VDU WGO ON AT RESID																			0.0060	0.0060
IL25RB1	5VDU LVGO ON ATRESID																			0.2703	0.2703
IH25RB1	5VDU HVGO ON ATRESID																			0.4593	0.4593
IAS5RB1	5VDU ASP ON AT RESID																			0.2645	0.2645
*	CHECK																			1	1
	VDU YIELDS (LV) AS PROPERTY OF ATM RESIDS EX CDU 1, 2 & 5																				
*	6 VDU AT 50 MBPD																				
IW16RA1	6VDU WGO ON ATRESID	0.0040	0.0040	0.0050	0.0050					0.005	0.005	0.005	0.005	0.006	0.006						
IL16RA1	6VDU LVGO ON ATRESID	0.2462	0.2462	0.2362	0.2362					0.236225	0.236225	0.267768	0.267768	0.256768	0.256768						
IH16RA1	6VDU HVGO ON ATRESID	0.4362	0.4362	0.4402	0.4402					0.440166	0.440166	0.449519	0.448519	0.455519	0.455519						
IV16RA1	6VDU VR ON AT RESID	0.3136	0.3136	0.3186	0.3186					0.318609	0.318609	0.278713	0.278713	0.281713	0.281713						
*	CHECK	1	1	1	1					1	1	1	1	1	1						
*																					
*	VAC YLD AS PROPERTY OF VAC RESID (3, 4 CDU)																				
*	5 VDU Normal mode																				
IW15VA3	5VDU WGO ON V. RESID					0.0080	0.0080									0	0				
IL15VA3	5VDU LVGO ON V.RESID					0.0675	0.0675									0.09926	0.09926				
IH15VA3	5VDU HVGO ON V.RESID					0.5086	0.5086	0.3653	0.3653							0.518779	0.518779	0.388033	0.388033		
IV15VA3	5VDU VR ON V. RESID					0.4158	0.4158	0.6347	0.6347							0.381962	0.381962	0.611967	0.611967		
*	Check					1	1	1	1							1	1	1	1		
*	6 VDU AT 65 MBPD																				
IW26VA3	6VDU WGO ON V. RESID																				
IL26VA3	6VDU LVGO ON V.RESID					0.0565	0.0565									0.068011	0.068011				
IH26VA3	6VDU HVGO ON V.RESID					0.4846	0.4846	0.2903	0.2903							0.516282	0.516282	0.32379	0.32379		
IV26VA3	6VDU VR ON V. RESID					0.4588	0.4588	0.7097	0.7097							0.415707	0.415707	0.67621	0.67621		
*	Check					1	1	1	1							1	1	1	1		
*	ATMOS CUT PROPERTIES																				
												Properties in BLNGASO : RVP, 230, FPI, OLE, LSR									
*	LS1 – CDU LSR																				
ISPGLS1	LSR – Spec Grav	0.6563	0.6563	0.6666	0.6666	0.6663	0.6663	0.6762	0.6762	0.6695	0.6695	0.6674	0.6674	0.6639	0.6639	0.6626	0.6626	0.6795	0.6795	0.6662	0.6662
ISULLS1	LSR – Sulfur	0.018	0.018	0.017	0.017	0.016	0.016	0.018	0.018	0.017	0.017	0.003	0.003	0.002	0.002	0.003	0.003	0.005	0.005	0.003	0.003
I130LS1	LSR – %VAP @ 130	50.0	50.0	56.0	56.0	60.0	60.0	38.0	38.0	51.0	51.0	40.0	40.0	43.0	43.0	51.0	51.0	24.0	24.0	42.0	42.0
IRONLS1	LSR-RON	61.5	61.5	61.5	61.5	63.5	63.5	54.5	54.5	61.5	61.5	61.0	61.0	61.0	61.0	64.5	64.5	51.5	51.5	61.0	61.0
IR30LS1	LSR – RON 3CC	82.5	82.5	82.5	82.5	84.5	84.5	75.5	75.5	82.5	82.5	82.0	82.0	82.0	82.0	85.5	85.5	72.4	72.4	82.0	82.0
IMONLS1	LSR – MON	60.5	60.5	60.5	60.5	62.5	62.5	53.5	53.5	60.5	60.5	60.0	60.0	60.0	60.0	63.5	63.5	50.5	50.5	60.0	60.0
IM30LS1	LSR – MON 3CC	81.5	81.5	81.5	81.5	83.5	83.5	74.5	74.5	81.5	81.5	81.0	81.0	81.0	81.0	84.5	84.5	71.4	71.4	81.0	81.0
IVLPLS1	LSR – VLPT	47.0	47.0	46.0	46.0	46.0	46.0	47.0	47.0	48.0	48.0	48.0	48.0	47.0	47.0	47.0	47.0	51.0	51.0	48.0	48.0
ISPGMN4	MSR – Spec Grav							0.7502	0.7456									0.7453	0.7415		
ISULMN4	MSR – Sulfur							0.037	0.025									0.016	0.014		
IRVPMN4	MSR – RVP							4.9	4.9									4.2	4.2		
IFPIMN4	MSR – Freeze Index							9	9									9	9		
IN2AMN4	MSR – N+2A							39.6	36.4									47.3	43.0		

Table 13-17
Continued

*	MODE	1CDU A1N	1CDU A1K	2CDU A2N	2CDU A2K	3CDU A3N	3CDU A3K	4CDU A4N	4CDU A4K	5CDU A5N	5CDU A5K	1CDU B1N	1CDU B1K	2CDU B2N	2CDU B2K	3CDU B3N	3CDU B3K	4CDU B4N	4CDU B4K	5CDU B5N	5CDU B5K
ISPGHN1	HSR – SG	0.7364	0.7322	0.7338	0.7300	0.7315	0.7284	0.7940		0.7348	0.7307	0.7331	0.7284	0.7302	0.7240	0.7282	0.7248	0.7936		0.7309	0.7263
ISULHN1	HSR – SUL	0.033	0.029	0.029	0.026	0.031	0.028	0.143		0.028	0.027	0.017	0.013	0.013	0.010	0.014	0.012	0.121		0.014	0.011
IRVPHN1	HSR – RVP	7.6	7.6	8.3	8.3	9.7	9.7	0.7		7.6	7.6	6.9	6.9	7.6	7.6	9.0	9.0	0.7		7.6	7.6
IN2AHN1	HSR – N + 2A	34.9	34.9	34.1	35.3	32.7	31.5	58.2		34.7	34.1	40.6	38.5	39.5	36.6	38.1	36.6	64.6		39.7	37.6
IDIIHN1	HSR – Diesel Index							66.4										66			
IPPIHN1	HSR – Pour Index							36										36			
I144HN1	HSR – 144 Index							−191										−191			
I154HN1	HSR – 154 Index							−175										−175			
IVBIHN1	HSR – H Value							−267										−282			
IFPIHN1	HSR – Freeze Index							32.3										28.9			
ISPGKN1	Kero – SG	0.7879	0.7865	0.7890	0.7893	0.7884	0.7873	0.8024	0.7880	0.7862	0.7855	0.7881	0.7865	0.7890	0.7892	0.7887	0.7874	0.7914	0.7894	0.7863	0.7857
ISULKN1	Kero – Sul	0.107		0.104		0.119		0.237	0.237	0.125		0.094		0.091		0.105		0.205	0.205	0.112	
IVBIKN1	Kero – H Value	−303		−291		−296		−222	−222	−310		−310		−303		−305		−233	−233	−315	
IFPIKN1	Kero – Freeze Index	33		34		34.1		75.7	75.7	29		29.6		30		29		65.5	65.5	26	
IPPIKN1	Kero – Pour Index	24	23	28	28	25	25	56	24	23	22	26	25	28	29	27	26	32	28	24	24
IDIIKN1	Kero – Diesel Index	67.1		67.2		67.2		63.2	63.2	67.6		66.4		66.4		66.4		61.8	61.8	66.9	
I144KN1	Kero – 144 Index	−196	−197	−195	−195	−196	−197	12	−196	−198	−198	−196	−197	−195	−195	−195	−196	−193	−195	−197	−198
I154KN1	Kero – 154 Index	−179	−181	−179	−178	−179	−180	8	−179	−181	−181	−179	−181	−179	−178	−179	−180	−177	−178	−181	−181
ISPGDN1	LDO – SG	0.8391	0.8391	0.8461	0.8461	0.8511	0.8511	0.8344	0.8344	0.8424	0.8424	0.8428	0.8428	0.8504	0.8504	0.8556	0.8556	0.8409	0.8409	0.8485	0.8485
ISULDN1	LDO – Sul	1.100	1.100	1.251	1.251	1.320	1.320	0.840	0.840	1.160	1.160	1.029	1.029	1.169	1.169	1.271	1.271	0.804	0.804	1.148	1.148
IPPIDN1	LDO – Pour Index	197	197	235	235	308	308	190	190	240	240	197	197	220	220	300	300	187	187	240	240
I144DN1	LDO – 144 Index	58	58	64	64	34	34	77	77	56	56	57	57	63	63	34	34	66	66	57	57
I154DN1	LDO – 154 Index	33	33	39	39	14	14	48	48	31	31	32	32	38	38	15	15	39	39	33	33
IDIIDN1	LDO – Diesel Index	57.9	57.9	56.6	56.6	57.1	57.1	57.4	57.4	57.2	57.2	55.3	55.3	53.7	53.7	52.5	52.5	54.5	54.5	54.2	54.2
IVBIDN1	LDO – H Value	−68	−68	−37	−37	−19	−19	−76	−76	−50	−50	−72	−72	−45	−45	−24	−24	−82	−82	−50	−50
ISPGMI4	M/ID – SG							0.8681	0.8681									0.8718	0.8718		
ISULMI4	M/ID – Sul							1.677	1.677									1.645	1.645		
IVBIMI4	M/ID – H Value							57	57									48	48		
IPPIMI4	M/ID – Pour Index							525	525									467	467		
IDIIMI4	M/ID – Diesel Index							51.5	51.5									49.2	49.2		
I144MI4	M/ID – 144 Index							67	67									62	62		
I154MI4	M/ID – 154 Index							42	42									39	39		
ISPGRA1	Atm Res – SG	0.9463	0.9463	0.9465	0.9465					0.9465	0.9465	0.9459	0.9459	0.9463	0.9463						
ISULRA1	Atm Res – Sul	3.14	3.14	3.17	3.17					3.17	3.17	3.38	3.38	3.41	3.41						
IVBIRA1	Atm Res – H Value	446.1	446.1	448.1	448.1					448.1	448.1	431.7	431.7	434.7	434.7						
ISPGHD3	HDO – SG					0.9058	0.9058	0.9113	0.9113							0.9088	0.9088	0.9144	0.9144		
ISULHD3	HDO – Sul					2.265	2.265	2.365	2.365							2.480	2.480	2.586	2.586		

IVBIHD3	HDO – H Value					244	244	271	271					234	234	261	261
IPPIHD3	HDO – Pour Index					1287	1287	1507	1507					1251	1251	1473	1473
I144HD3	HDO – 144 Index					58	58	57	57					57	57	55	55
I154HD3	HDO – 154 Index					38	38	37	37					37	37	36	36
ICONHD3	HDO – Con Carbon					0.04	0.04	0.07	0.07					0.04	0.04	0.07	0.07
ISPGVA3	Vac Res – SG					0.9672	0.9672	0.9965	0.9965					0.9697	0.9697	0.9993	0.9993
ISULVA3	Vac Res – Sul					3.522	3.522	4.030	4.030					3.835	3.835	4.389	4.389
IVBIVA3	Vac Res – H Value					542.0	542.0	661.2	661.2					540.0	540.0	670.7	670.7
ICONVA3	Vac Res – Con Carbon					8.93	8.93	13.4	13.4					10.63	10.63	15.3	15.3
I144VA3	Vac Res – 144 Index					39	39	31	31					38	38	29	29
I154VA3	Vac Res – 154 Index					25	25	20	20					24	24	19	19
IDIIVA3	Vac Res – Diesel Index					31.3	31.3	21.5	21.5					31.3	31.3	21.5	21.5
IPPIVA3	Vac Res – Pour Index					523.2	523.2	990.3	990.3					523.2	523.2	990.3	990.3
*	DEFERRED CUT PROPERTIES																
* 1VDU Normal mode																	
ISPGW11	1VDU WGO – SG	0.8554	0.8554	0.8536	0.8536					0.859	0.859	0.8568	0.8568				
ISULW11	1VDU WGO – Sul	1.408	1.408	1.373	1.373					1.369	1.369	1.332	1.332				
IPPIW11	1VDU WGO – Pour Ind	388	388	378	378					378	378	367	367				
IVBIW11	1VDU WGO – H Value	4	4	−4	−4					−4	−4	−12	−12				
IDIIW11	1VDU WGO – Diesel Ix	50.2	50.2	50.2	50.2					54.1	54.1	54.1	54.1				
ISPGL11	1VDU DGO – SG	0.9093	0.9093	0.9099	0.9099					0.9122	0.9122	0.9128	0.9128				
ISULL11	1VDU DGO – Sul	2.352	2.352	2.362	2.362					2.567	2.567	2.582	2.582				
IPPIL11	1VDU DGO – Pour Ind	1541	1541	1562	1562					1610	1610	1641	1641				
IVBIL11	1VDU DGO – H Value	261	261	264	264					252	252	255	255				
ISPGV11	1VDU VR – SG	1.0076	1.0076	1.0076	1.0076					1.0165	1.0165	1.0165	1.0165				
ISULV11	1VDU VR – Sul	4.029	4.029	4.029	4.029					4.640	4.640	4.639	4.639				
IVBIV11	1VDU VR – H Value	712	712	712	712					748	748	748	748				
ICONV11	1VDU VR – Con Carbon	16.6	16.6	16.6	16.6					20.1	20.1	20.1	20.1				
I144V11	1VDU VR – 144 Index	26	26	26	26					22	22	22	22				
I154V11	1VDU VR – 154 Index	17	17	17	17					15	15	15	15				
*																	
* 1VDU Asphalt Mode																	
ISPGW21	1VDU WGO – SG									0.8585	0.8585	0.846	0.846				
ISULW21	1VDU WGO – Sul									1.171	1.171	1.134	1.134				
IPPIW21	1VDU WGO – Pour Ind									277	277	268	268				
IVBIW21	1VDU WGO – H Value									−29	−29	−29	−29				
IDIIW21	1VDU WGO – Diesel Ind									54.1	54.1	54.7	54.7				
ISPGL21	1VDU DGO – SG									0.9174	0.9174	0.9183	0.9183				
ISULL21	1VDU DGO – Sul									2.691	2.691	2.712	2.712				
IPPIL21	1VDU DGO – Pour Ind									1941	1941	1972	1972				
IVBIL21	1VDU DGO – H Value									280	280	284	284				
1 VDU bsgo mode																	
ISPGW31	1VDU WGO – SG	0.8435	0.8435	0.843	0.843												
ISULW31	1VDU WGO – Sul	1.196	1.196	1.189	1.189												
IPPIW31	1VDU WGO – Pour Ind	260	260	276	276												
IVBIW31	1VDU WGO – H Value	−46	−46	−47	−47												
IDIIW31	1VDU WGO – Diesel Ind	54.1	54.1	54.7	54.7												
ISPGL31	1VDU DGO – SG	0.9149	0.9149	0.9156	0.9156					0.9149	0.9149	0.9156	0.9156				
ISULL31	1VDU DGO – Sul	2.46	2.46	2.471	2.471					2.46	2.46	2.471	2.471				

Table 13-17
Continued

*	MODE	1CDU A1N	1CDU A1K	2CDU A2N	2CDU A2K	3CDU A3N	3CDU A3K	4CDU A4N	4CDU A4K	5CDU A5N	5CDU A5K	1CDU B1N	1CDU B1K	2CDU B2N	2CDU B2K	3CDU B3N	3CDU B3K	4CDU B4N	4CDU B4K	5CDU B5N	5CDU B5K
IPPIL31	1VDU DGO – Pour Ind	1872	1872	1893	1893							1872	1872	1893	1893						
IVBIL31	1VDU DGO – H Value	289	289	292	292							289	289	292	292						
ISPGV31	1VDU VR – SG	1.0220	1.0220	1.0217	1.0217																
ISULV31	1VDU VR – Sul	4.249	4.249	4.245	4.245																
IVBIV31	1VDU VR – H Value	774	774	772	772																
ICONV31	1VDU VR – Con Carbon	21.5	21.5	21.4	21.4																
I144V31		0.22																			
I154V31		0.15																			
* 5 VDU Normal mode																					
ISPGW15	5VDU WGO – SG	0.8155	0.8155	0.8164	0.8164	0.8045	0.8045			0.8164	0.8164	0.8112	0.8112	0.8112	0.8112					0.8112	0.8112
ISULW15	5VDU WGO – Sul	0.460	0.460	0.530	0.530	0.259	0.259			0.530	0.530	0.385	0.385	0.385	0.385					0.385	0.385
IPPIW15	5VDU WGO – Pour Ind	61	61	61	61	93	93			61	61	56	56	56	56					56	56
IVBIW15	5VDU WGO – H Value	185	185	220	220	180	180			220	220	−200	−200	−200	−200					−200	−200
IDIIW15	5VDU WGO – Diesel Ind	62.5	62.5	62.6	62.6	62.1	62.1			64.4	64.4	60.8	60.8	61.3	61.3					64.2	64.2
I144W15	5VDU WGO – 144 Index	50	50	−101	−101	25	25			−101	−101	49	49	43	43					15	15
I154W15	5VDU WGO – 154 Index	24	24	−99	−99	4	4			−99	−99	23	23	18	18					−5	−5
ISPGL15	5VDU LVGO – SG	0.8716	0.8716	0.8703	0.8703	0.8854	0.8854			0.8703	0.8703	0.8756	0.8756	0.8742	0.8742	0.8856	0.8856			0.8742	0.8742
ISULL15	5VDU LVGO – Sul	1.932	1.932	1.910	1.910	2.158	2.158			1.910	1.910	1.939	1.939	1.916	1.916	2.161	2.161			1.916	1.916
IPPIL15	5VDU LVGO – Pour Index	560	560	560	560	779	779			560	560	540	540	525	525	722	722			525	525
IVBIL15	5VDU LVGO – H Value	78	78	72	72	145	145			72	72	67	67	63	63	128	128			63	63
IDIIL15	5VDU LVGO – Dsl Indx	51.2	51.2	51.5	51.5	48.4	48.4			51.5	51.5	48.4	48.4	48.7	48.7	44.8	44.8			48.7	48.7
I144L15	5VDU LVGO – 144	63.5	63.5	61.1	61.1	57.9	57.9			61	61	62.2	62.2	62.1	62.1	57.6	57.6			60	60
I154L15	5VDU LVGO – 154	39.7	39.7	37.6	37.6	36	36			38	38	38.9	38.9	38.7	38.7	35.8	35.8			37	37
*	DEFERRED CUT PROPERTIES																				
ISPGH15	5VDU HVGO – SG	0.9386	0.9386	0.9388	0.9388	0.9432	0.9432	0.9503	0.9503	0.9388	0.9388	0.9416	0.9416	0.9416	0.9416	0.9425	0.9425	0.9505	0.9505	0.9416	0.9416
ISULH15	5VDU HVGO – Sul	2.864	2.864	2.866	2.866	2.87	2.87	3.061	3.061	2.866	2.866	3.241	3.241	3.24	3.24	3.26	3.26	3.57	3.57	3.24	3.24
IPPIH15	5VDU HVGO – Pour Ind	2800	2800	2800	2800	2800	2800	3100	3100	2800	2800	3100	3100	3100	3100	3100	3100	3500	3500	3100	3100
IVBIH15	5VDU HVGO – H Value	401	401	402	402	404	404	439	439	402	402	400	400	400	400	404	404	446	446	400	400
I144H15	5VDU HVGO – 144	48	48	48	48	46	46	44	44	48	48	47	47	47	47	46	46	44	44	47	47
I154H15	5VDU HVGO – 154	31	31	31	31	30	30	29	29	31	31	30	30	30	30	30	30	29	29	30	30
ISPGV15	5VDU VR – SG	1.0252	1.0252	1.0249	1.0249	1.0208	1.0208	1.0201	1.0201	1.0249	1.0249	1.0402	1.0402	1.0372	1.0372	1.0298	1.0298	1.0275	1.0275	1.0372	1.0372
ISULV15	5VDU VR – Sul	4.49	4.49	4.49	4.49	4.48	4.48	4.39	4.39	4.49	4.49	4.91	4.91	4.91	4.91	4.88	4.88	4.86	4.86	4.91	4.91
IVBIV15	5VDU VR – VBI	782.0	782.0	781.0	781.0	776.0	776.0	770.0	770.0	781.0	781.0	814.0	814.0	813.0	813.0	804.0	804.0	800.0	800.0	813.0	813.0
ICONV15	5VDU VR – Con Carbon	21.9	21.9	21.5	21.5	21.2	21.2	20.2	20.2	21.5	21.5	26.4	26.4	26	26	25.3	25.3	23.9	23.9	26	26
I144V15	5VDU VR – 144 Index	23	23	23	23	24	24	24	24	23	23	18	18	18	18	21	21	21	21	18	18
I154V15	5VDU VR – 154 Index	15	15	15	15	16	16	16	16	15	15	11	11	12	12	13	13	14	14	12	12

*	6 VDU 65 mbpd																				
ISPGW26	6VDU (65) WGO – SG	0.82	0.82	0.8158	0.8158					0.8158	0.8158	0.8188	0.8188	0.8134	0.8134					0.8134	0.8134
ISULW26	6VDU (65) WGO – Sul	0.664	0.664	0.58	0.58					0.58	0.58	0.543	0.543	0.458	0.458					0.458	0.458
IPPIW26	6VDU (65) WGO – PI	92	92	79	79					79	79	81	81	68	68					68	68
IVBIW26	6VDU (65) WGO – H	−144	−144	−164	−164					−164	−164	−167	−167	−190	−190					−190	−190
IDIIW26	6VDU (65) WGO – DI	60.7	60.7	61.4	61.4					61.4	61.4	59.5	59.5	60.5	60.5					60.5	60.5
I144W26	6VDU (65) WGO – 144	−329	−329	−336	−336					−336	−336	−331	−331	−340	−340					−340	−340
I154W26	6VDU (65) WGO – 154	−286	−286	−292	−292					−292	−292	−288	−288	−296	−296					−296	−296
ISPGL26	6VDU (65) LVGO – SG	0.8705	0.8705	0.8703	0.8703	0.8822	0.8822			0.8703	0.8703	0.874	0.874	0.8737	0.8737	0.8861	0.8861			0.8737	0.8737
ISULL26	6VDU (65) LVGO – Sul	1.91	1.91	1.907	1.907	2.113	2.113			1.907	1.907	1.907	1.907	1.906	1.906	2.154	2.154			1.906	1.906
IPPIL26	6VDU (65) LVGO – PI	547	547	551	551	682	682			551	551	522	522	525	525	654	654			525	525
IVBIL26	6VDU (65) LVGO – H	72	72	72	72	127	127			72	72	61	61	61	61	117	117			61	61
IDIIL26	6VDU (65) LVGO – DI	51.4	51.4	51.5	51.5	48.9	48.9			51.5	51.5	48.7	48.7	48.8	48.8	46	46			48.8	48.8
I144L26	6VDU (65) LVGO – 144	63	63	62	62	65	65			62	62	61	61	60	60	63	63			60	60
I154L26	6VDU (65) LVGO – 154	39	39	39	39	41	41			39	39	38	38	37	37	41	41			37	37
ISPGH26	6VDU (65) HVGO – SG	0.9348	0.9348	0.9348	0.9348	0.933	0.933	0.9476	0.9476	0.9348	0.9348	0.9379	0.9379	0.9379	0.9379	0.9362	0.9362	0.9509	0.9509	0.9379	0.9379
ISULH26	6VDU (65) HVGO – Sul	2.795	2.795	2.795	2.795	2.757	2.757	3.009	3.009	2.795	2.795	3.159	3.159	3.163	3.163	3.119	3.119	3.440	3.440	3.163	3.163
IPPIH26	6VDU (65) HVGO – PI	2609	2609	2609	2609	2477	2477	3065	3065	2609	2609	2856	2856	2856	2856	2676	2676	3379	3379	2856	2856
IVBIH26	6VDU (65) HVGO – H	385	385	385	385	375	375	441	441	385	385	383	383	383	383	374	374	443	443	383	383
I144H26	6VDU (65) HVGO – 144	49	49	49	49	50	50	45	45	49	49	48	48	48	48	48	48	44	44	48	48
I154H26	6VDU (65) HVGO – 154	32	32	32	32	32	32	29	29	32	32	31	31	31	31	31	31	29	29	31	31
ISPGV26	6VDU (65) VR – SG	1.0254	1.0254	1.0252	1.0252	1.0210	1.0210	1.0173	1.0173	1.0252	1.0252	1.0326	1.0326	1.0325	1.0325	1.0279	1.0279	1.0232	1.0232	1.0325	1.0325
ISULV26	6VDU (65) VR – Sul	4.5	4.5	4.5	4.5	4.4	4.4	4.4	4.4	4.5	4.5	4.5	4.5	4.5	4.5	4.8	4.8	4.8	4.8	4.9	4.9
IVBIV26	6VDU (65) VR – H	782.0	782.0	781.0	781.0	764.0	764.0	748.0	748.0	781.0	781.0	816.0	816.0	816.0	816.0	795.0	795.0	774.0	774.0	816.0	816.0
ICONV26	6VDU (65) VR -Carbon	21.2	21.2	21.1	21.1	19.5	19.5	18.2	18.2	21.1	21.1	26	26	26	26	24	24	22.2	22.2	26	26
I144V26	6VDU (65) VR – 144	23	23	23	23	24	24	25	25	23	23	20	20	20	20	21	21	22	22	20	20
I154V26	6VDU (65) VR – 154	15	15	15	15	16	16	16	16	15	15	13	13	13	13	14	14	15	15	13	13

The fifth set of rows is used to represent CDUs capacities. There are some other rows in this set, starting with E, L, or G, showing that it is equal to, less than, or greater to that row. These are control rows. The row names in this set are seven characters long and span all crude unit submodels.

CRDDISTL TABLE

This table is used by the model to define the structure of a number of theoretical or logical crude distillation submodels. Each logical crude unit may be considered one of many modes of operation of a crude unit. This table provides mapping between the Assays Table data and these logical crude units. Table 13-18 is an example of this table. The column names are three character tags for various operation modes of a crude units. Where individual crudes are being segregated, it may be convenient to use crude tags as logical unit tags, but this is not a requirement.

The rows of this table are divided into four sets. The first set comprises row names ATMTWR and VACTWR. The entries in these rows are integer values indicating which physical crude distillation towers are used by various logical crude units.

The second set of rows are ESTxxx, where xxx is the crude tag. The entries in these rows define the estimated charge of each crude to each logical crude unit. These estimates need not be accurate. The charge estimates are either on a weight or volume basis, depending on how crude assay data are entered. Thus, if the crude assay data are on a volume basis, entries in this table should be in volume units. The LP model uses these data to calculate the properties of straight-run cuts from a predefined mix of crude.

The next set of rows in this table are named ATMuuu and VACuuu and these define the consumption of utilities (fuel, steam, electricity, cooling water etc) in atmospheric and vacuum units for each of logical crude units.

CRDCUTS TABLE

The CRDCUTS Table defines the crude distillation cutting scheme used in the model. The row names are the three-character stream tags of the straight-run distillation cuts (see Table 13-19).

The column TYPE is used to identify the type of cut, and each cut must be allocated one of the following numbers:

> Generally the vendor supplied LP packages have an in-built mechanism to generate an Atmospheric Distillation Unit sub model

Table 13-18
CRDDISTL Table

TABLE CRDDISTL	TEXT	LC1	LC2	LC3	LC4	LC5
	REFINERY CONFIGURATION					
ATMTWR	PHYSICAL CDU	1	2	3	4	5
VACTWR	PHYSICAL VDU			3	4	
	CRUDE TO UNIT ESTIMATE					
ESTA1N	Chg AB 1CDU Normal	1.000				
ESTA1K	Chg AB 1CDU Kero	1.000				
ESTB1N	Chg Bah 1CDU Normal	1.000				
ESTB1K	Chg Bah 1CDU Kero	17.000				
ESTA2N	Chg AB 2CDU Normal		1.000			
ESTA2K	Chg AB 2CDU Kero		1.000			
ESTB2N	Chg Bah 2CDU Normal		1.000			
ESTB2K	Chg Bah 2CDU Kero		17.000			
ESTA3N	Chg AB 3CDU Normal			61.000		
ESTA3K	Chg AB 3CDU Kero			1.000		
ESTB3N	Chg Bah 3CDU Normal			1.000		
ESTB3K	Chg Bah 3CDU Kero			1.000		
ESTA4N	Chg AB 4CDU Normal				100.000	
ESTA4K	Chg AB 4CDU Kero				1.000	
ESTB4N	Chg Bah 4CDU Normal				1.000	
ESTB4K	Chg Bah 4CDU Kero				1.000	
ESTA5N	Chg AB 5CDU Normal					28.000
ESTA5K	Chg AB 5CDU Kero					2.000
ESTB5N	Chg Bah 5CDU Normal					28.000
ESTB5K	Chg Bah 5CDU Kero					2.000
	TOTAL CRUDE EST 265.000	20.000	20.000	64.000	103.000	58.000
	UTILITY CONSUMPTIONS (MSCF NATURAL GAS)					
ATMFUL	FUL ATM TOWER	0.1683	0.1683	0.2049	0.1829	0.1683

NOTE: THIS TABLE DEFINES THE ATMOS/VACUUM LOGIC. ONLY INTEGRAL CRUDE/VACUUM UNITS NEED BE NOTED HERE (I.E., 3 AND 4).

using data supplied by user in tables Assay, CRDDISTL, and CRDCUTS. The type of cut naming convention used in the LP package must be used. The cut type or number used here (0, 1, 2, 3, 9) are as per PIMS LP Package.

Table 13-19
CRDCUTS Table

TABLE CRDCUTS	TEXT	TYPE	LC1	LC2	LC3	LC4	LC5*
GA1	OFF GASES (C3 MINUS)	1	1	1	1	1	1
C41	CDU C3/C4	1	1	1	1	1	1
LS1	CDU LSR	1	1	1	1	1	1
MN4	4A MSR Pool	1				4	
HN1	HSR 39.5 C FLASH	1	1	2	3	4	5
HA4	HSR for KRU (Arab)	1				4	
HB4	HSR for KRU (Bah)	1				4	
KN1	KERO 39.5C FLASH	1	1	2	3	4	5
KA1	Kero for KRU (Arab)	1	1	2	3		5
KB1	Kero for KRU (Bah)	1	1	2	3		5
DN1	CDU LIGHT DIESEL	1	1	1	1	4	1
MI4	M + I DSL FROM 4A	1				4	
RA1	ATM RESID 1/2/5 (Arab)	1	1	2	1	1	5
RB1	ATM RESID 1/2/5 (Bah)	1	1	2	1	1	5
HD3	HVY DSL 3 & 4 CDU	3			3	4	
VA3	VAC RES 3 & 4 CDU	3			3	4	
	DEFERRED CUTS						
W11	1 VDU WGO POOL	9	1	1			
L11	1 VDU LVGO POOL	9	1	1			
V11	1 VDU VAC RES POOL	9	1	1			
LW1	1 VDU LWGO POOL (Asp)	9	1	1			
W21	1 VDU LVGO POOL (Asp)	9	1	1			
L21	1 VDU VAC RES POOL (A)	9	1	1			
BS1	1 VDU BSGO	9	1	1			
AS1	1 VDU ASPHALT	9	1	1			
W31	1 VDU WGO POOL (BSGO)	9	1	1			
L31	1 VDU DGO POOL (BSGO)	9	1	1			
BS3	1 VDU BSGO (BSGO)	9	1	1			
V31	1 VDU VAC RES POOL (BSGO)	9	1	1			
W15	5 VDU WGO POOL	9	1	1	1		1
L15	5 VDU LVGO POOL	9	1	1	1		1
H15	5 VDU HVGO POOL	9	1	1	1	1	1
V15	5 VDU VAC. RESID POOL	9	1	1	1	1	1
W25	5 VDU WGO POOL (Asp)	9					1
L25	5 VDU LVGO POOL (Asp)	9					1
H25	5 VDU HVGO POOL (Asp)	9					1
AS5	5 VDU ASPHALT	9					1
W16	6 VDU WGO AT 50 MBPD	9	1	1			1
L16	6 VDU LVGO AT 50 MBPD	9	1	1	1		1

Table 13-19
Continued

TABLE CRDCUTS	TEXT	TYPE	LC1	LC2	LC3	LC4	LC5*
H16	6 VDU HVGO AT 50 MBPD	9	1	1	1	1	1
V16	6 VDU VR AT 50 MBPD	9	1	1	1	1	1
W26	6 VDU WGO AT 65 MBPD	9	1	1			1
L26	6 VDU LVGO AT 65MBPD	9	1	1	1		1
H26	6 VDU HVGO AT 65MBPD	9	1	1	1	1	1
V26	6 VDU VR AT 65 MBPD	9	1	1	1	1	1
W46	6 VDU WGO FROM W11	9	1	1			
L46	6 VDU LVGO FROM W11	9	1	1			
H46	6 VDU HVGO FROM W11	9	1	1			
W56	6 VDU WGO FROM W11	9	1	1			
L56	6 VDU LVGO FROM W11	9	1	1			
H56	6 VDU HVGO FROM W11	9	1	1			

0. This indicates that cut yield is identified in the assay data, but this cut is produced by downstream separation, in a saturated gas plant and not on a crude unit.
1. This indicates that this a straight-run atmospheric cut.
2. This indicates that it is atmospheric residuum, which may be used elsewhere or fractionated in a vacuum unit. If an atmospheric resid is always directed to a vacuum unit, it is omitted from the cutting scheme in this table.
3. This indicates that it is a straight-run vacuum cut.
9. This indicates it is a deferred cut that is not produced on a crude distillation unit but in a downstream unit.

The other column names in this table are the three-character tags for the logical crude units and must exactly match the column names of the CRDDISTL Table. The entry at each intersection is the pool number to which the cut from each logical crude unit is directed.

For example, if this table contains a cut ABC, this stream yields streams ABC in pool 1, AB2 in pool 2, and AB3 in pool 3. Thus, when a stream is segregated, the model automatically constructs additional stream tags to identify the segregated stream. The third character in the stream tag is replaced with a pool number for pool 2 onward.

SCRD TABLE

The SCRD Table (Table 13-20) gives the disposition of crudes processed by the refinery to various crude distillation units. But only those crudes with nonblank entries in the material balance rows for crudes are available as potential feed to that logical crude unit. Therefore, if a crude is omitted from the logical crude unit, it will not be processed on that unit. Thus, Table 13-20 shows that crude ABP (Arab light) is processed on crude units 3, 4, and 5, while crude BAH (Bahrain crude) is processed only on crude units 1 and 2.

Tables 13-17 to 13-20 contain all the data on the crudes processed, their disposition, yield and properties, data on the physical atmospheric and vacuum distillation units present, their mode of operation, unit capacities, and utility consumption. The primary cuts from various crude units are pooled or segregated as per pooling map in the CRDCUTS Table. Thus, all refinery atmospheric distillation data is maintained in these tables. The CDU models are automatically generated from them by the LP package.

CDU SUBMODEL STRUCTURE

It may be useful to review exactly how a matrix generator program processes data in the Assay Table to produce material and property balances for CDU streams. The Assay Table differs from other tables in that it is not included in the matrix but is used by the program to build one or more CDU tables, which are included in the matrix.

The name of actual CDU submodel tables built by the matrix generator program is typically concocted by adding a character S (to denote submodel) before the column names of the CRDDISTL Table. Thus, if the first column in this table is LC1, the first CDU table is named SLC1 (PIMS Refinery LP Package).

The column names in the crude submodel tables are the operation modes of this crude distillation unit. Thus, A1N and A1K are the two operation modes of crude distillation unit SLC1 while processing light Arab crude, and B1N and B1K are the two operation modes while processing Bahrain crude. In the matrix generated from this table, the column names are SLC1A1N, SLC1A1K, SLC1B1N, and SLC1B1K. The row names would be identical to the table CRDCUTS row names; for example, in this case, VBALGA1,VBALC41, VBALLS1, and so on. The coefficients at row/column intersections are the yields of various cuts of data entered in the Assay Table.

The atmospheric residuum from crude units becomes feed to VDUs. To avoid unnecessary multiplicity of streams, only one atmospheric

Table 13-20
Disposition of Crudes

TABLE SCRD	TEXT	A1N	A1K	A2N	A2K	A3N	A3K	A4N	A4K	A5N	A5K	B1N	B1K	B2N	B2K	B3N	B3K	B4N	B4K	B5N	B5K
VBALABP	AL CRUDE	1	1	1	1	1	1	1	1	1	1										
VBALBAH	BAH CRUDE											1	1	1	1	1	1	1	1	1	1
VBALA1N	AB to 1 CDU Normal	−1																			
VBALA1K	AB to 1 CDU KRR		−1																		
VBALA2N	AB to 2 CDU Normal			−1																	
VBALA2K	AB to 2 CDU KRR				−1																
VBALA3N	AB to 3 CDU Normal					−1															
VBALA3K	AB to 3 CDU KRR						−1														
VBALA4N	AB to 4 CDU Normal							−1													
VBALA4K	AB to 4 CDU KRR								−1												
VBALA5N	AB to 5 CDU Normal									−1											
VBALA5K	AB to 5 CDU KRR										−1										
VBALB1N	Bah to 1 CDU Normal											−1									
VBALB1K	Bah to 1 CDU KRR												−1								
VBALB2N	Bah to 2 CDU Normal													−1							
VBALB2K	Bah to 2 CDU KRR														−1						
VBALB3N	Bah to 3 CDU Normal															−1					
VBALB3K	Bah to 3 CDU KRR																−1				
VBALB4N	Bah to 4 CDU Normal																	−1			
VBALB4K	Bah to 4 CDU KRR																		−1		
VBALB5N	Bah to 5 CDU Normal																			−1	
VBALB5K	Bah to 5 CDU KRR																				−1

resid stream is produced from the operation of one crude distillation column, irrespective of number of crudes processed there or the number of operation modes of the CDUs. This is done by pooling the various resid streams from different modes of a crude column, as shown next for crude distillation column 1 (SLC1). Also the yield and properties of streams produced from the vacuum distillation of this resid are determined here, as these properties depend on the properties of the composite resid stream generated here, even though the vacuum distillation streams are not produced here. A part of Table SLC1 follows, presenting pooling of resids from various operation modes of crude distillation column SLC1:

	A1N	A1K	B1N	B1K	AR1
VBALAR1	−0.5224	−0.5224	−0.5330	−0.5330	
RBALAR1	−0.5224	−0.5224	−0.5330	−0.5330	1
ISPGAR1	0.9463	0.9463	0.9459	0.9459	
RSPGAR1	−0.4943	−0.4943	−0.5041	−0.5041	999
ISULAR1	3.14	3.14	3.38	3.38	
IVBIAR1	446.1	446.1	431.7	431.7	
RSULAR1	−1.6403	−1.6403	−1.8015	−1.8015	999
RVBIAR1	−233.042	−233.042	−230.096	−230.096	999

The properties of the pooled stream AR1 are calculated next.

For example, consider the specific gravity of the pooled resid stream AR1. This stream is generated by pooling volumes in the VBALAR1 row. Row RBALAR1, which has the same coefficients as row VBALAR1, represents how many barrels of resids from each operation mode of SLC1 are used to form the pool. The specific gravity of individual cuts is given by row ISPGAR1 in the Assay Table. Row RSPGAR1 coefficients are formed by multiplying coefficients in row ISPGAR1 with the coefficient in row RBALAR1. For example, the coefficient of row RSPGAR1 in column A1N is the product of 0.9463*−0.5224, or 0.4943. The activity of this row is the sum of SG*barrels for the pooled stream. The SG of the composite stream can be calculated by dividing the row activity of RSPGAR1 by the row activity of RBALAR1, if column activities are known. Other properties of the composite streams can similarly be calculated.

It may be noted that, even though the rows VBALAR1 and RBALAR1 exhibit the same coefficients in the SLC1 Table, there is an important distinction between them. VBALAR1 is a material balance row, representing the actual physical production of atmospheric residue. RBALAR1, on the other hand, is a property balance row, which provides the denominator for computation of the average specific gravity of the pooled resid.

DBAL Rows (Deferred Cuts)

In the Assays Table, the yields for all atmospheric cuts, such as LPG, light naphtha, kerosene, or diesel, are represented by entries in VBAL rows; for example, VBALLS1 for LSR, VBALKN1 for kerosene yield, and VBALAR1 for atmospheric resid streams. This is because these streams are produced directly from CDUs and their volumes as well as properties are determined in the CDUs.

On the other hand, the vacuum product streams (LVGO, HVGO, V. RESID) and certain other streams, which are not produced on the CDU, are represented by DBALxxx type rows; for example, DBALW11, DBALL11, and DBALH11 represent WGO, LVGO, and HVGO streams from VDU1 from the distillation of atmospheric resid from the CDU1. The presence of these deferred products in the Assays Table reflects the fact that their properties depend on the crude mix and CDU fractionation. That they are represented by DBAL and not VBAL rows reflects that the option of exactly what volume of each stream is to be actually produced lies with the vacuum distillation units.

In the Assays Table the properties of the various atmospheric cuts and deferred cuts are represented by the suffix I:

ISPGL11	Specific gravity of DGO.
ISULL11	Sulfur of DGO.
IVBIL11	Viscosity blending index of DGO.

Consider distillate gas oil stream DGO, produced from VDU 1. This stream has no VBAL rows in the Assays Table. To calculate the specific gravity of a blend of streams, the LP multiplies the RBALL11 row with ISPGL11 rows and again the resulting "gravity*barrels" are placed in row RSPGL11. Row RBALL11 has the same coefficients as row DBALAR1, representing how many barrels of DGO distillate from each operation mode of SLC1 are used to form the pool. RBALxxx is used as the denominator to calculate the average stream property. The numerator is provided by the activity of row RSPGL11.

	A1N	A1K	B1N	B1K	L11
DBALL11	−0.2544	−0.2544	−0.2709	−0.2709	
RBALL11	−0.2544	−0.2544	−0.2709	−0.2709	1
ISPGL11	0.9093	0.9093	0.9122	0.9122	
RSPGL11	−0.2313	−0.2313	−0.2471	−0.2471	999

VDU Yield as a Property of Vacuum Resids

In the Assay Table, there is a group of rows with special significance. This group of I rows specify a special set of properties of the atmospheric resid streams. These represent the yield of various products in the VDU from distillation of the atmospheric resids, for example, the following rows:

IW11AR1	Yield of WGO from atmospheric resid AR1 on VDU1.
IL11AR1	Yield of DGO from atmospheric resid AR1 on VDU1.
IV11AR1	Yield of V. RESID from atmospheric resid AR1 on VDU1.

Although treated as properties in the Assays Table, these are the yields of products from the vacuum distillation of atmospheric resid streams, called the *pseudoproperties* of atmospheric resid because they are presented as similar to the properties of the atmospheric resid in I-type rows.

Consider a portion of the Assays Table data for processing of Arab and Bahrain crudes on CDU1. The following data present potential yields and properties of various vacuum distillation cuts from processing atmospheric resids produced from various operation modes of atmospheric crude distillation column CDU1, on vacuum distillation column VDU6:

	A1N	A1K	B1N	B1K
VBALAR1	−0.5224	−0.5224	−0.5330	−0.5330
*				
DBALW26	−0.0020	−0.0020	−0.0030	−0.0030
DBALL26	−0.1280	−0.1280	−0.1210	−0.1210
DBALH26	−0.2220	−0.2220	−0.2210	−0.2210
DBALV26	−0.1700	−0.1700	−0.1710	−0.1710
*				
IW26AR1	0.0040	0.0040	0.0050	0.0050
IL26AR1	0.2452	0.2452	0.2352	0.2352
IH26AR1	0.4250	0.4250	0.4282	0.4282
IV26AR1	0.3256	0.3256	0.3316	0.3316
*				
ISPGW26	0.8200	0.8200	0.8188	0.8188
ISPGL26	0.8705	0.8705	0.8740	0.8740
ISPGH26	0.9348	0.9348	0.9379	0.9379
ISPGV26	1.0254	1.0254	1.0326	1.0326
*				

In this table,

W26 = wet gas oil from distillation of CDU1 atmospheric resid on VDU6.

L26 = LVGO from distillation of CDU1 atmospheric resid on 6VDU6.

H26 = HVGO from distillation of CDU1 atmospheric resid on 6VDU6.

V26 = vacuum resid from distillation of CDU1 atmospheric resid on VDU6.

The yields of vacuum cuts are presented by DBAL rows instead of VBAL rows, indicating that these cuts are not produced on the crude column and the yields indicate only the potential yield of cuts from the vacuum column.

The second set of rows IW26AR1, IL26AR1, IH26AR1, and IV26AR1 are the potential yields of WGO, LVGO, HVGO, and vacuum resid on the atmospheric resid feed (not on the crude) by the distillation of CDU1 atmospheric resids on VDU6. Here, VDU6 yields of various cuts from processing CDU1 resid are indicated as a property of the feed resid.

Rows ISPGW26, ISPGL26, and the like indicate the properties of cuts (WGO, LVGO) from VDU6.

Referring to these Assays Table data, the following structure determines the yield of various vacuum cuts from a vacuum column; for example, VDU6 from a resid coming from mix of crudes processed on CDU1:

	A1N	A1K	B1N	B1K	AR1
RBALAR1	−0.5224	−0.5224	−0.5330	−0.5330	1
*					
RW26AR1	−0.0021	−0.0021	−0.0027	−0.0027	999
RL26AR1	−0.1281	−0.1281	−0.1281	−0.1281	999
RH26AR1	−0.2220	−0.2220	−0.2282	−0.2282	999
RV26AR1	−0.1701	−0.1701	−0.1767	−0.1767	999

Here, the coefficients in row RW26AR1 represent the yield of wet gas oil (WGO) from VDU6. The coefficient in column A1N in this row is obtained by multiplying the corresponding coefficients in row RBALAR1 and row IW26AR1 ($-0.5224 \times 0.0040 = -0.0020$). Similarly, the coefficient in row RV26AR1, column A1N is formed as a product of corresponding coefficients in RBALAR1 and IV26AR1 ($-0.5224 \times 0.3256 = -0.1701$).

The overall yields of vacuum cuts W26 (WGO), L26 (LVGO), H26 (HVGO), and V26 (vacuum resid) from the distillation of composite atmospheric resid from crude unit 1 or SLC1 is determined by dividing the row activity of each RW26AR1, RL26AR1, RH26AR1, and RV26AR1 by the row activity of RBALAR1 once the column activities are known.

The 999 is a placeholder for the first guess of the yield of the vacuum products, required to start the recursion process.

The yields of vacuum cuts determined here are transmitted to various vacuum distillation submodels for use.

Properties of VDU Streams

The properties of a vacuum distillation stream depends on the composition of the atmospheric resids from which the VDU stream is produced. In the LP structure that follows, the specific gravities of W26 (WGO), L26 (LVGO), H26 (HVGO), and V26 (vacuum resid) from VDU6 are determined.

Here, the feed to VDU6 is atmospheric resid from crude distillation unit 1 (SLC1). Row DBALW26 gives the potential yield of WGO from different atmospheric resids from different modes of crude distillation column 1. The coefficients in row RBALW26 are identical to those in row DBALW26. The coefficient in row RSPGW26, for example, in column A1N, is formed by multiplying the coefficient in row RBALW26 with the coefficient in row ISPGW26 ($-0.0020 \times 0.8200 = -0.0016$). The specific gravity of the WGO stream is determined by dividing the activity of row RSPGW26 by the activity of row RBALW26 once the column activities are known:

- Specific gravity of stream W26, wet gas oil from VDU6 processing CDU1 atmospheric resid.

	A1N	A1K	B1N	B1K	W26
DBALW26	−0.0020	−0.0020	−0.0030	−0.0030	
RBALW26	−0.0020	−0.0020	−0.0030	−0.0030	1
RSPGW26	−0.0016	−0.0016	−0.0025	−0.0025	999

- Specific gravity of stream L26, LVGO from VDU6 processing CDU1 atmospheric resid.

	A1N	A1K	B1N	B1K	L26
RBALL26	−0.1280	−0.1280	−0.1210	−0.1210	1
RSPGL26	−0.1114	−0.1114	−0.1058	−0.1058	999

- Specific gravity of stream H26, HVGO from VDU6 processing CDU1 atmospheric resid.

	A1N	A1K	B1N	B1K	H26
RBALH26	−0.2220	−0.2220	−0.2210	−0.2210	1
RSPGH26	−0.2075	−0.2075	−0.2073	−0.2073	999

- Specific gravity of stream V26, vacuum resid from VDU6 processing CDU1 atmospheric resid.

	A1N	A1K	B1N	B1K	V26
RBALV26	−0.1700	−0.1700	−0.1710	−0.1710	1
RSPGV26	−0.1743	−0.1743	−0.1766	−0.1766	999

Other properties (sulfur, viscosity blend index, pour index) of various vacuum cuts are determined in an identical manner.

We see that both the properties and potential yields of vacuum cuts are determined in an atmospheric crude distillation model, where the feed to vacuum distillation unit originates and relative ratios of various crudes or their operating modes of CDU are available. The yield and property data of vacuum cuts can be retrieved in VDU submodels to give the proper yield of the vacuum stream from each atmospheric resid charged to a given vacuum unit.

The advantage of modeling vacuum yields as pseudoproperties of atmospheric resids in the Assays Table is that all vacuum unit yield data are maintained in a single table, the Assays Table. The use of pseudo-properties allows the flexibility to add unlimited number of crudes to any CDU and still obtain the correct yields in VDU submodels without modifying the VDU submodels in any way.

Although the use of deferred distillation rows and a number of pseudo-property rows in the Assays Table appears to complicate the model, this additional structure in the table enormously simplifies the modeling

VDUs as well as the addition of crude to the model and model maintenance in general. The advantages of deferred distillation becomes apparent in VDU submodels.

VACUUM DISTILLATION UNIT MODELING

In refineries in which each CDU has its own associated VDU, modeling a composite crude and vacuum unit becomes easy, and yields from the vacuum unit are treated exactly like those from CDU. However, in refineries where atmospheric residuum from a given CDU can go to a number of VDUs, one cannot treat CDUs and VDUs as a single combination, since one does not know in advance in which VDU a given atmospheric residue will be processed nor which set of vacuum streams will be produced.

Consider the following VDU6 model. Here, the feed is AR1, the atmospheric resid from CDU1. The product streams are

W62	Wet gas oil.
L62	LVGO.
H62	HVGO.
V62	Vacuum resid.

In the SVD6 Table,

	AR1	W26	L26	H26	V26
VBALAR1	1				
VBALW26		1			
VBALL26			1		
VBALH26				1	
VBALV26					1
VBALWG6		−1			
VBALVM6			−1		
VBALVH6				−1	
VBALVR6					−1
EW26AR1	−999	1			
EL26AR1	−999		1		
EH26AR1	−999			1	
EV26AR1	−999				1

In rows EW26AR1, EL26AR1, EH26AR1, and EV26, the second to fourth characters match the names of pseudoproperties of AR1, which were defined in the Assays Table to represent the yields on VDU6 of WGO, LVGO, HVGO, and vacuum resid, respectively. In the SVD6 Table, these E (equality) rows all have entries of −999 in column AR1. The matrix generator program will substitute, for −999, the current yield of each vacuum product, as calculated in the Assays Table for CDU1 atmospheric resid. Thus, the correct vacuum yields for a mix of crudes processed on CDU1 is calculated in CDU1 submodel and transmitted to VDU6 submodel in column AR1.

Now, let us see how the yields transmitted into AR1 are utilized. For example, consider the vacuum resid yields represented in row EV26AR1. The program will substitute for −999 in column AR1, the yield of vacuum resids from AR1.

Suppose this is 0.33 or 33%. Then, −0.33 is substituted for −999. Further suppose that 1 mb (million barrels) of CDU1 atmospheric resid AR1 is charged to VDU6; then, the volume placed into row EV26AR1 in column AR1 is as follows:

$$1\,\text{mb AR1} \times 0.33\ \text{(yield of vacuum resid)} = -0.33\,\text{mb vacuum resid}$$

The negative sign denotes production. The only other entry in row EV26AR1 is a positive 1 in column V26. Since this is an equality row that must balance, column or variable V26 takes on an activity equal to the volume of vacuum resid contained in CDU1 atmospheric resid charged to VDU6. The activities of the other columns W26, L26, H26 are similarly determined by these E rows.

Other entries shown in column V26 are −1, representing production, in row VBALVR6 +1, representing consumption, in row VBALV26. Therefore we are actually producing not stream V26, which is VDU6 vacuum resid derived from CDU1 atmospheric resid, but VR6, the overall vacuum resid produced on VDU6 from all atmospheric resids from different CDUs. This is consistent with the physical reality of the refinery, where all atmospheric resids are processed as a mixture to produce a single set of vacuum products from a VDU.

A vacuum distillation unit may have more than atmospheric resid feed. Let us add one more resid, AR3 coming from CDU3, to the preceding structure, as shown next, this time showing only vacuum resid production:

	AR1	V26	AR3	V36	VR6
VBALAR1	1				
VBALAR3			1		
VBALV26		1			
VBALV36				1	
VBALVR6		−1		−1	
EV26AR1	−999	1			
EV36AR3			−999	1	
*					
RBALVR6		−1		−1	1
RSPGVR6		−999		−999	999
RSULVR6		−999		−999	999
RVB1VR6		−999		−999	999

We see that the structure in this table for AR3 is exactly parallel to that for AR1. Column V36 has activity equal to the volume of VDU6-derived vacuum resid derived from CDU3, just as the activity of column V26 represents the VDU6 vacuum resid derived from CDU1.

Note that both columns V26 and V36 have entries of −1 in row VBALVR6, showing that these resids coming from different CDUs combine into a single vacuum resid stream, VR6. Columns V26 and V36 have +1 entries in material balance rows VBALV26 and VBALV36, indicating that these streams are consumed to form stream VR6.

To estimate the properties of the composite vacuum resid stream, row RBALVR6 shows the components that form the stream, in the form of columns that have −1 entries in this row, that is, V26 and V36. The properties (SG, SUL, VBI, etc.) of V26 and V36 are calculated in the CDU1 and CDU3 submodels and transmitted here to replace the placeholders −999. Placeholder 999, in column VR6, is the first guess of the pooled stream to start the recursion process.

To summarize, the advantage of modeling vacuum distillation with deferred cuts and the stream's pseudoproperties are as follows:

1. The calculation of all VDU yields and stream properties are done in the Assay Table and transmitted to the respective VDU model. The VDU models themselves contain no yield data. All data maintenance is only through the Assays Table.
2. Only one atmospheric resid stream is produced for each CDU, irrespective of the number of crudes processed in it. In the absence

of deferred distillation, one atmos resid is generated for every crude.

3. For each VDU, only one set of vacuum streams is produced, rather than a separate set for each crude/CDU/VDU combination.
4. The streams produced in the model correspond much more closely to streams physically produced in the refinery.
5. All vacuum yields are specified in the Assays Table, eliminating the need to maintain vacuum yield data in VDU submodels.

SINGLE-PRODUCT LP BLENDER

Refinery blending of a cargo is done with following objectives:

1. Meet the requirement of a specific shipment with respect to volume and product specification.
2. Optimum blending; that is, there is no giveaway on product quality and minimum cost to the refinery.
3. Minimize the inventory of stocks by giving priority to reblending tank heels and other unwanted stocks.

To meet these objectives, many refineries use an on-line single-product LP system. This system interfaces with the refinery's LAB system for tank test results and product specifications. The refinery on-line LAB system records, monitors, and reports on streams and product tank test results done by the refinery's laboratory, forming the database for a single-product LP blender.

The on-line single-product LP blender aims to provide minimum-cost blends within user-defined volumetric and specification constraints, using a linear programming algorithm. Optimization of the blend is done on the basis of the prices of blend streams provided by the user. The LP blender system minimizes the cost of the blend produced while meeting product specifications.

For the optimization step, the prices of the blend stocks must be provided. Generally, a very low cost is assigned to stocks such as tank heels and other unwanted stocks to maximize their blending. If the objective is to minimize inventory, arbitrary prices may be used; the assumed price of a stock in this case is inversely proportional to the available inventory.

The user enters only the tank numbers allowed in the blend, the target specifications, and the total blend volume required. Latest tank quality data, such as specific gravity, sulfur, or flash point, are retrieved from the LAB system, once the tank number is entered. The LP next works out the optimum blend composition that meets all given specifications.

The optimum blend composition thus worked out is used for actual blending of cargoes, instead of using past experience with similar blends, manual calculations, or tedious trial and error procedures.

EXAMPLE 13-5

We want to blend 100 mb of a gasoline grade with the following specifications:

RON	MINIMUM	95
MON	MINIMUM	83.5
RVP	kPa MAX	62.0
RELATIVE DENSITY	MIN/MAX	0.7200/0.775
% EVAPORATED AT 70°C	MINIMUM	12

The blending is to be done in tank 74, which contains 8000 bbl of previously shipped gasoline cargo. Determine the most economic blend, using the available stocks. The available blending components and assumed prices are as follows:

STOCK	TANK NO.	PRICE, $/bbl
FCCU LIGHT NAPHTHA	75	17.5
FCCU HEAVY NAPHTHA	77	17.1
REFORMATE	54	19.0
MTBE	79	26.5
COKER LIGHT NAPHTHA	72	15.6
ALKYLATE	73	22.5
GASOLINE (TANK HEEL)	74	15.0

Using a single-product blender system, the tank volume and property data are retrieved from the LAB system, which contains the latest test results done on these tanks. The user insert only the blend component prices for optimization calculations and the total volume of the blend required. The following table is the result:

STOCK	TANK NO.	AVAILABLE VOL, mb	SG	RON, CLEAR	MON, CLEAR	RVP, kPa	% DISTILLED, 70°C
FCCU LIGHT NAPHTHA	75	60.5	0.7095	93	82.7	63	51.0
FCCU HEAVY NAPHTHA	77	40.0	0.8390	93.8	83.2	0	0
REFORMATE	54	20.0	0.7830	96.5	86.4	54.6	10.5
MTBE	79	15.5	0.7450	112.5	100.0	56.7	100
COKER LIGHT NAPHTHA	72	25.4	0.7030	74	65.8	53.2	30
ALKYLATE	73	10.3	0.6845	97.8	94.5	51.8	10.0
GASOLINE (TANK HEEL)	74	8.0	0.7550	93	83	62.0	15.0

The problem is converted into an LP model as presented next. Here, T75, T77, T54, BLN, and the like are variables and their values are found subject to the following conditions. Find

$$\text{T75}, \text{T77}, \text{T54}, \text{T79}, \text{T72}, \text{T73}, \text{T74}, \text{BLN} \geq 0$$

$$\text{BLN} = 100$$

such that

$$\text{T75} \leq 60.5$$

$$\text{T77} \leq 40.0$$

$$\text{T54} \leq 20.0$$

$$\text{T79} \leq 15.5$$

$$\text{T72} \leq 25.4$$

$$\text{T73} \leq 10.3$$

$$\text{T74} = 8.0$$

and the following constraints

	T75	T77	T54	T79	T72	T73	T74	BLN	
MAT. BAL.	1	1	1	1	1	1	1	−1	= 0
SG (MIN)	0.7095	0.8390	0.7830	0.7450	0.7030	0.6845	0.7550	−0.7200	≥ 0
SG (MAX)	0.7095	0.8390	0.7830	0.7450	0.7030	0.6845	0.7550	−0.7750	≤ 0
RON	93.0	93.8	96.5	112.5	74.0	97.8	91.0	−95.0	≥ 0
MON	82.7	83.2	86.4	100.0	65.8	94.5	81.0	−83.5	≥ 0
RVP	63.0	0	54.6	56.7	53.2	51.8	62.0	−62.0	≤ 0
DISTILLED, 70°C	51.0	0	10.5	100.0	30.0	10.0	15.0	−12.0	≤ 0

and such that

$$Y = 17.5^{*}(T75) + 17.1^{*}(T77) + 19.0^{*}(T54) + 26.5^{*}(T79) + 15.6^{*}(T72) + 22.5^{*}(T73) + 15.0^{*}(T74) \text{ is minimum}$$

The blend composition and properties are found on solution of the preceding problem, as follows:

BLEND COMPONENT	VOLUME, mb
LIGHT CAT NAPHTHA	41.69
HEAVY CAT NAPHTHA	30.00
REFORMATE	13.74
MTBE	6.56
COKER LIGHT NAPHTHA	0
ALKYLATE	0
TANK HEEL	8.00
TOTAL BLEND	100.00

Gasoline blend properties are as follows:

PROPERTY	SPECIFICATION	BLEND
SG	0.720–0.775	0.764
RON	95, MIN	95.2
MON	83.5, MIN	84.8
RVP	62 kPa, MAX	42.5
% DISTILLED, 70°C	12% (VOL), MIN	30.5

NOTES

1. For example, IBM MPSX/370 or CPLEX Optimisation Inc.
2. Aspen Technologies Company, originally developed by Bechtal Inc.
3. J. S. Bonner, "Advance Pooling Techniques in Refinery Modeling." Bonner and Moore Associates GmbH seminar Wiesbaden, Germany, Oct 19–21, 1987.
4. G. B. Dantzig, *Linear Programming and Extensions.* Princeton, NJ: Princeton University Press, 1946.
5. The LP packages from different vendors may have different row and column naming conventions. The one described here is from the PIMS from Aspen Technology.

CHAPTER FOURTEEN

Pricing Petroleum Products

NETBACK AND FORMULA PRICING FOR CRUDE OIL

The netback pricing of crude oil set a crude oil price on the basis of the product market. Netback and other formula techniques seek to provide reduced market risk and reasonable return to the refiner during extreme market fluctuation, and they make long-term contracts between crude producers and refiners possible.

In the past, violent price fluctuations created huge trading losses for some companies; and this accelerated the shift to netback and other formula pricing as a tactic to minimize risk in place of outright sale at a negotiated price.

There are four basic components of any netback deal: yield of the finished products from refining the crude in question, product prices, timing, processing fees, and transportation cost.

YIELD

The yield is the portion of each refined product that, when combined with refinery fuel and loss, adds up to the whole barrel of crude. A specific spot product price reference point (as monitored by an agreed-on published source) is selected for each portion of yield to determine the total value of crude oil. The processing fees include refining cost, freight, and other elements that are deducted to arrive at a "net" value of the crude "back" at the point of origin (i.e., the netback). The basic method of calculating the netback price follows.

PRODUCT PRICES

Spot prices quotes from any reporting service can be used, although Platts spot price quotes are most popular. The netback price reference is based on the "high," "mean," or "low" point of the price range reported. The choice has a substantial influence on the resulting crude price. In almost all cases, the product pricing base used is that at the intended destination refining center, with few exceptions. For example, an Iranian crude processed in Singapore uses Singapore Market quotes—high, mean or low, as per the terms of the agreement.

TIMING

Typically, the timing component is expressed as a certain number of days (usually between 0 and 60) after bill of lading date. This implies that the prices to be used for calculating the gross product worth are the spot quotes (high, mean, or low) prevailing exactly on the agreed number of days after the crude is loaded. Timing can sometimes create an incentive for the buyer to rush or slow vessel streaming. For example, if prices are falling, the buyer has an incentive to speed up the vessel to process the crude and sell the products at the highest possible prices.

In some netback deals, the product quotes are averaged over a number of days to avoid the chance that the product quote deviates substantially from the prevailing market levels on any single day and unduly distorts the netback prices. The period for averaging may be 5–10 days or more and is agreed to in the netback deal.

The product prices for the indicated time period are multiplied by the yield according to set percentages (on either a weight or volume basis). The total of the calculation gives the overall value or "gross product worth" at the refinery location. The processing fees are then deducted from this total, giving netback or the price of the crude.

PROCESSING FEES

The processing fees consist of the operating cost to the refinery per barrel of the crude processed plus minimum refinery profit. The refining cost include cost of utilities (fuel, electricity, water, etc.), catalysts and chemicals, and personnel incurred in processing one barrel of crude.

Refining costs do not include the amortization and depreciation costs of fixed refinery assets. It is closely related to the refinery bottom-line profit margin. The variation in processing fees are large, ranging from as high as $2 to as low as $0.5 for a simple refinery. The processing fees are usually related to the complexity of the refinery. For a refinery with an extensive conversion facility, the fees may be many times more than for one with only basic crude topping facilities.

TRANSPORTATION COST

For crude transport, it is the cost of chartering an appropriately sized vessel (see Table 14-1) on the spot market for a single voyage. The transport cost is set by Worldscale, a trade association that publishes a flat base rate for voyages between each oil loading and receiving port. Daily tanker market fluctuations are measured in Worldscale "points," which are a percentage of standard flat rate.

For example, the cost of chartering a ship, at Worldscale 35, would be 35% of the flat rate for spot cost of transport to Singapore from the Arabian Gulf:

Flat rate per long ton (VLCC class) = $25.32

Flat rate per barrel = $3.38

Converted at 7.49 bbl/ton of 34 API crude.

Table 14-1
Crude and Product Tanker-Size Classification

CLASS	SHIP, dwt (long tons)
GP	16,000–24,999
MR	25,000–44,999
LR1	45,000–79,999
LR2	80,000–159,999
VLCC	160,000–319,999
ULCC	320,000–OVER

$$\text{Transport cost per barrel at WS } 35 = 3.38 \times 35/100$$
$$= \$1.183$$

EXAMPLE 14-1

Determine the netback price for light Arabian crude refined in Singapore with its product yield sold in the spot market. Assume April 1998 mean prices. The product yield from this crude (LV%) (LV = Liquid volume %) is presented below;

PRODUCT	PRICES SINGAPORE, MEAN, $	PRODUCT YIELD, VOL%	VALUE OF YIELD, $
NAPHTHA	17.50	14.149	2.48
PREMIUM GASOLINE	25.90	5.557	1.44
JET A-1	22.30	16.140	3.60
DIESEL	22.50	31.687	7.13
FUEL OIL, 3.5% SULFUR	13.4	32.467	4.35
LOSSES	0	0	0
GROSS PRODUCT WORTH			18.99

1. The gross product worth (GPW) of crude is calculated first as shown in table.
2. By subtracting the refining cost, freight, other costs (insurance, loss, financing, duties, etc.), the spot product prices are translated into an equivalent crude oil value at the crude loading port of origin, or the "FOB netback." Therefore,

$$\text{Total Arabian light product (GPW)} = \$18.99$$
$$\text{Less incremental refining fees} = \$1.30$$
$$\text{Less spot freight cost} = \$1.18$$
$$\text{Less insurance, loss, etc.} = \$0.25$$
$$\text{Implied netback of the Arabian light crude} = \$16.26$$

EXAMPLE 14-2

A refinery in the Arabian Gulf sells 40,000 barrels of topped light Arabian crude to a refiner in Singapore for further processing in the Singapore refinery. Establish the FOB price from the Arabian Gulf as per netback principles. Assume a refining cost of $2.10 per barrel.

The dollar per barrel revenue obtained from the topped crude (300°F + cut) is established by multiplying the calculated yield of the product obtained when this feedstock is processed in the Singapore refinery by the market prices prevailing for these products in the month in question.

PRODUCT	YIELD, LV%	Mean of Platts (MOP), $	BARREL COST, $
NAPHTHA	2.0	28.35	0.567
KEROSENE	18.0	33.10	5.958
DIESEL	30.0	28.75	8.625
FUEL OIL	46.0	21.70	9.982
TOTAL	96.0		25.132

From the dollar per barrel revenue just calculated, the actual refining cost and per barrel freight for an Arabian Gulf/Singapore trip in an MR class vessel (Table 14-1) is deducted, to obtain netback price for the topped crude:

$$\begin{aligned}\text{Sales realization per barrel topped crude in Singapore} &= \$25.13\\ \text{Less refining cost} &= \$2.10\\ \text{Less per barrel freight, Singapore/Arabian Gulf} &= \$0.75\\ \text{Estimated price per barrel topped crude, FOB Arabian Gulf} &= \$22.28\end{aligned}$$

FORMULA PRICING

Formula pricing techniques do not rely on the product yield value to determine the crude price nor do they secure any margin for the refiner. In formula pricing, the crude price is linked to another crude or group of crudes for which price quotes are regularly available. The usual practice is to select a popular spot crude of comparable quality (API, sulfur, etc.) or a basket of crudes from the same producing area. For example,

Mexican crudes sold in U.S. markets at one time were tied to the prices of three U.S. domestic crudes plus an element of residual fuel oil (33% West Texas sour + 33% West Texas intermediate + 33% Alaskan North slope, less 3% fuel). It provided for price determination based on U.S. quotes for 5 days around the bill of lading date.

Formula pricing is inflexible compared to the spot market, to which oil companies are becoming accustomed. The buyers are reluctant to tie themselves to long-term arrangements through complex formulas that offer no advantage, unless the prices are very attractive over long-term contracts, and do not offer secure margins to refiners, as is in the case of netback deals, which will remain popular so long as market volatility persists.

PRICING PETROLEUM PRODUCTS AND INTERMEDIATE STOCKS

For the purpose of economic settlement between the participants, it becomes necessary to estimate the dollar value of the petroleum product and intermediate process stocks inventory in the refinery tanks at a given time. Market quotes for a few finished products, which are regularly traded in bulk, are published regularly and thus easily available. These become the reference prices for computing the prices of other petroleum products or intermediate stocks for which no market price quotes are available.

Product properties produced by a refinery may be quite different from the quality of the reference products for which price quotes are available. For example, a refinery in the Arabian Gulf region may produce 95 octane gasoline. As no price quotes for gasoline are available in the Arabian Gulf market, available gasoline quotes for the Singapore or Mediterranean market may be used as reference. Price adjustments are required, however, for quality variations. Process or intermediate stocks, for the purpose of price determination, are considered a blend of two or more stocks whose price quotes are available.

REFERENCE MARKETS

The appropriate reference market is chosen depending on the location of the refinery where its products are most likely to be sold. Market quotations from the following markets are regularly published by Platt's Oilgram Service or Platt's Marketscan and other reporting services and are used as reference prices for estimating the price of other products:

U.S. Gulf Coast.
Northwestern Europe/Rotterdam (NWE).
Mediterranean.
Arabian Gulf (AG).
Singapore.

To estimate the prices of products and intermediate streams, average quotes over a period of time should be used. The following examples illustrate the mechanism of product pricing from the reference price data.

EXAMPLE 14-3

Determine the prices of motor gasolines, unleaded (RON 91 and 95), for an Arabian Gulf coast location for September 1998 from the following data:

PRODUCT	MARKET	AVERAGE PRICE, $/TON	BARRELS PER TON	SPECIFIC GRAVITY
NAPHTHA	AG	143.55	9.00	0.699
NAPHTHA	NWE	157.23	8.90	0.707
MOGAS REG, 91	NWE	169.34	8.46	0.744
MOGAS PREM, 95	NWE	174.67	8.46	0.755

Price quotations for gasoline grades in the Arabian Gulf Markets are not available as gasolines are not traded in this market.

Gasoline prices for Arabian Gulf locations are, therefore, calculated from NWE quotes as follows:

$$\begin{aligned}\text{MOGAS (RON 91)} &= \text{Arabian Gulf naphtha price} + (\text{NWE, regular unleaded price} - \text{NWE naphtha price})\\ &= [143.55 + (169.34 - 157.23)]\\ &= \$155.66/\text{ton}\end{aligned}$$

$$\begin{aligned}\text{MOGAS (RON 95)} &= \text{Arabian Gulf naphtha price} + (\text{NWE premium unleaded price} - \text{NWE naphtha price})\\ &= [143.55 + (174.67 - 157.23)]\\ &= \$160.99/\text{ton}\end{aligned}$$

EXAMPLE 14-4, REFORMATE 96 RON PRICING

Calculate the price of catalytic reformate (RON 96) from the regular and premium gasoline prices for the Arabian Gulf market, calculated in the preceding example.

Reformate 96 prices are estimated on the basis of RON parity with premium and regular gasolines blends:

PRODUCT	RON	vol fraction	SG	wt%	PRICE, $/ton
MOGAS REG	91	−0.2500	0.744	−24.5	155.66
MOGAS PREM	95	1.2500	0.755	124.5	160.99
REFORMATE	96	1.0000	0.758	100.0	162.29

EXAMPLE 14-5, LIGHT CAT NAPHTHA

Calculate the price of light cat naphtha from the FCCU unit, with following properties:

$$\text{RON} = 92.8$$

$$\text{Specific gravity} = 0.788$$

$$\text{Barrels/ton} = 7.985$$

As in the case of reformate, the light cat naphtha prices are estimated on the basis of RON parity with premium and regular gasoline blends as follows:

PRODUCT	RON	vol fraction	SG	wt%	PRICE, $/ton
MOGAS 91	91.0	0.5500	0.744	0.5460	155.66
MOGAS 95	95.0	0.4500	0.755	0.4540	160.98
LIGHT CAT NAPHTHA	92.8	1.0000	0.749	1.0000	158.06

This price is for a product with a specific gravity of 0.749 and it must be corrected for the specific gravity of the required product:

$$\text{The corrected Light Cat Naptha (LCN) price} = 158.06 \times (0.749/0.788)$$
$$= \$150.28/\text{ton}$$

GAS OIL PRICING

Gas oil pricing is done on the basis of cloud point parity; then, correction is applied for specific gravity and sulfur, if required.

EXAMPLE 14-6

Estimate the price of winter grade gas oil (NWE market) with the following properties:

$$\text{Cloud point} = -7°\text{C}$$
$$\text{Specific gravity} = 0.8330$$
$$\text{Sulfur} = 0.19 \text{ wt\%}$$

PRODUCT	CLOUD POINT, °C	CLOUD POINT, °F	CLOUD POINT INDEX[1]	wt fraction	SULFUR	SG	PRICE, $/ton, NWE
KEROSENE	−50	−58	1.15	0.5516	0.10	0.783	155.28
GAS OIL	5	41	38.52	0.4484	0.50	0.845	142.20
NWE GAS OIL	−7	19.4	17.91	1.0000	0.28	0.811	149.42

$$\text{Specific gravity correction} = 149.42 \times (0.811/0.833)$$
$$= \$145.43/\text{ton}$$

Sulfur correction is applied next. This is based on market quotes published on the gas oil sulfur differential ($/% sulfur) for the period in question:

$$\text{Sulfur differential} = 2.4/\text{per percent Sulfur}$$
$$\text{Sulfur correction} = (0.28 - 0.19) \times 2.4$$
$$= 0.22\ \$/\text{ton}$$
$$\text{Gas oil price} = \$149.42 + 0.22$$
$$\text{NWE with Sulfur correction} = 149.64\ \$/\text{ton}$$
$$\text{NWE GAS oil price with S.G correction} = 149.64 \times \frac{0.811}{0.833}$$
$$= 145.69\ \$/\text{ton}$$

PRICING FUEL OILS

Fuel oil pricing is done on viscosity parity basis with a blend of 180 cst, 3.5% sulfur fuel oil and diesel. The prices of these reference fuel oil grades and diesel are available from market quotes.

EXAMPLE 14-7

Estimate the price of 350 cst and 4% sulfur fuel oil on the basis of Arabian Gulf market prices:

Fuel oil sulfur = 4%

Fuel oil viscosity = 380 cst at 50°C.

PRODUCT	VISCOSITY, cst	VISCOSITY BLEND INDEX[2]	wt fraction	SULFUR, wt%	PRICE, \$/ton
GAS OIL	2.5	13.55	−0.09	0.50	142.2
FUEL OIL	180	34.93	1.09	3.5	78.01
FUEL OIL	380	36.88	1.00	3.77	72.17

The sulfur differential is based on Mediterranean and Singapore market quotes for a differential between 3.5% and 1.0% sulfur fuel oils:

$$\text{Sulfur differential} = 5.199\ \$/\text{per}\ \%\ \text{sulfur}$$

$$\text{Sulfur correction} = 5.199 \times (4.0 - 3.77) = \$1.18/\text{ton}$$

$$\text{Fuel oil price, after sulfur adjustment} = 72.17 - 1.18 = 70.99\ \$/\text{ton}$$

PRICING VACUUM GAS OILS

Vacuum gas oil (VGO) can be considered a blend of 180 cst, 3.5% fuel oil and diesel to match the viscosity and sulfur of the VGO.

EXAMPLE 14-8

Calculate the price of desulfurized heavy vacuum gas oil (HVGO) from the distillate hydrocracker (bleed stream) with following properties:

$$\text{Viscosity at } 50^\circ\text{C, cst} = 12.7$$
$$\text{Sulfur, wt\%} = 0.15$$

PRODUCT	VISCOSITY, cst	VISCOSITY BLEND INDEX	wt fraction	SULFUR, wt%	PRICE $/ton
GAS OIL	2.5	13.55	0.470	0.50	142.20
FUEL OIL	180	34.93	0.530	3.50	78.01
VACUUM GAS OIL	12.7	24.88	1.000	2.089	108.20

$$\text{Sulfur differential} = \$5.199/\% \text{ sulfur}$$
$$\text{Sulfur adjustment} = 5.199 \times (2.089 - 0.15)$$
$$= \$10.08/\text{ton}$$
$$\text{Final VGO price} = \$118.28/\text{ton}$$

PRICING CUTTER STOCKS FOR FUEL OILS

A typical pricing basis for cutter stocks is viscosity parity or with a blend of kerosene and diesel.

EXAMPLE 14-9

Calculate the price of cutter stock with 1.22 cst viscosity at 50°C for fuel oil blending on the basis of kerosene and diesel properties that follow:

PRODUCT	VISCOSITY, cst	VISCOSITY BLEND INDEX	wt fraction	SULFUR, wt%	PRICE $/ton
KEROSENE	1.0	3.25	0.750	0.20	155.28
DIESEL	2.5	13.55	0.250	0.50	142.20
CUTTER STOCK	1.22	5.85	1.000	0.28	151.97

The reference prices for diesel and kerosene used here are the market quotes. Here, the price is computed for a cutter viscosity of 1.22 cst.

COST OF ENERGY

Energy is a major cost item in any refinery. Energy may be used in many forms, such as natural or associated gas used as refinery fuel, steam generated in the refinery, distilled and cooling water, and electricity or feed to the hydrogen plant.

The energy cost is usually derived from the cost of heavy fuel oil (380 cst and 3.5% sulfur) as per the Mean of Platts (MOP) published price at any reference time. The calorific value of this grade is around 38 million Btu per ton. The cost of energy-intensive utilities, such as electricity, steam, cooling water, or distilled water, is a function of the energy cost and can be expressed as follows:

UTILITY	UNIT	COST, $/UNIT
ELECTRICITY	kWhr	$A \times E + B$
STEAM	mmBtu	$C \times E + D$
COOLING WATER	m^3	$K \times E + L$
DISTILLED WATER	m^3	$M \times E + N$

m^3 = METER CUBE.

where E is the energy cost per million Btu and A, B, C, D, K, L, and the like are constants.

For example, if the heavy fuel oil price is $76/ton, the corresponding energy cost would be 76/38 or $2.0/million Btu.

The values of the constants can be derived from the refinery's operating data over a period of time.

ASSIGNED CRUDE YIELDS

A refinery may process more than one crude oil at a time. The crude may be received under a netback or similar agreement, in which the refiner pays the crude supplier the gross product worth of the product produced and receives only a processing fee plus a premium. Therefore, it becomes necessary to determine the yield of the products from a crude before the gross product worth of a crude can be determined.

If the refinery is processing only one crude received under netback agreement, the determination of yields from the crude is determined simply from refinery stock balance. However, the refinery may be processing more than one crude at the same time and one of the crudes processed is received under netback arrangement. The determination of the actual yield from the netback crude in the refinery is complicated. To determine the yields, a crude received by refinery under netback or similar arrangement is called an *assigned crude*.

DETERMINATION OF ASSIGNED CRUDE YIELDS

Single-Ownership Refineries

In a single-ownership refinery, the assigned crude yield is determined as follows (the steps are shown in Table 14-2):

1. Two separate LP models are set up for simulating processing of assigned and other crudes processed in the refinery. Only the balancing-grade products, no fixed-grade products, are produced. Also, no drawdown or buildup of process stocks is allowed in the LP. Identical product prices of balancing grade are used to drive the two LP models.
2. The maximum available processing unit capacities to be used in the assigned crude LP model is determined from the "assigned crude ratio" (assigned crude/total refinery crude) and the refinery's maximum available processing unit capacities. For example, if the assigned crude ratio is 0.6 and the total crude unit capacity of the refinery is 100 mbpd, the crude unit capacity to be used in the assigned crude LP model is (100×0.6) or 60 mbpd. The remaining CDU capacity (i.e., 40 mbpd) is used in the second LP model, processing the other crude. Downstream processing unit capacities

Table 14-2
Assigned Crude Yields

GRADES	ARAB CRUDE LP PRODUCT	BAHRAIN CRUDE LP PRODUCT	TOTAL LP* PRODUCT	REFINERY PRODUCTION	INITIAL DELTA	DELTA OFFSET	NET DELTA	DELTA CRUDE RATIO	YIELD FROM ARAB CRUDE	% VOL/ VOL YIELD
	(1)	(2)	(3 = 1 + 2)	(4)	(5 = 4 − 3)	(6)	(7 = 5 + 6)	(8)	(9 = 1 + 8)	(10)
I-150	19115	0	19115	3753	−15362	−86	−15447	−12758	6357	0.10%
I-201	77502	0	77502	59656	−17846	−100	−17946	−14822	62681	1.02%
I-210	1042001	197199	1239200	1230148	−9052	−51	−9102	−7517	1034483	16.89%
I-397	300496	52811	353307	534659	181352	−1013	180339	148942	449438	7.34%
I-440	686546	117938	804484	1227704	423220	−2363	420857	347586	1034132	16.89%
I-888	2385691	576836	2962526	2476700	−485826	−2713	−488539	−403485	1982206	32.37%
I-961	1576524	337949	1914473	1844702	−69771	−390	−70161	−57946	1518579	24.80%
TOTAL	6087875	1282732	7370607	7377322	6715	−6715	0	0		99.41%
CRUDE	6124230	1291296	7415526						6124230	

NOTES:

CRUDE	BARRELS
LIGHT ARAB CRUDE	6,124,230
BAHRAIN CRUDE	1,291,296
TOTAL REFINERY CRUDE	7,415,526

ASSIGNED CRUDE RATIO (ARAB) = C.8259.
COLUMN 7 = (−1)*(*ABSOLUTE VALUE OF COLUMN 6/SUM OF ABSOLUTE VALUES OF DELTAS IN COLUMN 6*)**SUM OF COLUMN.*
COLUMN 8 = COLUMN 6 + COLUMN 7.
COLUMN 9 = COLUMN 8 × *CRUDE RATIO.*
WHERE CRUDE RATIO = ASSIGNED CRUDE/TOTAL REFINERY CRUDE.
SUM OF ABSOLUTE VALUES OF DELTAS 1202429.
*LP = LINEAR PROGRAM.

are similarly split between the two LP models for processing the two crudes.

3. The models are run to determine the balancing grade product yields. The LP optimizes the production of balancing grades on the basis of their prices.
4. The balancing grade production is adjusted to reflect unaccounted losses. The total unaccounted losses are determined as a certain percentage of crude processed. A figure of 0.6% by volume may be used if no refinery data are available. These losses are spread over the balancing grades in the ratio of their production.
5. The revised LP products are adjusted to reflect the actual refinery blending. The adjustment factors (initial deltas) are established by deducting the sum of the revised LP production from the total refinery production (expressed in balancing grades by means of product equivalencies).
6. Ideally, the sum of initial deltas should equal zero, which in practice is not so. The sum of the initial deltas is distributed over all balancing grades in the ratio of their production disregarding their signs; that is, (absolute value of initial delta)/(sum of absolute values of initial deltas). This factor, called the *delta offset*, is deducted from the initial delta to get the net delta.
7. The net deltas established are multiplied by the assigned crude ratio and added to the revised (assigned crude) LP products to determine the overall assigned crude yield.

Joint-Ownership Refineries

In joint-ownership refineries, the assigned crude yield is determined as follows (Table 14-3 shows various steps):

1. An LP model simulating the assigned crude processing is set up containing the actual unit yields, unit capacities, and so forth available for processing the assigned crude. The driving force behind the model is the mean value of Platts (MOP) product prices for the products that prevailed during the month.
2. The maximum available processing unit capacities to be used in the LP model for processing assigned crude are determined from the assigned crude ratio (assigned crude/total refinery crude) and maximum available processing unit capacities. For example, if the assigned crude ratio is 0.6 and the total crude unit capacity of the

Table 14-3
Assigned Crude Yields

GRADES	AOC LP PRODUCT	BOC LP PRODUCT	TOTAL LP PRODUCT FROM L.ARAB	%	LESS 0.6% CRUDE	REVISED LP PROD. YIELD	TOTAL RETROSPECTIVE DOP	TOTAL ALLOCATED PRODUCTION	INITIAL DELTA	%	DELTA OFFSET*	NET DELTA	DELTA* CRUDE RATIO	TOTAL REVISED LP PRODUCTS	LIGHT ARAB CRUDE YIELDS	CRUDE, %
	(1)	(2)	(3 = 1 + 2)	(4)	(5)	(6 = 3 − 5)	(7)	(8)	(9 = 8 − 7)	(10)	(11)	(12 = 9 + 11)	(13)	(6)	(14 = 13 + 6)	(15)
I-150	19230	0	19230	0.31%	115	19115	3161	3753	0	0	0	0	0	19115	19115	0.31%
I-201	0	77970	77970	1.27%	468	77502	316228	59656	−256572	16.59%	−3618	−260190	−214891	77502	−137389	−2.24%
I-210	928560	119730	1048290	17.12%	6289	1042001	976069	1230148	254079	16.43%	−3583	250496	206885	1042001	1248886	20.39%
I-397	194100	108210	302310	4.94%	1814	300496	521905	534659	12754	0.82%	−180	12574	10385	300496	310881	5.08%
I-440	550620	140070	690690	11.28%	4144	686546	825186	1227704	402518	26.02%	−5677	396842	327751	686546	1014297	16.56%
I-888	1921110	478980	2400090	39.19%	14399	2385691	2982594	2476700	−505894	32.71%	−7135	−513029	−423710	2385690	1961980	32.04%
I-961	1315830	270210	1586040	25.90%	9516	1576524	1729774	1844702	114928	7.43%	−1621	113307	93580	1576524	1670104	27.27%
TOTAL	4929450	1195170	6124620	100.00%	36745	6087875	7354917	7377322	21814	100.00%	−21814	0	0	6087874	6087874	99.41%

NOTES:

CRUDE	PARTICIPANT AOC	PARTICIPANT BOC	TOTAL, bbl
LIGHT ARAB	4924230	1200000	6124230
BAHRAIN	1291296		1291296
TOTAL REFINERY	6215526	1200000	7415526

ASSIGNED[1] (LIGHT ARAB) CRUDE RATIO = 0.8259
SUM OF ABSOLUTE TOTAL DELTAS 1546746
ASSIGNED CRUDE RATIO IS THE RATIO OF CRUDE PROCESSED UNDER ASSIGNED CRUDE AGREEMENT TO TOTAL REFINERY CRUDE.
*DELTA OFFSET (COLUMN 11) = SUM OF COLUMN 9 MULTIPLIED WITH (COLUMN 10) AND (−1).

refinery is 100 mbpd, the crude unit capacity to be used in the LP model is (100×0.6) or 60 mbpd. Downstream processing unit capacities are similarly determined.

3. The model is run to determine the balancing-grade production yields. The balancing grade production is adjusted to reflect unaccounted losses. The total unaccounted losses are determined as a certain percentage of crude processed. A figure of 0.6% by volume may be used if no refinery data are available. These losses are spread over the balancing grades in the ratio of their production.
4. The revised LP products are adjusted to reflect the actual refinery blending. The adjustment factors (initial deltas) are established by deducting the balancing grade equivalents of the "retrospective definitive operating program (DOP)"[3] production from the total refinery "allocated production" (expressed in balancing grades).
5. As described earlier, the sum of initial deltas should equal zero, which in practice is not so. The sum of the initial deltas is distributed over all balancing grades in the ratio of their production, disregarding their signs; that is, (absolute value of initial delta)/(sum of absolute values of initial deltas). This factor, called the *delta offset*, is deducted from the initial delta to get the net delta.
6. The net deltas established are multiplied by the assigned crude ratio and added to the revised LP products to determine the overall assigned crude yield.

NOTES

1. For the cloud point index for blending on a weight basis, refer to Table 11-15.
2. Based on weight based viscosity blend index I,

$$I = 23.097 + 33.468 \log_{10} \log_{10}(\nu + 0.8)$$

where ν = viscosity in centistokes (refer to Table 11-13).

3. Refer to Chapter 16, "Product Allocation" for procedures to determine the retrospective DOP and allocated production.

CHAPTER FIFTEEN

Definitive Operating Plan

Before the start of an operating period, usually a month, the planner submits an operating plan, termed the *definitive operating plan* (DOP), to the refinery Operations Department, defining how the refinery is to be run during the month.

The DOP for a given month contains information on the following:

1. The crude oils to be processed during the month and their quantities.
2. Product grades and the quantities to be produced during the month.
3. Intermediate stocks to be drawn from the inventory to make the refinery's product slate.
4. Inventory of intermediate stocks to be built up during the month.
5. Process unit capacities available, based on the refinery's latest shutdown schedule and estimated process unit capacities utilized to make the product slate.
6. Composition of feed to all conversion units.
7. Product prices of the balancing-grade products and energy price used to optimize product blending.
8. Blending instructions for critical stocks, where alternate blending is possible.
9. Additives to be used for blending products and their estimated consumption. Only additives used for product blending (for example, dyes for gasoline, antistatic and anti-icing additives for jet fuels, pour point depressants for diesels) must be listed.
10. Expected production rate for all product groups during the month.

The DOP is prepared using a refinery linear programming model. Refinery material balance spreadsheet programs can also be used, but LP models are preferred as no product blending optimization is possible with spreadsheet programs. The first step in preparing the DOP for a given month is to update the refinery LP model by incorporating the following:

- *Refinery crude run.* The crude processing rate is estimated from the total crude to be run during the month.
- *Available processing units' capacity.* Every refinery processing unit has an agreed-on minimum and maximum nominal capacity in mb (thousand barrels) per stream day. The data are based on previous test runs on the unit. Also, the refinery maintains a unit shutdown schedule for the next 12 or more months to perform routine maintenance, major repairs, catalyst change or regeneration, and the like. This document is constantly updated. Thus, from the latest shutdown schedule, it is possible to find out as to how many days during the month a given unit would be available to process feedstock. From such data, two on-stream factors (OSF) are worked out:

$$\text{OSF1} = (\text{number of days unit is likely to be available}) / (\text{total number of days in the month})$$

  Here, OSF is the on-stream factor for scheduled maintenance. To take into account unit unavailability due to possible unscheduled shutdowns, another factor (OSF2) is used. This is based on the past experience in operating the unit.

$$\text{Available (max) unit capacity} = \text{nominal max capacity} \times \text{OSF1} \times \text{OSF2}$$

  For example, if the nominal throughput of a unit is 50 mbpd, OSF1 is 0.70, and OSF2 is 0.985, available unit capacity $= 50 \times 0.7 \times 0.985$, or 34.475 mb per stream days during the month.
- *Product grades to be made during the month and their quantities.* This is based on the expected shipment of products during the month. The schedule of ships likely to arrive at the refinery marine terminal and the product to be loaded on them is known several weeks ahead of their actual arrival. Keeping in view the product

inventory likely to be available in refinery tanks, the production rates of all fixed product grades are worked out to meet the expected shipments. As the crude rate is already fixed, any blending stocks left after blending fixed grades is blended into balancing-grade products for spot sale by the refinery.

- *Product prices for all balancing-grade products.* The LP model optimize the production of balancing-grade products based on their prices. Prices are entered only for balancing-grade products whose production is not fixed. The LP optimizes the production of balancing-grade products based on their prices and maximizes refinery profit at the same time.
- *Updated specifications for all product to be made.* Although the LP model has built-in data on all product grades the refinery usually produces, updating may be necessary if the specifications for a product grade to be made are different.
- *Updated blend map.* Every LP model has a blend map that specifies which blend streams are to be allowed in blending a product grade. For example, if a refinery has a surplus heavy vacuum gas oil (HVGO) after meeting the requirements of conversion units, the model would dump all surplus HVGO in fuel oil. Disallowing HVGO in fuel oil in the blending map forces the LP to produce HVGO as an export stream.
- *Buildup or drawdown of intermediate and process stocks from inventory.* Inventories of all process stocks are closely monitored and regulated. Drawdown of intermediate or process stocks is done to ensure that processing units run to their full capacity or to meet given product demand. Buildup of stocks is sometimes done to ensure adequate stock ahead of a large shipment. For example, a refinery may build up reformate stock ahead of a large gasoline shipment or buildup HVGO ahead of a large export shipment of HVGO.

The format of an actual refinery DOP is shown in Table 15-1. The crude to be processed and product produced in the DOP become the refinery targets for the month (see Table 15-2). Also, feed rates and composition of all conversion units is kept close to DOP figures. If a product grade requires additives, such as military jet fuels, these are specified in the DOP and the refinery follows these directions while blending such products.

Table 15-1
Definitive Operating Plan of a Refinery

	bpcd	mb
CRUDE RUN		
ARAB LIGHT CRUDE	201000	6030
BAHRAIN CRUDE	42000	1260
CONDENSATE IMPORT	3600	108
TOTAL	246600	7398
INTERMEDIATE STOCK INVENTORY CHANGES		
LIGHT CAT NAPHTHA	−2000	−60
POLYMER GASOLINE	−1000	−30
CAT REFORMATE 90R	−600	−18
CAT REFORMATE 95R	−2000	−60
TOTAL	−5600	−168
PRODUCTION GRADES		
I-150 LPG	600	18
LIGHT NAPHTHA	3600	108
I-210 W.S.R.NAPHTHA	39483	1184
I-220 LIGHT NAPHTHA	29	1
I-390 GASOLINE 90R	2560	77
I-395 GAS 95 R	0	0
I-397L GAS 97 R	19951	599
I-398E GAS 98 R	1470	44
I-419 DUAL-PURPOSE KEROSENE	14000	420
JP-4	8000	240
I-440 KEROSENE JET A-1	24301	729
I-800 +6 POUR DIESEL	14000	420
I-876 −3 POUR DIESEL	10000	300
I-876ZP −18 POUR DIESEL	8000	240
I-885 −3 POUR, 0.5S DIESEL	0	0
I-888 −3 POUR 1%S DIESEL	47098	1413
I-928 2.8 %S 180 cst FO	12000	360
I-961 3.5 %S, 180 cst FO	43513	1305
I-1138 ASPHALT	3553	107
TOTAL	252158	7565

NOTES:
MONTH = APRIL 2000
NEGATIVE ENTRY IN THE TABLE INDICATES A DRAWDOWN.
POSITIVE ENTRY INDICATES A BUILDUP OF STOCK.

(DOP continued)

Table 15-2
Unit Capacity Utilization

UNITS	TOTAL AVAILABLE CAPACITY, bpcd	TOTAL UTILIZED bpcd	%UTILIZED bpcd
CRUDE UNITS 1–5	243000	243000	100.00%
VACUUM UNITS 1, 5, 6	95000	92108	96.96%
KEROSENE RERUN UNIT	15421	15421	100.00%
VISBREAKER	20000	19173	95.87%
ASPHALT CONVERTER	4800	2961	61.69%
KEROSENE TREATING UNIT	45000	41987	93.30%
No. 1 HDU DIESEL HDS	20000	19998	99.99%
No. 2 HDU, HYDROCRACKER	50000	49005	98.01%
FCCU	39000	37833	97.01%
CAT REFORMER	15000	3625	24.17%

NOTES:
THE MODEL BLENDS WSR (WHOLE STRAIGHT RUN) TO SG = 0.6918
DISPOSITION OF CRITICAL STOCKS:

		bpcd	mb
HEAVY CAT NAPTHA TO	MOGAS	0	0
	DIESEL	863	25.89
	FUEL	1358	40.74
	TOTAL	2221	66.63
KEROSENE TO	DIESEL	2351	70.53
	FUEL	0	0
LCGO TO	HDU 1	6463	193.89
LIMITING SPECS		DIESEL	DI/SULFUR
		FUEL	VISCOSITY
FCCU FEED COMPOSITION			
	MEDIUM/HEAVY ISOMATE		91%
	FRESH FEED (HVGO)		9%
	VACUUM RESID		0%
	TOTAL		100%

The refinery prepares its own estimates of production rates, based on a refinery spreadsheet model and revised every week on the basis of refinery operating conditions. At the end of the month, the total crude processed should be close to that stipulated in the DOP. The product produced would, of course, be different because of the following:

- Blending of products in the refinery is done to meet specific shipments and not to maximize profit, as in the LP.
- Production rates of blend stocks can be different from that in the LP due to unscheduled shutdown of a processing unit, or yields and properties may be different due to a change in the operating conditions of a processing unit.

DOPS IN JOINT-OWNERSHIP REFINERIES

In joint-ownership refineries, the crude distillation unit (CDU) and downstream processing unit capacities are split between the two participants in the ratio of their equity in the refinery. Thus, the refinery is divided into two "virtual" refineries. Each participant is free to utilize its share of the refinery as it wishes.

In joint-ownership refineries, each participant prepares a DOP for its share of refinery capacity, stating the crude to be processed during the month and product to be produced. The DOP format is identical to that for single-ownership refineries.

To ensure that the participant DOPs are on a common basis, a reference LP model of the refinery is prepared by the refinery, reflecting the total CDU and process unit capacities likely to be available to the participants during the month, taking into account the refinery's shutdown schedule. Process yields and utilities, catalyst, and chemical consumption for all units; product qualities of all streams; and product specifications are updated, if required. The driving force behind the LP is product price for the agreed-on balancing grades, which are usually the mean of Platt published price (MOP) for each product prevailing during the previous month.

Two LP models are prepared from the reference refinery model by incorporating the unit capacity available to each participant as though these were independent refineries. Each participant updates its refinery model for preparation of its DOP by incorporating the following data:

1. The anticipated crude run during the month and any process stock drawn down or built up to meet its product slate.
2. The product grades and quantities it wishes to make during the month.
3. The maximum and minimum unit capacity of each processing unit available during the month, taking into account unit shutdowns, if any.

The LP model is run and the results of the LP feasible and optimum run are incorporated in the DOP.

REFINERY DOP

The refinery combines the two DOPs to make a refinery DOP before the start of the month. Examples of participant DOPs and combined refinery DOP is shown in Tables 15-3 to 15-8. The combined DOP becomes the

Table 15-3
DOP of Participant AOC

	bpcd	mb	%
FEED STOCKS			
ARABIAN LIGHT CRUDE	161000	4,830.0	
BAHRAIN CRUDE	42000	1,260.0	
TOTAL CRUDE	203000	6,090.0	
INTERMEDIATE STOCK INVENTORY CHANGES			
LIGHT CAT NAPHTHA	−2000		
POLYMER GASOLINE	−1000		
REFORMATE 95R	−2000		
REFORMATE 90R	−600		
TOTAL REFINERY FEED	208600	6,258.0	
FINISHED PRODUCTS			
I-150 LPG	600	18.0	0.29
I-201 LSR	0	0.0	0.00
I-210 WSR	31845	955.4	15.26
I-220 LSR	0	0.0	0.00
I-390 GASOLINE	2560	76.8	1.23
I-395 GASOLINE	0	0.0	0.00
I-397L GASOLINE	17242	517.3	8.26
I-398E GASOLINE	1470	44.1	0.70
I-419 DUAL KEROSENE	14000	420.0	6.71
I-434 JP-4	8000	240.0	3.83
I-440 JET A-1	17401	522.0	8.34
I-800 1.0%S +6 POUR	14000	420.0	6.71
I-876 1.0%S −6 POUR	10000	300.0	4.79
I-876ZP 0.4%S −18 POUR	8000	240.0	3.83
I-885 .5%S	0	0.0	0.00
I-888 1.0%S −3 POUR	33348	1000.4	15.99
I-928 2.8%S 180 CST	12000	360.0	5.75
I-961 3.5%S 180 CST	36153	1084.6	17.33
I-1138 BULK ASPHALT	2000	60.0	0.96
TOTAL PRODUCTS	208619	6258.6	100.00
LIQUID RECOVERY	100.01%		

(DOP of Participant AOC continued)

Table 15-4
Unit Capacity Utilization of Participant AOC

UNITS	AVAILABLE, bpcd	UTILIZED, bpcd	UTILIZED, mb	UTILIZED, %
CRUDE UNITS	203000	203000	6,090.0	100.0%
VACUUM UNITS	82086	76408	2,292.2	93.1%
NO.6 KEROSENE RERUN	23809	10521	315.6	44.2%
VISBREAKER	15873	15873	476.2	100.0%
ASPHALT CONVERTER	3968	1265	38.0	31.9%
KEROSENE TREATING	35087	35087	1,052.6	100.0%
HDU NO. 1	16708	16708	501.2	100.0%
HDU NO. 2	40935	40935	1,228.1	100.0%
FCCU	31603	31603	948.1	100.0%
CAT REFORMER	14285	2984	89.5	20.9%

NOTES:
THE MODEL BLENDS WSR TO SG 0.6917
DISPOSITION OF CRITICAL STOCKS:

		bpcd	mb
HEAVY CAT NAPHTHA	MOGAS	0	0.0
	DIESEL	793	23.8
	FUEL	1206	36.2
		1999	60.0
KEROSINE TO	DIESEL	1981	59.4
	FUEL	0	0.0
LCGO TO	1HDU	5255	157.7
LIMITING SPECS	DIESEL	DI/SULFUR	
	FUEL	VISCOSITY	

FCCU FEED COMPOSITION	
MEDIUM/HEAVY ISOMATE	92.35%
FRESH FEED (HVGO)	7.65%
4 A VACUUM RESID	0.00%
	100.00%

operating plan of the refinery, and the Joint Operating Committee (JOC) tries to run the refinery as close as possible to that plan. In case of any unscheduled shutdown of a major process unit or other emergency, like a major storage tank out of service or problem at marine or shipping terminal, the JOC can ask the participants to resubmit their DOPs after taking into consideration the latest situation.

In joint-ownership refineries, participants DOPs, apart from serving as a refinery operating plan, are used as the basis of product allocation, or establishing the ownership of stocks. The mechanism of product allocation is explained later on in the book. The concepts of fixed- and balancing-grade products, used both in the LP models for DOP and product allocation, is explained here. Fixed-grade product equivalencies are used later in the product allocation program.

Table 15-5
DOP of Participant BOC

	bpcd	mb	%
FEED STOCKS			
ARABIAN LIGHT CRUDE	40000	1,200.0	
BAHRAIN CRUDE	0	0.0	
CONDENSATE IMPORT	3600	108.0	
TOTAL CRUDES	43600	1,308.0	
INITIAL STOCK INVENTORY CHANGES			
STOCK			
LIGHT CAT NAPHTHA	0		
POLYMER GASOLINE	0		
REFORMATE 95R	0		
REFORMATE 90R	0		
TOTAL REFINERY FEED	43600	1,308.0	
FINISHED PRODUCTS			
I-150 LPG	0	0.0	0.00
I-201 LSR	3600	108.0	8.27
I-210 WSR	7638	229.1	17.54
I-220 LSR	29	0.9	0.07
I-390 GASOLINE	0	0.0	0.00
I-395 GASOLINE	0	0.0	0.00
I-397L GASOLINE	2709	81.3	6.22
I-398E GASOLINE	0	0.0	0.00
I-419 DUAL KEROSENE	0	0.0	0.00
I-434 JP-4	0	0.0	0.00
I-440 JET A-1	6900	207.0	15.85
I-800 1.0%S +6 POUR	0	0.0	0.00
I-876 1.0%S −6 POUR	0	0.0	0.00
I-876ZP 0.4%S −18 POUR	0	0.0	0.00
I-885 .5% S	0	0.0	0.00
I-888 1.0%S −3 POUR	13750	412.5	31.58
I-928 2.8%S 180 CST	0	0.0	0.00
I-961 3.5%S 180 CST	7360	220.8	16.90
I-1138 BULK ASPHALT	1553	46.6	3.57
TOTAL PRODUCTS	43539	1306.17	100.00
LIQUID RECOVERY	99.86%		

(DOP of Participant BOC continued)

Table 15-6
Unit Capacity Utilization of Participant BOC

UNITS	AVAILABLE, bpcd	UTILIZED, bpcd	UTILIZED, mb	UTILIZED, %
CRUDE UNITS 1–5	40000	40000	1,200.0	100.0%
VACUUM UNITS 1, 5, 6	15700	15700	471.0	100.0%
KEROSENE RERUN NO. 6	4900	4900	147.0	100.0%
VISBREAKER	3300	3300	99.0	100.0%
ASPHALT CONVERTER	4800	1696	50.9	35.3%
KEROSENE TREATING	6900	6900	207.0	100.0%
HDU NO. 1	3290	3290	98.7	100.0%
HDU NO. 2	8070	8070	242.1	100.0%
FCCU	6230	6230	186.9	100.0%
CAT REFORMER	2800	641	19.2	22.9%

NOTES:
DISPOSITION OF CRITICAL STOCKS:

			bpcd	mb
HEAVY CAT NAPHTHA TO	MOGAS		0	0.0
	DIESEL		70	2.1
	FUEL		152	4.6
			222	6.7
KEROSINE TO	DIESEL		370	11.1
	FUEL		0	0.0
LIGHT CYCLE GAS OIL TO	1HDU		1208	36.2
LIMITING SPECS	DIESEL	D.I.		
	FUEL	VISCOSITY		
FCCU FEED COMPOSITION			V/V	
MEDIUM/HEAVY ISOMATE			90.00%	
FRESH FEED (HVGO)			10.00%	
VACUUM RESID			0.00%	
			100.00%	

Table 15-7
Combined DOP of Participants

	AOC, bpcd	BOC, bpcd	TOTAL, bpcd	TOTAL, mb
PRODUCTION GRADE				
I-150	600	0	600	18.0
I-201	0	3600	3600	108.0
I-210	31845	7638	39483	1184.5
I-220	0	29	29	0.9
I-390	2560	0	2560	76.8
I-395	0	0	0	0.0
I-397L	17242	2709	19951	598.5
I-398E	1470	0	1470	44.1
I-419	14000	0	14000	420.0
I-434	8000	0	8000	240.0
I-440	17401	6900	24301	729.0
I-800	14000	0	14000	420.0
I-876	10000	0	10000	300.0
I-876ZP	8000	0	8000	240.0
I-885	0	0	0	0.0
I-888	33348	13750	47098	1412.9
I-928	12000	0	12000	360.0
I-961	36153	7360	43513	1305.4
I-1138	2000	1553	3553	106.6
TOTAL	208619	43539	252158	7564.7
INTERMEDIATE STOCK CHANGES				
LCN	−2000	0	−2000	−60.0
CPOR	−1000	0	−1000	−30.0
CR90	−600	0	−600	−18.0
CR95	−2000	0	−2000	−60.0
TOTAL	−5600	0	−5600	−168.0

NOTES:
NEGATIVE ENTRY IN THE TABLE INDICATES A DRAWDOWN.
POSITIVE ENTRY INDICATES A BUILDUP OF STOCK.

CRUDE RUN	AOC bpcd	BOC bpcd	TOTAL bpcd
ARABIAN LIGHT	161000	40000	201000
BAHRAIN	42000	0	42000
CONDENSATE IMPORT	0	3600	3600
TOTAL	203000	43600	246600

Table 15-8
Combined Unit Capacity Utilization

UNITS	TOTAL AVAILABLE, bpcd	UTILIZED AOC, bpcd	UTILIZED BOC, bpcd	TOTAL UTILIZED, bpcd	UTILIZED, bpcd%
CRUDE UNITS 1–5	243000	203000	40000	243000	100.0
VACUUM UNITS 1, 5, 6	95000	76408	15700	92108	97.0
NO. 6 KEROSENE RERUN	15421	10521	4900	15421	100.0
VISBREAKER	20000	15873	3300	19173	95.9
ASPHALT CONVERTER	4800	1265	1696	2961	61.7
KEROSENE TREATING	45000	35087	6900	41987	93.3
HDU NO. 1	20000	16708	3290	19998	100.0
HDU NO. 2	50000	40935	8070	49005	98.0
FCCU	39000	31603	6230	37833	97.0
CAT REFORMER	15000	2984	641	3625	24.2

NOTES: THE MODEL BLENDS WSR to S.G. = 0.6918
DISPOSITION OF CRITICAL STOCKS:

		bpcd	mb
HEAVY CAT NAPTHA TO	MOGAS	0	0
	DIESEL	863	25.89
	FUEL	1358	40.74
	TOTAL	2221	66.63
KEROSENE TO	DIESEL	2351	70.53
	FUEL	0	0
LCGO TO	1HDU	6463	193.89
LIMITING SPECS		DIESEL	DI/SULFUR
		FUEL	VISCOSITY

FCCU FEED COMPOSITION

MEDIUM/HEAVY ISOMATE	91%
FRESH FEED (HVGO)	9%
VACUUM RESID	0%
TOTAL	100%

EXAMPLE 15-1

A joint ownership refinery has processing unit capacities as follows. Participant AOC has a 60% share of the refinery and Participant BOC has a 40% share of the refinery. Estimate the maximum crude the two participants can process within their capacity rights.

Crude distillation units = 200 mbpcd
Catalytic reformer = 20 mbpcd
FCCU = 40 mbpcd
Distillate hydrocracker = 50 mbpcd

The processing unit capacities available to each participant for processing their feed stock follows next:

PROCESSING UNIT	REFINERY CAPACITY	AOC, 60%	BOC, 40%
CRUDE DISTILLATION UNITS	200 mbpcd	120	80
CATALYTIC REFORMER	20 mbpcd	12	8
FCCU	40 mbpcd	24	16
DISTILLATE HYDROCRACKER	50 mbpcd	30	20

The maximum crude run of a participant is dictated by the crude unit capacity available to it. Therefore, the maximum crude run by participant company AOC is 120 mbpcd. Also, process unit capacities utilized for processing this crude must not exceed the capacities available to it.

FIXED- AND BALANCING-GRADE PRODUCTS

Refinery products grades may be divided into two categories: fixed grades and balancing grades. Fixed grades are those products produced to meet a definite export commitment. The volumes required and product specifications are known before beforehand (see Table 15-9), and these are blended first. The balance of blending components in a product group, after meeting fixed grades demand, is blended to a balancing-grade specification. Balancing-grade production cannot be fixed. It depends on the volumes and specifications of fixed grades as well as its own product specification.

One grade in every product group (naphtha, gasoline, kerosene, diesel, fuel oil, etc.) is nominated as the balancing grade. In participants' DOPs, the production of balancing grades is optimized based on their price.

For the operation of the joint refinery, each participant can have different product specifications for its fixed grades that give it the flexibility to produce products to suit its market requirements. It is essential, however, that both participants have common balancing-grade products.

Table 15-9
Refinery Product Grades

GRADE	PRODUCT GROUP	KEY SPECIFICATIONS
I-150*	**LPG**	**RVP** 275 – 690 **KPA**
I-151	LPG	RVP 485 MAX
I-201*	**LSR NAPHTHA**	**LSR, RVP 95 KPA MAX., SG MIN 0.654**
I-220	LSR NAPHTHA	LSR, RVP 95 KPA MAX., SG MIN 0.654-0.702
I-222	LSR NAPHTHA	LSR, RVP 95 KPA MAX., SG MIN 0.654-0.702
I-206	NAPHTHA	WSR, RVP 80 KPA, SG 0.702 MAX.
I-210*	**NAPHTHA**	**WSR, RVP 75 KPA, SG 0.69 – 0.735 MAX.**
I-211	NAPHTHA	WSR, RVP 75 KPA, SG 0.70 – 0.735 MAX.
I-213	NAPHTHA	WSR, RVP 65 KPA MAX.
I-253	GASOLINE	FCCU LIGHT NAPHTHA, RON 88 MIN, RVP 70 MAX.
I-387	GASOLINE	RON 87, RVP 75 MAX.
I-390	GASOLINE	RON 90, RVP 70 MAX.
I-393	GASOLINE	RON 93, RVP 75 MAX.
I-395	GASOLINE	RON 95, RVP 65 MAX.
I-397*	**GASOLINE**	**RON 97, RVP 85 MAX.**
I-400	KEROSENE	ILLUMINATING KERO, SMOKE 21, FLASH 35 0C MIN.
I-411	KEROSENE	ILLUMINATING KERO, SMOKE 25, FLASH 35 MIN.
I-419	KEROSENE	DPK, 38°C FLASH, −47°C FREEZE.
I-434	KEROSENE	JP-4, −58°C FREEZE, 25 VOL % AROMATICS.
I-440*	**KEROSENE**	**JET FUEL, 25 SMOKE, −47°C FREEZE**
I-710	HVGO	HEAVY LUBE DISTILLATE, VISCOSITY 14 CST AT 100°C
I-711	HVGO	HVGO, VIS. 6–9 CST AT 100°C
I-725	HVGO	FCCU CHARGE STOCK, SULFUR MAX 3.0%
I-800	DIESEL	DIESEL INDEX 45, POUR + 6°C
I-803	DIESEL	CETANE INDEX 45, CLOUD POINT 3°C
I-808	DIESEL	CETANE INDEX 50, CLOUD POINT 2°C
I-875	DIESEL	CETANE INDEX 47, CLOUD POINT −4°C
I-876	DIESEL	CETANE INDEX 46, CLOUD POINT −1°C
I-876Z	DIESEL	CETANE INDEX 46, CLOUD POINT −12°C
I-884	DIESEL	CETANE INDEX 47, CLOUD POINT −6°C
I-885	DIESEL	CETANE INDEX 50, POUR −3°C, 0.5% SULFUR
I-888*	**DIESEL**	**CETANE INDEX 50, POUR** −3°C
I-890	MARINE DSL	CETANE INDEX 45, POUR −9°C

NOTE:
AN ASTERISK (*) DENOTES BALANCING GRADE PRODUCTS.

Table 15-9
Continued

I-892	MARINE DSL	CETANE INDEX 43, POUR 3°C, 2% CON CARBON
I-893	MARINE DSL	CETANE INDEX 45, POUR −9°C, 0.08% CON CARBON
I-903	FUEL OIL	VISCOSITY 80 CST AT 50°C, 0.3% SULFUR
I-925	FUEL OIL	VISCOSITY 180 CST AT 50°C, 2.5% SULFUR
I-928	FUEL OIL	VISCOSITY 180 CST AT 50°C, 2.8% SULFUR
I-933	FUEL OIL	VISCOSITY 80 CST AT 50°C, 3.5% SULFUR
I-934	FUEL OIL	VISCOSITY 120 CST AT 50°C, 3.0% SULFUR
I-955	FUEL OIL	VISCOSITY 75 CST AT 50°C, 3.5% SULFUR
I-957	FUEL OIL	VISCOSITY 225 CST AT 50°C, 3.5% SULFUR
I-960	FUEL OIL	VISCOSITY 120 CST AT 50°C, 3.5% SULFUR
I-961*	**FUEL OIL**	**VISCOSITY 180 CST AT 50°C, 3.5% SULFUR**
I-962	FUEL OIL	VISCOSITY 280 CST AT 50°C, 3.5% SULFUR
I-963	FUEL OIL	VISCOSITY 180 CST AT 50°C, 3.8% SULFUR
I-964	FUEL OIL	VISCOSITY 140 CST AT 50°C, 3.5% SULFUR
I-965	FUEL OIL	VISCOSITY 280 CST AT 50°C, 3.5% SULFUR
I-966	FUEL OIL	VISCOSITY 280 CST AT 50°C, 4.0% SULFUR
I-967	FUEL OIL	VISCOSITY 180 CST AT 50°C, 3.5% SULFUR
I-971	FUEL OIL	VISCOSITY 380 CST AT 50°C, 4.0% SULFUR
I-975	FUEL OIL	VISCOSITY 530 CST AT 50°C, 3.5% SULFUR
I-1110B	ASPHALT	BULK ASPHALT, PENETRATION 180–220
I-1110D	ASPHALT	DRUMMED ASPHALT, PENETRATION 180–220
I-1129B	ASPHALT	BULK ASPHALT, PENETRATION 85–100
I-1129D	ASPHALT	DRUMMED ASPHALT, PENETRATION 85–100
I-1138B	ASPHALT	BULK ASPHALT, PENETRATION 60–70
I-1138D	ASPHALT	DRUMMED ASPHALT, PENETRATION 60–70
I-1149B	ASPHALT	BULK ASPHALT, PENETRATION 40–50
I-1149D	ASPHALT	DRUMMED ASPHALT, PENETRATION 40–50

GRADE/CODE	PROCESS STOCK
PGAS	GASOLINE BASE STOCK
PLCGAS	TREATED LIGHT CAT NAPHTHA
PLLCN	UNTREATED LIGHT CAT NAPHTHA
PLSRBS	LSR UNTREATED
PMSR	UNTREATED MEDIUM ST. RUN NAPHTHA
PP90R	CAT REFORMATE, 90 RON
PP95R	CAT REFORMATE, 95 RON
PP97R	CAT REFORMATE, 97 RON
PSMCN	SWEET MEDIUM CAT NAPHTHA
PSWMSR	TREATED MSR NAPHTHA
PPOLY	RAW POLYMER GASOLINE
PUFCHG	UNIFINER CHARGE
PWCN	WHOLE CAT NAPHTHA

Table 15-9
Continued

GRADE/CODE	PROCESS STOCK
PSKERO	SWEET KEROSENE DISTILLATE
PDSDSL	DESULFURISED DIESEL STOCK
PDSL	UNTREATED DIESEL STOCK
PLTISO	LIGHT DIESEL EX HYDROCRACKER
PMIDSL	MEDIUM/INTER DIESEL STOCK
PBFUEL	BURNER FUEL OIL
PCTISO	DESULFURISED HVGO EX MILD HYDROCRACKER
PCTTR	FUEL OIL CUTTER STOCK
PFCCF	FCCU CHARGE, HVGO
PFCCO	FCCU CHARGE, DESULFURISED HVGO
PFDISO	HVGO FEED TO HYDROCRACKER
PLANTS	PLANTS AND LINE CONTENT
PMEISO	MEDIUM ISOMATE CUTTER
PRESID	ATM RESID/VDU FEED
PSLOPD	SLOP DISTILLATE
PSLOPO	SLOP OIL

EXAMPLE 15-2

We want to estimate the production of balancing grade fuel oil (I-961). The production of fuel oil blending components in a refinery and their properties and its fixed-grade production are as follows:

STREAM	VOL, mb	VBI*	SULFUR, wt%	SPECIFIC GRAVITY
VACUUM RESID	1112	706	4.28	1.0170
CUTTERS	581	140	2.40	0.9684
HEAVY CAT NAPHTHA	95	−250	0.40	0.8256
TOTAL	1788	471.2	3.46	0.9910

* VISCOSITY BLENDING INDEX

To estimate the production of balancing-grade fuel oil (I-961), the average properties of the fixed fuel oil grades is estimated, as shown next:

FIXED GRADES	REQUIREMENTS, mb	VBI	SULFUR, wt%
I-928	250	460	2.0
I-957	150	460	3.4
I-971	300	516	4.0
TOTAL	700	470	3.2

Out of 1788 mb of available fuel oil components, 700 mb go to blending fixed fuel oil grades. The remaining 1088 mb have a viscosity blending index of 472.

For disposal of this material, we have two options: to blend off this material to the specifications of balancing grade fuel oil or to keep this material in inventory as process stock for product blending later.

In option 1, this material must be blended to fuel oil balancing-grade (BG) specifications I-961 by adding finished diesel (I-888) to the remaining fuel oil pool to lower its viscosity blending index to the I-961 specs of 458.

STREAM	VOL, mb	VBI	SULFUR, wt%
FIXED GRADE	700	469	3.2
REMAINING POOL	1088	472	3.4
ADD DIESEL	30	−30	0.97
TOTAL I-961 (BG)	1118	458	3.50

Process stocks have no product specifications to meet. The product quality depends on the process conditions of the refinery units from which they originate. These are treated as fixed-grade products, even though they are not blended products.

From among the various product specifications for a product pool, we consider for possible nomination to balancing grade only those that can be met by average pool properties. For example, consider fuel oil pool (vacuum resids, cutter stocks, and the like) from a refinery with following properties:

Sulfur	3.50%
Viscosity	400 cst
Con carbon	15.0%

A fuel oil balancing grade must have properties close to the pool properties. For example, if the fuel oil balancing grade has a viscosity of 100 cst, this product could be blended only by downgrading a large amount of diesel to the fuel oil pool to reduce the pool viscosity from 400 cst to 100 cst. Similarly, if a balancing-grade fuel oil had a sulfur content of 1.5% instead of 3.5%, the pool sulfur could be lowered only by inclusion of a large amount of diesel and a large economic penalty.

In the allocation program, balancing grades are required as reference grades, in terms of which all fixed grades and process stock equivalencies are expressed. All settlements between the participants for overlifting or underlifting is done by exchange of balancing-grade products, as these are produced by both participants.

In the final allocation LP models for each participant, production of balancing grades is optimized on the basis of their price after producing all the fixed stocks and inventory changes as per the actual productions and available unit capacities.

It is essential that both the allocation program and allocation LP models for both participants have identical balancing grades.

PRODUCT EQUIVALENCIES

The concept of a fixed-grade product equivalencies makes it possible to express any fixed grade product into its balancing grade equivalent, in term of its volume. Note that each participant in a joint-ownership refinery may produce different product fixed grades within a product group, but both participants produce same balancing grade within the group. The concept of product equivalencies is fundamental to the mechanism of valuation and exchange of fixed-grade products with different specifications. In a joint-ownership refinery, all outstanding claims between the two participants are settled by exchanging balancing-grade products. The concept of equivalency allows participants to produce and ship products with the different specifications required for its sale domain without causing any accounting problems.

A joint-ownership refinery produces a large number of product grades and process stocks for its operations. Table 15-10 shows the product grades produced by one such refinery. Both participants are free to produce any number of fixed grades in a group, but they must also produce the balancing grade of the group. Fixed-grade volumes are converted into balancing-grade volumes by means of product equivalency. As balancing

Table 15-10
Fixed-Grade Product Equivalencies

PRODUCT	CLASS	GROUP	I-150	I-201	I-210	I-397	I-440	I-888	I-961
I-1110B	ASPHALT	961	0.0000	0.0000	0.0000	0.0000	0.0000	−0.7287	1.7287
I-1110D	ASPHALT	961	0.0000	0.0000	0.0000	0.0000	0.0000	−0.7287	1.7287
I-1129B	ASPHALT	961	0.0000	0.0000	0.0000	0.0000	0.0000	−0.7287	1.7287
I-1129D	ASPHALT	961	0.0000	0.0000	0.0000	0.0000	0.0000	−0.7287	1.7287
I-1138B	ASPHALT	961	0.0000	0.0000	0.0000	0.0000	0.0000	−0.7287	1.7287
I-1138D	ASPHALT	961	0.0000	0.0000	0.0000	0.0000	0.0000	−0.7287	1.7287
I-1149B	ASPHALT	961	0.0000	0.0000	0.0000	0.0000	0.0000	−0.7287	1.7287
I-1149D	ASPHALT	961	0.0000	0.0000	0.0000	0.0000	0.0000	−0.7287	1.7287
I-150	LPG	150	1.0000	0.0000	0.0000	0.0000	0.0000	0.0000	0.0000
I-151	LPG	150	1.0000	0.0000	0.0000	0.0000	0.0000	0.0000	0.0000
I-201	LSR	201	0.0000	1.0000	0.0000	0.0000	0.0000	0.0000	0.0000
I-210	WSR	210	0.0000	0.0000	1.0000	0.0000	0.0000	0.0000	0.0000
I-211	WSR	210	0.0245	−0.3110	1.2865	0.0000	0.0000	0.0000	0.0000
I-213	WSR	210	−0.0400	−0.1680	1.2080	0.0000	0.0000	0.0000	0.0000
I-220	WSR	201	0.0000	1.0000	0.0000	0.0000	0.0000	0.0000	0.0000
I-222	WSR	201	0.0000	1.0000	0.0000	0.0000	0.0000	0.0000	0.0000
I-253	CAT NAPHTHA	397	−0.0078	0.1634	0.0000	0.8444	0.0000	0.0000	0.0000
I-387	GASOLINE	397	0.0054	0.6160	0.0000	0.3786	0.0000	0.0000	0.0000
I-390	GASOLINE	397	−0.0068	0.4543	0.0000	0.5525	0.0000	0.0000	0.0000
I-393	GASOLINE	397	0.0103	0.2916	0.0000	0.6981	0.0000	0.0000	0.0000
I-395	GASOLINE	397	0.0119	0.1834	0.0000	0.8047	0.0000	0.0000	0.0000
I-397	GASOLINE	397	0.0000	0.0000	0.0000	1.0000	0.0000	0.0000	0.0000
I-398	GASOLINE	397	−0.0003	0.0216	0.0000	0.9787	0.0000	0.0000	0.0000
I-400	KEROSENE	440	0.0000	0.0000	0.0000	0.0000	1.0000	0.0000	0.0000
I-411	KEROSENE	440	0.0000	0.0000	0.0000	0.0000	1.0000	0.0000	0.0000
I-419	KEROSENE	440	0.0000	0.0000	0.0000	0.0000	1.0000	0.0000	0.0000
I-434	JP-4	210	−0.0350	0.0000	0.5230	0.0000	0.5120	0.0000	0.0000
I-440	KEROSENE	440	0.0000	0.0000	0.0000	0.0000	1.0000	0.0000	0.0000
I-710	HVGO	961	0.0000	0.0000	0.0000	0.0000	0.0000	0.2500	0.7500
I-711	HVGO	961	0.0000	0.0000	0.0000	0.0000	0.0000	0.2500	0.7500
I-725	HVGO	961	0.0000	0.0000	0.0000	0.0000	0.0000	0.3468	0.6532
I-800	DIESEL	888	0.0000	0.0000	0.0000	0.0000	−0.5681	1.5681	0.0000

Table 15-10
Continued

PRODUCT	CLASS	GROUP	I-150	I-201	I-210	I-397	I-440	I-888	I-961
I-803	DIESEL	888	0.0000	0.0000	0.0000	0.0000	0.1478	0.8522	0.0000
I-808	DIESEL	888	0.0000	0.0000	0.0000	0.0000	0.0000	1.0000	0.0000
I-875	DIESEL	888	0.0000	0.0000	0.0000	0.0000	0.1478	0.8522	0.0000
I-876	DIESEL	888	0.0000	0.0000	0.0000	0.0000	0.1478	0.8522	0.0000
I-876ZP	DIESEL	888	0.0000	0.0000	0.0000	0.0000	0.5768	0.4232	0.0000
I-884	DIESEL	888	0.0000	0.0000	0.0000	0.0000	0.2754	0.7246	0.0000
I-885	DIESEL	888	0.0000	0.0000	0.0000	0.0000	0.0000	1.0000	0.0000
I-888	DIESEL	888	0.0000	0.0000	0.0000	0.0000	0.0000	1.0000	0.0000
I-892	BULK DIESEL	888	0.0000	0.0000	0.0000	0.0000	0.0000	0.7320	0.2680
I-893S	BULK DIESEL	888	0.0000	0.0000	0.0000	0.0000	0.0260	0.8800	0.0940
I-893W	BULK DIESEL	888	0.0000	0.0000	0.0000	0.0000	0.0000	0.9963	0.0037
I-903	FUEL OIL	961	0.0000	0.0000	0.0000	0.0000	0.0000	0.0000	1.0000
I-925	FUEL OIL	961	0.0000	0.0000	0.0000	0.0000	0.0000	0.0015	0.9985
I-928	FUEL OIL	961	0.0000	0.0000	0.0000	0.0000	0.0000	−0.0204	1.0204
I-933	FUEL OIL	961	0.0000	0.0000	0.0000	0.0000	0.0000	0.1271	0.8729
I-934	FUEL OIL	961	0.0000	0.0000	0.0000	0.0000	0.0000	0.0501	0.9499
I-955	FUEL OIL	961	0.0000	0.0000	0.0000	0.0000	0.0000	0.1392	0.8608
I-957	FUEL OIL	961	0.0000	0.0000	0.0000	0.0000	0.0000	0.2168	0.7832
I-960	FUEL OIL	961	0.0000	0.0000	0.0000	0.0000	0.0000	0.0542	0.9458
I-961	FUEL OIL	961	0.0000	0.0000	0.0000	0.0000	0.0000	0.0000	1.0000
I-962	FUEL OIL	961	0.0000	0.0000	0.0000	0.0000	0.0000	−0.0548	1.0548
I-964	FUEL OIL	961	0.0000	0.0000	0.0000	0.0000	0.0000	0.0360	0.9640
I-966	FUEL OIL	961	0.0000	0.0000	0.0000	0.0000	0.0000	−0.0719	1.0719
I-971	FUEL OIL	961	0.0000	0.0000	0.0000	0.0000	0.0000	−0.1042	1.1042
I-975	FUEL OIL	961	0.0000	0.0000	0.0000	0.0000	0.0000	−0.1454	1.1454

PBFUEL	BURNER FUEL OIL	961	0.0000	0.0000	0.0000	0.0000	0.0000	0.0000	1.0000
PCTISO	ISOMATE CUTTER	961	0.0000	0.0000	0.0000	0.0000	0.0000	0.3360	0.6640
PCTTR	FCC CUTTER	961	0.0000	0.0000	0.0000	0.0000	0.0000	0.6438	0.3562
PDSDSL	DSL EX 1HDU	888	0.0000	0.0000	0.0000	0.0000	0.0000	1.0000	0.0000
PDSL	UNDESULF.DSL	888	0.0000	0.0000	0.0000	0.0000	0.0000	1.0000	0.0000
PFCCF	HVGO	961	0.0000	0.0000	0.0000	0.0000	0.0000	0.2406	0.7594
PFCOO	H.ISOM	961	0.0000	0.0000	0.0000	0.0000	0.0000	0.2697	0.7303
PFDISO	HVGO FEED TO HCR.	961	0.0000	0.0000	0.0000	0.0000	0.0000	0.2406	0.7594
PGABS	GASOLINE B.S	397	−0.0914	0.5571	0.0000	0.5343	0.0000	0.0000	0.0000
PKERO	TKE/SOUR KEROSENE	440	0.0000	0.0000	0.0000	0.0000	1.0000	0.0000	0.0000
PLANTS	PLANT & LINE CONT.	961	0.0103	0.1040	0.1105	0.1518	0.0886	0.1854	0.3494
PLCGAS	LIGHT CAT NAPTHA, TREATED	397	−0.0075	0.0327	0.0000	0.9748	0.0000	0.0000	0.0000
PLLCN	LIGHT CAT NAPTHA, UNTR.	397	−0.0075	0.0327	0.0000	0.9748	0.0000	0.0000	0.0000
PLTISO	LIGHT ISOMATE	888	0.0000	0.0000	0.0000	0.0000	0.0000	1.0000	0.0000
PMEISO	MEDIUM ISOMATE	961	0.0000	0.0000	0.0000	0.0000	0.0000	0.3360	0.6640
PMIDSL	4A M/I DSL	888	0.0000	0.0000	0.0000	0.0000	−0.4551	1.4551	0.0000
PMSR	THN/SOUR MSR	210	−0.1140	0.4690	1.5830	0.0000	0.0000	0.0000	0.0000
PPOLY	POLY GASOLINE	397	−0.0154	−0.1076	0.0000	1.1230	0.0000	0.0000	0.0000
PP90R	REFORMATE 90 R	397	−0.0227	−0.0263	0.0000	1.0490	0.0000	0.0000	0.0000
PP95R	REFORMATE 95 R	397	−0.0206	−0.1615	0.0000	1.1821	0.0000	0.0000	0.0000
PRESID	6VDU FEED	961	0.0000	0.0000	0.0000	0.0000	0.0000	0.2218	0.7782
PSKERO	SWEET KEROSENE	440	0.0000	0.0000	0.0000	0.0000	1.0000	0.0000	0.0000
PSLOPD	SLOP DISTILLATE	961	0.0000	0.1000	0.1000	0.1400	0.1900	0.3700	0.1000
PSLOPO	SLOP OIL	961	0.0000	0.1000	0.1000	0.1400	0.1900	0.3700	0.1000
PSLOPT	SLOP	397	−0.0123	0.1414	0.0000	0.4936	0.0488	0.3285	0.0000
PSMCN	MCN	397	−0.1742	0.3624	0.0000	0.8118	0.0000	0.0000	0.0000
PSWMSR	SWEET MSR	210	−0.1140	−0.4690	1.5830	0.0000	0.0000	0.0000	0.0000
PUFCHG	UNIFINER CHG.	210	−0.1140	−0.4690	1.5830	0.0000	0.0000	0.0000	0.0000
PWCN	NAPTHA FROM VISBREAKER	397	−0.1742	0.3624	0.0000	0.8118	0.0000	0.0000	0.0000

grades are produced by both participants, any overlifting by one participant is settled in terms of balancing grade products.

PRODUCT EQUIVALENCY DETERMINATION

To work out the product equivalency of a fixed grade in any group, the basis of product equivalencies must be identified:

Naphtha. Calculations are based on density and Reid vapor pressure (RVP). The density and RVP of a fixed grade naphtha, whose equivalency is to be determined, must match those of a hypothetical blend of two balancing-grade products (LPG and naphtha).

Gasoline. The RVP and RON of the fixed-grade gasoline are matched with those of a blend of a balancing-grade naphtha, a balancing grade gasoline, and balancing grade LPG.

Diesel. The fixed-grade diesel equivalencies are based on pour point, cetane index, or the sulfur content of the grade.

Vacuum gas oils. VGO is an important intermediate stock. Its product equivalency is determined on the basis of its viscosity, in terms of diesel and fuel oil balancing grades.

Fuel oils. The key properties are the viscosity, sulfur, and Conradson carbon specs, which must be matched.

Representative values of the balancing-grade properties are used to determine the fixed-grade product equivalencies (see Table 15-11).

Table 15-11
Balancing-Grade Properties

GRADE	GROUP	SPECIFIC GRAVITY	RVP, psia	RON	POUR INDEX	VBI	SULFUR, wt%	CON. CARBON
I-150	LPG	0.5740	60.00	98.0				
I-201	LSR	0.6698	13.97	55.2				
I-210	WSR	0.7052	8.80					
I-397	GASOLINE	0.7279	10.29	97.2				
I-440	KEROSENE	0.7890	0		38.7		0.30	
I-888	DIESEL				250.0	−45	0.97	
I-961	FUEL OIL					463	3.35	13.4

EXAMPLE 15-3

Calculate the product equivalency of I-434 (JP-4), a military jet fuel, blended from kerosene, naphtha, and LPG. The I-434 target blending specs follow:

$$\text{Density} = 0.7527\ \text{gm/cc}$$
$$\text{RVP} = 2.50\ \text{psia}$$

In the I-434 blend, let

$$\text{I-150} = x$$
$$\text{I-210} = y$$
$$\text{I-440} = z$$

where x, y, and z are the volume fraction of blend components required to blend grade I-434.

Thus,

	x	y	z	$= \text{I} - 434$
MATERIAL BALANCE	1	1	1	= 1
DENSITY	0.5740	0.7052	0.7890	= 0.7527
RVP	60	8.8	0	= 2.5

Solving for three equations for three unknowns gives the following result:

$$\text{I-150} = -0.035$$
$$\text{I-210} = 0.512$$
$$\text{I-440} = 0.521$$

So, for accounting purposes, one barrel of JP-4 is equivalent to 0.521 bbl I-440 plus 0.512 bbl I-210 minus 0.033 bbl I-150. Therefore. the equivalency of JP-4 (I-434) is written as follows:

	I-150	I-210	I-440
I-434	−0.0350	0.5120	0.5210

EXAMPLE 15-4

Determine the equivalency of a diesel grade I-876Z blended from diesel and kerosene. The target blending specs of I-876Z are as follows:

$$\text{Pour point} = -0.4°\text{F}$$
$$\text{Sulfur} = 0.4\,\text{wt\%}$$

I-876 ZP could be considered a blend of kerosene and diesel to match the pour point −0.4°F. The limiting specification is the I-876Z pour point. As sulfur is not limiting, this property is not considered.

GRADE	PRODUCT	VOLUME	POUR POINT, °F	POUR INDEX
I-440	KEROSENE	0.5768	−50.0	45.68
I-888	DIESEL	0.4232	26.60	388.70
I-876ZP	DIESEL	1.0000	−0.40	190.42

Hence, the equivalency of I-876Zp becomes

$$1\,\text{bbl I-876ZP} = 0.5768\,\text{bbl kerosene} + 0.4232\,\text{bbl diesel}$$

EXAMPLE 15-5

Determine the equivalency of a marine diesel, grade I-892, blended from diesel and fuel oil. The target blending specs of I-892 are as follows:

$$\text{Viscosity} = 3.0 - 9.0\,\text{cst}$$
$$\text{Sulfur} = 1.6\,\text{wt\%}$$
$$\text{Con carbon} = 2.0\,\text{wt\%}$$

Marine diesel grade I-892 could be considered a blend of diesel and fuel oil to achieve a Con carbon content of 2.0 wt%, as follows:

GRADE	PRODUCT	VOLUME	CON CARBON, wt%	VBI
I-888	DIESEL	0.8582	0	−45
I-961	FUEL OIL	0.1418	13.4	458
I-892	MARINE DIESEL	1.0000	1.90	26.3

We see that the viscosity index of the blend is 26.3, which corresponds to a viscosity of 4.1 cst, a value that falls within the specification range. Therefore, the equivalency of I-892 becomes

1 bbl I-892 = 0.8582 bbl I-888 + 0.14218 bbl I-961

EXAMPLE 15-6

Determine the product equivalency of fuel oil grade I-971 (380 cst viscosity, 4% sulfur) in terms of balancing grades.

I-971 could be considered a hypothetical blend of fuel oil I-961 and diesel I-888. The viscosity of I-971 could be achieved by removing some diesel from the fuel oil grade, as follows:

GRADE	PRODUCT	VOLUME	VBI	SULFUR, wt%
I-888	DIESEL	−0.1044	−45	0.97
I-961	FUEL OIL	1.1044	458	3.35
I-971	FUEL OIL	1.0000	511	3.76

Hence, the equivalency of I-971 becomes

1 bbl I-971 = −0.1044 bbl I-888 + 1.1044 bbl I-961

EXAMPLE 15-7

Determine the equivalency of asphalt, I-1138 (viscosity blending index 814), in terms of balancing grade products.

I-1138 (asphalt) could be considered a hypothetical blend of fuel oil and diesel. The viscosity of I-1138 could be achieved by removing some diesel from the fuel oil grade, as follows:

GRADE	PRODUCT	VOLUME	VBI	SULFUR, wt%
I-888	DIESEL	−0.7287	−45	0.97
I-961	FUEL OIL	1.7287	458	3.35
I-1138	FUEL OIL	1.0000	825	5.34

Hence the equivalency of I-1138 becomes

$$1 \text{ bbl I-1138} = -0.7287 \text{ bbl I-888} + 1.7287 \text{ bbl I-961}$$

CRUDE OIL EQUIVALENCY

Crude oil equivalency is determined for a joint ownership refinery by conducting a series of test runs on the crude distillation unit by gradually increasing CDU feed from its nominal design capacity to the hydraulic limits of the crude distillation unit. The incremental crude yields are recorded and used to convert the incremental crude run to equivalent barrels of balancing-grade products (see Table 15-12). Crude oil equivalency is required for preliminary allocation and have virtually no effect on final allocation.

Table 15-12
Incremental Yield of Balancing Grade from Crude

CRUDE RUN, mbpcd	I-201	I-210	I-440	I-888	I-961	LOSS
200	0.0010	0.1830	0.1000	0.3270	0.3740	0.0170
260	−0.0100	0.1800	0.1000	0.3100	0.4030	0.0170
270	−0.0150	0.2090	0.1000	0.2760	0.4110	0.0190
280	−0.0290	0.2360	0.1000	0.2620	0.4150	0.0160
290	−0.0400	0.2350	0.1000	0.1150	0.5740	0.0160

EQUIVALENCY OF SLOP

Slop is the off-specification product produced by various processing units during start-up, shutdown, or emergency situations. This material is of variable composition and reprocessed either in the CDU or in product blending. The equivalency of the slop is determined from its ASTM distillation to determine the potential volumes of LPG, naphtha, kerosene, diesel, and fuel oil content likely to be obtained reprocessing it in a crude distillation unit. As the slop composition changes with time, the analysis is frequently updated for its use in determining the product equivalency. A typical slop equivalency is as follows:

$$\begin{aligned}1\,\text{bbl slop} = {} & 0.0\,\text{bbl I-150} + 0.10\,\text{bbl I-201} + 0.1\,\text{bbl I-210} \\ & + 0.14\,\text{bbl I-397} + 0.19\,\text{bbl I-440} + 0.37\,\text{bbl I-888} \\ & + 0.10\,\text{bbl I-61}\end{aligned}$$

As product equivalencies are defined in terms of balancing-grade products, if either the balancing grades chosen or their properties undergoes a change, the equivalencies of fixed grades also change. Product equivalencies are not unique. These are refinery specific. It is sometimes possible to define a fixed-grade equivalency in more than one way. The equivalency adopted must be acceptable to both participants. For product allocation, the equivalency of every fixed grade and process stock, including slop, must be determined.

CHAPTER SIXTEEN

Product Allocation

In a joint-ownership refinery, the participants can use their share of the refinery in any way they wish, provided it is not detrimental to the other participant or refinery as a whole. Therefore, at any given time, the participants will be using different proportions of their capacity entitlement on the crude and downstream units to produce a wide range of finished products, some of which are common to both participants and others produced for one participant only. Also, the inventory of intermediate or process stocks will change constantly.

At some stage, the ownership of various products and intermediate stocks has to be determined. In fact, the allocation of feedstocks and production is an exercise required under the processing agreement. This chapter describes how this is done. Product allocation for an operating period is routinely done at the end of the operating period. For example, if the operating period is a month, say January, product allocation will determine the ownership of products and process stocks at the end of January 31. Also, product allocation requires all operating data for the entire month, which would be available only after January 31. The actual ownership of products and process stocks in the refinery inventory on January 31, may be available only some weeks later, after the completion of the product allocation exercise for the month of January.

Product allocation is carried out in two distinct stages: preliminary allocation and final allocation. Essentially, the preliminary allocation consists of collecting data from all sources, feeding the data into an allocation program, making forecaster changes, and cross-checking the input data. Preliminary allocation allocates only the fixed-product grades and process stocks. Final allocation allocates the refinery production of balancing-grade products to the participants in the ratio of their allocation linear programs (LP) production of balancing grades.

INPUT DATA

The following input data are required:

1. Each participant's definitive operating plan (DOP) for the month.
2. The participant's opening inventories of finished products and intermediate stocks. The previous month's preliminary and final allocation reports. The previous month closing inventories become the opening inventories for current month.
3. Each participant's product lifting, showing actual product shipped, local sales, pipeline transfers, etc. during the month.
4. Refinery closing inventories and consumption of finished products and intermediate stocks.
5. Actual crude throughput of the refinery for the month.
6. Product equivalencies for all fixed grades and process stocks.

FORECASTER CHANGES

The actual refinery production during the month of all fixed grades is compared with the production of the fixed grades in the combined or refinery DOP (Table 15-7). The differences between actual production and the DOP for fixed grades and process stocks are allocated to the participants as per a set of rules.

These differences, called *deltas*, between actual production and the combined DOP is allocated to the participants in such a way that the sum of two revised DOPs is made equal to actual production for all fixed grades and process stocks. This allocation of fixed-grade and process-stock deltas to the participants is called *forecaster assigned changes* or simply *forecaster changes*. Note that these changes are made only to fixed grades and process stocks and not to balancing grades. The objective is to revise the participants' DOPs in such a way as to bring all fixed-grade production in the combined DOP equal to the actual production.

RULES FOR FORECASTER CHANGES

First, the cause of difference between the actual refinery production and combined DOP is investigated and the delta allocated to the participants according to following rules.

Fixed Grades

1. *For grades where only one participant requests production in its DOP*. The delta is allocated to the participant with the DOP request as shown in following example. Consider a fixed grade I-390. Comparing the combined DOP for I-390 with its actual refinery production shows a delta of 7879 barrels. The entire delta is allocated to participant AOC, as BOC did not request production of this grade in its DOP:

PARTICIPANT	DOP, bbl	DELTA, bbl
AOC	55300	7879
BOC	0	0
TOTAL	55300	7879
ACTUAL PRODUCTION	63179	
DELTA	7879	

2. *For a grade where both participants request production in their DOP*. Delta is allocated in the ratio of their DOP requests. However, if allocation by this rule causes negative closing inventory to a participant, the delta should be changed:

PARTICIPANT	DOP	DOP RATIO	DELTA	LIFTING	OPENING INVENTORY	CLOSING INVENTORY
AOC	69185	0.8279	20126	98000	5260	−3429
BOC	14384	0.1721	4184	0	3400	21968
TOTAL	83569	1.0000	24310		8660	18539
PRODUCTION	107879					
DELTA	24310					

In this case, splitting the delta in the DOP ratio has caused negative closing inventory due to overlifting by one participant. Here, AOC's delta is made equal to (lifting – opening inventory – DOP) to eliminate the negative closing inventory as follows:

PARTICIPANT	DOP	DELTA	LIFTING	OPENING INVENTORY	CLOSING INVENTORY
AOC	69185	23555	98000	5260	0
BOC	14384	755	0	3400	18539
TOTAL	83569	24310		8660	18539
PRODUCTION	107879				
DELTA	24310				

3. *No DOP request but there was an actual positive production.* In this case, the delta is allocated in the crude run ratio:

PARTICIPANT	DOP	CRUDE RATIO	DELTA
AOC	0	0.7112	3200
BOC	0	0.2888	1300
TOTAL	0	1.0000	4500
PRODUCTION	4500		
DELTA	4500		

4. *No DOP request but there was an actual negative production due to reblending of a product grade.* In this case, the delta is allocated to participants in the ratio of their opening inventories:

PARTICIPANT	DOP	OPENING INVENTORY	DELTA
AOC	0	8290	−990
BOC	0	4350	−520
TOTAL	0	12640	−1510
PRODUCTION	−1510		
DELTA	−1510		

Process and Intermediate Stocks

1. *One participant requests a buildup and there actually is a buildup close to the request.* In this case, the delta is allocated to the requesting participant as follows:

PARTICIPANT	DOP, bbl	DELTA, bbl
AOC	25000	2230
BOC	0	0
TOTAL	25000	2230
ACTUAL PRODUCTION	27230	
DELTA	2230	

2. *One participant requests a buildup and there actually is a buildup but much larger than the request.* In this case, the delta is allocated in the ratio of crude run of the participants:

PARTICIPANT	DOP, bbl	CRUDE RATIO	DELTA, bbl
AOC	19000	0.6250	14896
BOC	0	0.3750	8937
TOTAL	19000	1.0000	23833
ACTUAL PRODUCTION	42833		
DELTA	23833		

3. *One participant requests a buildup and there actually is a buildup but much smaller than the request.* In this case, the delta is allocated in the ratio of DOP of the participants:

PARTICIPANT	DOP, bbl	DELTA, bbl
AOC	40000	−30000
BOC	0	0
TOTAL	40000	−30000
ACTUAL PRODUCTION	10000	
DELTA	−30000	

4. *One participant requests a buildup and there actually is a drawdown that is small.* In this case, all drawdown (not delta) is allocated to the nonrequesting participant:

PARTICIPANT	DOP, bbl	DELTA, bbl
AOC	25000	−25000
BOC	0	−1580
TOTAL	25000	−26580
ACTUAL PRODUCTION	−1580	
DELTA	−26580	

5. *One participant requests a buildup, and there actually is a drawdown that is large.* In this case, all the drawdown (not delta) is allocated in the ratio of the opening inventories of the participants:

PARTICIPANT	DOP, bbl	OPENING INV RATIO	DRAWDOWN, bbl	DELTA, bbl
AOC	20000	0.3342	−6528	−26528
BOC	0	0.6658	−13005	−13005
TOTAL	20000	1.0000	−19533	−39533
ACTUAL PRODUCTION	−19533			
DELTA	−39533			

6. *Both participants request a buildup and there actually is a buildup.* In this case, the delta is allocated in the ratio of DOP of the participants as follows:

PARTICIPANT	DOP, bbl	DOP RATIO	DELTA, bbl
AOC	21333	0.3983	−6033
BOC	32230	0.6017	−9114
TOTAL	53563	1.0000	−15147
ACTUAL PRODUCTION	38416		
DELTA	−15147		

7. *Both participants request a buildup and there actually is a drawdown.* In this case, the drawdown (not delta) is allocated in the ratio of opening inventories of the participants:

PARTICIPANT	DOP, bbl	OP. INV RATIO	DRAWDOWN, bbl	DELTA, bbl
AOC	20000	0.3341	−8353	−28353
BOC	41000	0.6659	−16647	−57647
TOTAL	61000	1.0000	−25000	−86000
ACTUAL PRODUCTION	−25000			
DELTA	−86000			

8. *Only one participant requests a drawdown and there actually is a drawdown close to that request.* In this case, the delta is allocated to the requesting participant as follows:

PARTICIPANT	DOP, bbl	DELTA, bbl
AOC	0	0
BOC	−36000	−3610
TOTAL	−36000	−3610
ACTUAL PRODUCTION	−39610	
DELTA	−3610	

9. *Only one participant requests a drawdown and there actually is a drawdown much larger than the request.* In this case, the delta is allocated to participants in the ratio of opening inventories, as follows:

PARTICIPANT	DOP, bbl	OPENING INV RATIO	DELTA, bbl
AOC	0	0.4355	−12126
BOC	−19480	0.5645	−15717
TOTAL	−19480	1.0000	−27843
ACTUAL PRODUCTION	−47323		
DELTA	−27843		

10. *Only one participant requests a drawdown and there actually is a drawdown much smaller than the request.* In this case, the delta is allocated to participants in the ratio of DOPs, as follows:

PARTICIPANT	DOP, bbl	DELTA, bbl
AOC	0	0
BOC	−20788	19212
TOTAL	−20788	19212
ACTUAL PRODUCTION	−1576	
DELTA	19212	

11. *No DOP request but there actually is a drawdown.* In this case, the delta is allocated to participants in the ratio of opening inventories, as follows:

PARTICIPANT	DOP, bbl	OPENING INV RATIO	DELTA, bbl
AOC	0	0.6814	−2418
BOC	0	0.3185	−1130
TOTAL	0	1.0000	−3548
ACTUAL PRODUCTION	−3548		
DELTA	−3548		

12. *One participant requests a buildup, the other asks for a drawdown; there actually is a buildup that is small.* In this case, protect the participant who asks for a drawdown:

PARTICIPANT	DOP, bbl	DELTA, bbl
AOC	−20000	20000
BOC	15000	−10345
TOTAL	−5000	9655
ACTUAL PRODUCTION	4655	
DELTA	9655	

13. *One participant requests a buildup, the other asks for a drawdown; there actually is a buildup that is large.* In this case, protect the participant who asks for a drawdown (delta 1), then split the remaining delta (delta 2) in the ratio of crude run of the participants:

PARTICIPANT	DOP	CRUDE RATIO	DELTA 1, bbl	DELTA 2, bbl	TOTAL DELTA, bbl
AOC	−6500	0.3971	6500	10376	16876
BOC	9230	0.6029		15754	15754
TOTAL	2730	1.0000	6500	26130	32630
ACTUAL PRODUCTION	35360				
DELTA	32630				

14. *One participant requests a buildup, the other asks for a drawdown; there actually is a drawdown that is small.* In this case, protect the participant who asks for a buildup, then allocate the remaining delta to the other participant:

PARTICIPANT	DOP, bbl	DELTA, bbl
AOC	−21740	18600
BOC	15000	−15000
TOTAL	−6740	3600
ACTUAL PRODUCTION	−3140	
DELTA	3600	

15. *One participant requests a buildup, the other asks for a drawdown; there actualy is a drawdown that is large.* In this case, protect the participant who asks for a buildup (delta 1), then split the remaining delta in the ratio of opening inventories:

PARTICIPANT	DOP, bbl	OPENING INV RATIO	DELTA 1, bbl	DELTA 2, bbl	TOTAL DELTA, bbl
AOC	12681	0.3189	−12681	−7019	−19700
BOC	−6730	0.6811		−14992	−14992
TOTAL	5951	1.0000	−12681	−22011	−34692
ACTUAL PRODUCTION	−28741				
DELTA	−34692				

16. *Only one participant requests a drawdown; there actually is a buildup that is small.* In this case, allocate the buildup (not delta) to the nonrequesting participant:

PARTICIPANT	DOP, bbl	ALLOCATION, bbl	DELTA, bbl
AOC	0	1499	1499
BOC	−27471	0	27471
TOTAL	−27471	1499	28970
ACTUAL PRODUCTION	1499		
DELTA	28970		

17. *Only one participant requests a drawdown; there actually is a buildup that is large.* In this case, allocate the buildup (not delta) on a crude run ratio basis. Use this ratio to allocate the delta:

PARTICIPANT	DOP, bbl	CRUDE RATIO	ALLOCATION, bbl	DELTA, bbl
AOC	0	0.3972	10777	10777
BOC	−29491	0.6028	16356	45847
TOTAL	−29491	1.0000	27133	56624
ACTUAL PRODUCTION	27133			
DELTA	56624			

18. *Both participants request a drawdown; there actually is a drawdown.* In this case, allocate delta in the ratio of DOP:

PARTICIPANT	DOP, bbl	DOP RATIO	DELTA, bbl
AOC	−17320	0.4471	4578
BOC	−21417	0.5529	5662
TOTAL	−38737	1.0000	10240
ACTUAL PRODUCTION	−28497		
DELTA	10240		

19. *Both participants request a drawdown; there actually is a buildup.* In this case, allocate the buildup (not delta) in the ratio of crude run. Use this ratio to allocate the delta:

PARTICIPANT	DOP, bbl	CRUDE RATIO	ALLOCATION, bbl	DELTA, bbl
AOC	−17600	0.6032	12215	29815
BOC	−10100	0.3968	8035	18135
TOTAL	−27700	1.0000	20250	47950
ACTUAL PRODUCTION	20250			
DELTA	47950			

20. *No DOP request but there actually is a buildup.* In this case, the delta is allocated to participants in the ratio of their crude run ratio, as follows:

PARTICIPANT	DOP, bbl	CRUDE RUN RATIO	DELTA, bbl
AOC	0	0.6032	11530
BOC	0	0.3968	7585
TOTAL	0	1.0000	19115
ACTUAL PRODUCTION	19115		
DELTA	19115		

We see that the allocation of deltas between DOP requests and actual production involves considerable judgment. However, if a wrong decision is made, compensating changes to the balancing grade, which is equal in magnitude but opposite in sign, ensure that the participants' fixed grade is merely exchanged with equal barrels of balancing-grade products.

Allocation of Cat Reformate

Allocation of cat reformate, of different octane numbers, from the catalytic reformer unit is done using a different procedure. The actual cat reformer production during the month is split in the ratio of the DOP request of the participants. The reformate used by a participant in blending its gasoline cargoes is deducted from it. The remaining reformate is its allocation. A more-detailed procedure follows:

1. From the refinery records, for the reference month, determine the total reformer feed and the split for each mode of (severity) operation.
2. Split the estimated reformate production in the ratio of the DOPs.
3. Make an estimate of reformate usage of either severity in the blending of gasoline shipments during the month by each participant and use this ratio to split the actual reformate used in blending gasoline grades.
4. Deduct the reformate used from the reformate production of each participant to estimate its allocation.

The following example make the procedure clear.

EXAMPLE 16-1

During a given month, the cat reformer ran for 16 days at 8000 barrels per calendar day and processed 128,000 bpcd of feed. The split of the feed into different operation modes was as follows:

SEVERITY	FEED	REFORMATE YIELD	REFORMATE PRODUCTION
RON 90	48000	0.8200	39000
RON 95	80000	0.7700	62000
TOTAL	108000		101000

1. Split the estimated production of reformate, in the ratio of the DOP of the participants, as follows:

PARTICIPANT	DOP	DOP RATIO	90R REFORMATE	95R REFORMATE
AOC	65296	0.7167	27951	44435
BOC	25816	0.2833	11049	17565
TOTAL	91112	1.0000	39000	62000

2. Estimates of actual reformate usage in different shipments by both the participants are next prepared:

GRADE	AOC 90R, bbl	AOC 95R, bbl	BOC 90R, bbl	BOC 95R, bbl
I-393	0	2500	0	0
I-395	9400	4350	4600	2150
I-397	11500	500	0	0
I-397M	18000	36900	0	0
I-398	4000	0	0	0
TOTAL	42900	49250	4600	2150

3. Actual reformate usage in gasoline blending during the month from the refinery records, follows. This is split between the participants

in the ratio of their actual usage. The reformate actually used in gasoline blending is

$$\text{Reformate 90R} = 54635\,\text{bbls}$$
$$\text{Reformate 95R} = 47298\,\text{bbls}$$

4. Split this actual refinery usage of reformate, in the ratio of participants' 90R and 95R usage in their shipments. Allocation is remaining production of a participant; that is (production – usage) as follows. For reformate 90R,

PARTICIPANT	SHIPMENT RECORD, bbl	SHIPMENT RATIO, bbl	ACTUAL USAGE, bbl	REFORMER PRODUCTION, bbl	ALLOCATION, bbl
AOC	42900	0.9032	49344	27951	−21393
BOC	4600	0.0968	5291	11049	5758
TOTAL	47500	1.0000	54635	39000	−15635

For reformate 95R,

PARTICIPANT	SHIPMENT RECORD, bbl	SHIPMENT RATIO, bbl	ACTUAL USAGE, bbl	REFORMER PRODUCTION, bbl	ALLOCATION, bbl
AOC	49250	0.9582	45320	44435	−885
BOC	2150	0.0418	1978	17565	15587
TOTAL	51400	1.0000	47298	62000	14702

FORECASTER COMPENSATING CHANGES

Every forecaster change results in an equal and opposite change to the balancing grades, depending on the product equivalency of that grade. Thus, any production of fixed grade by a participant over and above what was indicated in the participant's DOP has to come at the expense of balancing grade from a given crude run.

EXAMPLE 16-2

Suppose forecaster change for I-1138 (bulk asphalt) for participant AOC is -9604 bbl and the equivalency for asphalt is as follows:

$$\begin{aligned} \text{I-1138} &= 1.0000 \\ \text{I-888} &= -0.7287 \\ \text{I-961} &= 1.7287 \end{aligned}$$

or one barrel of asphalt is equivalent to 1.7287 bbl I-961 (fuel oil) minus 0.7287 barrels of I-888 (diesel).

As such, -9604 bbl asphalt (I-1138) would be equivalent to

$$\begin{aligned} \text{I-961} &= 1.7287 \times (-9604) \\ &= -16602\,\text{bbl} \\ \text{I-888} &= -0.7287 \times (-9604) \\ &= 6998\,\text{bbl} \end{aligned}$$

Thus, a fixed-grade I-1138 (asphalt) production decrease of 9604 barrels results in a balancing-grade fuel oil (I-961) production increase of 16,602 barrels and a balancing-grade diesel (I-888) production decrease of 6998 barrels. So, participant AOC is compensated for 9604 bbl asphalt loss with +16602 bbl I-961 (fuel oil) and -6998 bbl I-888 (diesel), as follows:

FIXED GRADE	FORECASTER CHANGE, bbl	FORECASTER COMPENSATING CHANGES, bbl		TOTAL FORECASTER COMPENSATING CHANGE, bbl
		I-961	I-888	
I-1138	−9604	−6998	16602	9604

Compensating changes to the balancing grade are equal in magnitude but opposite in sign to forecaster change.

The forecaster compensating changes are added algebraically to the balancing grade production of the participant.

CRUDE OIL CHANGES

The actual crude run during a month cannot be exactly the same as the participants' DOP estimates, even though the joint-operating company (JOC) tries to keep the crude run as close as possible to the participants' DOP.

Any delta between the actual crude run of a participant and its DOP rude run is reflected as forecaster change The crude oil delta is converted into balancing grade products by means of crude oil equivalency. The crude oil equivalency is the yields of balancing-grade products from the incremental crude run, generally determined by actual test runs on the refinery at various levels of crude throughput (Table 15-12).

EXAMPLE 16-3

The actual crude run of the refinery during a month was 125,526 barrels more than the combined DOP of the participants as follows:

CRUDE RUN	DOP, bbl	ACTUAL, bbl	DELTA, bbl
AOC	6,090,000	6,215,526	125,526
BOC	1,200,000	1,200,000	0
TOTAL REFINERY	7,290,000	7,415,526	125,526

Any change in the crude run is reflected in the production of balancing grades by means of crude oil equivalency. Here refinery crude run is 243 mbpcd therefore from table 15-12, crude oil equivalency for crude run 200–259 mbpcd is used.

Thus, additional processing of 125,526 bbl crude oil results in following balancing-grade production for participant AOC:

I-201 = −126 bbl
I-210 = 22,971 bbl
I-440 = 12,553 bbl
I-888 = 41,047 bbl
I-961 = 46,947 bbl
Loss = 2,134 bbl
Total = 125,526 bbl

DOP "LOSS" ADJUSTMENT

During processing in the conversion units of the refinery (FCCU, hydrocracker, etc.) the stocks expand in volume due to reduction in density, resulting in volume gain or negative loss. A situation can arise in which the DOP of one participant shows volume gain (negative loss) while the other participant's DOP shows a positive loss. The allocation of actual refinery "losses" on the basis of retrospective DOPs becomes unrealistic, as the following example shows:

Retrospective DOP loss, AOC = −5,000
Retrospective DOP loss, BOC = 9,600
Total refinery retrospective losses = 4,600
Actual refinery losses = 30,000
Loss delta = 25,400

Splitting the loss delta in the ratio of retrospective DOP losses yields the following results:

AOC delta = −27,607 bbl
BOC delta = 53,007 bbl
Total retrospective delta = 25,400 bbl
AOC allocated losses = −32,607 bbl
BOC allocated losses = 62,607 bbl
Total refinery losses = 30,000 bbl

This allocation is unrealistic. Therefore, to make the losses positive in both retrospective DOPs, the balancing-grade production of each is reduced by 0.6 vol%. The actual refinery losses are next allocated in the ratio of the revised retrospective DOPs, after deducting 0.6% from balancing-grade production volumes.

The figure of 0.6%, which roughly represents unaccounted losses of the refinery during crude processing, is chosen to make the losses positive in both retrospective DOPs or, in other words, a little more than the volume gain encountered in the retrospective DOPs.

RETROSPECTIVE DOP

By applying forecaster changes, forecaster compensating changes, and crude changes, the original DOP of the participants has been amended to bring their fixed grades, process stock, and crude run equal to the actual production of fixed grades, process stocks, and crude. These amended DOPs of the participants are called *retrospective DOPs*, or *retro DOPs*, which are combined to make a refinery retro DOP.

Note that, in the refinery retro DOP,

- All fixed grade and process stock production equals their actual refinery production.
- The total crude run in the retro DOP equals the actual crude run.
- The delta between the actual production and retro DOP of the refinery remains on balancing grades.
- The delta between the actual production and retro DOP of the refinery remains on losses.

The retrospective DOP is used for the allocation of balancing grades and actual refinery losses.

ALLOCATION OF BALANCING GRADES

We see that, in the retrospective DOP, all deltas on fixed grades, intermediate stocks, and crude run have been eliminated. The deltas between actual production and retro DOP remain only in balancing grades and refinery loss (actual refinery losses minus retro DOP losses). In the allocation procedure, loss is treated as a balancing-grade product.

The balancing-grade deltas (actual production minus retro production) are allocated by following procedure:

1. The retro DOP of the participants is expressed in balancing-grade equivalents using product equivalencies.
2. The total delta (actual production minus retro DOP) for each balancing grade and loss is determined. The sum of these deltas must equal zero.
3. Next, the deltas are allocated to participants in proportion to their respective shares of production of that product in the combined retro DOP (expressed in balancing grades).
4. By the nature of this calculation, the sum of each participant's delta allocation will not be balanced between the participants but equal in amount and of opposite plus and minus signs, since the sum of such imbalances must equal zero. Accordingly, a further step is required to eliminate these imbalances. This is done by reverse allocation, as described next.

REVERSE ALLOCATION

5. The absolute values of all balancing-grade (including loss) deltas are added to obtain the absolute value of combined total deltas (AVCTD).
6. The ratio of the absolute value of delta for each product and AVCTD is multiplied by the total imbalance (sum of delta allocations) to obtain the reverse allocation of that product for the other participant. The sum of such reverse allocation for each participant is equal to and of opposite sign to that participant's total imbalance (sum of delta allocations), accordingly eliminating the imbalance.
7. The total allocation of each participant is the sum of the retrospective DOP plus the sum of allocations for such product made under steps 3 and 4.

The preliminary product allocation is now complete.

EXAMPLE 16-4, PRELIMINARY ALLOCATION

Two participants, AOC and BOC, run a 250 mbpd oil refinery. Participant AOC has a 210 mbpd refinery capacity and participant BOC has a 40 mbpd refinery capacity. The refinery is operated as a joint-ownership refinery as per their processing agreement (see the Appendix). The entire product allocation procedure for effecting their preliminary allocation is shown in Tables 16-1 to 16-8.

Table 16-1 shows the DOP of participants AOC and BOC for a given month. The combined production of two DOPs make up the refinery DOP. Tables 16-2 and 16-3 present procedure for adjusting the balancing grade for refinery losses, by participant. The total adjustment is 0.6% (LV) of the crude run. This is spread over the balancing grades in proportion to the participant's production, as shown in these tables.

Table 16-4 compares the actual refinery production of every fixed grade and process stock with that in the combined DOP. The delta between these is the total "forecaster" change. The total forecaster change is allocated to the participants as per the rules for allocation of fixed grades and process stocks explained earlier. Table 16-5 shows data for the actual refinery crude run and that, in the combined DOP, allocated to the participants as per the crude distillation unit capacity available to each.

Tables 16-6 and 16-7 show calculation of the forecaster compensating changes for each participant. The forecaster compensating changes for a fixed product grade or process stock are calculated by converting every forecaster change to its balancing-grades equivalent by means of a product equivalency and multiplying it by −1. The forecaster compensating change for crude is also calculated by converting to balancing-grades equivalent by means of a crude oil equivalency. However, unlike the fixed grades, the forecaster compensating changes for crude are of the same sign as the forecaster change for crude oil.

Table 16-8 shows the retrospective DOP of the participants (columns 6 and 11) and the overall refinery retrospective DOP (column 12). We see that

1. The fixed grade production in the participant's retrospective DOP is, in fact, its allocation. The sum of the participants' retrospective DOPs for any fixed grade or process stock equals the actual refinery production for that grade. Thus, in the retrospective DOP, the actual refinery production of all fixed grades and process stocks has been split or "allocated" to the participants.
2. The delta between the actual refinery production and the total retro DOP (column 12) remains only on balancing grades and losses.

Table 16-1
Refinery DOP (All Figures in Barrels)

PRODUCT	AOC	BOC	TOTAL
1138B	60000	46590	106590
1149B	0	0	0
150	17839	0	17839
201	0	864	864
210	946831	227578	1174409
220	0	0	0
383	0	0	0
390	76800	0	76800
395	0	0	0
397	512647	80716	593363
397E	0	0	0
398	44100	0	44100
411	0	0	0
419	420000	0	420000
434	240000	0	240000
440	517375	205589	722964
711	0	0	0
725	0	0	0
800	420000	0	420000
876	300000	0	300000
876ZP	240000	0	240000
888	991519	409686	1401205
892	0	0	0
928	360000	0	360000
961	1074918	219295	1294213
PBFUEL	0	0	0
PCTTR	0	0	0
PDSDSL	0	0	0
PDSL	0	0	0
PFCOO	0	0	0
PFDISO	0	0	0
PKERO	0	0	0
PLCGAS	−60000	0	−60000
PLLCN	0	0	0
PLTISO	0	0	0
PMEISO	0	0	0
PMIDSL	0	0	0
PMSR	0	0	0
PPOLY	−30000	0	−30000
PP90R	−18000	0	−18000
PP95R	−60000	0	−60000
PRESID	0	0	0
PSKERO	0	0	0
PSLOPD	0	0	0
PSLOPO	0	0	0
PSMCN	0	0	0
PSWMSR	0	0	0
PUFCHG	0	0	0
PWCN	0	0	0
PROD + STOCK	6054029	1190318	7244347
LOSSES	35971	9682	45653
CRUDE	6090000	1200000	7290000

Table 16-2
Adjustment of Losses, Participant AOC

BALANCING GRADES	DOP, bbls	PERCENT	ADJUSTMENT, bbls	ADJUSTED DOP, bbls
I-150	18000	0.4	161	17839
I-201	0	0.0	0	0
I-210	955350	23.3	8519	946831
I-397	517260	12.6	4613	512647
I-440	522030	12.7	4655	517375
I-888	1000440	24.4	8921	991519
I-961	1084590	26.5	9672	1074918
	4097670	100.0	36541	4061129

NOTES:
TOTAL CRUDE RUN DURING THE MONTH = 6090000 bbls
DEDUCT 0.6% FOR UNACCOUNTED LOSSES = 36540 bbls

Table 16-9 shows the retro DOP of the participants converted to balancing grade equivalents. This is used to allocate balancing grades and losses (loss is considered as balancing grade).

Table 16-10 shows the allocation of balancing grade deltas to participants as per procedure described earlier. Table 16-11 shows the complete preliminary product allocation, in which the actual refinery production of all fixed and balancing grades has been split (allocated) to the participants, AOC and BOC.

Table 16-3
Adjustment of Losses, Participant BOC

BALANCING GRADE	DOP, bbls	PERCENT	ADJUSTMENT, bbls	ADJUSTED DOP, bbls
I-150	0	0	0	0
I-201	870	0.1	6	864
I-210	229140	19.9	1562	227578
I-397	81270	7.1	554	80716
I-440	207000	18	1411	205589
I-888	412500	35.7	2811	409689
I-961	220800	19.2	1505	219295
	1151580	100.0	7849	1143731

NOTES:
CRUDE RUN FOR BOC = 1308000 bbl
DEDUCT 0.6% FOR UNACCOUNTED LOSSES = 7848 bbl

Table 16-4
Forecaster Changes for Fixed Grades and Process Stocks
(All Figures in Barrels)

PRODUCTS	ACTUAL PRODUCTION	COMBINED DOP	TOTAL FORCASTER CHANGE	FORECASTER CHANGE, AOC	FORECASTER CHANGE, BOC
GRADES					
1138B	85638	106590	−20952	−9604	−11348
1149B	25568	0	25568	13451	12117
150	19553	17839	BG		
201	0	864	BG		
210	1029704	1174409	BG		
220	575	0	575	0	575
383	17686	0	17686	17686	0
390	88039	76800	11239	11239	0
395	104597	0	104597	87671	16926
397	289195	593363	BG		
397E	3384	0	3384	0	3384
398	67508	44100	23408	23408	0
411	100	0	100	100	0
419	220278	420000	−199722	−199722	0
434	309908	240000	69908	69908	0
440	944758	722964	BG		
711	−28374	0	−28374	0	−28374
725	100752	0	100752	100752	0
800	378169	420000	−41831	−41831	0
876	−130601	300000	−430601	−430601	0
876ZP	232098	240000	−7902	−7902	0
888	2002599	1401205	BG		
892	−66	0	−66	−38	−28
928	403652	360000	43652	43652	0
961	1345430	1294213	BG		
PROCESS STOCKS					
PBFUEL	−1301	0	−1301	−657	−644
PCTTR	−8731	0	−8731	−7294	−1437
PDDSL	1470	0	1470	1232	238
PDSL	33564	0	33564	28133	5431
PFCOO	−25764	0	−25764	−17798	−7966
PFDISO	−196533	0	−196533	−188993	−7540
PKERO	−19637	0	−19637	−14004	−5633
PLCGAS	−18227	−60000	41773	41773	0
PLLCN	2674	0	2674	2241	433
PLTISO	10800	0	10800	9052	1748
PMEISO	35309	0	35309	29595	5714
PMIDSL	−26688	0	−26688	−20269	−6419
PMSR	19029	0	19029	15950	3079

Table 16-4
Continued

PPOLY	3652	−30000	33652	30000	3652
PP90R	1363	−18000	19363	18000	1363
PP95R	36857	−60000	96857	90893	5964
PRESID	793	0	793	665	128
PSKERO	13473	0	13473	11293	2180
PSLOPD	−4861	0	−4861	−4045	−816
PSLOPO	−4719	0	−4719	−3874	−845
PSMCN	9387	0	9387	7867	1520
PSWMSR	8909	0	8909	7467	1442
PUFCHG	−3099	0	−3099	−2590	−509
PWCN	−547	0	−547	−451	−96
TOTAL	7377323	7244347	−292859	−287645	−5761
LOSSES	38204	45650			
CRUDE	7415526	7290000	125526	125526	0

NOTE: BG = BALANCING GRADE

FINAL ALLOCATION

In the preliminary allocation, the allocation of balancing grades is unsatisfactory because it is based on the participants' DOPs. A participant DOP is prepared from the refinery LP model using its share of processing unit capacity. The DOPs are prepared before the start of the reference month. At that time, the exact refinery unit capacity available for sharing between the participants, during the month, is not known. Also, the average balancing-grade product prices prevailing during the month, which are used to drive the LP, are not known. Participants may be using approximate data on processing unit capacity available to them and balancing-grade product prices in their refinery LP model. The result is that the optimization of balancing-grades production is not on the same basis in the two participants' LPs.

Table 16-5
Crude Changes (All Figures in Barrels)

	ACTUAL CRUDE	COMBINED DOP	DELTA	FORECASTER, AOC	FORECASTER, BOC
CRUDE RUN	7415526	729000	125526	125526	0

Table 16-6
Forecaster Compensating Changes, AOC (All Figures in Barrels)

GRADE	FORECASTER CHANGE	FORECASTER COMPENSATING AOC								
		I-150	I-201	I-210	I-397	I-440	I-888	I-961	LOSS	CHECK
1138B	−9604	0	0	0	0	0	−6998	16602		9604
1149B	13451	0	0	0	0	0	9802	−23253		−13451
150										
201										
210										
220	0	0	0	0	0	0	0	0		0
383	17686	478	−14738	0	−3426	0	0	0		−17686
390	11239	76	−5106	0	−6210	0	0	0		−11239
395	87671	−1043	−16079	0	−70549	0	0	0		−87671
397										
397E	0	0	0	0	0	0	0	0		0
398	23408	7	−506	0	−22909	0	0	0		−23408
411	100	0	0	0	0	−100	0	0		−100
419	−199722	0	0	0	0	199722	0	0		199722
434	69908	2447	0	−36562	0	−35793	0	0		−69908
440										
711	0	0	0	0	0	0	0	0		0
725	100752	0	0	0	0	0	−34941	−65811		−100752
800	−41831	0	0	0	0	−23764	65595	0		41831
876	−430601	0	0	0	0	63643	366958	0		430601
876ZP	−7902	0	0	0	0	4558	3344	0		7902
888										
892	−38	0	0	0	0	0	28	10		38
928	43652	0	0	0	0	0	891	−44543		−43652
961										
Process Stocks										
PBFUEL	−657	0	0	0	0	0	0	657		657
PCTTR	−7294	0	0	0	0	0	4696	2598		7294
PDDSL	1232	0	0	0	0	0	−1232	0		−1232
PDSL	28133	0	0	0	0	0	−28133	0		−28133
PFCOO	−17798	0	0	0	0	0	4800	12998		17798
PFDISO	−188993	0	0	0	0	0	45472	143521		188993
PKERO	−14004	0	0	0	0	14004	0	0		14004
PLCGAS	41773	313	−1366	0	−40720	0	0	0		−41773
PLLCN	2241	17	−73	0	−2185	0	0	0		−2241
PLTISO	9052	0	0	0	0	0	−9052	0		−9052
PMEISO	29595	0	0	0	0	0	−9944	−19651		−29595
PMIDSL	−20269	0	0	0	0	−9224	29493	0		20269
PMSR	15950	1818	7481	−25249	0	0	0	0		−15950
PPOLY	30000	462	3228	0	−33690	0	0	0		−30000
PP90R	18000	409	473	0	−18882	0	0	0		−18000
PP95R	90893	1872	14679	0	−107445	0	0	0		−90893
PRESID	665	0	0	0	0	0	−147	−518		−665
PSKERO	11293	0	0	0	0	−11293	0	0		−11293
PSLOPD	−4045	0	405	405	566	769	1497	405		4045
PSLOPO	−3874	0	387	387	542	736	1433	387		3874
PSMCN	7867	1370	−2851	0	−6386	0	0	0		−7867
PSWMSR	7467	851	3502	−11820	0	0	0	0		−7467
PUFCHG	−2590	−295	−1215	4100	0	0	0	0		2590
PWCN	−451	−79	163	0	366	0	0	0		451
TOTAL	−287645	8704	−11615	−68739	−310927	203257	443561	23404		287645
CRUDE	125526	0	−1255	22595	0	12553	38913	50587	2134	125526

Table 16-7
Forecaster Compensating Changes, BOC (All Figures in Barrels)

GRADE	FORECASTER CHANGE	FORECASTER COMPENSATING CHANGES TO BALANCING GRADES								
		I-150	I-201	I-210	I-397	I-440	I-888	I-961	LOSS	CHECK
1138B	−11348	0	0	0	0	0	−8269	19617		11348
1149B	12117	0	0	0	0	0	8830	−20947		−12117
150										
201										
210										
220	575	0	−575	0	0	0	0	0		−575
383	0	0	0	0	0	0	0	0		0
390	0	0	0	0	0	0	0	0		0
395	16926	−201	−3104	0	−13620	0	0	0		−16926
397										
397E	3384	0	0	0	−3384	0	0	0		−3384
398	0	0	0	0	0	0	0	0		0
411	0	0	0	0	0	0	0	0		0
419	0	0	0	0	0	0	0	0		0
434	0	0	0	0	0	0	0	0		0
440										
711	−28374	0	0	0	0	0	7094	21281		28374
725	0	0	0	0	0	0	0	0		0
800	0	0	0	0	0	0	0	0		0
876	0	0	0	0	0	0	0	0		0
876ZP	0	0	0	0	0	0	0	0		0
888										
892	−28	0	0	0	0	0	20	8		28
928	0	0	0	0	0	0	0	0		0
961										
Process Stocks										
PBFUEL	−644	0	0	0	0	0	0	644		644
PCTTR	−1437	0	0	0	0	0	925	512		1437
PDDSL	238	0	0	0	0	0	−238	0		−238
PDSL	5431	0	0	0	0	0	−5431	0		−5431
PFCOO	−7966	0	0	0	0	0	2148	5818		7966
PFDISO	−7540	0	0	0	0	0	1814	5726		7540
PKERO	−5633	0	0	0	0	5633	0	0		5633
PLCGAS	0	0	0	0	0	0	0	0		0
PLLCN	433	3	−14	0	−422	0	0	0		−433
PLTISO	1748	0	0	0	0	0	−1748	0		−1748
PMEISO	5714	0	0	0	0	0	−1920	−3794		−5714
PMIDSL	−6419	0	0	0	0	−2921	9340	0		6419
PMSR	3079	351	1444	−4874	0	0	0	0		−3079
PPOLY	3652	56	393	0	−4101	0	0	0		−3652
PP90R	1363	31	36	0	−1430	0	0	0		−1363
PP95R	5964	123	963	0	−7050	0	0	0		−5964
PRESID	128	0	0	0	0	0	−28	−100		−128
PSKERO	2180	0	0	0	0	−2180	0	0		−2180
PSLOPD	−816	0	82	82	114	155	302	82		816
PSLOPO	−845	0	85	85	118	161	313	85		845
PSMCN	1520	265	−551	0	−1234	0	0	0		−1520
PSWMSR	1442	164	676	−2283	0	0	0	0		−1442
PUFCHG	−509	−58	−239	806	0	0	0	0		509
PWCN	−96	−17	35	0	78	0	0	0		96
TOTAL	−5761	717	−770	−6185	−30931	847	13152	28930		5761
CRUDE	0	0	0	0	0	0	0	0	0	0

The final allocation is done after the end of the reference month. At that time, the monthly average unit capacity of each processing unit available to participants is known with certainty. Also, the monthly average balancing grade products published prices are known. Therefore, modeling the refinery unit capacities available to each participant can be accurate. Both models are driven by an average "Mean of Platt" or other published product price quotes prevailing during the month. Crude run, fixed grade productions, and process stocks drawdown or buildup are entered in the LP models from the preliminary allocation of the participants.

The final product allocation process has two main parts to it: running the allocation LPs, then rerunning the allocation procedure.

ALLOCATION LPs

Allocation LPs are the most important part of the allocation procedure. Two refinery LP models are prepared, modeling AOC's and BOC's shares of the refinery for the reference month. For example, if the joint refinery had available crude distillation capacity of 200,000 barrels per day and the equity of the two participants is 60% and 40%, then AOC will have 120,000 bpd and BOC will have 80,000 bpd crude distillation capacity. Capacity of the downstream units will be similarly split up. The unit capacities used in the LP models are the available unit capacities during the month (not the actually used capacities, which could be less than that available for certain units). All unit yields, stream properties, blending, and product specifications are accurately modeled as per actual refinery operations during the reference month.

PROCESS UNIT CAPACITIES

Data on the unit capacity available on each day of its operation during the month is maintained for every refinery unit and an average available unit capacity is determined at the end of the month (Table 16-12). These available capacities for every unit are split in the ratio of equity to model AOC's and BOC's shares of refinery capacity.

Table 16-8
Retrospective DOP (All Figures in Barrels)

PRODUCT GRADE (1)	AOC DOP (2)	AOC FORECASTER (3)	AOC FC COMPENS. (4)	AOC CRUDE (5)	AOC RETRO DOP (6)	BOC DOP (7)	BOC FORE-CASTER (8)	BOC FC COMPENS (9)	BOC CRUDE (10)	BOC RETRO DOP (11)	TOTAL RETRO (12)	ACTUAL PROD-UCTION (13)	DELTA (14)
1138B	60000	−9604			50396	46590	−11348			35242	85638	85638	0
1149B	0	13451			13451	0	12117			12117	25568	25568	0
150	17839		8704	0	26543	0		717	0	717	27260	19553	−7707
201	0		−11615	−1255	−12870	864		−770	0	94	−12776	0	12776
210	946831		−68739	22595	900687	227578		−6185	0	221393	1122080	1029704	−92376
220	0	0			0	0	575			575	575	575	0
383	0	17686			17686	0	0			0	17686	17686	0
390	76800	11239			88039	0	0			0	88039	88039	0
395	0	87671			87671	0	16926			16926	104597	104597	0
397	512647		−310927	0	201720	80716		−30931	0	49785	251505	289195	37690
397E	0	0			0	0	3384			3384	3384	3384	0
398	44100	23408			67508	0	0			0	67508	67508	0
411	0	100			100	0	0			0	100	100	0
419	420000	−199722			220278	0	0			0	220278	220278	0
434	240000	69908			309908	0	0			0	309908	309908	0
440	517375		203257	12553	733184	205589		847	0	206436	939621	944758	5137
711	0	0			0	0	−28374			−28374	−28374	−28374	0
725	0	100752			100752	0	0			0	100752	100752	0
800	420000	−41831			378169	0	0			0	378169	378169	0
876	300000	−430601			−130601	0	0			0	−130601	−130601	0
876ZP	240000	−7902			232098	0	0			0	232098	232098	0
888	991519		443561	38913	1473993	409686		13152	0	422838	1896831	2002599	105768
892	0	−38			−38	0	−28			−28	−66	−66	0
928	360000	43652			403652	0	0			0	403652	403652	0
961	1074918		23404	50587	1148909	219295		28930	0	248225	1397134	1345430	−51704
PBFUEL	0	−657			−657	0	−644			−644	−1301	−1301	0

Table 16-8
Continued

PRODUCT GRADE (1)	AOC DOP (2)	AOC FORECASTER (3)	AOC FC COMPENS. (4)	AOC CRUDE (5)	AOC RETRO DOP (6)	BOC DOP (7)	BOC FORE-CASTER (8)	BOC FC COMPENS (9)	BOC CRUDE (10)	BOC RETRO DOP (11)	TOTAL RETRO (12)	ACTUAL PROD-UCTION (13)	DELTA (14)
PCTTR	0	−7294			−7294	0	−1437			−1437	−8731	−8731	0
PDDSL	0	1232			1232	0	238			238	1470	1470	0
PDSL	0	28133			28133	0	5431			5431	33564	33564	0
PFCOO	0	−17798			−17798	0	−7966			−7966	−25764	−25764	0
PFDISO	0	−188993			−188993	0	−7540			−7540	−196533	−196533	0
PKERO	0	−14004			−14004	0	−5633			−5633	−19637	−19637	0
PLCGAS	−60000	41773			−18227	0	0			0	−18227	−18227	0
PLLCN	0	2241			2241	0	433			433	2674	2674	0
PLTISO	0	9052			9052	0	1748			1748	10800	10800	0
PMEISO	0	29595			29595	0	5714			5714	35309	35309	0
PMIDSL	0	−20269			−20269	0	−6419			−6419	−26688	−26688	0
PMSR	0	15950			15950	0	3079			3079	19029	19029	0
PPOLY	−30000	30000			0	0	3652			3652	3652	3652	0
PP90R	−18000	18000			0	0	1363			1363	1363	1363	0
PP95R	−60000	90893			30893	0	5964			5964	36857	36857	0
PRESID	0	665			665	0	128			128	793	793	0
PSKERO	0	11293			11293	0	2180			2180	13473	13473	0
PSLOPD	0	−4045			−4045	0	−816			−816	−4861	−4861	0
PSLOPO	0	−3874			−3874	0	−845			−845	−4719	−4719	0
PSMCN	0	7867			7867	0	1520			1520	9387	9387	0
PSWMSR	0	7467			7467	0	1442			1442	8909	8909	0
PUFCHG	0	−2590			−2590	0	−509			−509	−3099	−3099	0
PWCN	0	−451			−451	0	−96			−96	−547	−547	0
LOSSES	35971			2133.9	38104.94	9682				9682	47786.94	38204	−9582.9
CRUDE	6090000	125526			6215526	1200000	0			1200000	7415526	7415526	0
COL TOT	6054029	−287645	287645				−5761	5761	0				0

NOTE: COLUMNS 4 AND 9 ARE FORECASTER COMPENSATING CHANGES TO BALANCING GRADES.

Table 16-9
Retrospective DOP in Terms of Balancing Grades

BALANCING GRADES	AOC		BOC		TOTAL, bbl
	%	bbl	%	bbl	
I-150	11460	100.00	0	0.00	11460
I-201	46017	98.16	864	1.84	46881
I-210	1094946	82.79	227578	17.21	1322524
I-397	416254	83.76	80716	16.24	496970
I-440	1016978	83.18	205589	16.82	1222567
I-888	1995196	84.15	375736	15.85	2370932
I-961	1596571	84.19	299835	15.81	1896406
LOSS	38105	79.74	9682	20.26	47787
TOTAL	6215526	83.82	1200000	16.18	7415526

FEED AND PRODUCTS

1. Crude run in the LPs is per the participant's retrospective DOPs.
2. In the LP, the fixed grade products to be made are the allocated production for the fixed grades per the preliminary allocation. Negative production of any fixed grade is not entered into the LP and has to be taken care of by a forecaster change in the allocation program.
3. Negative allocation of process stock in the preliminary allocation is reflected as drawdown and positive allocation is reflected as a buildup in the LP models.
4. Allocation of slop (drawdown or buildup) is not modeled in the LPs. This is reflected as a forecaster change in the final allocation.

PRODUCT PRICES

The product prices driving the LPs are the agreed-on prices for balancing-grade products (Table 16-13). Normally, these are the average mean of Platts (MOP) published prices for the month in question. The location of the joint refinery dictates which market price quotes are to be used. Similarly, energy prices for refinery fuel and the like are based on MOP published data for the month. The LP models optimize the production of balancing-grade products based on balancing-grade prices in the two LPs.

Table 16-10
Allocation of Balancing-Grade Deltas (All Figures in Barrels)

							FIRST STEP		REVERSE ALLOCATION		SUM OF DELTAS ALLOCTION	
PRODUCTS	TOTAL DELTA (1)	AOC, IN TERMS OF BG (2)	BOC, IN TERMS OF BG (3)	TOTAL (4)	% AOC (5)	% BOC (6)	AOC DELTA (7)	BOC DELTA (8)	AOC (9)	BOC (10)	AOC TOTAL DELTA (11)	BOC TOTAL DELTA (12)
I-150	−7707	11460	0	11460	1.0000	0.0000	−7707	0	−48	49	−7756	49
I-201	12776	46017	864	46881	0.9816	0.0184	12540	235	−80	80	12460	316
I-210	−92376	1094946	227578	1322524	0.8279	0.1721	−76480	−15896	−581	581	−77061	−15315
I-397	37690	416254	80716	496970	0.8376	0.1624	31568	6121	−237	237	31331	6359
I-440	5137	1016978	205589	1222567	0.8318	0.1682	4273	864	−32	32	4241	896
I-888	105768	1995196	375736	2370932	0.8415	0.1585	89007	16762	−666	666	88341	17427
I-961	−51704	1596571	299835	1896406	0.8419	0.1581	−43529	−8175	−325	325	−43855	−7849
LOSS	−9583	38105	9682	47787	0.7974	0.2026	−7641	−1942	−60	60	−7702	−1881
TOTAL	0						2031	−2031	−2031	2031	0	0

NOTES:
COLUMN 1 BALANCING-GRADE DELTAS BETWEEN ACTUAL PRODUCTION AND COMBINED RETROSPECTIVE DOP.
COLUMNS 2 AND 3 RETRO DOP OF AOC AND BOC EXPRESSED IN TERMS OF BALANCING GRADES.
COLUMNS 7 AND 8 TOTAL DELTA IN COLUMN 1 SPLIT IN THE RATIO OF RETRO DOP EXPRESSED IN TERMS OF BALANCING GRADES.
AVCTD THE ABSOLUTE VALUES OF COMBINED TOTAL DELTAS IN COLUMN 1, =322,740.
COLUMNS 9 AND 10 REVERSE ALLOCATION OF DELTAS TO MAKE THE SUM OF AOC AND BOC DELTAS INDIVIDUALLY EQUAL TO ZERO.
COLUMN 11 = COLUMN 7 + COLUMN 9.

Table 16-11
Preliminary Product Allocation (All Figures in Barrels)

PRODUCT	AOC RETRO	BOC RETRO	AOC BG DELTA	BOC BG DELTA	AOC ALLOCATION	BOC ALLOCATION	ACTUAL PRODUCTION
1138B	50396	35242			50396	35242	85638
1149B	13451	12117			13451	12117	25568
150	26543	717	−7756	49	18787	766	19553
201	−12870	94	12460	316	−410	410	0
210	900687	221393	−77061	−15315	823626	206079	1029704
220	0	575			0	575	575
383	17686	0			17686	0	17686
390	88039	0			88039	0	88039
395	87671	16926			87671	16926	104597
397	201720	49785	31331	6359	233051	56144	289195
397E	0	3384			0	3384	3384
398	67508	0			67508	0	67508
411	100	0			100	0	100
419	220278	0			220278	0	220278
434	309908	0			309908	0	309908
440	733184	206436	4241	896	737425	207333	944758
711	0	−28374			0	−28374	−28374
725	100752	0			100752	0	100752
800	378169	0			378169	0	378169
876	−130601	0			−130601	0	−130601
876ZP	232098	0			232098	0	232098
888	1473993	422838	88341	17427	1562334	440265	2002599
892	−38	−28			−38	−28	−66
928	403652	0			403652	0	403652
961	1148909	248225	−43855	−7849	1105054	240376	1345430
PBFUEL	−657	−644			−657	−644	−1301
PCTTR	−7294	−1437			−7294	−1437	−8731
PDDSL	1232	238			1232	238	1470
PDSL	28133	5431			28133	5431	33564
PFCOO	−17798	−7966			−17798	−7966	−25764
PFDISO	−188993	−7540			−188993	−7540	−196533
PKERO	−14004	−5633			−14004	−5633	−19637
PLCGAS	−18227	0			−18227	0	−18227
PLLCN	2241	433			2241	433	2674
PLTISO	9052	1748			9052	1748	10800
PMEISO	29595	5714			29595	5714	35309
PMIDSL	−20269	−6419			−20269	−6419	−26688
PMSR	15950	3079			15950	3079	19029
PPOLY	0	3652			0	3652	3652
PP90R	0	1363			0	1363	1363
PP95R	30893	5964			30893	5964	36857
PRESID	665	128			665	128	793
PSKERO	11293	2180			11293	2180	13473
PSLOPD	−4045	−816			−4045	−816	−4861
PSLOPO	−3874	−845			−3874	−845	−4719
PSMCN	7867	1520			7867	1520	9387
PSWMSR	7467	1442			7467	1442	8909
PUFCHG	−2590	−509			−2590	−509	−3099
PWCN	−451	−96			−451	−96	−547
LOSSES	38105	9682	−7702	−1881	30403	7801	38204
COL TOTAL					6215526	1200001	7415527
CRUDE	6215526	1200000			6215526	1200000	7415526

SUMMARY OF PRIMARY DATA INPUT FOR LPs

1. Balancing-grade product prices.
2. Available capacity of units during the month.
3. Actual crude run of each participant.
4. Fixed-grade product requirements as per preliminary allocation of that participant (negative allocations of fixed grades and slop allocation, positive or negative, are not put in the LPs).
5. Intermediate stocks drawdown and buildup as per the preliminary allocation data.

EXAMINATION OF LP RESULTS

The two LPs are run and the results examined for the following points:

1. The solution should be optimum and converged.
2. A data check should be conducted to ensure that the crude run, fixed grade, and process stock production are entered in the LP as per the preliminary allocation report.
3. The utilization of key units (crude distillation, FCCU, hydrocracker, and other conversion units) should be close to maximum or to the actual refinery performance during the month. If any key conversion unit is grossly under utilized, the cause should be investigated.
4. LP product blending should be examined to ensure there is no unnecessary giveaway and no dumping of distillate streams in fuel oils. For example,

   - If the FCCU is running full in one participant's LP but not in the other, adjustment would be made to the buildup or drawdown of process stock HVGO (FCCU feed) in an attempt to get the FCCU to run full in both LPs or at the same percentage of their capacity utilization.
   - If the price differential between naphtha and gasoline is such that the catalytic reformer is running at unrealistically high rates, then limit the unit to what it actually ran during the month.
   - From time to time, a situation arises where there is no production of a balancing grade, usually gasoline. If adjusting the inventory

Table 16-12
Calculation of Available Processing Unit Capacities for Allocation LPS

PROCESS UNITS DAY	NOMINAL CAPACITY	AVAILABLE PROCESSING UNITS CAPACITIES DURING THE MONTH, MBPSD														
		(1)	(2)	(3)	(4)	(5)	(6)	(7)	(8)	(9)	(10)	(11)	(12)	(13)	(14)	(15)
1CDU	20	22	22	22	22	22	22	22	22	22	22	22	22	22	22	22
2CDU	20	22	22	22	22	22	22	22	22	22	22	22	22	22	22	22
3CDU	64	69	69	69	69	69	69	69	69	69	69	69	69	69	69	69
4CDU	93	104	104	104	104	104	104	104	104	104	104	104	104	104	104	104
5CDU	46	50	50	50	50	50	50	50	50	50	50	50	50	50	50	50
TOTAL CDU	243	267	267	267	267	267	267	267	267	267	267	267	267	267	267	267
1VDU	9.6	9.1	9.1	9.1	9.1	9.1	9.1	9.1	9.1	9.1	9.1	9.1	9.1	9.1	9.1	9.1
5VDU	20	33	33	33	33	33	33	33	33	33	33	33	33	33	33	33
6VDU	52.63	67	67	67	50	50	50	50	50	67	67	67	67	67	67	67
TOTAL VDU	82.23	109.1	109.1	109.1	109.1	109.1	109.1	109.1	109.1	109.1	109.1	109.1	109.1	109.1	109.1	109.1
DIESEL HDS	20	22	22	22	22	22	22	22	22	22	22	22	22	22	22	22
HYDROCRACKER	50	52	52	52	42	42	42	42	52	47	52	52	52	52	52	52
VISBREAKER	20	22	22	22	22	22	22	22	22	22	22	22	22	22	22	22
FCCU	39	45	45	45	45	45	0	0	0	0	0	0	45	45	45	45
CAT REFORMER	18	15.4	15.4	15.4	15.4	15.4	15.4	15.4	15.4	15.4	15.4	15.4	15.4	15.4	15.4	15.4
KERO TREATER	42	42	42	42	42	42	42	42	42	42	42	42	42	42	42	42
SULFUR PLANT	0.165	0.165	0.165	0.165	0.165	0.165	0.165	0.165	0.165	0.165	0.165	0.165	0.165	0.165	0.165	0.165
H2 PLANT	11.1	11.1	11.1	11.1	11.1	11.1	11.1	11.1	11.1	11.1	11.1	11.1	11.1	11.1	11.1	11.1

PROCESS UNITS DAY	NOMINAL CAPACITY	AVAILABLE PROCESSING UNITS CAPACITIES DURING THE MONTH, MBPSD															
		(16)	(17)	(18)	(19)	(20)	(21)	(22)	(23)	(24)	(25)	(26)	(27)	(28)	(29)	(30)	AVERAGE
1CDU	20	22	22	22	22	22	22	22	22	22	22	22	22	22	22	22	22.0
2CDU	20	22	22	22	22	22	22	22	22	22	22	22	22	22	22	22	22.0
3CDU	64	69	69	69	69	69	69	69	69	69	69	69	69	69	69	69	69.0
4CDU	93	104	104	104	104	104	104	104	104	104	104	104	0	0	0	0	90.1
5CDU	46	50	50	50	50	50	50	50	50	50	50	50	50	50	50	50	50.0
TOTAL CDU	243	267	267	267	267	267	267	267	267	267	267	267	163	163	163	163	253.1
1VDU	9.6	9.1	9.1	9.1	9.1	9.1	9.1	9.1	9.1	9.1	9.1	9.1	9.1	9.1	9.1	9.1	9.1
5VDU	20	33	33	33	33	33	33	33	33	33	33	33	33	33	33	33	33.0
6VDU	52.63	67	67	67	67	67	67	67	67	67	67	67	67	67	67	67	64.2
TOTAL VDU	82.23	109.1	109.1	109.1	109.1	109.1	109.1	109.1	109.1	109.1	109.1	109.1	109.1	109.1	109.1	109.1	109.1
DIESEL HDS	20	22	22	22	22	22	22	22	22	22	22	22	22	22	22	22	22.0
HYDROCRACKER	50	52	52	52	52	52	52	52	52	52	52	52	52	52	52	52	50.5
VISBREAKER	20	22	22	22	22	22	22	22	22	22	22	22	22	22	22	22	22.0
FCCU	39	45	45	45	45	45	45	45	45	45	45	45	45	45	45	45	36.0
CAT REFORMER	18	15.4	15.4	15.4	15.4	15.4	15.4	15.4	15.4	15.4	15.4	15.4	15.4	15.4	15.4	15.4	15.4
KERO TREATER	42	42	42	42	42	42	42	42	42	42	42	42	42	42	42	42	42.0
SULFUR PLANT	0.165	0.165	0.165	0.165	0.165	0.165	0.165	0.165	0.165	0.165	0.165	0.165	0.165	0.165	0.165	0.165	0.2
H2 PLANT	11.1	11.1	11.1	11.1	11.1	11.1	11.1	11.1	11.1	11.1	11.1	11.1	11.1	11.1	11.1	11.1	11.1

NOTES:
SULFUR PLANT CAPACITY IN M TONS/DAY SULFUR.
H2 PLANT CAPACITY IN MMSCF H2/DAY.
ALL OTHER CAPACITIES IN M BPSD.
CAPACITIES USED IN ALLOCATION LPS ARE THE AVERAGE OF AVAILABLE UNIT CAPACITIES FOR THE MONTH.

changes of the blendstocks does not rectify the situation, then the balancing grade should be changed.
- The hydrogen balance listed in the LPs sometimes is not accurate. If a hydrogen consuming unit, such as a desulfurizer or hydrocracker, is constrained by the hydrogen supply, verify that the constraint is real or increase the hydrogen plant capacity by, say, 10% to allow conversion unit to run at full capacity.

Other points to look for are

- Are the specifications of all products correct?
- Which specifications are limiting? Are they reasonable? For example, if the flash point is the only limiting specs on diesel, there could be a problem elsewhere.
- For specifications that are not limiting, how much giveaway is there?
- Is the way the LP has chosen to run reasonable? Has it picked up an acceptable mode of operation of a particular unit?
- Are there any streams in the refinery fuel pool that should not be there?

These examples do not cover all the problems that could be encountered. It may be necessary to make additional LP runs until the programmer is satisfied with the results of both LPs for use in the allocation program.

INFEASIBILITIES

Sometimes, the LP cannot meet all the fixed grade product demand as per the preliminary allocation report and it returns an infeasible solution. In this case, the product or feedstock causing the infeasibility should be identified and its production in the DOP reduced to make the solution feasible. The delta between the actual production and the DOP for this product is entered as a forecaster change in the final allocation.

Gross underutilization of any conversion unit capacity can be handled similarly. For example, if an FCCU is grossly underutilized because of a buildup of VGO feed in the LP model, the VGO buildup could be appropriately reduced to allow fuller utilization of the FCCU and the rest of VGO buildup reflected as a forecaster change in the final allocation.

Table 16-13
Calculation of LP Driving Prices (Example)

DAY	AG LPG, $/ton	AG NAPHTHA, $/ton	AG KEROSENE, $/bbl	AG DIESEL, $/bbl	AG FUEL OIL, $/Month	MED. NAPHTHA, $/ton	MED. PREM., $/ton GAS
1							
2	111.50	120.50	13.90	13.20	47.00	112.00	171.00
3	111.50	112.00	13.80	13.00	46.50	112.00	168.00
4	111.50	106.50	13.80	13.00	46.50	105.00	160.00
5	111.50	105.50	13.80	12.40	46.00	106.00	158.00
6	111.50	105.50	13.35	12.40	45.00	106.00	155.00
7							
8							
9	111.50	105.50	13.35	12.40	45.00	104.00	150.50
10	111.50	105.50	13.35	12.40	44.00	103.50	150.00
11	111.50	105.50	13.35	11.95	43.00	103.50	150.00
12	111.50	105.50	13.35	11.95	43.00	103.50	150.00
13	111.50	105.50	13.35	11.95	42.00	103.50	150.00
14							
15							
16	111.50	105.50	13.35	11.95	42.00	103.50	150.00
17	101.00	105.50	12.35	11.95	42.00	103.50	150.50
18	101.00	107.00	12.25	11.95	42.00	102.00	150.50
19	101.00	107.00	12.25	11.95	42.00	102.00	150.50
20	101.00	106.50	12.25	11.85	42.00	100.50	150.50
21							
22							
23	101.00	105.50	12.25	11.75	42.00	100.50	150.50
24	101.00	100.50	12.15	11.65	42.50	100.50	147.00
25	101.00	100.50	12.05	11.35	42.50	100.50	146.00
26	101.00	95.50	12.05	11.25	42.50	100.50	143.50
27	101.00	93.50	12.05	11.25	42.50	100.50	143.00
28							
29							
30	101.00	92.50	12.05	11.25	43.00	95.50	140.50
31							
MONTH AVERAGE	106.50	104.62	12.88	12.04	43.48	103.26	151.67

NOTES:
AG = ARABIAN GULF MARKET QUOTE (MEAN OF PLATT).
MED = MEDITERRANEAN MARKET QUOTES (MEAN OF PLATT).
GASOLINE AG MARKET PRICE IS EXTIMATED AS FOLLOWS, AS THERE ARE NO AG GASOLINE PRICES.

AVERAGE MEAN OF PLATT FOR MED PREM GASOLINE = 151.69 $/Ton
AVERAGE MEAN OF PLATT FOR MED NAPHTHA = 103.26 $/Ton
RATIO OF MED MOGAS /MED NAPHTHA = 1.4690
AVERAGE MEAN OF PLATT FOR AG GASOLINE = 103.36*; 1.4690
= 153.68 $/Ton

FINAL ALLOCATION CYCLE

INITIAL DOP (FINAL ALLOCATION)

After the two allocation LPs are satisfactorily run, the LP productions of fixed grades and process stocks are entered as the initial DOP of the final allocation cycle (Tables 16-14 to 16-16). Balancing-grade production from the LP results are reduced by 0.6% to account for unaccounted losses, as was done in the case of preliminary allocation, before entering in the initial DOP.

FORECASTER CHANGES

Forecaster changes are next applied. There should be almost no forecaster change for most fixed grade and process stocks

All productions are entered in the LP in units of thousand barrels per day. Actual LP production in thousand barrels per day are reconverted into barrels per month before entering this data in the initial DOP of the final allocation. These conversions can cause small rounding errors between the preliminary allocation figures and those entered in the initial DOP of final allocation as shown next.

Consider for example the preliminary allocation of fixed grade I-390. The allocation for participant AOC is 88,039 bbls. This is entered in the AOC allocation LP as 88,039/(1000*30), or 2.934 thousand bbl/day.

In the initial DOP of the final allocation, this figure is again converted into barrels/month as 2.934*1000*30 or 88,020 bbl. To make it equal to the preliminary allocation figure of 88,039, a forecaster change of (88,039 – 88,020) or 19 bbl is introduced for AOC.

Similarly, any rounding error on the crude run is entered in units of thousand barrels/day in the LPs is corrected as a forecaster change, to make it equal to actual crude run of the participant as per preliminary allocation.

For all fixed grades and process stocks that were not modeled in the LPs, the initial DOP (final allocation) would be zero and forecaster changes should be the same as in the preliminary allocation.

A situation can arise in which the allocated production of a process stock in the preliminary allocation, when reflected in the participants' LP, causes serious underutilization of a key conversion unit, such as the hydrocracker or FCCU. In this case, the LP production values are sufficiently reduced and the remaining allocated production reflected as forecaster changes, as shown by following example.

EXAMPLE 16-5

The allocated production of a feedstock PFDISO (HVGO feed to hydrocracker) in the preliminary allocation is as follows:

$$\text{AOC PFDISO} = 39{,}816\,\text{bbl}$$
$$\text{BOC PFDISO} = 7839\,\text{bbl}$$
$$\text{Refinery buildup} = 47{,}655\,\text{bbl}$$

As this buildup of 39,816 and 7839 bbl in the participants' allocation LPs causes underutilization of the hydrocracker unit (due to lack of feed), the buildup is reduced as follows:

	BUILDUP, bbl	DELTA, bbl
AOC	17,010	22,806
BOC	1,830	6,009
TOTAL	18,840	28,815

The remaining deltas are applied as forecaster change in the final allocation.

RETROSPECTIVE DOP (FINAL ALLOCATION)

By applying forecaster changes, forecaster compensating changes, and crude changes, the initial DOP based on allocation LP production is amended to bring fixed grades, process stock and crude run equal to the actual production of the participants. These amended participant DOPs are called *retrospective DOPs* or *retro DOPs*. These two retro DOPs of the participants are combined to make a refinery retro DOP. The procedure is identical to that followed for the retro DOP in the preliminary allocation. The retrospective DOP is used for the allocation of balancing grades and actual refinery losses.

Table 16-14
Refinery DOP (bbl)

PRODUCT	AOC	BOC	TOTAL
1138B	50396	35242	85638
1149B	13451	12117	25568
150	19067	0	19067
201	204504	50906	255410
210	622971	151556	774527
220	0	570	570
383	17700	0	17700
390	88050	0	88050
395	87660	16920	104580
397	215748	58918	274666
397E	0	3390	3390
398	67500	0	67500
411	90	0	90
419	220290	0	220290
434	310170	0	310170
440	376971	144019	520990
711	0	−28380	−28380
725	100740	0	100740
800	378180	0	378180
876	0	0	0
876ZP	232110	0	232110
888	1943011	450586	2393597
892	0	0	0
928	403650	0	403650
961	985664	242584	1228248
PBFUEL	0	0	0
PCTTR	−7290	−1440	−8730
PDSDSL	0	1470	1470
PDSL	0	33570	33570
PFCOO	−17790	−7980	−25770
PFDISO	−189000	−7530	−196530
PKERO	−19637	0	−19637
PLCGAS	−18227	0	−18227
PLLCN	2237	420	2657
PLTISO	10800	0	10800
PMEISO	29610	5700	35310
PMIDSL	−20280	−6420	−26700
PMSR	15950	3079	19029
PPOLY	0	3660	3660
PP90R	0	1350	1350
PP95R	30900	5970	36870
PRESID	660	120	780
PSKERO	−13	13470	13457
PSLOPD	0	0	0
PSLOPO	0	0	0
PSMCN	7860	1530	9390
PSWMSR	7467	1442	8909
PUFCHG	−3107	9	−3098
PWCN	0	0	0
PROD + STOCK	6168063	1186848	7354911
LOSSES	47457	13152	60609
CRUDE	6215520	1200000	7415520

NOTE:
THE TWO LP PRODUCTIONS ARE COMBINED TO GIVE AN INITIAL DOP. THE BALANCING GRADE PRODUCTION HAS BEEN ADJUSTED FOR UNACCOUNTED LOSSES.

Table 16-15
Adjustment of Losses, Participant AOC

BALANCING GRADE	DOP, bbl	PERCENT	ADJUSTMENT, bbl	ADJUSTED PRODUCTION, bbl
I-150	19230	0.44	163	19067
I-201	206250	4.68	1746	204504
I-210	628290	14.26	5319	622971
I-397	217590	4.94	1842	215748
I-440	380190	8.63	3219	376971
I-888	1959600	44.48	16589	1943011
I-961	994080	22.57	8416	985664
	4405230	100.00	37293	4367937

NOTES:
CRUDE RUN FOR AOC IN THE LP MODEL = 6215520 bbl
DEDUCT 0.6% FOR UNACCOUNTED LOSSES = 37293 bbl

Table 16-16
Adjustment of Losses, Participant BOC

BALANCING GRADE	DOP, bbl	PERCENT	ADJUSTMENT, bbl	ADJUSTED PRODUCTION, bbl
I-150	0	0.00	0	0
I-201	51270	4.63	364	50906
I-210	152640	13.80	1084	151556
I-397	59340	5.36	422	58918
I-440	145050	13.11	1031	144019
I-888	453810	41.02	3224	450586
I-961	244320	22.08	1736	242584
	1106430	100.00	7861	1098569

NOTES:
CRUDE RUN FOR BOC IN THE LP MODEL = 1310160 bbl
DEDUCT 0.6% FOR UNACCOUNTED LOSSES = 7861 bbl

ALLOCATION OF BALANCING GRADES (FINAL ALLOCATION)

In the retrospective DOP, all deltas on fixed grades, process stocks, and crude run have been eliminated. Deltas between the actual production and retro DOP remain only in balancing grades and refinery loss (actual refinery losses minus retro DOP losses).

The procedure for allocating the balancing grades in the final allocation is identical to that followed for preliminary allocation. The allocation of fixed grades and process stocks also remain unchanged from that in the preliminary allocation. Only the balancing-grade allocations have changed from the preliminary allocation.

EXAMPLE 16-6 (FINAL ALLOCATION)

Continuing from Example 16-4, the entire final product allocation procedure is shown in Tables 16-12 to 16-24.

It may be recalled that the allocated productions of fixed grades and process stocks from the preliminary product allocation was input in the respective LP models of the participants. The participants' allocation LPs are run, and the LP production of all grades for both participants is entered as the initial DOP for the final allocation. The crude run of the participants is fixed as per the preliminary allocation. The allocation LPs are used to optimize the production of balancing grades for a given fixed grades/process stock slate.

Table 16-14 shows participants' AOC and BOC initial DOP for final allocation. The combined production of the two DOPs make a refinery DOP. Tables 16-15 and 16-16 show the adjustment of balancing grades for unaccounted refinery losses. The adjustment is 0.6% (LV) of the crude run of the participant. These are spread over the balancing grades in proportion to their production as shown in these tables.

Table 16-17 compares the actual refinery production of every fixed grade and process stock with that in the combined refinery DOP. The delta between these is the total "forecaster" change. There should be no forecaster change for fixed grades, process stocks, and crude run for all those grades whose allocated production as per preliminary allocation is reflected in the allocation LPs (except for small rounding error, as explained earlier). For all those grades that were not reflected in the allocation LPs, such as fixed grades with negative production or process stocks such as slop, the forecaster change is exactly as per the preliminary allocation.

Table 16-18 shows data for actual refinery crude run and that in the combined DOP. The small forecaster change is due to rounding error generated by converting LP figures in thousand barrels per day to barrels in the initial DOP of the participants.

Tables 16-19 and 16-20 show the calculation of forecaster compensating changes for the participants. The procedure is identical to that followed in the preliminary allocation.

Table 16-21 shows the retrospective DOP of the participants (columns 6 and 11) and the overall refinery retrospective DOP (column 12). We see that

1. Fixed grade production in each participant's retrospective DOP is, in fact, its allocation. The sum of the participants' retrospective DOPs for any fixed grade or process stock equals the actual refinery production for that grade.
2. The delta between the actual refinery production and the total retro DOP (column 12) remains only on balancing grades and losses.

Table 16-22 shows the retro DOP of the participants converted to balancing-grade equivalents. This is used for allocation of balancing grades and losses (loss is considered as balancing grade). Table 16-23 shows allocation of balancing grade deltas to participants, as per the procedure described earlier.

Table 16-24 shows the complete final product allocation, in which the actual refinery production of all fixed and balancing grades is split, or "allocated," to participants AOC and BOC.

ALLOCATION SPREADSHEET PROGRAM

The product allocation procedures can be aided by a spreadsheet program. The program is used for both the preliminary and final allocations. An example in the use of the allocation program follows.

PRELIMINARY ALLOCATION

Data are entered only in the shaded cells shown in the screens.

1. The participants' submitted DOPs, including crude run, is entered on Screen 1 of the program. The participants' DOP balancing-grade production must be reduced by 0.6% of their crude run to account for unaccounted losses.
2. The participants' opening inventories of all product grades, process stocks, product lifting, and local sales are entered on Screens 2 and 3. The source of opening inventory data is the previous month's

final allocation report. The previous month closing inventories become the opening inventories of the current month.

3. The refinery's physical closing inventory of all products, process stocks, and consumption are entered on Screen 4. The source of the data is refinery stock report. The program calculates the production of a product from the following relationship:

$$\text{Production} = \text{closing inventory} - \text{opening inventory} + \text{product lifted} + \text{local sales} + \text{refinery consumption}$$

   where the refinery's opening inventory, product lifted, and local sales is the sum of the participants' opening inventories, product lifted, and local sales data entered in Screen 2 and 3.

4. The total forecaster change for any fixed grade is computed by the program as the delta between the total refinery production and the sum of participants' DOPs.
5. The total forecaster change for a fixed grade or process stock is split between the participants as per the rules of allocation of fixed grades and process stocks and entered on Screen 1.
6. The refinery actual (total) crude run during the month is entered on Screen 1.
7. Product equivalencies of all fixed grades, process stocks, and crude are entered on Screen 5. Unless a new grade is introduced or a balancing grade is changed, no data need be entered on this screen.

The program calculates the following:

1. The retrospective DOP of the two participants (Screen 6).
2. The retrospective DOP of the overall refinery by combining the participants' DOPs and the retro DOP expressed in terms of balancing grades (Screens 7 and 8).
3. The delta on the balancing grades between actual refinery production and the combined DOP production of the participants is split between the participants, and allocation of balancing-grade products is done by the program and presented on Screen 9.

Table 16-17
Forecaster Changes for Fixed Grades and Process Stocks (bbl)

PRODUCT	ACTUAL PRODUCTION	REFINERY DOP*	TOTAL FORCASTER CHANGE	FORECASTER AOC	FORECASTER BOC
1138B	85638	85638	0	0	0
1149B	25568	25568	0	0	0
150	19553	0	BG		
201	0	50906	BG		
210	1029704	151556	BG		
220	575	570	5	0	5
383	17686	0	17686	−14	17700
390	88039	0	88039	−11	88050
395	104597	16920	87677	11	87666
397	289195	58918	BG		
397E	3384	3390	−6	0	−6
398	67508	0	67508	8	67500
411	100	0	100	10	90
419	220278	0	220278	0	220278
434	309908	0	309908	−262	310170
440	944758	144019	BG		
711	−28374	−28380	6	0	6
725	100752	0	100752	12	100740
800	378169	0	378169	−11	378180
876	−130601	0	−130601	−130601	0
876ZP	232098	0	232098	−12	232110
888	2002599	450586	BG		
892	−66	0	−66	−38	−28
928	403652	0	403652	2	403650
961	1345430	242584	BG		
Process Stocks					
PBFUEL	−1301	0	−1301	−657	−644
PCTTR	−8731	−1440	−7291	−4	−7287
PDDSL	1470	1470	0	0	0
PDSL	33564	33570	−6	0	−6
PFCOO	−25764	−7980	−17784	5	−17789
PFDISO	−196533	−7530	−189003	7	189010
PKERO	−19637	0	−19637	0	−19637
PLCGAS	−18227	0	−18227	0	−18227
PLLCN	2674	420	2254	4	2250
PLTISO	10800	0	10800	0	10800
PMEISO	35309	5700	29609	−15	29624
PMIDSL	−26688	−6420	−20268	11	−20279
PMSR	19029	3079	15950	0	15950
PPOLY	3652	3660	−8	0	−8
PP90R	1363	1350	13	0	13

Table 16-17
Continued

PRODUCT	ACTUAL PRODUCTION	REFINERY DOP*	TOTAL FORCASTER CHANGE	FORECASTER AOC	FORECASTER BOC
PP95R	36857	5970	30887	−7	30894
PRESID	793	120	673	5	668
PSKERO	13473	13470	3	13	−10
PSLOPD	−4861	0	−4861	−4045	−816
PSLOPO	−4719	0	−4719	−3874	−845
PSMCN	9387	1530	7857	−3	7860
PSWMSR	8909	1442	7467	0	7467
PUFCHG	−3099	9	−3108	8	−3116
PWCN	−547	0	−547	−451	−96
TOTAL	7377323	1250695	1593958	−139909	1733867
LOSSES	38204	6164825			
CRUDE	7415526	7415520	6	6	0

* SUM OF ALLOCATION LPS.

4. A summary of the allocation of all product grades and process stocks is determined by the program and presented on Screen 10.
5. The allocated production of the participants is transmitted to Screens 2 and 3, and the program calculates the closing inventories of the participants.

FINAL ALLOCATION

In the final allocation cycle, the following data remain unchanged from the preliminary allocation: the allocation of fixed grades and process

Table 16-18
Crude Changes (bbl)

	DOP CRUDE	ACTUAL CRUDE	DELTA
AOC	6215520	6215526	6
BOC	1200000	1200000	0
TOTAL REFINERY	7415520	7415526	6

Table 16-19
Forecaster Compensating Changes, AOC (bbl)

PRODUCT	FORECASTER CHANGE	FORECASTER COMPENSATING CHANGES TO BALANCING GRADES								CHECK
		I-150	I-201	I-210	I-397	I-440	I-888	I-961	LOSS	
1138B	0	0	0	0	0	0	0	0		0
1149B	0	0	0	0	0	0	0	0		0
150	BG									
201	BG									
210	BG									
220	0	0	0	0	0	0	0	0		0
383	−14	0	12	0	3	0	0	0		14
390	−11	0	5	0	6	0	0	0		11
395	11	0	−2	0	−9	0	0	0		−11
397	BG									
397E	0	0	0	0	0	0	0	0		0
398	8	0	0	0	−8	0	0	0		−8
411	10	0	0	0	0	−10	0	0		−10
419	0	0	0	0	0	0	0	0		0
434	−262	−9	0	137	0	134	0	0		262
440	BG									
711	0	0	0	0	0	0	0	0		0
725	12	0	0	0	0	0	−4	−8		−12
800	−11	0	0	0	0	−6	17	0		11
876	−130601	0	0	0	0	19303	111298	0		130601
876ZP	−12	0	0	0	0	7	5	0		12
888	BG									
892	−38	0	0	0	0	0	28	10		38
928	2	0	0	0	0	0	0	−2		−2
961	BG									
Process Stocks										
PBFUEL	−657	0	0	0	0	0	0	657		657
PCTTR	−4	0	0	0	0	0	3	1		4
PDDSL	0	0	0	0	0	0	0	0		0
PDSL	0	0	0	0	0	0	0	0		0
PFCOO	5	0	0	0	0	0	−1	−4		−5
PFDISO	7	0	0	0	0	0	−2	−5		−7
PKERO	0	0	0	0	0	0	0	0		0
PLCGAS	0	0	0	0	0	0	0	0		0
PLLCN	4	0	0	0	−4	0	0	0		−4
PLTISO	0	0	0	0	0	0	0	0		0
PMEISO	−15	0	0	0	0	0	5	10		15
PMIDSL	11	0	0	0	0	5	−16	0		−11
PMSR	0	0	0	0	0	0	0	0		0
PPOLY	0	0	0	0	0	0	0	0		0
PP90R	0	0	0	0	0	0	0	0		0
PP95R	−7	0	−1	0	8	0	0	0		7
PRESID	5	0	0	0	0	0	−1	−4		−5
PSKERO	13	0	0	0	0	−13	0	0		−13
PSLOPD	−4045	0	405	405	566	769	1497	405		4045
PSLOPO	−3874	0	387	387	542	736	1433	387		3874
PSMCN	−3	−1	1	0	2	0	0	0		3
PSWMSR	0	0	0	0	0	0	0	0		0
PUFCHG	8	1	4	−13	0	0	0	0		−8
PWCN	−451	−79	163	0	366	0	0	0		451
TOTAL	−139909	−88	973	916	1474	20924	114262	1448		139909
CRUDE	6	0	0	1	0	1	2	2	0	6

Table 16-20
Forecaster Compensating Changes, BOC (bbl)

		FORECASTER COMPENSATING CHANGES TO BALANCING GRADES								
PRODUCT	FORECASTER CHANGE	I-150	I-201	I-210	I-397	I-440	I-888	I-961	LOSS	CHECK
1138B	0	0	0	0	0	0	0	0		0
1149B	0	0	0	0	0	0	0	0		0
150										
201										
210										
220	5	0	−5	0	0	0	0	0		−5
383	0	0	0	0	0	0	0	0		0
390	0	0	0	0	0	0	0	0		0
395	6	0	−1	0	−5	0	0	0		−6
397										
397E	−6	0	0	0	6	0	0	0		6
398	0	0	0	0	0	0	0	0		0
411	0	0	0	0	0	0	0	0		0
419	−12	0	0	0	0	12	0	0		12
434	0	0	0	0	0	0	0	0		0
440										
711	6	0	0	0	0	0	−2	−5		−6
725	0	0	0	0	0	0	0	0		0
800	0	0	0	0	0	0	0	0		0
876	0	0	0	0	0	0	0	0		0
876ZP	0	0	0	0	0	0	0	0		0
888										
892	−28	0	0	0	0	0	20	8		28
928	0	0	0	0	0	0	0	0		0
961										
PROCESS STOCKS										
PBFUEL	−644	0	0	0	0	0	0	644		644
PCTTR	3	0	0	0	0	0	−2	−1		−3
PDDSL	0	0	0	0	0	0	0	0		0
PDSL	−6	0	0	0	0	0	6	0		6
PFCOO	1	0	0	0	0	0	0	−1		−1
PFDISO	−10	0	0	0	0	0	2	8		10
PKERO	0	0	0	0	0	0	0	0		0
PLCGAS	0	0	0	0	0	0	0	0		0
PLLCN	13	0	0	0	−13	0	0	0		−13
PLTISO	0	0	0	0	0	0	0	0		0
PMEISO	14	0	0	0	0	0	−5	−9		−14
PMIDSL	1	0	0	0	0	0	−1	0		−1
PMSR	0	0	0	0	0	0	0	0		0
PPOLY	−8	0	−1	0	9	0	0	0		8
PP90R	13	0	0	0	−14	0	0	0		−13
PP95R	−6	0	−1	0	7	0	0	0		6
PRESID	8	0	0	0	0	0	−2	−6		−8
PSKERO	3	0	0	0	0	−3	0	0		−3
PSLOPD	−816	0	82	82	114	155	302	82		816
PSLOPO	−845	0	85	85	118	161	313	85		845
PSMCN	0	0	0	0	0	0	0	0		0
PSWMSR	0	0	0	0	0	0	0	0		0
PUFCHG	−9	−1	−4	14	0	0	0	0		9
PWCN	−96	−17	35	0	78	0	0	0		96
TOTAL	−2413	−18	189	180	301	325	632	803		2413
CRUDE	0	0	0	0	0	0	0	0	0	0

stock and all data on opening inventories, refinery closing inventory, product lifted, local sales, and refinery consumption (Screens 2 to 4) remain unchanged.

The preliminary allocation file is saved under a different name and the initial DOP of the participants in Screen 1 is updated. Instead of the participants' submitted DOPs, the production from the allocation LPs of the participants are entered on Screen 1, subject to following exceptions:

- Only positive production of fixed grades as per allocation LPs and process stock buildup or drawdown as reflected in the LPs are entered.
- Actual crude run of the participants as per the allocation LPs is entered in Screen 1.
- Ideally there should be no forecaster changes for all fixed grades and process stocks reflected in the allocation LPs, except for small rounding errors due to conversion of LP figures in thousand barrels per day to DOP figures in barrels per month. The rounding error is the difference between the allocated production as per preliminary allocation of that grade and the DOP figure of the participant entered on Screen 1.
- If process stocks are not modeled in the LPs, the forecaster change for the product or stock is identical to that in the preliminary allocation.
- If any process stock production, positive or negative, cannot be fully reflected in the allocation LP (due to an LP convergence problem or underutilization of a key conversion unit), the difference between the allocated production of the grade and the initial DOP of the participant (Screen 1) is reflected as forecaster change for that participant.

The program calculates the rest of the data, such as allocation of balancing grades, in a manner similar to that for the preliminary allocation.

We see that, in the final allocation, only the allocation of balancing grades and refinery losses change from the preliminary allocation. The rest of the report is identical to the preliminary allocation report.

PRODUCT ALLOCATION PROBLEMS

DUMPING OF KEROSENE INTO DIESEL

Due to constraints on the storage capacity or available ullage to sustain the refinery production rate of middle distillates, the refinery is some-

Table 16-21
Retrospective DOP (bbl)

PRODUCT (1)	AOC DOP (2)	AOC FORECASTER (3)	AOC FC COMPENS. (4)	AOC CRUDE (5)	AOC RETRO DOP (6)	BOC DOP (7)	BOC FORECASTER (8)	BOC FC COMPENS. (9)	BOC CRUDE (10)	BOC RETRO DOP (11)	REFINERY RETRO. DOP (12)	ACTUAL PRODUCTION (13)	DELTA (14)
1138B	50396	0			50396	35242	0			35242	85638	85638	0
1149B	13451	0			13451	12117	0			12117	25568	25568	0
150	19067		−88	0	18979	0		−18	0	−18	18961	19553	592
201	204504		973	0	205477	50906		189	0	51095	256572	0	−256572
210	622971		916	1	623888	151556		180	0	151736	775625	1029704	254079
220	0	0			0	570	5			575	575	575	0
383	17700	−14			17686	0	0			0	17686	17686	0
390	88050	−11			88039	0	0			0	88039	88039	0
395	87660	11			87671	16920	6			16926	104597	104597	0
397	215748		1474	0	217222	58918		301	0	59219	276441	289195	12754
397E	0	0			0	3390	−6			3384	3384	3384	0
398	67500	8			67508	0	0			0	67508	67508	0
411	90	10			100	0	0			0	100	100	0
419	220290	0			220290	0	−12			−12	220278	220278	0
434	310170	−262			309908	0	0			0	309908	309908	0
440	376971		20924	1	397896	144019		325	0	144344	542240	944758	402518
711	0	0			0	−28380	6			−28374	−28374	−28374	0
725	100740	12			100752	0	0			0	100752	100752	0
800	378180	−11			378169	0	0			0	378169	378169	0
876	0	−130601			−130601	0	0			0	−130601	−130601	0
876ZP	232110	−12			232098	0	0			0	232098	232098	0
888	1943011		114262	2	2057275	450586		632	0	451218	2508492	2002599	−505893
892	0	−38			−38	0	−28			−28	−66	−66	0
928	403650	2			403652	0	0			0	403652	403652	0
961	985664		1448	2	987114	242584		803	0	243387	1230502	1345430	114928
PBFUEL	0	−657			−657	0	−644			−644	−1301	−1301	0
PCTTR	−7290	−4			−7294	−1440	3			−1437	−8731	−8731	0

Table 16-21
Continued

PRODUCT (1)	AOC DOP (2)	AOC FORECASTER (3)	AOC FC COMPENS. (4)	AOC CRUDE (5)	AOC RETRO DOP (6)	BOC DOP (7)	BOC FORECASTER (8)	BOC FC COMPENS. (9)	BOC CRUDE (10)	BOC RETRO DOP (11)	REFINERY RETRO. DOP (12)	ACTUAL PRODUCTION (13)	DELTA (14)
PDDSL	0	0			0	1470	0			1470	1470	1470	0
PDSL	0	0			0	33570	−6			33564	33564	33564	0
PFCOO	−17790	5			−17785	−7980	1			−7979	−25764	−25764	0
PFDISO	−189000	7			−188993	−7530	−10			−7540	−196533	−196533	0
PKERO	−19637	0			−19637	0	0			0	−19637	−19637	0
PLCGAS	−18227	0			−18227	0	0			0	−18227	−18227	0
PLLCN	2237	4			2241	420	13			433	2674	2674	0
PLTISO	10800	0			10800	0	0			0	10800	10800	0
PMEISO	29610	−15			29595	5700	14			5714	35309	35309	0
PMIDSL	−20280	11			−20269	−6420	1			−6419	−26688	−26688	0
PMSR	15950	0			15950	3079	0			3079	19029	19029	0
PPOLY	0	0			0	3660	−8			3652	3652	3652	0
PP90R	0	0			0	1350	13			1363	1363	1363	0
PP95R	30900	−7			30893	5970	−6			5964	36857	36857	0
PRESID	660	5			665	120	8			128	793	793	0
PSKERO	−13	13			0	13470	3			13473	13473	13473	0
PSLOPD	0	−4045			−4045	0	−816			−816	−4861	−4861	0
PSLOPO	0	−3874			−3874	0	−845			−845	−4719	−4719	0
PSMCN	7860	−3			7857	1530	0			1530	9387	9387	0
PSWMSR	7467	0			7467	1442	0			1442	8909	8909	0
PUFCHG	−3107	8			−3099	9	−9			0	−3099	−3099	0
PWCN	0	−451			−451	0	−96			−96	−547	−547	0
LOSSES	47457			0.102	47457.102	13152				13152	60609.1	38204	−22405
CRUDE	6215520	6			6215526	1200000	0			1200000	7415526	7415526	0
COL TOTAL	6168063	−139909	139909	6	6168069	1186848	−2413	2413	0	1186848	7354917	7377323	0

NOTES:
COLUMN 3 = FORECASTER CHANGE AOC
COLUMN 4 = FORECASTER COMPENSATING CHANGES TO BALANCING GRADES, AOC.
COLUMN 5 = CHANGES TO BALANCING GRADES DUE TO CRUDE CHANGES.

times forced to dump kerosene into diesel. Such situations can arise from failure to lift its production due to depressed kerosene sales by one or both participants. The effect of dumping of kerosene by one participant can affect the allocation of balancing grades of both the participants, and it is important that the effect of kerosene dumping be correctly modeled in the allocation and the effect confined only to the participant that caused the problem.

EXAMPLE 16-7

During the month, a refinery's kerosene inventory was becoming critical and the refinery had two options to contain the kerosene inventory: cut the crude rate of the refinery or dump kerosene to diesel.

As the problem was caused by participant BOC's nonlifting of its kerosene inventory and lack of ullage in its allocated tankage capacity, it was decided that BOC's kerosene be dumped into diesel until the end of the month.

Estimation of amount of kerosene dumped into diesel was complicated because part of this dumping was achieved by undercutting kerosene on the crude distillation unit itself. The amount of the kerosene to be dumped into diesel was estimated by determining the pour point of diesel during the period of dumping.

Table 16-22
Retrospective DOP in Terms of Balancing Grades

BALANCING GRADES	AOC		BOC		
	bbl	%	bbl	%	TOTAL, bbl
150	11460	100.00	0	0.00	11460
201	47147	98.20	864	1.80	48011
210	1095322	82.80	227578	17.20	1322900
397	416254	83.76	80716	16.24	496970
440	1016977	83.18	205589	16.82	1222566
888	1997330	84.17	375739	15.83	2373069
961	1592931	84.16	299835	15.84	1892766
LOSS	38105	79.74	9679	20.26	47784
TOTAL	6215526	83.82	1200000	16.18	7415526

Table 16-23
Allocation of Balancing-Grade Deltas (bbl)

							FIRST STEP		REVERSE ALLOCATION		SUM OF TOTAL ALLOCATION DELTAS	
PRODUCTS	TOTAL DELTA (1)	AOC IN TERMS OF BG (2)	BOC IN TERMS OF BG (3)	TOTAL (4)	AOC, % (5)	BOC, % (6)	AOC DELTA (7)	BOC DELTA (8)	AOC (9)	BOC (10)	AOC (11)	BOC (12)
I-150	592	3956	−795	3161	1.2514	−0.2514	740	−149	8	−8	748	−157
I-201	−256572	264599	51629	316228	0.8367	0.1633	−214683	−41889	3471	−3472	−211211	−45361
I-210	254079	817342	158727	976069	0.8374	0.1626	212761	41318	3438	−3438	216199	37880
I-397	12754	431747	90158	521905	0.8273	0.1727	10551	2203	173	−173	10723	2031
I-440	402518	664775	160411	825186	0.8056	0.1944	324271	78247	5446	−5446	329717	72801
I-888	−505893	2550864	431730	2982594	0.8553	0.1447	−432665	−73228	6845	−6845	−425821	−80073
I-961	114928	1434786	294988	1729774	0.8295	0.1705	95329	19599	1555	−1555	96884	18044
LOSS	−22405	47457	13152	60609	0.7830	0.2170	−17543	−4862	303	−303	−17240	−5165
TOTAL	1						−21239	21240	21239	−21240	0	0

NOTES:
COLUMN 1 BALANCING GRADES DELTAS BETWEEN ACTUAL PRODUCTION AND COMBINED RETROSPECTIVE DOP.
COLUMNS 2 AND 3 RETRO DOP OF AOC AND BOC EXPRESSED IN TERMS OF BALANCING GRADES.
COLUMNS 7 AND 8 TOTAL DELTA IN COLUMN 1 SPLIT IN THE RATIO OF RETRO DOP EXPRESSED IN TERMS OF BALANCING GRADES.
AVCTD THE ABSOLUTE VALUE OF COMBINED TOTAL DELTAS IN COLUMN 1, = 1,569,742.
COLUMNS 9 AND 10 REVERSE ALLOCATION OF DELTAS TO MAKE THE SUM OF AOC AND BOC DELTAS INDIVIDUALLY EQUAL TO ZERO.

Table 16-24
Final Product Allocation (bbl)

PRODUCT	AOC RETRO	BOC RETRO	AOC DELTA	BOC DELTA	AOC ALLOCATION	BOC ALLOCATION	ACTUAL PRODUCTION
1138B	50396	35242			50396	35242	85638
1149B	13451	12117			13451	12117	25568
150	18979	−18	748	−157	19727	−174	19553
201	205477	51095	−211211	−45361	−5734	5734	0
210	623888	151736	216199	37880	840087	189617	1029704
220	0	575			0	575	575
383	17686	0			17686	0	17686
390	88039	0			88039	0	88039
395	87671	16926			87671	16926	104597
397	217222	59219	10723	2031	227945	61250	289195
397E	0	3384			0	3384	3384
398	67508	0			67508	0	67508
411	100	0			100	0	100
419	220290	−12			220290	−12	220278
434	309908	0			309908	0	309908
440	397896	144344	329717	72801	727613	217145	944758
711	0	−28374			0	−28374	−28374
725	100752	0			100752	0	100752
800	378169	0			378169	0	378169
876	−130601	0			−130601	0	−130601
876ZP	232098	0			232098	0	232098
888	2057275	451218	−425821	−80073	1631454	371145	2002599
892	−38	−28			−38	−28	−66
928	403652	0			403652	0	403652
961	987114	243387	96884	18044	1083998	261432	1345430
PBFUEL	−657	−644			−657	−644	−1301
PCTTR	−7294	−1437			−7294	−1437	−8731
PDDSL	0	1470			0	1470	1470
PDSL	0	33564			0	33564	33564
PFCOO	−17785	−7979			−17785	−7979	−25764
PFDISO	−188993	−7540			−188993	−7540	−196533
PKERO	−19637	0			−19637	0	−19637
PLCGAS	−18227	0			−18227	0	−18227
PLLCN	2241	433			2241	433	2674
PLTISO	10800	0			10800	0	10800
PMEISO	29595	5714			29595	5714	35309
PMIDSL	−20269	−6419			−20269	−6419	−26688
PMSR	15950	3079			15950	3079	19029
PPOLY	0	3652			0	3652	3652
PP90R	0	1363			0	1363	1363
PP95R	30893	5964			30893	5964	36857
PRESID	665	128			665	128	793
PSKERO	0	13473			0	13473	13473
PSLOPD	−4045	−816			−4045	−816	−4861
PSLOPO	−3874	−845			−3874	−845	−4719
PSMCN	7857	1530			7857	1530	9387
PSWMSR	7467	1442			7467	1442	8909
PUFCHG	−3099	0			−3099	0	−3099
PWCN	−451	−96			−451	−96	−547
LOSSES	47457	13152	−17240	−5165	30217	7987	38204
COL. TOTAL					6215526	1200000	7415527
CRUDE	6215526	1200000			6215526	1200000	7415526

First, at the month's end, diesel tank 701 contained 120 mb I-885 inventory with a pour point of −9 °C. The volume of kerosene that could be backed out to bring the diesel pour point to its normal value of −3°C is estimated, as follows, at 33 mb.

PRODUCT GRADE	POUR POINT, °C	POUR POINT, °F	POUR BLEND INDEX (PI)	BLEND VOLUME, mb	PI*VOL
I-885	−3.0	26.6	387	87.0	33669.0
I-440	−45	−49	47.1	33.0	1554.3
T-701	−9	15.8	295.2	120.0	35424.0
BLEND					

During the same period, a total of 118 mb of diesel (I-885) was blended and shipped with a −6°C pour point. The volume of kerosene that could be backed out to bring pour point of diesel to −3° was estimated at 17 mb, as follows:

PRODUCT GRADE	POUR POINT, °C	POUR POINT, °F	POUR BLEND INDEX (PI)	BLEND VOLUME, mb	PI*VOL
I-885	−3.0	26.6	387	101.0	39087.0
I-440	−45	−49	47.1	17.0	800.7
BLEND	−6	21.2	336.3	17.0	39683.0

Thus, the total downgrading of kerosene to diesel equals (33 + 17) or 50 mb.

Suppose that the downgrading of kerosene to diesel was caused by a lack of ullage in participant BOC kerosene tankage. This is reflected in the allocation as follows:

1. The allocation program is first run with correct accounting closing inventories of I-440 (KERO) and I-888 (DIESEL).
2. The allocation program is next run with closing inventory of I-440 increased by 50 mb and I-888 decreased by 50 mb.

The difference in the allocated production for the two cases is recorded as follows. This table shows the effect of dumping 50,000 bbl kerosene into diesel on the allocated production of the participants.

PRODUCT GRADE	AOC, bbl	BOC, bbl	TOTAL, bbl
I-150	177	−177	0
I-201	50	−50	0
I-210	394	−394	0
I-397	133	−133	0
I-440	−45063	−4937	−50000
I-888	43226	6774	50000
I-961	951	−951	0
TOTAL	−132	132	0

Participant AOC was originally allocated 45,063 bbl of 50,000 bbl kerosene downgrading (instead of 0) and BOC only −4937 bbl instead of 50,000 bbl. Also, there is net transfer of volume (132 bbl) from AOC to BOC.

To correct this, the following hand adjustments were made to the allocated production of the participants.

PRODUCT GRADE	AOC, bbl	BOC, bbl	TOTAL, bbl
I-150	−177	177	0
I-201	−50	50	0
I-210	−394	394	0
I-397	−133	133	0
I-440	45063	−45063	0
I-888	−43226	43226	0
I-961	−951	951	0
TOTAL	132	−132	0

The allocation delta, for the two cases after adjustment, became

PRODUCT GRADE	AOC, bbl	BOC, bbl	TOTAL, bbl
I-150	0	0	0
I-201	0	0	0
I-210	0	0	0
I-397	0	0	0
I-440	0	−50000	−50000
I-888	0	50000	50000
I-961	0	0	0
TOTAL	0	0	0

Screen 1
Initial DOPs and Forecaster Changes (Preliminary Allocation, November 1999)

GRADE	PRODUCT DESCRIPTION	INITIAL DOP BOC BARRELS	INITIAL DOP AOC BARRELS	TOTAL BARRELS	ACTUAL PROD BARRELS	TOTAL DELTA BARRELS	FORECASTER CHANGES BOC DELTA BARRELS	FORECASTER CHANGES AOC DELTA BARRELS
I-1138	Asphalt 60/70	0	36000	36000	37467	1467	0	1467
I-1149	Asphalt 40/50	0	0	0	880	880	352	528
I-11xx	Asphalt	0	0	0	0	0	0	0
I-11xx	Asphalt	0	0	0	0	0	0	0
I-150	LPG	0	24000	24000	23160	−840	0	0
I-151	LPG	0	0	0	0	0	0	0
I-1xx	LPG	0	0	0	0	0	0	0
I-1xx	LPG	0	0	0	0	0	0	0
I-201	LSR Naphtha	0	0	0	0	0	0	0
I-210	WSR Naphtha	362670	671320	1033990	800332	−233658	0	0
I-220	LSR Naphtha	144000	0	144000	198553	54553	0	0
I-2xx	Naphtha	0	0	0	0	0	0	0
I-2xx	Naphtha	0	0	0	0	0	0	0
I-2xx	Naphtha	0	0	0	0	0	0	0
I-2xx	Naphtha	0	0	0	0	0	0	0
I-253	Whole Cat Naphtha	0	0	0	0	0	0	0
I-383	Mogas (83 RON, 0.84 Pb)	0	0	0	−8268	−8268	−3147	−5121
I-387	Mogas (87 RON, 0.84 Pb)	0	0	0	29039	29039	11616	17423
I-387R	Mogas (87 RON, 0.40 Pb)	0	0	0	0	0	0	0
I-387S		0	0	0	0	0	0	0
I-390E		0	0	0	0	0	0	0
I-390J	Mogas	0	105000	105000	140488	35488	0	35488
I-390R		0	0	0	0	0	0	0
I-390S	Mogas (0.40 Pb, 8 RVP)	0	0	0	0	0	0	0
I-390X		0	0	0	0	0	0	0
I-390Z	0.4 Pb, 9 RVP	207990	0	207990	209046	1056	1056	0

Screen 1
Continued

GRADE	PRODUCT DESCRIPTION	INITIAL DOP			FORECASTER CHANGES			
		BOC BARRELS	AOC BARRELS	TOTAL BARRELS	ACTUAL PROD BARRELS	TOTAL DELTA BARRELS	BOC DELTA BARRELS	AOC DELTA BARRELS
I-393	Mogas (93 RON, 0.84 Pb)	0	0	0	0	0	0	0
I-393S	Mogas (93 RON, 0.40 Pb)	0	0	0	0	0	0	0
I-393X	Mogas (93 RON, 0.60 Pb)	0	0	0	0	0	0	0
I-395L	Mogas (95 RON, 0.84 Pb)	0	0	0	– 72896	– 72896	0	– 72896
I-395LL	Mogas (95 RON, 0.15 Pb)	0	0	0	31462	31462	330	31132
I-395M	Mogas (95 RON, 0.40 Pb)	0	68400	68400	0	– 68400	0	0
I-395S	Mogas	0	0	0	0	0	0	0
I-397		0	0	0	0	0	0	0
I-397C	Mogas (97 RON, 0.6 Pb)	0	0	0	0	0	0	0
I-397E	Mogas (97 RON, 0.84 Pb)	0	0	0	0	0	0	0
I-397LL	Mogas (97 RON, 0.15 Pb)	27000	42000	69000	33551	– 35449	– 6950	– 28499
I-397R	Mogas (97 RON, 0.4 Pb)	0	0	0	0	0	0	0
I-397S	Mogas (0.4 Pb, 9 RVP)	0	0	0	0	0	0	0
I-3975		0	0	0	0	0	0	0
I-398E	Mogas (98 RON, 0.84 Pb)	0	0	0	249558	249558	0	249558
I-398M	Mogas	0	57900	57900	146705	88805	0	88805
I-398S	Mogas (0.4 Pb, 10 RVP)	0	99990	99990	0	– 99990	0	– 99990
I-398X	Mogas (98 RON, 0.60 Pb)	0	0	0	0	0	0	0
I-3xx	Mogas	0	0	0	0	0	0	0
I-3xx	Mogas	0	0	0	0	0	0	0
I-3xx	Mogas	0	0	0	0	0	0	0
I-3xx	Mogas	0	0	0	0	0	0	0
I-3xx	Mogas	0	0	0	0	0	0	0
I-3xx	Mogas	0	0	0	0	0	0	0

I-3xx	Mogas	0	0	0	0	0	0	0
I-3xx	Mogas	0	0	0	0	0	0	0
I-3xx	Mogas	0	0	0	0	0	0	0
I-3xx	Mogas	0	0	0	0	0	0	0
I-400	Special Kerosene	0	0	0	0	0	0	0
I-411	Standard White Kerosene	0	9000	9000	0	– 9000	0	– 9000
I-419	Dual Purpose Kerosene	0	0	0	0	0	0	0
I-434	Jet Fuel JP-4	0	0	0	– 6	– 6	0	– 6
I-440	Jet A-1 Fuel	576000	855000	1431000	1476919	45919	0	0
I-440IM		0	0	0	0	0	0	0
I-4xx	Kerosene	0	0	0	0	0	0	0
I-4xx	Kerosene	0	0	0	0	0	0	0
I-4xx	Kerosene	0	0	0	0	0	0	0
I-4xx	Kerosene	0	0	0	0	0	0	0
I-4xx	Kerosene	0	0	0	0	0	0	0
I-711	Lube Distillate	0	0	0	0	0	0	0
I-725	FCCU Charge Stock	0	0	0	0	0	0	0
I-726	L.S. FCCU Charge Stock	0	0	0	0	0	0	0
I-730	Arabian Crude HVGO	0	0	0	0	0	0	0
I-7xx	Heavy Gasoil	0	0	0	0	0	0	0
I-7xx	Heavy Gasoil	0	0	0	0	0	0	0
I-7xx	Heavy Gasoil	0	0	0	0	0	0	0
I-800	High Speed Diesel	0	0	0	0	0	0	0
I-80001	Zero Pour HSD	210000	681000	891000	928278	37278	8786	28492
I-875	Diesel (47 CI, -6 Pour)	0	0	0	0	0	0	0
I-876	Diesel (46 CI, -6 Pour)	0	204000	204000	320348	116348	0	116348
I-876ZP	Diesel (46 CI, -18 Pour)	0	0	0	0	0	0	0
I-884	Diesel (47 CI, 0.5 Sulf)	0	0	0	0	0	0	0
I-885	Diesel (50 CI, 0.5 Sulf)	24990	66990	91980	42505	– 49475	– 19755	– 29720
I-888	Diesel (50 CI, -3 Pour)	599130	471930	1071060	1095206	24146	0	0
I-88852		0	0	0	0	0	0	0
I-8xx	Diesel	0	0	0	0	0	0	0
I-8xx	Diesel	0	0	0	0	0	0	0
I-8xx	Diesel	0	0	0	0	0	0	0
I-8xx	Diesel	0	0	0	0	0	0	0
I-8xx	Diesel	0	0	0	0	0	0	0

Screen 1
Continued

GRADE	PRODUCT DESCRIPTION	INITIAL DOP			FORECASTER CHANGES			
		BOC BARRELS	AOC BARRELS	TOTAL BARRELS	ACTUAL PROD BARRELS	TOTAL DELTA BARRELS	BOC DELTA BARRELS	AOC DELTA BARRELS
I-8xx	Diesel	0	0	0	0	0	0	0
I-8xx	Diesel	0	0	0	0	0	0	0
I-8xx	Diesel	0	0	0	0	0	0	0
I-892	Marine Diesel Oil	0	0	0	−185	−185	−104	−81
I-928	Fuel Oil (2.8 S, 180 cSt)	0	0	0	0	0	0	0
I-933	Fuel Oil (3.5 S, 80 cSt)	0	0	0	0	0	0	0
I-955	Fuel Oil (3.5 S, 75 cSt)	0	0	0	0	0	0	0
I-957	Fuel Oil (3.5 S, 225 SUS)	0	0	0	0	0	0	0
I-960	Fuel Oil (3.5 S, 120 cSt)	0	0	0	0	0	0	0
I-96001		0	0	0	0	0	0	0
I-9601		0	0	0	0	0	0	0
I-9602		0	0	0	0	0	0	0
I-961	Fuel Oil (3.5 S, 180 cSt)	724230	582030	1306260	1218169	−88091	0	0
I-96107	Fuel Oil (4.0 S, 180 cSt)	0	420990	420990	291431	−129559	0	−129559
I-962	Fuel Oil (3.5 S, 280 cSt)	0	0	0	0	0	0	0
I-96201		0	0	0	0	0	0	0
I-96202		0	0	0	0	0	0	0
I-971	Fuel Oil (4.0 S, 380 cSt)	0	0	0	0	0	0	0
I-97103		0	387990	387990	381455	−6535	0	−6535
I-9xx	Fuel Oil	0	0	0	0	0	0	0
I-9xx	Fuel Oil	0	0	0	0	0	0	0
I-9xx	Fuel Oil	0	0	0	0	0	0	0
I-9xx	Fuel Oil	0	0	0	0	0	0	0
I-9xx	Fuel Oil	0	0	0	0	0	0	0
I-9xx	Fuel Oil	0	0	0	0	0	0	0

I-9xx	Fuel Oil	0	0	0	0	0	0	0
I-9xx	Fuel Oil	0	0	0	0	0	0	0
I-9xx	Fuel Oil	0	0	0	0	0	0	0
I-9xx	Fuel Oil	0	0	0	0	0	0	0
Pxxxx	Intermediate stock	0	0	0	0	0	0	0
Pxxxx	Intermediate stock	0	0	0	0	0	0	0
Pxxxx	Intermediate stock	0	0	0	0	0	0	0
Pxxxx	Intermediate stock	0	0	0	0	0	0	0
Pxxxx	Intermediate stock	0	0	0	0	0	0	0
Pxxxx	Intermediate stock	0	0	0	0	0	0	0
PBFUEL	Burner Fuel Oil	0	0	0	– 846	– 846	– 255	– 591
PBHDSL	Diesel Oil ex Bombay High	0	0	0	0	0	0	0
PCTISO	Isomate Cutter	0	0	0	0	0	0	0
PCTTR	Cutter Stock	0	6000	6000	2868	– 3132	0	– 3132
PDSDSL	Desulfurized Diesel	0	9000	9000	80009	71009	28404	42605
PDSL	Diesel	0	14010	14010	88946	74936	29974	44962
PFCCF	HVGO to FCCU	0	0	0	0	0	0	0
PFCOO	Hvy Isomate to FCCU	– 30990	– 27990	– 58980	– 37350	21630	11365	10265
PFDISO	HVGO to 2HDU	– 51000	– 120000	– 171000	– 161672	9328	2782	6546
PGABS	Gasoline	0	0	0	0	0	0	0
PGPDSL	Diesel ex Gippsland	0	0	0	0	0	0	0
PISIMP	Isomate Import	0	0	0	0	0	0	0
PKERIM	Kerosene Import	0	0	0	0	0	0	0
PKERO	Kerosene	0	0	0	– 5572	– 5572	– 2391	– 3181
PLANTS	Plants & Lines Contents	0	0	0	0	0	0	0
PLCGAS	Lt Cat Naphtha	33990	69000	102990	115403	12413	4097	8316
PLLCN	Lt Cat Naphtha	0	0	0	1543	1543	617	926
PLSRBS	I-220 LSR	0	0	0	0	0	0	0
PLTISO	Lt. Isomate	2010	0	2010	– 25866	– 27876	– 7758	– 20118
PMEISO	Med. Isomate Cutter	0	0	0	– 9766	– 9766	– 4470	– 5296
PMIDSL	CDU 4A M/I Diesel	6990	9990	16980	– 3349	– 20329	– 13119	– 7210
PMSR	Sour MSR Naphtha	0	0	0	– 3136	– 3136	– 1239	– 1897
POMRSD	Oman Resid Import	0	0	0	0	0	0	0
PPOLY	Polymer Gasoline	3990	6990	10980	9393	– 1587	– 577	– 1010

Screen 1
Continued

GRADE	PRODUCT DESCRIPTION	BOC BARRELS	INITIAL DOP AOC BARRELS	TOTAL BARRELS	FORECASTER CHANGES: ACTUAL PROD BARRELS	TOTAL DELTA BARRELS	BOC DELTA BARRELS	AOC DELTA BARRELS
PP90R	90 RON Reformate	0	0	0	0	0	0	0
PP93R	93 RON Reformate	−990	−2010	−3000	−5867	−2867	−946	−1921
PP95R	95 RON Reformate	0	0	0	0	0	0	0
PP96R	96 RON Reformate	9990	18000	27990	24257	−3733	−1332	−2401
PRESID	Residuum	363960	15870	379830	340735	−39095	−37462	−1633
PRESMI	M/I Dsls in 6VDU Feed	0	0	0	0	0	0	0
PRES75	I-725 in 6VDU Feed	0	0	0	0	0	0	0
PSKERO	Sweet Kerosene	0	0	0	658	658	263	395
PSLOPD	Slop Distillate	0	0	0	−4714	−4714	−1860	−2854
PSLOPO	Slop Oil	0	0	0	−5518	−5518	−1901	−3617
PSLOPT	Slop ex Marine Terminal	0	0	0	0	0	0	0
PSMCN	Sweet Med. Cat Naphtha	−8010	−15000	−23010	−29517	−6507	−2265	−4242
PSWMSR	Sweet MSR	0	−6000	−6000	−100	5900	0	5900
PUFCHG	Unifiner Feed	−3990	0	−3990	−9419	−5429	−1859	−3570
PUJFBS	Sour Jet B.S.	0	0	0	0	0	0	0
PWCN	Whole Cat Naphtha	0	0	0	2999	2999	1200	1799
P440BS	Jet B.S.	0	0	0	0	0	0	0
TOTAL	Finished products	2876010	4783540		7573197	−86353	−7816	187834
TOTAL	Process stocks	325950	−22140		364119	60309	1268	59041
TOTAL	Finished products + Process stocks	3201960	4761400		7937316	−26044	−6548	246875
CRUDE	Crude oil run	3225000	4806000	8031000	8037551	6551	−9980	16531
LOSSES		23040	44600		100235			

Screen 2
Participant BOC Opening Inventory and Shipping Data

PRODUCT	OPENING INVENTORY bbls	ADJUST bbls	ALLOCATION bbls	TOTAL AVAILABLE bbls	LIFTED bbls	LOCAL SALES bbls	CONSUMED bbls	CLOSING INVENTORY bbls
I-1138	46862		0	46862	22624			24238
I-1149	251		352	603	5301			−4698
I-1138D	0		0	0				0
I-11xx	0		0	0				0
I-150	122442		−543	121899				121899
I-151	0		0	0				0
I-1xx	0		0	0				0
I-1xx	0		0	0				0
I-201	0		0	0				0
I-210	261907		273846	535753	240091			295662
I-220	124639	155202	192072	471913	490190			−18277
I-2xx	0		0	0				0
I-2xx	0		0	0				0
I-2xx	0		0	0				0
I-2xx	0		0	0				0
I-253	0		0	0				0
I-383	8718		−3147	5571				5571
I-387	0		11616	11616				11616
I-387R	0		0	0				0
I-387S	0		0	0				0
I-390E	0		0	0				0
I-390J	0		0	0				0
I-390R	0		0	0				0
I-390S	0		0	0				0
I-390X	0		0	0				0
I-390Z	17672		209046	226718	208480			18238
I-393	0		0	0				0
I-393S	0		0	0				0
I-393X	0		0	0				0
I-395L	0		0	0				0
I-395LL	0		330	330				330
I-395M	145602		48572	194174				194174
I-395S	0		0	0				0
I-397	0		0	0				0
I-397C	0		0	0				0
I-397E	0		0	0				0
I-397LL	6127		20050	26177	26177			0
I-397R	0		0	0				0
I-397S	0		0	0				0
I-3975	0		0	0				0
I-398E	0		0	0				0
I-398M	0		0	0				0
I-398S	0		0	0				0
I-398X	0		0	0				0
I-3xx	0		0	0				0
I-3xx	0		0	0				0
I-3xx	0		0	0				0
I-3xx	0		0	0				0
I-3xx	0		0	0				0
I-3xx	0		0	0				0
I-3xx	0		0	0				0
I-3xx	0		0	0				0
I-3xx	0		0	0				0

Screen 2
Continued

PRODUCT	OPENING INVENTORY bbls	ADJUST bbls	ALLOCATION bbls	TOTAL AVAILABLE bbls	LIFTED bbls	LOCAL SALES bbls	CONSUMED bbls	CLOSING INVENTORY bbls
I-3xx	0		0	0				0
I-400	0		0	0				0
I-411	0		0	0				0
I-419	0	237296	0	237296	237296			0
I-434	0		0	0				0
I-440	206732	−237296	591669	561105	83346			477759
I-440IM	0		0	0				0
I-4xx	0		0	0				0
I-4xx	0		0	0				0
I-4xx	0		0	0				0
I-4xx	0		0	0				0
I-4xx	0		0	0				0
I-711	0		0	0				0
I-725	0		0	0				0
I-726	0		0	0				0
I-730	0		0	0				0
I-7xx	0		0	0				0
I-7xx	0		0	0				0
I-7xx	0		0	0				0
I-800	0		0	0				0
I-80001	0		218786	218786	210332			8454
I-875	0		0	0				0
I-876	2847		0	2847				2847
I-876ZP	0		0	0				0
I-884	0		0	0				0
I-885	20332		5235	25567	25567			0
I-888	136994		650516	787510	698346	16932	314	71918
I-88852	0		0	0				0
I-8xx	0		0	0				0
I-8xx	0		0	0				0
I-8xx	0		0	0				0
I-8xx	0		0	0				0
I-8xx	0		0	0				0
I-8xx	0		0	0				0
I-8xx	0		0	0				0
I-8xx	0		0	0				0
I-892	15687		−104	15583				15583
I-928	0		0	0				0
I-933	0		0	0				0
I-955	0		0	0				0
I-957	0		0	0				0
I-960	0		0	0				0
I-96001	0		0	0				0
I-9601	0		0	0				0
I-9602	0		0	0				0
I-961	314131		636558	950689	1089965	1172		−140448
I-96107	0		0	0				0
I-962	0		0	0				0
I-96201	0		0	0				0
I-96202	0		0	0				0
I-971	0		0	0				0
I-97103	0		0	0				0
I-9xx	0		0	0				0

I-9xx	0		0	0				0
I-9xx	0		0	0				0
I-9xx	0		0	0				0
I-9xx	0		0	0				0
I-9xx	0		0	0				0
I-9xx	0		0	0				0
I-9xx	0		0	0				0
I-9xx	0		0	0				0
I-9xx	0		0	0				0
Pxxxx	0		0	0				0
Pxxxx	0		0	0				0
Pxxxx	0		0	0				0
Pxxxx	0		0	0				0
Pxxxx	0		0	0				0
Pxxxx	0		0	0				0
PBFUEL	3105		−255	2850				2850
PBHDSL	0		0	0				0
PCTISO	0		0	0				0
PCTTR	20931		0	20931				20931
PDSDSL	20447		28404	48851				48851
PDSL	30211		29974	60185				60185
PFCCF	0		0	0				0
PFCOO	69478		−19625	49853				49853
PFDISO	75257		−48218	27039				27039
PGABS	0		0	0				0
PGPDSL	0		0	0				0
PISIMP	0		0	0				0
PKERIM	0		0	0				0
PKERO	64261		−2391	61870				61870
PLANTS	27614		0	27614				27614
PLCGAS	45273		38087	83360				83360
PLLCN	1679		617	2296				2296
PLSRBS	0		0	0				0
PLTISO	21169		−5748	15421				15421
PMEISO	22667		−4470	18197				18197
PMIDSL	8407		−6129	2278				2278
PMSR	8780		−1239	7541				7541
POMRSD	0		0	0				0
PPOLY	13885		3413	17298				17298
PP90R	0		0	0				0
PP93R	45566		−1936	43630				43630
PP95R	0		0	0				0
PP96R	45550		8658	54208				54208
PRESID	5730		326498	332228				332228
PRESMI	0		0	0				0
PRES75	0		0	0				0
PSKERO	13398		263	13661				13661
PSLOPD	13218		−1860	11358				11358
PSLOPO	15147		−1901	13246				13246
PSLOPT	0		0	0				0
PSMCN	27559		−10275	17284				17284
PSWMSR	14008		0	14008				14008
PUFCHG	17971		−5849	12122				12122
PUJFBS	0		0	0				0
PWCN	858		1200	2058				2058
P440BS	0		0	0				0
TOT. PROCESS	632169	0	327218	959387	0	0	0	959387
TOT.FINISH PRD.	1430943	155202	2854854.18	4440999	3337715	18104	314	1084866.2
TOTAL	2063112	155202	3182072.18	5400386	3337715	18104	314	2044253.2

Screen 3
Participant AOC Opening Inventory and Shipping Data

PRODUCT	OPENING INVENTORY BARRELS	ADJUST BARRELS	ALLOCATION BARRELS	TOTAL AVAILABLE BARRELS	LIFTED BARRELS	LOCAL SALES BARRELS	CONSUMED BARRELS	CLOSING INVENTORY BARRELS
I-1138	−11689		37467	25778	17641	9491		−1354
I-1149	34130		528	34658	7952			26706
I-1138D	688		0	688				688
I-11xx	0		0	0				0
I-150	−110983		23703	−87280		22361		−109641
I-151	0		0	0				0
I-1xx	0		0	0				0
I-1xx	0		0	0				0
I-201	0		0	0				0
I-210	17406		526486	543892	249762			294130
I-220	107928		6481	114409				114409
I-2xx	0		0	0				0
I-2xx	0		0	0				0
I-2xx	0		0	0				0
I-2xx	0		0	0				0
I-253	0		0	0				0
I-383	14186		−5121	9065				9065
I-387	0		17423	17423				17423
I-387R	0		0	0				0
I-387S	0		0	0				0
I-390E	0		0	0				0
I-390J	32354		140488	172842		115302		57540
I-390R	0		0	0				0
I-390S	0		0	0				0
I-390X	0		0	0				0
I-390Z	0		0	0				0
I-393	0		0	0				0
I-393S	0		0	0				0
I-393X	0		0	0				0
I-395L	72896		−72896	0				0
I-395LL	218044		31132	249176	249176			0
I-395M	−145602		−48572	−194174				−194174
I-395S	0		0	0				0
I-397	0		0	0				0
I-397C	0		0	0				0
I-397E	0		0	0				0
I-397LL	105858		13501	119359	119359			0
I-397R	0		0	0				0
I-397S	0		0	0				0
I-3975	0		0	0				0
I-398E	0		249558	249558	34050			215508
I-398M	31538		146705	178243		68532		109711
I-398S	0		0	0				0
I-398X	0		0	0				0
I-3xx	0		0	0				0
I-3xx	0		0	0				0
I-3xx	0		0	0				0
I-3xx	0		0	0				0
I-3xx	0		0	0				0
I-3xx	0		0	0				0
I-3xx	0		0	0				0
I-3xx	0		0	0				0
I-3xx	0		0	0				0

I-3xx	0		0	0				0
I-400	0		0	0				0
I-411	12		0	12				12
I-419	0	841363	0	841363	841363			0
I-434	50163		−6	50157		8996		41161
I-440	432083	−841363	885250	475970	158164	5326		312480
I-440IM	0		0	0				0
I-4xx	0		0	0				0
I-4xx	0		0	0				0
I-4xx	0		0	0				0
I-4xx	0		0	0				0
I-4xx	0		0	0				0
I-711	0		0	0				0
I-725	0		0	0				0
I-726	0		0	0				0
I-730	0		0	0				0
I-7xx	0		0	0				0
I-7xx	0		0	0				0
I-7xx	0		0	0				0
I-800	0		0	0				0
I-80001	126709		709492	836201	542109			294092
I-875	0		0	0				0
I-876	280974		320348	601322	237453			363869
I-876ZP	0		0	0				0
I-884	0		0	0				0
I-885	92640		37270	129910	129910			0
I-888	51290		444690	495980	349946	67905	470	77659
I-88852	0		0	0				0
I-8xx	0		0	0				0
I-8xx	0		0	0				0
I-8xx	0		0	0				0
I-8xx	0		0	0				0
I-8xx	0		0	0				0
I-8xx	0		0	0				0
I-8xx	0		0	0				0
I-8xx	0		0	0				0
I-892	12293		−81	12212	1561			10651
I-928	0		0	0				0
I-933	0		0	0				0
I-955	0		0	0				0
I-957	0		0	0				0
I-960	0		0	0				0
I-96001	0		0	0				0
I-9601	0		0	0				0
I-9602	0		0	0				0
I-961	399679		581611	981290	395852			585438
I-96107	0		291431	291431	291431			0
I-962	0		0	0				0
I-96201	0		0	0				0
I-96202	0		0	0				0
I-971	0		0	0				0
I-97103	0		381455	381455				381455
I-9xx	0		0	0				0
I-9xx	0		0	0				0
I-9xx	0		0	0				0
I-9xx	0		0	0				0
I-9xx	0		0	0				0
I-9xx	0		0	0				0
I-9xx	0		0	0				0
I-9xx	0		0	0				0

Screen 3
Continued

PRODUCT	OPENING INVENTORY BARRELS	ADJUST BARRELS	ALLOCATION BARRELS	TOTAL AVAILABLE BARRELS	LIFTED BARRELS	LOCAL SALES BARRELS	CONSUMED BARRELS	CLOSING INVENTORY BARRELS
I-9xx	0		0	0				0
I-9xx	0		0	0				0
Pxxxx	0		0	0				0
Pxxxx	0		0	0				0
Pxxxx	0		0	0				0
Pxxxx	0		0	0				0
Pxxxx	0		0	0				0
Pxxxx	0		0	0				0
PBFUEL	7180		−591	6589				6589
PBHDSL	0		0	0				0
PCTISO	0		0	0				0
PCTTR	2712		2868	5580				5580
PDSDSL	44409		51605	96014				96014
PDSL	46110		58972	105082				105082
PFCCF	0		0	0				0
PFCOO	61850		−17725	44125				44125
PFDISO	202441		−113454	88987				88987
PGABS	0		0	0				0
PGPDSL	0		0	0				0
PISIMP	0		0	0				0
PKERIM	0		0	0				0
PKERO	85508		−3181	82327				82327
PLANTS	41421		0	41421				41421
PLCGAS	15093		77316	92409				92409
PLLCN	3234		926	4160				4160
PLSRBS	0		0	0				0
PLTISO	54893		−20118	34775				34775
PMEISO	26858		−5296	21562				21562
PMIDSL	4620		2780	7400				7400
PMSR	13445		−1897	11548				11548
POMRSD	0		0	0				0
PPOLY	7996		5980	13976				13976
PP90R	0		0	0				0
PP93R	41797		−3931	37866				37866
PP95R	0		0	0				0
PP96R	51489		15599	67088				67088
PRESID	87622		14237	101859				101859
PRESMI	0		0	0				0
PRES75	0		0	0				0
PSKERO	25555		395	25950				25950
PSLOPD	20285		−2854	17431				17431
PSLOPO	28826		−3617	25209				25209
PSLOPT	0		0	0				0
PSMCN	48855		−19242	29613				29613
PSWMSR	21738		−100	21638				21638
PUFCHG	34502		−3570	30932				30932
PUJFBS	0		0	0				0
PWCN	1715		1799	3514				3514
P440BS	0		0	0				0
PROCESS	980154	0	36901	1017055	0	0	0	1017055
FINISH PR	1812597	0	4718342.82	6530940	3625729	297913	470	2606827.8
TOTAL	2792751	0	4755243.82	7547995	3625729	297913	470	3623882.8

Screen 4
Refinery Opening and Closing Physical Inventory

PRODUCT	OPENING INVENTORY BARRELS	ADJUST BARRELS	PRODUCED BARRELS	TOT. AVAILABLE BARRELS	LIFTED BARRELS	LOCAL SALES BARRELS	CONSUMED BARRELS	CLOSING INVENTORY BARRELS
I-1138	35173	0	37467	72640	40265	9491	0	22884
I-1149	34381	0	880	35261	13253	0	0	22008
I-1138D	688	0	0	688	0	0	0	688
I-11xx	0	0	0	0	0	0	0	0
I-150	11459	0	23160	34619	0	22361	0	12258
I-151	0	0	0	0	0	0	0	0
I-1xx	0	0	0	0	0	0	0	0
I-1xx	0	0	0	0	0	0	0	0
I-201	0	0	0	0	0	0	0	0
I-210	279313	0	800332	1079645	489853	0	0	589792
I-220	232567	155202	198553	586322	490190	0	0	96132
I-2xx	0	0	0	0	0	0	0	0
I-2xx	0	0	0	0	0	0	0	0
I-2xx	0	0	0	0	0	0	0	0
I-2xx	0	0	0	0	0	0	0	0
I-253	0	0	0	0	0	0	0	0
I-383	22904	0	−8268	14636	0	0	0	14636
I-387	0	0	29039	29039	0	0	0	29039
I-387R	0	0	0	0	0	0	0	0
I-387S	0	0	0	0	0	0	0	0
I-390E	0	0	0	0	0	0	0	0
I-390J	32354	0	140488	172842	0	115302	0	57540
I-390R	0	0	0	0	0	0	0	0
I-390S	0	0	0	0	0	0	0	0
I-390X	0	0	0	0	0	0	0	0
I-390Z	17672	0	209046	226718	208480	0	0	18238
I-393	0	0	0	0	0	0	0	0
I-393S	0	0	0	0	0	0	0	0
I-393X	0	0	0	0	0	0	0	0
I-395L	72896	0	−72896	0	0	0	0	0
I-395LL	218044	0	31462	249506	249176	0	0	330
I-395M	0	0	0	0	0	0	0	0
I-395S	0	0	0	0	0	0	0	0
I-397	0	0	0	0	0	0	0	0
I-397C	0	0	0	0	0	0	0	0
I-397E	0	0	0	0	0	0	0	0
I-397LL	111985	0	33551	145536	145536	0	0	0
I-397R	0	0	0	0	0	0	0	0
I-397S	0	0	0	0	0	0	0	0
I-3975	0	0	0	0	0	0	0	0
I-398E	0	0	249558	249558	34050	0	0	215508
I-398M	31538	0	146705	178243	0	68532	0	109711
I-398S	0	0	0	0	0	0	0	0
I-398X	0	0	0	0	0	0	0	0
I-3xx	0	0	0	0	0	0	0	0
I-3xx	0	0	0	0	0	0	0	0
I-3xx	0	0	0	0	0	0	0	0
I-3xx	0	0	0	0	0	0	0	0
I-3xx	0	0	0	0	0	0	0	0
I-3xx	0	0	0	0	0	0	0	0
I-3xx	0	0	0	0	0	0	0	0
I-3xx	0	0	0	0	0	0	0	0
I-3xx	0	0	0	0	0	0	0	0

Screen 4
Continued

PRODUCT	OPENING INVENTORY BARRELS	ADJUST BARRELS	PRODUCED BARRELS	TOT. AVAILABLE BARRELS	LIFTED BARRELS	LOCAL SALES BARRELS	CONSUMED BARRELS	CLOSING INVENTORY BARRELS
I-3xx	0	0	0	0	0	0	0	0
I-400	0	0	0	0	0	0	0	0
I-411	12	0	0	12	0	0	0	12
I-419	0	1078659	0	1078659	1078659	0	0	0
I-434	50163	0	−6	50157	0	8996	0	41161
I-440	638815	−1078659	1476919	1037075	241510	5326	0	790239
I-440IM	0	0	0	0	0	0	0	0
I-4xx	0	0	0	0	0	0	0	0
I-4xx	0	0	0	0	0	0	0	0
I-4xx	0	0	0	0	0	0	0	0
I-4xx	0	0	0	0	0	0	0	0
I-4xx	0	0	0	0	0	0	0	0
I-711	0	0	0	0	0	0	0	0
I-725	0	0	0	0	0	0	0	0
I-726	0	0	0	0	0	0	0	0
I-730	0	0	0	0	0	0	0	0
I-7xx	0	0	0	0	0	0	0	0
I-7xx	0	0	0	0	0	0	0	0
I-7xx	0	0	0	0	0	0	0	0
I-800	0	0	0	0	0	0	0	0
I-80001	126709	0	928278	1054987	752441	0	0	302546
I-875	0	0	0	0	0	0	0	0
I-876	283821	0	320348	604169	237453	0	0	366716
I-876ZP	0	0	0	0	0	0	0	0
I-884	0	0	0	0	0	0	0	0
I-885	112972	0	42505	155477	155477	0	0	0
I-888	188284	0	1095206	1283490	1048292	84837	784	149577
I-88852	0	0	0	0	0	0	0	0
I-8xx	0	0	0	0	0	0	0	0
I-8xx	0	0	0	0	0	0	0	0
I-8xx	0	0	0	0	0	0	0	0
I-8xx	0	0	0	0	0	0	0	0
I-8xx	0	0	0	0	0	0	0	0
I-8xx	0	0	0	0	0	0	0	0
I-8xx	0	0	0	0	0	0	0	0
I-8xx	0	0	0	0	0	0	0	0
I-892	27980	0	−185	27795	1561	0	0	26234
I-928	0	0	0	0	0	0	0	0
I-933	0	0	0	0	0	0	0	0
I-955	0	0	0	0	0	0	0	0
I-957	0	0	0	0	0	0	0	0
I-960	0	0	0	0	0	0	0	0
I-96001	0	0	0	0	0	0	0	0
I-9601	0	0	0	0	0	0	0	0
I-9602	0	0	0	0	0	0	0	0
I-961	713810	0	1218169	1931979	1485817	1172	0	444990
I-96107	0	0	291431	291431	291431	0	0	0
I-962	0	0	0	0	0	0	0	0
I-96201	0	0	0	0	0	0	0	0
I-96202	0	0	0	0	0	0	0	0
I-971	0	0	0	0	0	0	0	0
I-97103	0	0	381455	381455	0	0	0	381455
I-9xx	0	0	0	0	0	0	0	0

I-9xx	0	0	0	0	0	0	0	0
I-9xx	0	0	0	0	0	0	0	0
I-9xx	0	0	0	0	0	0	0	0
I-9xx	0	0	0	0	0	0	0	0
I-9xx	0	0	0	0	0	0	0	0
I-9xx	0	0	0	0	0	0	0	0
I-9xx	0	0	0	0	0	0	0	0
I-9xx	0	0	0	0	0	0	0	0
I-9xx	0	0	0	0	0	0	0	0
Pxxxx	0	0	0	0	0	0	0	0
Pxxxx	0	0	0	0	0	0	0	0
Pxxxx	0	0	0	0	0	0	0	0
Pxxxx	0	0	0	0	0	0	0	0
Pxxxx	0	0	0	0	0	0	0	0
Pxxxx	0	0	0	0	0	0	0	0
PBFUEL	10285	0	−846	9439	0	0	0	9439
PBHDSL	0	0	0	0	0	0	0	0
PCTISO	0	0	0	0	0	0	0	0
PCTTR	23643	0	2868	26511	0	0	0	26511
PDSDSL	64856	0	80009	144865	0	0	0	144865
PDSL	76321	0	88946	165267	0	0	0	165267
PFCCF	0	0	0	0	0	0	0	0
PFCOO	131328	0	−37350	93978	0	0	0	93978
PFDISO	277698	0	−161672	116026	0	0	0	116026
PGABS	0	0	0	0	0	0	0	0
PGPDSL	0	0	0	0	0	0	0	0
PISIMP	0	0	0	0	0	0	0	0
PKERIM	0	0	0	0	0	0	0	0
PKERO	149769	0	−5572	144197	0	0	0	144197
PLANTS	69035	0	0	69035	0	0	0	69035
PLCGAS	60366	0	115403	175769	0	0	0	175769
PLLCN	4913	0	1543	6456	0	0	0	6456
PLSRBS	0	0	0	0	0	0	0	0
PLTISO	76062	0	−25866	50196	0	0	0	50196
PMEISO	49525	0	−9766	39759	0	0	0	39759
PMIDSL	13027	0	−3349	9678	0	0	0	9678
PMSR	22225	0	−3136	19089	0	0	0	19089
POMRSD	0	0	0	0	0	0	0	0
PPOLY	21881	0	9393	31274	0	0	0	31274
PP90R	0	0	0	0	0	0	0	0
PP93R	87363	0	−5867	81496	0	0	0	81496
PP95R	0	0	0	0	0	0	0	0
PP96R	97039	0	24257	121296	0	0	0	121296
PRESID	93352	0	340735	434087	0	0	0	434087
PRESMI	0	0	0	0	0	0	0	0
PRES75	0	0	0	0	0	0	0	0
PSKERO	38953	0	658	39611	0	0	0	39611
PSLOPD	33503	0	−4714	28789	0	0	0	28789
PSLOPO	43973	0	−5518	38455	0	0	0	38455
PSLOPT	0	0	0	0	0	0	0	0
PSMCN	76414	0	−29517	46897	0	0	0	46897
PSWMSR	35746	0	−100	35646	0	0	0	35646
PUFCHG	52473	0	−9419	43054	0	0	0	43054
PUJFBS	0	0	0	0	0	0	0	0
PWCN	2573	0	2999	5572	0	0	0	5572
P440BS	0	0	0	0	0	0	0	0
PROCESS	1612323	0	364119	1976442	0	0	0	1976442
FINISH PRD	3243540	155202	7573197	10971939	6963444	316017	784	3691694
TOTAL	4855863	155202	7937316	12948381	6963444	316017	784	5668136

Screen 5
Product Equivalencies

GRADE	DESCRIPTION	EQUIVALENCY IN BALANCING GRADES							CHECK
		I-150	I-220	I-210	I-395M	I-440	I-888	I-961	
I-1138	Asphalt 60/70	0.0000	0.0000	0.0000	0.0000	0.0000	−0.7287	1.7287	1.0000
I-1149	Asphalt 40/50	0.0000	0.0000	0.0000	0.0000	0.0000	−0.7287	1.7287	1.0000
I-11xx	Asphalt	0.0000	0.0000	0.0000	0.0000	0.0000	−0.7287	1.7287	1.0000
I-11xx	Asphalt	0.0000	0.0000	0.0000	0.0000	0.0000	−0.7287	1.7287	1.0000
I-150	LPG	1.0000	0.0000	0.0000	0.0000	0.0000	0.0000	0.0000	1.0000
I-151	LPG	1.0000	0.0000	0.0000	0.0000	0.0000	0.0000	0.0000	1.0000
I-1xx	LPG	1.0000	0.0000	0.0000	0.0000	0.0000	0.0000	0.0000	1.0000
I-1xx	LPG	1.0000	0.0000	0.0000	0.0000	0.0000	0.0000	0.0000	1.0000
I-201	LSR Naphtha	0.0000	1.0000	0.0000	0.0000	0.0000	0.0000	0.0000	1.0000
I-210	WSR Naphtha	0.0000	0.0000	1.0000	0.0000	0.0000	0.0000	0.0000	1.0000
I-220	LSR Naphtha	0.0000	1.0000	0.0000	0.0000	0.0000	0.0000	0.0000	1.0000
I-2xx	Naphtha	0.0000	0.0000	0.0000	0.0000	0.0000	0.0000	0.0000	0.0000
I-2xx	Naphtha	0.0000	0.0000	0.0000	0.0000	0.0000	0.0000	0.0000	0.0000
I-2xx	Naphtha	0.0000	0.0000	0.0000	0.0000	0.0000	0.0000	0.0000	0.0000
I-2xx	Naphtha	0.0000	0.0000	0.0000	0.0000	0.0000	0.0000	0.0000	0.0000
I-253	Whole Cat Naphtha	−0.0007	0.0279	0.0000	0.9728	0.0000	0.0000	0.0000	1.0000
I-383	Mogas (83 RON, 0.84 Pb)	−0.0215	0.8519	0.0000	0.1696	0.0000	0.0000	0.0000	1.0000
I-387	Mogas (87 RON, 0.84 Pb)	−0.0152	0.6049	0.0000	0.4103	0.0000	0.0000	0.0000	1.0000
I-387R	Mogas (87 RON, 0.40 Pb)	−0.0106	0.4207	0.0000	0.5899	0.0000	0.0000	0.0000	1.0000
I-387S		−0.0089	0.3528	0.0000	0.6561	0.0000	0.0000	0.0000	1.0000
I-390E		−0.0106	0.4198	0.0000	0.5908	0.0000	0.0000	0.0000	1.0000
I-390J	Mogas	−0.0066	0.2629	0.0000	0.7437	0.0000	0.0000	0.0000	1.0000
I-390R		−0.0078	0.3296	0.0000	0.6782	0.0000	0.0000	0.0000	1.0000
I-390S	Mogas (0.40 Pb, 8 RVP)	−0.0346	0.2619	0.0000	0.7727	0.0000	0.0000	0.0000	1.0000
I-390X		−0.0088	0.3408	0.0000	0.6680	0.0000	0.0000	0.0000	1.0000
I-390Z	0.4 Pb, 9 RVP	−0.0204	0.2624	0.0000	0.7580	0.0000	0.0000	0.0000	1.0000
I-393	Mogas (93 RON, 0.84 Pb)	−0.0059	0.2346	0.0000	0.7713	0.0000	0.0000	0.0000	1.0000
I-393S	Mogas (93 RON, 0.40 Pb)	−0.0027	0.1052	0.0000	0.8975	0.0000	0.0000	0.0000	1.0000
I-393X	Mogas (93 RON, 0.60 Pb)	−0.0043	0.1704	0.0000	0.8339	0.0000	0.0000	0.0000	1.0000
I-395L	Mogas (95 RON, 0.84 Pb)	−0.0028	0.1111	0.0000	0.8917	0.0000	0.0000	0.0000	1.0000
I-395LL	Mogas (95 RON, 0.15 Pb)	0.0040	−0.1570	0.0000	1.1530	0.0000	0.0000	0.0000	1.0000
I-395M	Mogas (95 RON, 0.40 Pb)	0.0000	0.0000	0.0000	1.0000	0.0000	0.0000	0.0000	1.0000
I-395S	Mogas	0.0040	−0.1570	0.0000	1.1530	0.0000	0.0000	0.0000	1.0000
I-397		0.0027	−0.1052	0.0000	1.1025	0.0000	0.0000	0.0000	1.0000
I-397C	Mogas (97 RON, 0.6 Pb)	0.0014	−0.0568	0.0000	1.0554	0.0000	0.0000	0.0000	1.0000
I-397E	Mogas (97 RON, 0.84 Pb)	0.0003	−0.0123	0.0000	1.0120	0.0000	0.0000	0.0000	1.0000
I-397LL	Mogas (97 RON, 0.15 Pb)	0.0049	−0.1943	0.0000	1.1894	0.0000	0.0000	0.0000	1.0000
I-397R	Mogas (97 RON, 0.4 Pb)	0.0027	−0.1052	0.0000	1.1025	0.0000	0.0000	0.0000	1.0000
I-397S	Mogas (0.4 Pb, 9 RVP)	−0.0111	−0.1057	0.0000	1.1168	0.0000	0.0000	0.0000	1.0000
I-3975		0.0033	−0.1315	0.0000	1.1282	0.0000	0.0000	0.0000	1.0000
I-398E	Mogas (98 RON, 0.84 Pb)	0.0019	−0.0741	0.0000	1.0722	0.0000	0.0000	0.0000	1.0000
I-398M	Mogas	0.0040	−0.1577	0.0000	1.1537	0.0000	0.0000	0.0000	1.0000
I-398S	Mogas (0.4 Pb, 10 RVP)	0.0045	−0.1577	0.0000	1.1532	0.0000	0.0000	0.0000	1.0000
I-398X	Mogas (98 RON, 0.60 Pb)	0.0029	−0.1136	0.0000	1.1107	0.0000	0.0000	0.0000	1.0000
I-3xx	Mogas	0.0000	0.0000	0.0000	1.0000	0.0000	0.0000	0.0000	1.0000
I-3xx	Mogas	0.0000	0.0000	0.0000	1.0000	0.0000	0.0000	0.0000	1.0000
I-3xx	Mogas	0.0000	0.0000	0.0000	1.0000	0.0000	0.0000	0.0000	1.0000
I-3xx	Mogas	0.0000	0.0000	0.0000	1.0000	0.0000	0.0000	0.0000	1.0000
I-3xx	Mogas	0.0000	0.0000	0.0000	1.0000	0.0000	0.0000	0.0000	1.0000
I-3xx	Mogas	0.0000	0.0000	0.0000	1.0000	0.0000	0.0000	0.0000	1.0000
I-3xx	Mogas	0.0000	0.0000	0.0000	1.0000	0.0000	0.0000	0.0000	1.0000
I-3xx	Mogas	0.0000	0.0000	0.0000	1.0000	0.0000	0.0000	0.0000	1.0000

I-3xx	Mogas	0.0000	0.0000	0.0000	1.0000	0.0000	0.0000	0.0000	1.0000
I-3xx	Mogas	0.0000	0.0000	0.0000	1.0000	0.0000	0.0000	0.0000	1.0000
I-400	Special Kerosene	0.0000	0.0000	0.0000	0.0000	1.0000	0.0000	0.0000	1.0000
I-411	Standard White Kerosene	0.0000	0.0000	0.0000	0.0000	1.0000	0.0000	0.0000	1.0000
I-419	Dual Purpose Kerosene	0.0000	0.0000	0.0000	0.0000	1.0000	0.0000	0.0000	1.0000
I-434	Jet Fuel JP-4	−0.0350	0.0000	0.5230	0.0000	0.5120	0.0000	0.0000	1.0000
I-440	Jet A-1 Fuel	0.0000	0.0000	0.0000	0.0000	1.0000	0.0000	0.0000	1.0000
I-440IM		0.0000	0.0000	0.0000	0.0000	0.0000	1.0000	0.0000	1.0000
I-4xx	Kerosene	0.0000	0.0000	0.0000	0.0000	1.0000	0.0000	0.0000	1.0000
I-4xx	Kerosene	0.0000	0.0000	0.0000	0.0000	1.0000	0.0000	0.0000	1.0000
I-4xx	Kerosene	0.0000	0.0000	0.0000	0.0000	1.0000	0.0000	0.0000	1.0000
I-4xx	Kerosene	0.0000	0.0000	0.0000	0.0000	1.0000	0.0000	0.0000	1.0000
I-4xx	Kerosene	0.0000	0.0000	0.0000	0.0000	1.0000	0.0000	0.0000	1.0000
I-711	Lube Distillate	0.0000	0.0000	0.0000	0.0000	0.0000	0.2500	0.7500	1.0000
I-725	FCCU Charge Stock	0.0000	0.0000	0.0000	0.0000	0.0000	0.3468	0.6532	1.0000
I-726	L.S. FCCU Charge Stock	0.0000	0.0000	0.0000	0.0000	0.0000	0.4000	0.6000	1.0000
I-730	Arabian Crude HVGO	0.0000	0.0000	0.0000	0.0000	0.0000	0.2709	0.7291	1.0000
I-7xx	Heavy Gasoil	0.0000	0.0000	0.0000	0.0000	0.0000	0.0000	0.0000	0.0000
I-7xx	Heavy Gasoil	0.0000	0.0000	0.0000	0.0000	0.0000	0.0000	0.0000	0.0000
I-7xx	Heavy Gasoil	0.0000	0.0000	0.0000	0.0000	0.0000	0.0000	0.0000	0.0000
I-800	High Speed Diesel	0.0000	0.0000	0.0000	0.0000	−0.5568	1.5568	0.0000	1.0000
I-80001	Zero Pour HSD	0.0000	0.0000	0.0000	0.0000	−0.1619	1.1619	0.0000	1.0000
I-875	Diesel (47 CI, -6 Pour)	0.0000	0.0000	0.0000	0.0000	0.1449	0.8551	0.0000	1.0000
I-876	Diesel (46 CI, -6 Pour)	0.0000	0.0000	0.0000	0.0000	0.1449	0.8551	0.0000	1.0000
I-876ZP	Diesel (46 CI, -18 Pour)	0.0000	0.0000	0.0000	0.0000	0.5653	0.4347	0.0000	1.0000
I-884	Diesel (47 CI, 0.5 Sulf)	0.0000	0.0000	0.0000	0.0000	0.2670	0.7330	0.0000	1.0000
I-885	Diesel (50 CI, 0.5 Sulf)	0.0000	0.0000	0.0000	0.0000	0.0000	1.0000	0.0000	1.0000
I-888	Diesel (50 CI, -3 Pour)	0.0000	0.0000	0.0000	0.0000	0.0000	1.0000	0.0000	1.0000
I-88852		0.0000	0.0000	0.0000	0.0000	0.0000	1.0000	0.0000	1.0000
I-8xx	Diesel	0.0000	0.0000	0.0000	0.0000	0.0000	1.0000	0.0000	1.0000
I-8xx	Diesel	0.0000	0.0000	0.0000	0.0000	0.0000	1.0000	0.0000	1.0000
I-8xx	Diesel	0.0000	0.0000	0.0000	0.0000	0.0000	1.0000	0.0000	1.0000
I-8xx	Diesel	0.0000	0.0000	0.0000	0.0000	0.0000	1.0000	0.0000	1.0000
I-8xx	Diesel	0.0000	0.0000	0.0000	0.0000	0.0000	1.0000	0.0000	1.0000
I-8xx	Diesel	0.0000	0.0000	0.0000	0.0000	0.0000	1.0000	0.0000	1.0000
I-8xx	Diesel	0.0000	0.0000	0.0000	0.0000	0.0000	1.0000	0.0000	1.0000
I-8xx	Diesel	0.0000	0.0000	0.0000	0.0000	0.0000	1.0000	0.0000	1.0000
I-892	Marine Diesel Oil	0.0000	0.0000	0.0000	0.0000	0.0000	0.7320	0.2680	1.0000
I-928	Fuel Oil (2.8 S, 180 cSt)	0.0000	0.0000	0.0000	0.0000	0.0000	0.0739	0.9261	1.0000
I-933	Fuel Oil (3.5 S, 80 cSt)	0.0000	0.0000	0.0000	0.0000	0.0000	0.1317	0.8683	1.0000
I-955	Fuel Oil (3.5 S, 75 cSt)	0.0000	0.0000	0.0000	0.0000	0.0000	0.1440	0.8560	1.0000
I-957	Fuel Oil (3.5 S, 225 SUS)	0.0000	0.0000	0.0000	0.0000	0.0000	0.2202	0.7798	1.0000
I-960	Fuel Oil (3.5 S, 120 cSt)	0.0000	0.0000	0.0000	0.0000	0.0000	0.0638	0.9362	1.0000
I-96001		0.0000	0.0000	0.0000	0.0000	0.0000	0.0638	0.9362	1.0000
I-9601		0.0000	0.0000	0.0000	0.0000	0.0000	0.0000	1.0000	1.0000
I-9602		0.0000	0.0000	0.0000	0.0000	0.0000	0.0000	1.0000	1.0000
I-961	Fuel Oil (3.5 S, 180 cSt)	0.0000	0.0000	0.0000	0.0000	0.0000	0.0000	1.0000	1.0000
I-96107	Fuel Oil (4.0 S, 180 cSt)	0.0000	0.0000	0.0000	0.0000	0.0000	0.0000	1.0000	1.0000
I-962	Fuel Oil (3.5 S, 280 cSt)	0.0000	0.0000	0.0000	0.0000	0.0000	−0.0638	1.0638	1.0000
I-96201		0.0000	0.0000	0.0000	0.0000	0.0000	−0.0844	1.0844	1.0000
I-96202		0.0000	0.0000	0.0000	0.0000	0.0000	−0.1029	1.1029	1.0000
I-971	Fuel Oil (4.0 S, 380 cSt)	0.0000	0.0000	0.0000	0.0000	0.0000	−0.1029	1.1029	1.0000
I-97103		0.0000	0.0000	0.0000	0.0000	0.0000	−0.1029	1.1029	1.0000
I-9xx	Fuel Oil	0.0000	0.0000	0.0000	0.0000	0.0000	0.0000	1.0000	1.0000
I-9xx	Fuel Oil	0.0000	0.0000	0.0000	0.0000	0.0000	0.0000	1.0000	1.0000
I-9xx	Fuel Oil	0.0000	0.0000	0.0000	0.0000	0.0000	0.0000	1.0000	1.0000
I-9xx	Fuel Oil	0.0000	0.0000	0.0000	0.0000	0.0000	0.0000	1.0000	1.0000
I-9xx	Fuel Oil	0.0000	0.0000	0.0000	0.0000	0.0000	0.0000	1.0000	1.0000
I-9xx	Fuel Oil	0.0000	0.0000	0.0000	0.0000	0.0000	0.0000	1.0000	1.0000
I-9xx	Fuel Oil	0.0000	0.0000	0.0000	0.0000	0.0000	0.0000	1.0000	1.0000
I-9xx	Fuel Oil	0.0000	0.0000	0.0000	0.0000	0.0000	0.0000	1.0000	1.0000

Screen 5
Continued

GRADE	DESCRIPTION	EQUIVALENCY IN BALANCING GRADES I-150	I-220	I-210	I-395M	I-440	I-888	I-961	CHECK
I-9xx	Fuel Oil	0.0000	0.0000	0.0000	0.0000	0.0000	0.0000	1.0000	1.0000
I-9xx	Fuel Oil	0.0000	0.0000	0.0000	0.0000	0.0000	0.0000	1.0000	1.0000
Pxxxx	Intermediate	0.0000	0.0000	0.0000	0.0000	0.0000	0.0000	0.0000	0.0000
Pxxxx	Intermediate	0.0000	0.0000	0.0000	0.0000	0.0000	0.0000	0.0000	0.0000
Pxxxx	Intermediate	0.0000	0.0000	0.0000	0.0000	0.0000	0.0000	0.0000	0.0000
Pxxxx	Intermediate	0.0000	0.0000	0.0000	0.0000	0.0000	0.0000	0.0000	0.0000
Pxxxx	Intermediate	0.0000	0.0000	0.0000	0.0000	0.0000	0.0000	0.0000	0.0000
Pxxxx	Intermediate	0.0000	0.0000	0.0000	0.0000	0.0000	0.0000	0.0000	0.0000
PBFUEL	Burner Fuel Oil	0.0000	0.0000	0.0000	0.0000	0.0000	0.0000	1.0000	1.0000
PBHDSL	Diesel Oil ex Bombay High	0.0000	0.0000	0.0000	0.0000	−1.3942	2.3942	0.0000	1.0000
PCTISO	Isomate Cutter	0.0000	0.0000	0.0000	0.0000	0.0000	0.3360	0.6640	1.0000
PCTTR	Cutter Stock	0.0000	0.0000	0.0000	0.0000	0.0000	0.6438	0.3562	1.0000
PDSDSL	Desulfurized Diesel	0.0000	0.0000	0.0000	0.0000	0.0000	1.0000	0.0000	1.0000
PDSL	Diesel	0.0000	0.0000	0.0000	0.0000	0.0000	1.0000	0.0000	1.0000
PFCCF	HVGO to FCCU	0.0000	0.0000	0.0000	0.0000	0.0000	0.2406	0.7594	1.0000
PFCOO	Hvy Isomate to FCCU	0.0000	0.0000	0.0000	0.0000	0.0000	0.2697	0.7303	1.0000
PFDISO	HVGO to 2HDU	0.0000	0.0000	0.0000	0.0000	0.0000	0.2406	0.7594	1.0000
PGABS	Gasoline	−0.0914	0.5571	0.0000	0.5343	0.0000	0.0000	0.0000	1.0000
PGPDSL	Diesel ex Gippsland	0.0000	0.0000	0.0000	0.0000	−1.3942	2.3942	0.0000	1.0000
PISIMP	Isomate Import	0.0000	0.0000	0.0000	0.0000	0.0000	0.2918	0.7082	1.0000
PKERIM	Kerosene Import	0.0000	0.0000	0.0000	0.0000	0.0000	1.0000	0.0000	1.0000
PKERO	Kerosene	0.0000	0.0000	0.0000	0.0000	1.0000	0.0000	0.0000	1.0000
PLANTS	Plants & Lines Contents	0.0103	0.1041	0.1105	0.1518	0.0886	0.1854	0.3494	1.0000
PLCGAS	Lt Cat Naphtha	0.0020	−0.0801	0.0000	1.0781	0.0000	0.0000	0.0000	1.0000
PLLCN	L. Lt Cat Naphtha	0.0020	−0.0801	0.0000	1.0781	0.0000	0.0000	0.0000	1.0000
PLSRBS	I-220 LSR	0.0000	1.0000	0.0000	0.0000	0.0000	0.0000	0.0000	1.0000
PLTISO	Lt. Isomate	0.0000	0.0000	0.0000	0.0000	0.0000	1.0000	0.0000	1.0000
PMEISO	Med. Isomate Cutter	0.0000	0.0000	0.0000	0.0000	0.0000	0.3360	0.6640	1.0000
PMIDSL	CDU 4A M/I Diesel	0.0000	0.0000	0.0000	0.0000	−0.4551	1.4551	0.0000	1.0000
PMSR	Sour MSR Naphtha	−0.1140	−0.4690	1.5830	0.0000	0.0000	0.0000	0.0000	1.0000
POMRSD	Oman Resid Import	0.0000	0.0000	0.0000	0.0000	0.0000	0.0245	0.9755	1.0000
PPOLY	Polymer Gasoline	−0.0102	−0.3091	0.0000	1.3193	0.0000	0.0000	0.0000	1.0000
PP90R	90 RON Reformate	−0.0272	−0.0458	0.0000	1.0730	0.0000	0.0000	0.0000	1.0000
PP93R	93 RON Reformate	−0.0246	−0.1503	0.0000	1.1749	0.0000	0.0000	0.0000	1.0000
PP95R	95 RON Reformate	−0.0228	−0.2200	0.0000	1.2428	0.0000	0.0000	0.0000	1.0000
PP96R	96 RON Reformate	−0.0219	−0.2548	0.0000	1.2767	0.0000	0.0000	0.0000	1.0000
PRESID	Residuum	0.0000	0.0000	0.0000	0.0000	0.0000	0.2218	0.7782	1.0000
PRESMI	M/I Dsls in 6VDU Feed	0.0000	0.0000	0.0000	0.0000	−0.4551	1.4551	0.0000	1.0000
PRES75	I-725 in 6VDU Feed	0.0000	0.0000	0.0000	0.0000	0.0000	0.3468	0.6532	1.0000
PSKERO	Sweet Kerosene	0.0000	0.0000	0.0000	0.0000	1.0000	0.0000	0.0000	1.0000
PSLOPD	Slop Distillate	0.0000	0.1000	0.1000	0.1400	0.1900	0.3700	0.1000	1.0000
PSLOPO	Slop Oil	0.0000	0.1000	0.1000	0.1400	0.1900	0.3700	0.1000	1.0000
PSLOPT	Slop ex Marine Terminal	−0.0123	0.1414	0.0000	0.4936	0.0488	0.3285	0.0000	1.0000
PSMCN	Sweet Med. Cat Naphtha	−0.1030	0.0870	0.0000	1.0160	0.0000	0.0000	0.0000	1.0000
PSWMSR	Sweet MSR	−0.1140	−0.4690	1.5830	0.0000	0.0000	0.0000	0.0000	1.0000
PUFCHG	Unifiner Feed	−0.1140	−0.4690	1.5830	0.0000	0.0000	0.0000	0.0000	1.0000
PUJFBS	Sour Jet B.S.	−0.1140	−0.4690	1.5830	0.0000	0.0000	0.0000	0.0000	1.0000
PWCN	Whole Cat Naphtha	−0.0032	0.1255	0.0000	0.8777	0.0000	0.0000	0.0000	1.0000
P440BS	Jet Basestock	0.0000	0.0000	0.0000	0.0000	1.0000	0.0000	0.0000	1.0000
CRUDE	Crude oil	0.0000	0.0100	0.1800	0.0000	0.1000	0.3100	0.4000	1.0000

Screen 6
Retrospective DOP

GRADE	PRODUCT	BOC					AOC				
		INITIAL DOP BOC	FORECASTER BOC	F.C COMPENS. CHANGES	CRUDE CHANGES BOC	RETRO DOP BOC	INITIAL DOP AOC	FORECASTER AOC	F.C COMPENS. AOC	CRUDE CHANGES AOC	RETRO DOP AOC
I-1138	Asphalt 60/70	0	0			0	36000	1467			37467
I-1149	Asphalt 40/50	0	352			352	0	528			528
I-11xx	Asphalt	0	0			0	0	0			0
I-11xx	Asphalt	0	0			0	0	0			0
I-150	LPG	0	0	−487	0	−487	24000	0	−690	0	23310
I-151	LPG	0	0			0	0	0			0
I-1xx	LPG	0	0			0	0	0			0
I-1xx	LPG	0	0			0	0	0			0
I-201	LSR Naphtha	0	0			0	0	0			0
I-210	WSR Naphtha	362670	0	5280	−1796	366154	671320	0	−35	2976	674260
I-220	LSR Naphtha	144000	0	−7234	−100	136666	0	0	9192	165	9357
I-2xx	Naphtha	0	0			0	0	0			0
I-2xx	Naphtha	0	0			0	0	0			0
I-2xx	Naphtha	0	0			0	0	0			0
I-2xx	Naphtha	0	0			0	0	0			0
I-253	Whole Cat Naphtha	0	0			0	0	0			0
I-383	Mogas (83 RON, 0.84 Pb)	0	−3147			−3147	0	−5121			−5121
I-387	Mogas (87 RON, 0.84 Pb)	0	11616			11616	0	17423			17423
I-387R	Mogas (87 RON, 0.40 Pb)	0	0			0	0	0			0
I-387S	Mogas	0	0			0	0	0			0
I-390E	Mogas	0	0			0	0	0			0
I-390J	Mogas	0	0			0	105000	35488			140488
I-390R	0.62 Pb	0	0			0	0	0			0
I-390S	Mogas (0.40 Pb, 8 RVP)	0	0			0	0	0			0
I-390X		0	0			0	0	0			0
I-390Z	0.4 Pb, 9 RVP	207990	1056			209046	0	0			0
I-393	Mogas (93 RON, 0.84 Pb)	0	0			0	0	0			0
I-393S	Mogas (93 RON, 0.40 Pb)	0	0			0	0	0			0
I-393X	Mogas (93 RON, 0.60 Pb)	0	0			0	0	0			0
I-395L	Mogas (95 RON, 0.84 Pb)	0	0			0	0	−72896			−72896
I-395LL	Mogas (95 RON, 0.15 Pb)	0	330			330	0	31132			31132

Screen 6
Continued

GRADE	PRODUCT	INITIAL DOP BOC	FORECASTER BOC	F.C COMPENS. CHANGES	CRUDE CHANGES BOC	RETRO DOP BOC	INITIAL DOP AOC	FORECASTER AOC	F.C COMPENS. AOC	CRUDE CHANGES AOC	RETRO DOP AOC
		BOC					AOC				
I-395M	Mogas (95 RON, 0.40 Pb)	0	0	3119	0	3119	68400	0	−224064	0	−155664
I-395S	Mogas	0	0			0	0	0			0
I-397	Mogas	0	0			0	0	0			0
I-397C	Mogas (97 RON, 0.6 Pb)	0	0			0	0	0			0
I-397E	Mogas (97 RON, 0.84 Pb)	0	0			0	0	0			0
I-397LL	Mogas (97 RON, 0.15 Pb)	27000	−6950			20050	42000	−28499			13501
I-397R	Mogas (97 RON, 0.4 Pb)	0	0			0	0	0			0
I-397S	Mogas (0.4 Pb, 9 RVP)	0	0			0	0	0			0
I-3975		0	0			0	0	0			0
I-398E	Mogas (98 RON, 0.84 Pb)	0	0			0	0	249558			249558
I-398M	Mogas	0	0			0	57900	88805			146705
I-398S	Mogas (0.4 Pb, 10 RVP)	0	0			0	99990	−99990			0
I-398X	Mogas (98 RON, 0.60 Pb)	0	0			0	0	0			0
I-3xx	Mogas	0	0			0	0	0			0
I-3xx	Mogas	0	0			0	0	0			0
I-3xx	Mogas	0	0			0	0	0			0
I-3xx	Mogas	0	0			0	0	0			0
I-3xx	Mogas	0	0			0	0	0			0
I-3xx	Mogas	0	0			0	0	0			0
I-3xx	Mogas	0	0			0	0	0			0
I-3xx	Mogas	0	0			0	0	0			0
I-3xx	Mogas	0	0			0	0	0			0
I-3xx	Mogas	0	0			0	0	0			0
I-400	Special Kerosene	0	0			0	0	0			0
I-411	Standard White Kerosene	0	0			0	9000	−9000			0
I-419	Dual Purpose Kerosene	0	0			0	0	0			0
I-434	Jet Fuel JP-4	0	0			0	0	−6			−6
I-440	Jet A-1 Fuel	576000	0	−1705	−998	573297	855000	0	−2509	1653	854144
I-440IM		0	0			0	0	0			0
I-4xx	Kerosene	0	0			0	0	0			0

I-4xx	Kerosene	0	0			0	0	0			0
I-4xx	Kerosene	0	0			0	0	0			0
I-4xx	Kerosene	0	0			0	0	0			0
I-4xx	Kerosene	0	0			0	0	0			0
I-711	Lube Distillate	0	0			0	0	0			0
I-725	FCCU Charge Stock	0	0			0	0	0			0
I-726	L.S. FCCU Charge Stock	0	0			0	0	0			0
I-730	Arabian Crude HVGO	0	0			0	0	0			0
I-7xx	Heavy Gasoil	0	0			0	0	0			0
I-7xx	Heavy Gasoil	0	0			0	0	0			0
I-7xx	Heavy Gasoil	0	0			0	0	0			0
I-800	High Speed Diesel	0	0			0	0	0			0
I-80001	Zero Pour HSD	210000	8786			218786	681000	28492			709492
I-875	Diesel (47 CI, −6 Pour)	0	0			0	0	0			0
I-876	Diesel (46 CI, −6 Pour)	0	0			0	204000	116348			320348
I-876ZP	Diesel (46 CI, −18 Pour)	0	0			0	0	0			0
I-884	Diesel (47 CI, 0.5 Sulf)	0	0			0	0	0			0
I-885	Diesel (50 CI, 0.5 Sulf)	24990	−19755			5235	66990	−29720			37270
I-888	Diesel (50 CI, −3 Pour)	599130	0	−14183	−3094	581853	471930	0	−156782	5125	320272
I-88852		0	0			0	0	0			0
I-8xx	Diesel	0	0			0	0	0			0
I-8xx	Diesel	0	0			0	0	0			0
I-8xx	Diesel	0	0			0	0	0			0
I-8xx	Diesel	0	0			0	0	0			0
I-8xx	Diesel	0	0			0	0	0			0
I-8xx	Diesel	0	0			0	0	0			0
I-8xx	Diesel	0	0			0	0	0			0
I-8xx	Diesel	0	0			0	0	0			0
I-892	Marine Diesel Oil	0	−104			−104	0	−81			−81
I-928	Fuel Oil (2.8 S, 180 cSt)	0	0			0	0	0			0
I-933	Fuel Oil (3.5 S, 80 cSt)	0	0			0	0	0			0
I-955	Fuel Oil (3.5 S, 75 cSt)	0	0			0	0	0			0
I-957	Fuel Oil (3.5 S, 225 SUS)	0	0			0	0	0			0
I-960	Fuel Oil (3.5 S, 120 cSt)	0	0			0	0	0			0
I-96001		0	0			0	0	0			0
I-9601		0	0			0	0	0			0
I-9602		0	0			0	0	0			0
I-961	Fuel Oil (3.5 S, 180 cSt)	724230	0	21759	−3992	741997	582030	0	128013	6612	716655
I-96107	Fuel Oil (4.0 S, 180 cSt)	0	0			0	420990	−129559			291431

Screen 6
Continued

GRADE	PRODUCT	BOC					AOC				
		INITIAL DOP BOC	FORECASTER BOC	F.C COMPENS. CHANGES	CRUDE CHANGES BOC	RETRO DOP BOC	INITIAL DOP AOC	FORECASTER AOC	F.C COMPENS. AOC	CRUDE CHANGES AOC	RETRO DOP AOC
I-962	Fuel Oil (3.5 S, 280 cSt)	0	0			0	0	0			0
I-96201		0	0			0	0	0			0
I-96202		0	0			0	0	0			0
I-971	Fuel Oil (4.0 S, 380 cSt)	0	0			0	0	0			0
I-97103		0	0			0	387990	−6535			381455
I-9xx	Fuel Oil	0	0			0	0	0			0
I-9xx	Fuel Oil	0	0			0	0	0			0
I-9xx	Fuel Oil	0	0			0	0	0			0
I-9xx	Fuel Oil	0	0			0	0	0			0
I-9xx	Fuel Oil	0	0			0	0	0			0
I-9xx	Fuel Oil	0	0			0	0	0			0
I-9xx	Fuel Oil	0	0			0	0	0			0
I-9xx	Fuel Oil	0	0			0	0	0			0
I-9xx	Fuel Oil	0	0			0	0	0			0
I-9xx	Fuel Oil	0	0			0	0	0			0
Pxxxx	Intermediate	0	0			0	0	0			0
Pxxxx	Intermediate	0	0			0	0	0			0
Pxxxx	Intermediate	0	0			0	0	0			0
Pxxxx	Intermediate	0	0			0	0	0			0
Pxxxx	Intermediate	0	0			0	0	0			0
Pxxxx	Intermediate	0	0			0	0	0			0
PBFUEL	Burner Fuel Oil	0	−255			−255	0	−591			−591
PBHDSL	Diesel Oil ex Bombay High	0	0			0	0	0			0
PCTISO	Isomate Cutter	0	0			0	0	0			0
PCTTR	Cutter Stock	0	0			0	6000	−3132			2868
PDSDSL	Desulfurized Diesel	0	28404			28404	9000	42605			51605
PDSL	Diesel	0	29974			29974	14010	44962			58972
PFCCF	HVGO to FCCU	0	0			0	0	0			0
PFCOO	Hvy Isomate to FCCU	−30990	11365			−19625	−27990	10265			−17725
PFDISO	HVGO to 2HDU	−51000	2782			−48218	−120000	6546			−113454

PGABS	Gasoline	0	0			0	0	0			0
PGPDSL	Diesel ex Gippsland	0	0			0	0	0			0
PISIMP	Isomate Import	0	0			0	0	0			0
PKERIM	Kerosene Import	0	0			0	0	0			0
PKERO	Kerosene	0	−2391			−2391	0	−3181			−3181
PLANTS	Plants & Lines Contents	0	0			0	0	0			0
PLCGAS	Lt Cat Naphtha	33990	4097			38087	69000	8316			77316
PLLCN	L. Lt Cat Naphtha	0	617			617	0	926			926
PLSRBS	I-220 LSR	0	0			0	0	0			0
PLTISO	Lt. Isomate	2010	−7758			−5748	0	−20118			−20118
PMEISO	Med. Isomate Cutter	0	−4470			−4470	0	−5296			−5296
PMIDSL	CDU 4A M/I Diesel	6990	−13119			−6129	9990	−7210			2780
PMSR	Sour MSR Naphtha	0	−1239			−1239	0	−1897			−1897
POMRSD	Oman Resid Import	0	0			0	0	0			0
PPOLY	Polymer Gasoline	3990	−577			3413	6990	−1010			5980
PP90R	90 RON Reformate	0	0			0	0	0			0
PP93R	93 RON Reformate	−990	−946			−1936	−2010	−1921			−3931
PP95R	95 RON Reformate	0	0			0	0	0			0
PP96R	96 RON Reformate	9990	−1332			8658	18000	−2401			15599
PRESID	Residuum	363960	−37462			326498	15870	−1633			14237
PRESMI	M/I Dsls in 6VDU Feed	0	0			0	0	0			0
PRES75	B725 in 6VDU Feed	0	0			0	0	0			0
PSKERO	Sweet Kerosene	0	263			263	0	395			395
PSLOPD	Slop Distillate	0	−1860			−1860	0	−2854			−2854
PSLOPO	Slop Oil	0	−1901			−1901	0	−3617			−3617
PSLOPT	Slop ex Marine Terminal	0	0			0	0	0			0
PSMCN	Sweet Med. Cat Naphtha	−8010	−2265			−10275	−15000	−4242			−19242
PSWMSR	Sweet MSR	0	0			0	−6000	5900			−100
PUFCHG	Unifiner Feed	−3990	−1859			−5849	0	−3570			−3570
PUJFBS	Sour Jet B.S.	0	0			0	0	0			0
PWCN	Whole Cat Naphtha	0	1200			1200	0	1799			1799
P440BS	Jet B.S.	0	0			0	0	0			0
	COLUMN SUM	3201960	−6548	6548	−9980	3191980	4761400	246875	−246875		4777931
CRUDE		3225000			−9980	3215020	4806000			16531	4822531
	FINISHED PRD.	2876010					4783540				
	PROCESS STOCKS	325950					−22140				
	PRD + PROCESS STOCKS	3201960									
LOSSES		23040				23040					44600

Screen 7
Retrospective DOP of BOC in Terms of Balancing Grades

GRADE	RETRO DOP	BALANCING GRADES EQUIVALENTS						
		I-150	I-220	I-210	I-395M	I-440	I-888	I-961
I-1138	0	0	0	0	0	0	0	0
I-1149	352	0	0	0	0	0	−257	609
I-11xx	0	0	0	0	0	0	0	0
I-11xx	0	0	0	0	0	0	0	0
I-150	−487	−487	0	0	0	0	0	0
I-151	0	0	0	0	0	0	0	0
I-1xx	0	0	0	0	0	0	0	0
I-1xx	0	0	0	0	0	0	0	0
I-201	0	0	0	0	0	0	0	0
I-210	366154	0	0	366154	0	0	0	0
I-220	136666	0	136666	0	0	0	0	0
I-2xx	0	0	0	0	0	0	0	0
I-2xx	0	0	0	0	0	0	0	0
I-2xx	0	0	0	0	0	0	0	0
I-2xx	0	0	0	0	0	0	0	0
I-253	0	0	0	0	0	0	0	0
I-383	−3147	68	−2681	0	−534	0	0	0
I-387	11616	−177	7027	0	4766	0	0	0
I-387R	0	0	0	0	0	0	0	0
I-387S	0	0	0	0	0	0	0	0
I-390E	0	0	0	0	0	0	0	0
I-390J	0	0	0	0	0	0	0	0
I-390R	0	0	0	0	0	0	0	0
I-390S	0	0	0	0	0	0	0	0
I-390X	0	0	0	0	0	0	0	0
I-390Z	209046	−4265	54854	0	158457	0	0	0
I-393	0	0	0	0	0	0	0	0
I-393S	0	0	0	0	0	0	0	0
I-393X	0	0	0	0	0	0	0	0
I-395L	0	0	0	0	0	0	0	0
I-395LL	330	1	−52	0	380	0	0	0
I-395M	3119	0	0	0	3119	0	0	0
I-395S	0	0	0	0	0	0	0	0
I-397	0	0	0	0	0	0	0	0
I-397C	0	0	0	0	0	0	0	0
I-397E	0	0	0	0	0	0	0	0
I-397LL	20050	98	−3896	0	23847	0	0	0
I-397R	0	0	0	0	0	0	0	0
I-397S	0	0	0	0	0	0	0	0
I-3975	0	0	0	0	0	0	0	0
I-398E	0	0	0	0	0	0	0	0

I-398M	0	0	0	0	0	0	0	0
I-398S	0	0	0	0	0	0	0	0
I-398X	0	0	0	0	0	0	0	0
I-3xx	0	0	0	0	0	0	0	0
I-3xx	0	0	0	0	0	0	0	0
I-3xx	0	0	0	0	0	0	0	0
I-3xx	0	0	0	0	0	0	0	0
I-3xx	0	0	0	0	0	0	0	0
I-3xx	0	0	0	0	0	0	0	0
I-3xx	0	0	0	0	0	0	0	0
I-3xx	0	0	0	0	0	0	0	0
I-3xx	0	0	0	0	0	0	0	0
I-3xx	0	0	0	0	0	0	0	0
I-400	0	0	0	0	0	0	0	0
I-411	0	0	0	0	0	0	0	0
I-419	0	0	0	0	0	0	0	0
I-434	0	0	0	0	0	0	0	0
I-440	573297	0	0	0	0	573297	0	0
I-440IM	0	0	0	0	0	0	0	0
I-4xx	0	0	0	0	0	0	0	0
I-4xx	0	0	0	0	0	0	0	0
I-4xx	0	0	0	0	0	0	0	0
I-4xx	0	0	0	0	0	0	0	0
I-4xx	0	0	0	0	0	0	0	0
I-711	0	0	0	0	0	0	0	0
I-725	0	0	0	0	0	0	0	0
I-726	0	0	0	0	0	0	0	0
I-730	0	0	0	0	0	0	0	0
I-7xx	0	0	0	0	0	0	0	0
I-7xx	0	0	0	0	0	0	0	0
I-7xx	0	0	0	0	0	0	0	0
I-800	0	0	0	0	0	0	0	0
I-80001	218786	0	0	0	0	−35421	254207	0
I-875	0	0	0	0	0	0	0	0
I-876	0	0	0	0	0	0	0	0
I-876ZP	0	0	0	0	0	0	0	0
I-884	0	0	0	0	0	0	0	0
I-885	5235	0	0	0	0	0	5235	0
I-888	581853	0	0	0	0	0	581853	0
I-88852	0	0	0	0	0	0	0	0
I-8xx	0	0	0	0	0	0	0	0
I-8xx	0	0	0	0	0	0	0	0
I-8xx	0	0	0	0	0	0	0	0
I-8xx	0	0	0	0	0	0	0	0
I-8xx	0	0	0	0	0	0	0	0
I-8xx	0	0	0	0	0	0	0	0
I-8xx	0	0	0	0	0	0	0	0
I-8xx	0	0	0	0	0	0	0	0
I-892	−104	0	0	0	0	0	−76	−28

Screen 7
Continued

GRADE	RETRO DOP	BALANCING GRADES EQUIVALENTS I-150	I-220	I-210	I-395M	I-440	I-888	I-961
I-928	0	0	0	0	0	0	0	0
I-933	0	0	0	0	0	0	0	0
I-955	0	0	0	0	0	0	0	0
I-957	0	0	0	0	0	0	0	0
I-960	0	0	0	0	0	0	0	0
I-96001	0	0	0	0	0	0	0	0
I-9601	0	0	0	0	0	0	0	0
I-9602	0	0	0	0	0	0	0	0
I-961	741997	0	0	0	0	0	0	741997
I-96107	0	0	0	0	0	0	0	0
I-962	0	0	0	0	0	0	0	0
I-96201	0	0	0	0	0	0	0	0
I-96202	0	0	0	0	0	0	0	0
I-971	0	0	0	0	0	0	0	0
I-97103	0	0	0	0	0	0	0	0
I-9xx	0	0	0	0	0	0	0	0
I-9xx	0	0	0	0	0	0	0	0
I-9xx	0	0	0	0	0	0	0	0
I-9xx	0	0	0	0	0	0	0	0
I-9xx	0	0	0	0	0	0	0	0
I-9xx	0	0	0	0	0	0	0	0
I-9xx	0	0	0	0	0	0	0	0
I-9xx	0	0	0	0	0	0	0	0
I-9xx	0	0	0	0	0	0	0	0
I-9xx	0	0	0	0	0	0	0	0
Pxxxx	0	0	0	0	0	0	0	0
Pxxxx	0	0	0	0	0	0	0	0
Pxxxx	0	0	0	0	0	0	0	0
Pxxxx	0	0	0	0	0	0	0	0
Pxxxx	0	0	0	0	0	0	0	0
Pxxxx	0	0	0	0	0	0	0	0
PBFUEL	−255	0	0	0	0	0	0	−255
PBHDSL	0	0	0	0	0	0	0	0
PCTISO	0	0	0	0	0	0	0	0
PCTTR	0	0	0	0	0	0	0	0
PDSDSL	28404	0	0	0	0	0	28404	0
PDSL	29974	0	0	0	0	0	29974	0
PFCCF	0	0	0	0	0	0	0	0
PFCOO	−19625	0	0	0	0	0	−5293	−14332
PFDISO	−48218	0	0	0	0	0	−11601	−36617
PGABS	0	0	0	0	0	0	0	0

PGPDSL	0	0	0	0	0	0	0	0
PISIMP	0	0	0	0	0	0	0	0
PKERIM	0	0	0	0	0	0	0	0
PKERO	−2391	0	0	0	0	−2391	0	0
PLANTS	0	0	0	0	0	0	0	0
PLCGAS	38087	76	−3051	0	41062	0	0	0
PLLCN	617	1	−49	0	665	0	0	0
PLSRBS	0	0	0	0	0	0	0	0
PLTISO	−5748	0	0	0	0	0	−5748	0
PMEISO	−4470	0	0	0	0	0	−1502	−2968
PMIDSL	−6129	0	0	0	0	2789	−8918	0
PMSR	−1239	141	581	−1961	0	0	0	0
POMRSD	0	0	0	0	0	0	0	0
PPOLY	3413	−35	−1055	0	4503	0	0	0
PP90R	0	0	0	0	0	0	0	0
PP93R	−1936	48	291	0	−2275	0	0	0
PP95R	0	0	0	0	0	0	0	0
PP96R	8658	−190	−2206	0	11054	0	0	0
PRESID	326498	0	0	0	0	0	72417	254081
PRESMI	0	0	0	0	0	0	0	0
PRES75	0	0	0	0	0	0	0	0
PSKERO	263	0	0	0	0	263	0	0
PSLOPD	−1860	0	−186	−186	−260	−353	−688	−186
PSLOPO	−1901	0	−190	−190	−266	−361	−703	−190
PSLOPT	0	0	0	0	0	0	0	0
PSMCN	−10275	1058	−894	0	−10439	0	0	0
PSWMSR	0	0	0	0	0	0	0	0
PUFCHG	−5849	667	2743	−9259	0	0	0	0
PUJFBS	0	0	0	0	0	0	0	0
PWCN	1200	−4	151	0	1053	0	0	0
P440BS	0	0	0	0	0	0	0	0
COL SUM	3191980	−2997.95	188052.6	354557.4	235131.8	537821.9	937304.1	942110.3

It would have been impossible to incorporate all this kerosene dumping into diesel in Participant BOC's LP model, as all the diesels were close to their flash point limit.

USE OF POUR POINT DEPRESSANTS

While processing certain waxy crudes, it is common practice to add pour point depressants to diesel produced to prevent or minimize kerosene dumping into diesel.

For example, kerosene from Bombay high crude may have a pour point of +9°C. Therefore, a significant amount of kerosene might be required

Screen 8
Retrospective DOP of Participant
AOC in Terms of Balancing Grades

GRADE	RETRO DOP	BALANCING GRADES EQUIVALENTS						
		I-150	I-220	I-210	I-395M	I-440	I-888	I-961
I-1138	37467	0	0	0	0	0	−27302	64769
I-1149	528	0	0	0	0	0	−385	913
I-11xx	0	0	0	0	0	0	0	0
I-11xx	0	0	0	0	0	0	0	0
I-150	23310	23310	0	0	0	0	0	0
I-151	0	0	0	0	0	0	0	0
I-1xx	0	0	0	0	0	0	0	0
I-1xx	0	0	0	0	0	0	0	0
I-201	0	0	0	0	0	0	0	0
I-210	674260	0	0	674260	0	0	0	0
I-220	9357	0	9357	0	0	0	0	0
I-2xx	0	0	0	0	0	0	0	0
I-2xx	0	0	0	0	0	0	0	0
I-2xx	0	0	0	0	0	0	0	0
I-2xx	0	0	0	0	0	0	0	0
I-253	0	0	0	0	0	0	0	0
I-383	−5121	110	−4363	0	−869	0	0	0
I-387	17423	−265	10539	0	7149	0	0	0
I-387R	0	0	0	0	0	0	0	0
I-387S	0	0	0	0	0	0	0	0
I-390E	0	0	0	0	0	0	0	0
I-390J	140488	−927	36934	0	104481	0	0	0
I-390R	0	0	0	0	0	0	0	0
I-390S	0	0	0	0	0	0	0	0
I-390X	0	0	0	0	0	0	0	0
I-390Z	0	0	0	0	0	0	0	0
I-393	0	0	0	0	0	0	0	0
I-393S	0	0	0	0	0	0	0	0
I-393X	0	0	0	0	0	0	0	0
I-395L	−72896	204	−8099	0	−65001	0	0	0
I-395LL	31132	125	−4888	0	35895	0	0	0
I-395M	−155664	0	0	0	−155664	0	0	0
I-395S	0	0	0	0	0	0	0	0
I-397	0	0	0	0	0	0	0	0
I-397C	0	0	0	0	0	0	0	0
I-397E	0	0	0	0	0	0	0	0
I-397LL	13501	66	−2623	0	16058	0	0	0
I-397R	0	0	0	0	0	0	0	0
I-397S	0	0	0	0	0	0	0	0

I-3975	0	0	0	0	0	0	0	0
I-398E	249558	474	−18492	0	267576	0	0	0
I-398M	146705	587	−23135	0	169254	0	0	0
I-398S	0	0	0	0	0	0	0	0
I-398X	0	0	0	0	0	0	0	0
I-3xx	0	0	0	0	0	0	0	0
I-3xx	0	0	0	0	0	0	0	0
I-3xx	0	0	0	0	0	0	0	0
I-3xx	0	0	0	0	0	0	0	0
I-3xx	0	0	0	0	0	0	0	0
I-3xx	0	0	0	0	0	0	0	0
I-3xx	0	0	0	0	0	0	0	0
I-3xx	0	0	0	0	0	0	0	0
I-3xx	0	0	0	0	0	0	0	0
I-3xx	0	0	0	0	0	0	0	0
I-400	0	0	0	0	0	0	0	0
I-411	0	0	0	0	0	0	0	0
I-419	0	0	0	0	0	0	0	0
I-434	−6	0	0	−3	0	−3	0	0
I-440	854144	0	0	0	0	854144	0	0
I-440IM	0	0	0	0	0	0	0	0
I-4xx	0	0	0	0	0	0	0	0
I-4xx	0	0	0	0	0	0	0	0
I-4xx	0	0	0	0	0	0	0	0
I-4xx	0	0	0	0	0	0	0	0
I-4xx	0	0	0	0	0	0	0	0
I-711	0	0	0	0	0	0	0	0
I-725	0	0	0	0	0	0	0	0
I-726	0	0	0	0	0	0	0	0
I-730	0	0	0	0	0	0	0	0
I-7xx	0	0	0	0	0	0	0	0
I-7xx	0	0	0	0	0	0	0	0
I-7xx	0	0	0	0	0	0	0	0
I-800	0	0	0	0	0	0	0	0
I-80001	709492	0	0	0	0	−114867	824359	0
I-875	0	0	0	0	0	0	0	0
I-876	320348	0	0	0	0	46418	273930	0
I-876ZP	0	0	0	0	0	0	0	0
I-884	0	0	0	0	0	0	0	0
I-885	37270	0	0	0	0	0	37270	0
I-888	320272	0	0	0	0	0	320272	0
I-88852	0	0	0	0	0	0	0	0
I-8xx	0	0	0	0	0	0	0	0
I-8xx	0	0	0	0	0	0	0	0
I-8xx	0	0	0	0	0	0	0	0
I-8xx	0	0	0	0	0	0	0	0

Screen 8
Continued

GRADE	RETRO DOP	BALANCING GRADES EQUIVALENTS						
		I-150	I-220	I-210	I-395M	I-440	I-888	I-961
I-8xx	0	0	0	0	0	0	0	0
I-8xx	0	0	0	0	0	0	0	0
I-8xx	0	0	0	0	0	0	0	0
I-8xx	0	0	0	0	0	0	0	0
I-892	−81	0	0	0	0	0	−59	−22
I-928	0	0	0	0	0	0	0	0
I-933	0	0	0	0	0	0	0	0
I-955	0	0	0	0	0	0	0	0
I-957	0	0	0	0	0	0	0	0
I-960	0	0	0	0	0	0	0	0
I-96001	0	0	0	0	0	0	0	0
I-9601	0	0	0	0	0	0	0	0
I-9602	0	0	0	0	0	0	0	0
I-961	716655	0	0	0	0	0	0	716655
I-96107	291431	0	0	0	0	0	0	291431
I-962	0	0	0	0	0	0	0	0
I-96201	0	0	0	0	0	0	0	0
I-96202	0	0	0	0	0	0	0	0
I-971	0	0	0	0	0	0	0	0
I-97103	381455	0	0	0	0	0	−39252	420707
I-9xx	0	0	0	0	0	0	0	0
I-9xx	0	0	0	0	0	0	0	0
I-9xx	0	0	0	0	0	0	0	0
I-9xx	0	0	0	0	0	0	0	0
I-9xx	0	0	0	0	0	0	0	0
I-9xx	0	0	0	0	0	0	0	0
I-9xx	0	0	0	0	0	0	0	0
I-9xx	0	0	0	0	0	0	0	0
I-9xx	0	0	0	0	0	0	0	0
I-9xx	0	0	0	0	0	0	0	0
Pxxxx	0	0	0	0	0	0	0	0
Pxxxx	0	0	0	0	0	0	0	0
Pxxxx	0	0	0	0	0	0	0	0
Pxxxx	0	0	0	0	0	0	0	0
Pxxxx	0	0	0	0	0	0	0	0
Pxxxx	0	0	0	0	0	0	0	0
PBFUEL	−591	0	0	0	0	0	0	−591
PBHDSL	0	0	0	0	0	0	0	0
PCTISO	0	0	0	0	0	0	0	0
PCTTR	2868	0	0	0	0	0	1846	1022

PDSDSL	51605	0	0	0	0	0	51605	0
PDSL	58972	0	0	0	0	0	58972	0
PFCCF	0	0	0	0	0	0	0	0
PFCOO	−17725	0	0	0	0	0	−4780	−12945
PFDISO	−113454	0	0	0	0	0	−27297	−86157
PGABS	0	0	0	0	0	0	0	0
PGPDSL	0	0	0	0	0	0	0	0
PISIMP	0	0	0	0	0	0	0	0
PKERIM	0	0	0	0	0	0	0	0
PKERO	−3181	0	0	0	0	−3181	0	0
PLANTS	0	0	0	0	0	0	0	0
PLCGAS	77316	155	−6193	0	83354	0	0	0
PLLCN	926	2	−74	0	998	0	0	0
PLSRBS	0	0	0	0	0	0	0	0
PLTISO	−20118	0	0	0	0	0	−20118	0
PMEISO	−5296	0	0	0	0	0	−1779	−3517
PMIDSL	2780	0	0	0	0	−1265	4045	0
PMSR	−1897	216	890	−3003	0	0	0	0
POMRSD	0	0	0	0	0	0	0	0
PPOLY	5980	−61	−1848	0	7889	0	0	0
PP90R	0	0	0	0	0	0	0	0
PP93R	−3931	97	591	0	−4619	0	0	0
PP95R	0	0	0	0	0	0	0	0
PP96R	15599	−342	−3975	0	19915	0	0	0
PRESID	14237	0	0	0	0	0	3158	11079
PRESMI	0	0	0	0	0	0	0	0
PRES75	0	0	0	0	0	0	0	0
PSKERO	395	0	0	0	0	395	0	0
PSLOPD	−2854	0	−285	−285	−400	−542	−1056	−285
PSLOPO	−3617	0	−362	−362	−506	−687	−1338	−362
PSLOPT	0	0	0	0	0	0	0	0
PSMCN	−19242	1982	−1674	0	−19550	0	0	0
PSWMSR	−100	11	47	−158	0	0	0	0
PUFCHG	−3570	407	1674	−5651	0	0	0	0
PUJFBS	0	0	0	0	0	0	0	0
PWCN	1799	−6	226	0	1579	0	0	0
P440BS	0	0	0	0	0	0	0	0
COL SUM	4777931	26145	−15753	664798	467541	780412	1452090	1402698

to meet the usual pour requirement of −3 to −6°C. The addition of a pour point depressant in the ppm range lowers the pour point of diesel by 3–6°C, reducing the kerosene dumping into diesel.

Screen 9
Allocation of Balancing Grades

BALANCING GRADES	RETRO DOP EXPRESSED IN BALANCING GRADES		RETRO DOP								REVERSE DELTA			
	DOP BOC	DOP AOC	DOP BOC	DOP AOC	DOP TOTAL	REFINERY PRODUCTION	TOTAL DELTA	DELTA BOC	DELTA 1 AOC	BOC	AOC	ALLOCATION BOC	ALLOCATION AOC
COL NO	(1)	(2)	(3)	(4)	(5)	(6)	(7)	(8)	(9)	(10)	(11)	(12)	(13)
I-150	−2998	26145	−487	23310	22823	23160	337	−44	381	−12	12	−543	23703
I-220	188053	−15753	136666	9357	146024	198553	52529	57332	−4803	−1926	1926	192072	6481
I-210	354557	664798	366154	674260	1040414	800332	−240082	−83507	−156576	−8801	8801	273846	526486
I-395M	235132	467541	3119	−155664	−152545	0	152545	51045	101500	−5592	5592	48572	−48572
I-440	537822	780412	573297	854144	1427441	1476919	49478	20186	29292	−1814	1814	591669	885250
I-888	937304	1452090	581853	320272	902125	1095206	193081	75741	117340	−7078	7078	650516	444690
I-961	942110	1402698	741997	716655	1458652	1218169	−240483	−96623	−143861	−8816	8816	636558	581611
LOSSES	23040	44600	23040	44600	67640	100235	32595	11103	21492	−1195	1195	32948	67287
TOTAL							0	35234	−35234	−35234	35234		
ABSOLUTE SUM OF COMBINED TOTAL DELTAS (COLUMN 7) =							961131						

Screen 10
Allocated Production

GRADE	PRODUCT	BOC ALLOCATION	AOC ALLOCATION	TOTAL PRODUCED
I-1138	Asphalt 60/70	0	37467	37467
I-1149	Asphalt 40/50	352	528	880
I-1138D	Asphalt	0	0	0
I-11xx	Asphalt	0	0	0
I-150	LPG	−543	23703	23160
I-151	LPG	0	0	0
I-1xx	LPG	0	0	0
I-1xx	LPG	0	0	0
I-201	LSR Naphtha	0	0	0
I-210	WSR Naphtha	273846	526486	800332
I-220	LSR Naphtha	192072	6481	198553
I-2xx	Naphtha	0	0	0
I-2xx	Naphtha	0	0	0
I-2xx	Naphtha	0	0	0
I-2xx	Naphtha	0	0	0
I-253	Whole Cat Naphtha	0	0	0
I-383	Mogas (83 RON, 0.84 Pb)	−3147	−5121	−8268
I-387	Mogas (87 RON, 0.84 Pb)	11616	17423	29039
I-387R	Mogas (87 RON, 0.40 Pb)	0	0	0
I-387S	Mogas 87 RON	0	0	0
I-390E	Mogas 90 RON	0	0	0
I-390J	Mogas	0	140488	140488
I-390R	Mogas	0	0	0
I-390S	Mogas (0.40 Pb, 8 RVP)	0	0	0
I-390X	Mogas	0	0	0
I-390Z	0.4 Pb, 9 RVP	209046	0	209046
I-393	Mogas (93 RON, 0.84 Pb)	0	0	0
I-393S	Mogas (93 RON, 0.40 Pb)	0	0	0
I-393X	Mogas (93 RON, 0.60 Pb)	0	0	0
I-395L	Mogas (95 RON, 0.84 Pb)	0	−72896	−72896
I-395LL	Mogas (95 RON, 0.15 Pb)	330	31132	31462
I-395M	Mogas (95 RON, 0.40 Pb)	48572	−48572	0
I-395S	Mogas	0	0	0
I-397	Mogas	0	0	0
I-397C	Mogas (97 RON, 0.6 Pb)	0	0	0
I-397E	Mogas (97 RON, 0.84 Pb)	0	0	0
I-397LL	Mogas (97 RON, 0.15 Pb)	20050	13501	33551
I-397R	Mogas (97 RON, 0.4 Pb)	0	0	0
I-397S	Mogas (0.4 Pb, 9 RVP)	0	0	0
I-3975	Mogas	0	0	0
I-398E	Mogas (98 RON, 0.84 Pb)	0	249558	249558

Screen 10
Continued

GRADE	PRODUCT	BOC ALLOCATION	AOC ALLOCATION	TOTAL PRODUCED
I-398M	Mogas	0	146705	146705
I-398S	Mogas (0.4 Pb, 10 RVP)	0	0	0
I-398X	Mogas (98 RON, 0.60 Pb)	0	0	0
I-3xx	Mogas	0	0	0
I-3xx	Mogas	0	0	0
I-3xx	Mogas	0	0	0
I-3xx	Mogas	0	0	0
I-3xx	Mogas	0	0	0
I-3xx	Mogas	0	0	0
I-3xx	Mogas	0	0	0
I-3xx	Mogas	0	0	0
I-3xx	Mogas	0	0	0
I-3xx	Mogas	0	0	0
I-400	Special Kerosene	0	0	0
I-411	Standard White Kerosene	0	0	0
I-419	Dual Purpose Kerosene	0	0	0
I-434	Jet Fuel JP-4	0	−6	−6
I-440	Jet A-1 Fuel	591669	885250	1476919
I-440IM		0	0	0
I-4xx	Kerosene	0	0	0
I-4xx	Kerosene	0	0	0
I-4xx	Kerosene	0	0	0
I-4xx	Kerosene	0	0	0
I-4xx	Kerosene	0	0	0
I-711	Lube Distillate	0	0	0
I-725	FCCU Charge Stock	0	0	0
I-726	L.S. FCCU Charge Stock	0	0	0
I-730	Arabian Crude HVGO	0	0	0
I-7xx	Heavy Gasoil	0	0	0
I-7xx	Heavy Gasoil	0	0	0
I-7xx	Heavy Gasoil	0	0	0
I-800	High Speed Diesel	0	0	0
I-80001	Zero Pour HSD	218786	709492	928278
I-875	Diesel (47 CI, −6 Pour)	0	0	0
I-876	Diesel (46 CI, −6 Pour)	0	320348	320348
I-876ZP	Diesel (46 CI, −18 Pour)	0	0	0
I-884	Diesel (47 CI, 0.5 Sulf)	0	0	0
I-885	Diesel (50 CI, 0.5 Sulf)	5235	37270	42505
I-888	Diesel (50 CI, −3 Pour)	650516	444690	1095206
I-88852	Diesel	0	0	0
I-8xx	Diesel	0	0	0

I-8xx	Diesel	0	0	0
I-8xx	Diesel	0	0	0
I-8xx	Diesel	0	0	0
I-8xx	Diesel	0	0	0
I-8xx	Diesel	0	0	0
I-8xx	Diesel	0	0	0
I-8xx	Diesel	0	0	0
I-892	Marine Diesel Oil	−104	−81	−185
I-928	Fuel Oil (2.8 S, 180 cSt)	0	0	0
I-933	Fuel Oil (3.5 S, 80 cSt)	0	0	0
I-955	Fuel Oil (3.5 S, 75 cSt)	0	0	0
I-957	Fuel Oil (3.5 S, 225 SUS)	0	0	0
I-960	Fuel Oil (3.5 S, 120 cSt)	0	0	0
I-96001		0	0	0
I-9601		0	0	0
I-9602		0	0	0
I-961	Fuel Oil (3.5 S, 180 cSt)	636558	581611	1218169
I-96107	Fuel Oil (4.0 S, 180 cSt)	0	291431	291431
I-962	Fuel Oil (3.5 S, 280 cSt)	0	0	0
I-96201		0	0	0
I-96202		0	0	0
I-971	Fuel Oil (4.0 S, 380 cSt)	0	0	0
I-97103		0	381455	381455
I-9xx	Fuel Oil	0	0	0
I-9xx	Fuel Oil	0	0	0
I-9xx	Fuel Oil	0	0	0
I-9xx	Fuel Oil	0	0	0
I-9xx	Fuel Oil	0	0	0
I-9xx	Fuel Oil	0	0	0
I-9xx	Fuel Oil	0	0	0
I-9xx	Fuel Oil	0	0	0
I-9xx	Fuel Oil	0	0	0
I-9xx	Fuel Oil	0	0	0
Pxxxx	Intermediate	0	0	0
Pxxxx	Intermediate	0	0	0
Pxxxx	Intermediate	0	0	0
Pxxxx	Intermediate	0	0	0
Pxxxx	Intermediate	0	0	0
Pxxxx	Intermediate	0	0	0
PBFUEL	Burner Fuel Oil	−255	−591	−846
PBHDSL	Diesel Oil ex Bombay High	0	0	0
PCTISO	Isomate Cutter	0	0	0
PCTTR	Cutter Stock	0	2868	2868
PDSDSL	Desulfurized Diesel	28404	51605	80009
PDSL	Diesel	29974	58972	88946
PFCCF	HVGO to FCCU	0	0	0

Screen 10
Allocated Production

GRADE	PRODUCT	BOC ALLOCATION	AOC ALLOCATION	TOTAL PRODUCED
PFCOO	Hvy Isomate to FCCU	−19625	−17725	−37350
PFDISO	HVGO to 2HDU	−48218	−113454	−161672
PGABS	Gasoline	0	0	0
PGPDSL	Diesel ex Gippsland	0	0	0
PISIMP	Isomate Import	0	0	0
PKERIM	Kerosene Import	0	0	0
PKERO	Kerosene	−2391	−3181	−5572
PLANTS	Plants & Lines Contents	0	0	0
PLCGAS	Lt Cat Naphtha	38087	77316	115403
PLLCN	L. Lt Cat Naphtha	617	926	1543
PLSRBS	B-220 LSR	0	0	0
PLTISO	Lt. Isomate	−5748	−20118	−25866
PMEISO	Med. Isomate Cutter	−4470	−5296	−9766
PMIDSL	CDU 4A M/I Diesel	−6129	2780	−3349
PMSR	Sour MSR Naphtha	−1239	−1897	−3136
POMRSD	Oman Resid Import	0	0	0
PPOLY	Polymer Gasoline	3413	5980	9393
PP90R	90 RON Reformate	0	0	0
PP93R	93 RON Reformate	−1936	−3931	−5867
PP95R	95 RON Reformate	0	0	0
PP96R	96 RON Reformate	8658	15599	24257
PRESID	Residuum	326498	14237	340735
PRESMI	M/I Dsls in 6VDU Feed	0	0	0
PRES75	B725 in 6VDU Feed	0	0	0
PSKERO	Sweet Kerosene	263	395	658
PSLOPD	Slop Distillate	−1860	−2854	−4714
PSLOPO	Slop Oil	−1901	−3617	−5518
PSLOPT	Slop ex Marine Terminal	0	0	0
PSMCN	Sweet Med. Cat Naphtha	−10275	−19242	−29517
PSWMSR	Sweet MSR	0	−100	−100
PUFCHG	Unifiner Feed	−5849	−3570	−9419
PUJFBS	Sour Jet B.S.	0	0	0
PWCN	Whole Cat Naphtha	1200	1799	2999
P440BS	Jet Fuel Base Stock	0	0	0
LOSS		32948	67287	100235
	TOTAL	3215020	4822531	8037551
CRUDE		3215020	4822531	8037551

The following procedure is adopted in the allocation calculations:

1. Estimate the amount of diesel produced during the month.
2. Determine, from the refinery records, the quantity of pour point depressant actually blended during the month.
3. Estimate the pour point dosage actually used.
4. The pour point lowering actually achieved by the preceding pour point dosage is next determined using the correlation between pour point depressant dosage and pour point actually lowered.

The pour point of the diesel with pour point depressant should be used in the allocation LPs. For example, if a diesel pour point without additive is +9°C and the addition of 120 ppm pour point depressant lowers the pour point to 6°C, then the lowered pour point of 6°C or the corresponding pour index should be used in the Bombay high diesel properties, to blend the diesel to the required pour point specification.

"IN TANK" SALE AND PURCHASE BETWEEN THE PARTICIPANTS

Sale of products from one partner to another is referred to as *in-tank sales*, since no physical movement of the product is called for. Adjustment is done to the opening inventories of both participants to reflect the transaction.

EXAMPLE 16-8

Participant AOC has agreed to sell 50 mb of RON 97 gasoline (I-397) to participant BOC during the month. This transaction is reflected in the allocation report of the month, as follows.

For participant AOC, the opening Inventory of I-397 is adjusted by −50,000 bbl by entering this figure in the "adjust" column. For participant BOC, the opening Inventory of I-397 is adjusted by +50,000 bbl by entering this figure in the "adjust" column. The overall result of these entries would be that AOC's closing inventory of I-397 decreases by 50,000 bbl and that of BOC increases by 50,000 bbl.

REBLENDING FINISHED PRODUCTS

Minor reprocessing and reblending of stocks is also reflected in the inventories tables as adjustments, as shown by the following examples.

1. AOC reprocessed 62,247 bbl of I-434 (jet fuel) during a month, as reflected in the "adjustment" column of the inventory table, on the basis of I-434 equivalency:

GRADE	ADJUSTMENT, bbl
I-434	−62247
I-150	−2179
I-201	32555
I-440	31871
TOTAL	62247

2. Participant BOC transferred 1375 barrels of asphalt (I-1138D) from leaking drums to an asphalt tank (I-1138B), as reflected in the inventory table as follows:

GRADE	ADJUSTMENT, bbl
I-1138D	−1375
I-1138B	1375
TOTAL	0

ALLOCATION OF SULFUR

Allocation of sulfur follows the general principles of allocation. At the end of the month, the figures for actual production of sulfur are available. This is split in the ratio of LP production of sulfur in the final allocation LPs of the participants.

EXAMPLE 16-9

The total sulfur production processing of 200 mbpcd light Arabian crude and 45 mbpcd Arabian medium crude is 5200 tons during a month.

AOC's allocation LP shows a sulfur production of 4500 tons and BOC's allocation LP estimates sulfur production at 1700 tons. The actual sulfur production is 5200 tons.

The actual sulfur production is allocated in the ratio of sulfur production in the participants' allocation LPs as follows:

	Tons/Month	Weight %
AOC LP sulfur production	4500	72.58
BOC LP sulfur production	1700	27.42
Total LP estimate	6200	100.00
Actual refinery sulfur production	5200	
AOC sulfur allocation	3774	
BOC sulfur allocation	1426	

ALLOCATION OF REFINERY FUEL CONSUMPTION

Many refineries burn natural gas as fuel, apart from the off gases produced by the processing units. The refinery purchase of natural gas from outside sources and the actual consumption of gas must be allocated to the participants in the ratio of their actual usage of the gas.

Consumption of purchased gas can be split into two parts: base load consumption and process unit consumption.

- The base load consumption represents the energy demand when all the process units are temporarily shutdown. It is the energy demand from off-site (utilities generation, storage tanks, etc.) and other miscellaneous usage (marine terminal consumption, lighting for refinery and offices, etc.).
- Process unit consumption is the fuel used in the fired heaters of the processing units, or as a feedstock for hydrogen production.

The actual refinery consumption of natural gas is allocated to the participants in the following manner:

1. The total base load of the refinery is known from the historical data, and it is assumed to remain constant. It is unaffected by the refinery crude run. This base load consumption is allocated to the participant in the ratio of its crude run.

2. The processing unit gas consumption for a participant is known from its allocation LP.
3. The sum of these two gives the estimated gas consumption of a participant.
4. The actual refinery gas consumption is allocated to the participant in the ratio of the participants gas consumption as estimated in Step 3.

An example of the gas allocation follows:

EXAMPLE 16-10

Estimate the natural gas allocation for a joint ownership refinery between its two participants AOC and BOC. During the month in question, net natural gas import was 2738 million standard cubic feet. The refinery has an average base load demand of 1023 million standard cubic feet. The crude run of the two participants during the month was 84.26 and 15.74% of the total crude run.

PARTICIPANT	CRUDE RATIO, (1)	LP GAS IMPORT, mmscf (2)	BASE LOAD, mmscf (3)	BASE LOAD+LP, mmscf (4)	REFINERY GAS IMPORTS, mmscf (5)	GAS ALLOCATION, mmscf (6)
AOC	0.8426	1907	862	2769		2204
BOC	0.1574	510	161	671		534
TOTAL	1.0000	2417	1023	3440	2738	2738

Here, the refinery base load is split between the participants in the ratio of their crude run (column 3). Allocation LPs gas imports of the participants is shown in column 2. Actual refinery gas consumption is allocated to the participants in the ratio of their (LP + base load) estimate, as per column 4.

EXAMPLE 16-11, SIMULATION OF REDUCTION IN A CONVERSION UNIT SEVERITY

During the processing of a waxy crude, the mild hydrocracker unit severity was reduced, for 8 days, by lowering the feed rector inlet temparature by 10°F. The normal unit operating data follows:

	LV%	bpcd
FEED	1.0000	50,000
PRODUCTS		
NAPHTHA	0.0170	
KEROSENE	0.0014	
DIESEL	0.2700	
HVGO	0.7116	

The objective is to reflect this change in the participants' allocation.

The net result of a drop in reactor temperature is that conversion of HVGO feed to diesel is reduced. Here, we first estimate the total loss in diesel barrels during the unit's 8 days operation at reduced severity.

As per the catalyst supplier's manual, the change in conversion with a drop in reactor inlet temperature is as follows:

REACTOR TEMPERATURE, °F	CONVERSION LV%
730	27.0
725	24.5
715	19.0

The loss in diesel yield is, by interpolation of these data, estimated at 5.25% by volume.

$$\text{Potential diesel lost to fuel oil} = 50{,}000 \times 0.0525 \times 8$$
$$= 21{,}000\,\text{bbl}$$

This diesel loss to fuel oil is simulated in the allocation LPs of the participants as follows. The diesel lost to fuel oil is split in the ratio of the equity of the participants:

PARTICIPANT	EQUITY RATIO	DIESEL LOSS, bbl
AOC	0.6000	12600
BOC	0.4000	8400
TOTAL	1.0000	21000

In the allocation LPs of the participants, the following amounts of diesel from the mild hydrocracker unit is forced into balancing-grade fuel oil (I-961):

$$\begin{aligned} \text{AOC} &= 12600/30 \\ &= 420\,\text{bbl/day} \\ \text{BOC} &= 8400/30 \\ &= 280\,\text{bbl/day} \end{aligned}$$

The Allocation LPs are run again to simulate this diesel dumping into fuel oil.

ELIMINATION OF NEGATIVE INVENTORY

By the very nature of allocation procedures, it has been found that, in certain product grades, the participants develop negative inventories, particularly, in the those products whose total physical inventory in the refinery is zero and those produced or sold very infrequently by the respective participant.

We want to bring product allocation book inventories in line with the physical inventories. This cleanup facilitates the accounting inventories by eliminating long-outstanding static balances and correcting inconsistencies.

Mechanism of Adjustment

Consider a fixed-grade product, say, I-211 (naphtha) for which the inventory situation is as follows:

PARTICIPANT	INVENTORY, bbl	REFINERY PHYSICAL INVENTORY, bbl
AOC	43691	
BOC	−43691	
TOTAL	0	0

This implies that participant BOC has overlifted 43,691 bbl of I-211, which actually belonged to AOC.

To eliminate the negative inventory of BOC, AOC gives BOC, 43,961 bbl of I-211. After this transfer, the situation is as follows:

PARTICIPANT	INVENTORY, bbl	ADJUSTMENT, bbl	AFTER ADJUSTMENT, bbl
AOC	43691	−43691	0
BOC	−43691	43691	0
TOTAL	0		0

As AOC has given 43,691 bbl of grade I-211 to BOC, it must receive from BOC an equivalent volume of balancing-grade products, based on product equivalency of I-211.

The product equivalency of I-211 is as follows:

I-211 = 1.0000

I-150 = 0.0245

I-201 = −0.3110

I-210 = 1.2865

Thus, the balancing grades are adjusted as follows:

PARTICIPANT	I-150, bbl	I-201,bbl	I-210, bbl
AOC	1070	−13588	56208
BOC	−1070	13588	−56208

This procedure is repeated for all fixed grade products to eliminate their negative inventories.

Sometimes, it is possible to eliminate the negative inventories of a participant for a particular grade by offsetting against those of some other grade in the same product group. This is possible only if the equivalency of all grades under consideration is identical as shown next. Consider participants AOC and BOC's asphalt grade inventories, in barrels, as follows.

The equivalency of all product grades (asphalt) is identical.

GRADE	INVENTORY		
	AOC	BOC	REFINERY, PHYSICAL
I-1129D	−14857	16710	1853
I-1138D	10199	15808	26007
I-1149D	4697	−4697	0
TOTAL	39	27821	27860

1. Consider that AOC buys 14,857 bbl grade I-1129D from BOC. After this transaction, I-1129D inventories become as follows:

   AOC = 0

   BOC = 1853 bbl

   Physical = 1853 bbl

2. Now BOC buys 4697 bbls of I-1149D from AOC to offset its negative inventory. After this, I-1149D inventories become as follows:

   AOC = 0

   BOC = 0

   Physical = 0

3. BOC buys another 10,160 bbl of grade I-1138D from AOC. This makes the sale and purchases of both participants exactly equal to each other; that is, 14,857 bbl. The effect on the inventory of I-1138D is as follows:

   AOC = 39 bbl

   BOC = 25,968 bbl

   Physical = 26,007 bbl

Thus, we have eliminated negative inventories of I-1129D and I-1149D, and the position of inventories is as follows:

GRADE	AOC, bbl	BOC, bbl	PHYSICAL, bbl
I-1129D	0	1853	1853
I-1138D	39	25968	26007
I-1149D	0	0	0
TOTAL	39	27821	27860

There is no change in the participants' overall asphalt inventories as a result of elimination of negative inventories.

Balancing-Grade Products

The balancing-grade products generally do not develop negative inventories, as both participants regularly produce these grades. In some rare instances, a particular balancing grade may develop negative inventory if one participant chooses not to produce that particular grade. The solution is either to change the balancing grade or effect an "in-tank" sale or purchase of the product from the other participant, who has positive inventory of that grade. For example, consider the inventory position of the participants, for balancing-grade naphtha I-201, as follows:

$$\text{AOC} = -25{,}000\,\text{bbl}$$
$$\text{BOC} = 80{,}000\,\text{bbl}$$
$$\text{Total} = 55{,}000\,\text{bbl}$$

AOC's negative inventory can be eliminated by an "in-tank" sale of 25,000 bbl I-201 from BOC, after which the position becomes as follows:

$$\text{AOC} = 0$$
$$\text{BOC} = 55{,}000\,\text{bbl}$$
$$\text{Total} = 55{,}000\,\text{bbl}$$

CHAPTER SEVENTEEN

Available Tankage Capacity

In an export refinery, the estimates of the total tankage capacity available for storage of various product groups are required for preparing inventory availability forecasts for product shipping. In single-ownership refineries, the entire tankage capacity is available to sustain the export operation. In joint-ownership refineries, the available tankage capacity for every product group is split between the participants in the ratio of their equity in the refinery. Each participant then uses its share of tankage capacity to hold its inventory for its product shipping.

The procedure for estimating the total available tankage capacity and its allocation to the participants is described in this chapter.

Each participant must maintain its product inventory levels within the storage capacity allocated to it. The tankage capacity allocation is necessary to prepare inventory forecasts of the participants, which is discussed in later chapters.

ESTIMATION OF TOTAL AVAILABLE CAPACITY

The procedure for estimating the total available tankage capacity for a product group is as follows:

1. All available tanks in the given product group service are listed. Only tanks in service for the given product group are considered. The tanks in service of a given product group keep changing. Every time a tank goes out of service for scheduled or unscheduled maintenance or a new tank is added in the service of a given product group, the tankage list must be updated.

2. Tanks in service in each group are listed along with the following data:

   Maximum gross volume (the physical volume of the tank).
   Maximum net volume (the total gross volume multiplied by a factor to compute the maximum net for a group).
   Working ullage (the space required in tanks for operational reasons, such as correcting tank composition or disengaging vapors; a working ullage is specified for a group of tanks).
   Minimum heel (the volume of tank below minimum gauge).
   Available working stock (the volume above the minimum gauge to provide suction head to the pump transferring products from the tank).

MINIMUM AND MAXIMUM INVENTORY (LI AND HI)

Each tank has some dead storage volume, which is necessary to separate any solid entrainment and water and so forth. This inventory or holdup is not available for export but, nevertheless, is necessary for operation of the tank. The total minimum inventory volume, which must be maintained for operational reasons, is termed LI. A tank LI is calculated as follows:

$$\mathrm{LI} = R + S$$

where R is the minimum heel and S is the available working stock.

The tank volume between the operational maximum level and tank's physical volume is termed HI. This volume is used to correct the tank blend composition, if so required. A tank HI is calculated as follows:

$$\mathrm{HI} = P - Q$$

where P is the maximum net volume of tank and Q is the working ullage.

TANKAGE CAPACITY AVAILABLE FOR PRODUCT STORAGE

The tank volume between LI and HI is actually available for storage of the products. Therefore,

$$\text{Available capacity} = \mathrm{HI} - \mathrm{LI}$$

PRODUCT GROUPS

Refinery product grades are combined into groups. For example, all light naphtha grades with a specific range of density, RVP, or distillation specifications constitute the LSR group, and the different grades of gasolines come under the gasoline group. A typical refinery might group products as follows:

GROUP	PRODUCT
LSR	LIGHT NAPHTHAS
WSR	WHOLE RANGE NAPHTHAS
KEROSENE	KEROSENE/JET FUELS
DIESEL	AUTOMOTIVE DIESELS
BDSL	BLACK OR MARINE DIESELS
HVGO	HEAVY VACUUM GAS OILS
FO	FUEL OILS
ASPHALT	ASPHALT GRADES

ALLOCATION OF TANKAGE CAPACITY

In joint-ownership refineries, HI and LI are computed for every group of tanks, and the HI and LI for the group are split in the ratio of the participants' equity in the refinery.

EXAMPLE 17-1

In a joint-ownership refinery, participant AOC has a 60% share and participant BOC has 40% share. The tanks in service for each product group and the tank data are shown in Tables 17-1 to 17-10. Estimate the tankage capacity available to each participant.

The solution is

$$\text{Participant AOC equity in refinery} = 40\%$$
$$\text{Participant BOC equity} = 60\%$$

A summary of HI and LI is made from the tankage data. The HI and LI are split in the ratio of the participants' equity. The capacity available to

Table 17-1
LPG Group Tanks

TANK NO.	MAXIMUM GROSS (1) O	MAXIMUM NET (2) $P = O \times 0.9570$	WORKING ULLAGE (3) Q	MAX. NET WORKING INVENTORY (4) $HI = P - Q$	MINIMUM HEEL (5) R	AVAILABLE WORKING STOCK (6) S	MINIMUM NET WORKING ULLAGE (7) $LI = R + S$
080	3.7	3.541			0.100		
081	3.7	3.541			0.100		
082	2.3	2.201			0.100		
083	2.3	2.201			0.100		
084	3.7	3.541			0.100		
085	3.7	3.541			0.100		
TOTAL	19.4	18.566	4.000	14.566	0.600	3.400	4.000

NOTES: *LI IS MINIMUM NET WORKING ULLAGE.*
HI IS MAXIMUM NET WORKING INVENTORY.
CALCULATION:
CALCULATE THE TOTAL HI AND LI FOR EACH PRODUCT GROUP.
THE CALCULATIONS OF HI AND LI PRESENTED HERE ARE SELF-EXPLANATORY.
ALL FIGURES ARE IN THOUSAND BARRELS.
ONLY TANKS IN SERVICE ARE CONSIDERED FOR COMPUTATION OF HI AND LI.
MAXIMUM NET IS COMPUTED FROM MAXIMUM GROSS BY MULTIPLYING WITH A FACTOR,
AND THIS FACTOR, 95.7% FOR GROUP LPG, MAY BE DIFFERENT FOR EVERY PRODUCT GROUP.
COLUMNS 4 AND 7 SHOW HI AND LI COMPUTATIONS.

Table 17-2
Light Naphtha Group Tanks

TANK NO.	MAXIMUM GROSS (1) O	MAXIMUM NET (2) $P = O \times 0.9766$	WORKING ULLAGE (3) Q	MAX. NET WORKING INVENTORY (4) $HI = P - Q$	MINIMUM HEEL (5) R	AVAILABLE WORKING STOCK (6) S	MINIMUM NET WORKING ULLAGE (7) $LI = R + S$
356	19	18.555			2.000		
457	14	13.672			2.000		
459	12	11.719			2.000		
460	12	11.719			2.000		
476	19	18.555			2.000		
944	191	186.531			20.000		
945	187	182.624			21.000		
946	165	161.139			22.000		
TOTAL	619	604.515	50.000	554.515	73.000	50.000	123.000

NOTES: *LI IS MINIMUM NET WORKING ULLAGE.*
HI IS MAXIMUM NET WORKING INVENTORY.
CALCULATION:
CALCULATE THE TOTAL HI AND LI FOR EACH PRODUCT GROUP.
THE CALCULATIONS OF HI AND LI PRESENTED HERE ARE SELF-EXPLANATORY.
ALL FIGURES ARE IN THOUSAND BARRELS.
ONLY TANKS IN SERVICE ARE CONSIDERED FOR COMPUTATION OF HI AND LI.
MAXIMUM NET IS COMPUTED FROM MAXIMUM GROSS BY MULTIPLYING WITH A FACTOR, AND THIS FACTOR IS 97.66% FOR THE GROUP LSR.
COLUMNS 4 AND 7 SHOW HI AND LI COMPUTATIONS.

Table 17-3
Whole Range Naphtha Group Tanks

TANK NO.	MAXIMUM GROSS (1) O	MAXIMUM NET (2) $P = O \times 0.9766$	WORKING ULLAGE (3) Q	MAX. NET WORKING INVENTORY (4) $HI = P - Q$	MINIMUM HEEL (5) R	AVAILABLE WORKING STOCK (6) S	MINIMUM NET WORKING ULLAGE (7) $LI = R + S$
965	527	514.668			41.000		
760	518	505.879			35.000		
925	120	117.192			16.000		
TOTAL 1	1165	1137.739	80.000	1057.739	92.000	50.000	142.000
TOTAL 2	638	623.071	55.000	568.071	51.000	30.000	81.000

NOTES: LI IS MINIMUM NET WORKING ULLAGE.
HI IS MAXIMUM NET WORKING INVENTORY.
CALCULATION:
CALCULATE THE TOTAL HI AND LI FOR EACH PRODUCT GROUP.
THE CALCULATIONS OF HI AND LI PRESENTED HERE ARE SELF-EXPLANATORY.
ALL FIGURES ARE IN THOUSAND BARRELS.
TANK 965 ALLOCATED FOR PARTICIPANT BOC USAGE ONLY.
ONLY TANKS IN SERVICE ARE CONSIDERED FOR COMPUTATION OF HI AND LI.
MAXIMUM NET IS COMPUTED FROM MAXIMUM GROSS BY MULTIPLYING WITH A FACTOR,
AND THIS FACTOR IS 97.66% FOR THE GROUP WSR.
COLUMNS 4 AND 7 SHOW HI AND LI COMPUTATIONS.
TOTAL 1 INCLUDES TANK 965 CAPACITY.
TOTAL 2 EXCLUDES TANK 965 CAPACITY.

Table 17-4
Gasoline Group Tanks

TANK NO.	MAXIMUM GROSS (1) *O*	MAXIMUM NET (2) $P = O \times 0.9570$	WORKING ULLAGE (3) *Q*	MAX. NET WORKING INVENTORY (4) $HI = P - Q$	MINIMUM HEEL (5) *R*	AVAILABLE WORKING STOCK (6) *S*	MINIMUM NET WORKING ULLAGE (7) $LI = R + S$
129	32	30.624			6.000		
250	12	11.484			1.000		
469	7	6.699			1.000		
470	7	6.699			1.000		
710	126	120.582			16.000		
905	127	121.539			16.000		
906	117	111.969			9.000		
915	121	115.797			17.000		
916	122	116.754			16.000		
924	115	110.055			16.000		
933	123	117.711			15.000		
934	121	115.797			15.000		
935	121	115.797			15.000		
TOTAL	1151	1101.507	325.000	776.507	144.000	155.000	299.000

NOTES: *LI IS MINIMUM NET WORKING ULLAGE.*
HI IS MAXIMUM NET WORKING INVENTORY.
CALCULATION:
CALCULATE THE TOTAL HI AND LI FOR EACH PRODUCT GROUP.
THE CALCULATIONS OF HI AND LI PRESENTED HERE ARE SELF-EXPLANATORY.
ALL FIGURES ARE IN THOUSAND BARRELS.
ONLY TANKS IN SERVICE ARE CONSIDERED FOR COMPUTATION OF HI AND LI.
MAXIMUM NET IS COMPUTED FROM MAXIMUM GROSS BY MULTIPLYING WITH A FACTOR, AND THIS FACTOR IS 95.70% FOR THE GROUP GASO.
COLUMNS 4 AND 7 SHOW HI AND LI COMPUTATIONS.

Table 17-5
Kerosene Group Tanks

TANK NO.	MAXIMUM GROSS (1) O	MAXIMUM NET (2) $P = O \times 0.9821$	WORKING ULLAGE (3) Q	MAX. NET WORKING INVENTORY (4) $HI = P - Q$	MINIMUM HEEL (5) R	AVAILABLE WORKING STOCK (6) S	MINIMUM NET WORKING ULLAGE (7) $LI = R + S$
105	88	86.425			10.000		
114	75	73.658			15.000		
115	84	82.496			16.000		
730	124	121.780			16.000		
740	134	131.601			21.000		
751	449	440.963			55.000		
901	73	71.693			11.000		
902	77	75.622			11.000		
TOTAL	1104	1084.238	180.000	904.238	155.000	200.000	355.000

NOTES: *LI IS MINIMUM NET WORKING ULLAGE.*
HI IS MAXIMUM NET WORKING INVENTORY.
CALCULATION:
CALCULATE THE TOTAL HI AND LI FOR EACH PRODUCT GROUP.
THE CALCULATIONS OF HI AND LI PRESENTED HERE ARE SELF-EXPLANATORY.
ALL FIGURES ARE IN THOUSAND BARRELS.
ONLY TANKS IN SERVICE ARE CONSIDERED FOR COMPUTATION OF HI AND LI.
MAXIMUM NET IS COMPUTED FROM MAXIMUM GROSS BY MULTIPLYING WITH A FACTOR,
AND THIS FACTOR IS 98.21% FOR THE GROUP KERO.
COLUMNS 4 AND 7 SHOW HI AND LI COMPUTATIONS.

Table 17-6
Automotive Diesel Group Tanks

TANK NO.	MAXIMUM GROSS (1) O	MAXIMUM NET (2) $P=O\times 0.9801$	WORKING ULLAGE (3) Q	MAX. NET WORKING INVENTORY (4) $HI=P-Q$	MINIMUM HEEL (5) R	AVAILABLE WORKING STOCK (6) S	MINIMUM NET WORKING ULLAGE (7) $LI=R+S$
701	125	122.513			3.000		
702	126	123.493			3.000		
706	604	591.980			8.000		
711	125	122.513			3.000		
721	124	121.532			3.000		
722	131	128.393			3.000		
731	123	120.552			3.000		
732	85	83.309			2.000		
913	126	123.493			3.000		
923	126	123.493			3.000		
452	7	6.861			0.200		
455	4	3.920			1.000		
41	1	0.980			0.000		
910	5	4.901			0.000		
917	1	0.980			0.000		
TOTAL	1713	1678.911	240.000	1438.911	35.200	175.000	210.200

NOTES: *LI IS MINIMUM NET WORKING ULLAGE.*
HI IS MAXIMUM NET WORKING INVENTORY.
CALCULATION:
CALCULATE THE TOTAL HI AND LI FOR EACH PRODUCT GROUP.
THE CALCULATIONS OF HI AND LI PRESENTED HERE ARE SELF-EXPLANATORY.
ALL FIGURES ARE IN THOUSAND BARRELS.
ONLY TANKS IN SERVICE ARE CONSIDERED FOR COMPUTATION OF HI AND LI.
MAXIMUM NET IS COMPUTED FROM MAXIMUM GROSS BY MULTIPLYING WITH A FACTOR, AND THIS FACTOR IS 98.01% FOR THE GROUP DSL.
COLUMNS 4 AND 7 SHOW HI AND LI COMPUTATIONS.

Table 17-7
Black or Marine Diesel Group Tanks

TANK NO.	MAXIMUM GROSS (1) O	MAXIMUM NET (2) $P = O \times 0.9842$	WORKING ULLAGE (3) Q	MAX. NET WORKING INVENTORY (4) $HI = P - Q$	MINIMUM HEEL (5) R	AVAILABLE WORKING STOCK (6) S	MINIMUM NET WORKING ULLAGE (7) $LI = R + S$
912	28	27.558			1.000		
TOTAL	28	27.558	5.000	22.558	1.000	5.000	6.000

NOTES: *LI IS MINIMUM NET WORKING ULLAGE.*
HI IS MAXIMUM NET WORKING INVENTORY.
CALCULATION:
CALCULATE THE TOTAL HI AND LI FOR EACH PRODUCT GROUP.
THE CALCULATIONS OF HI AND LI PRESENTED BELOW ARE SELF EXPLANATORY.
ALL FIGURES ARE IN THOUSAND BARRELS.
ONLY TANKS IN SERVICE ARE CONSIDERED FOR COMPUTATION OF HI AND LI.
MAXIMUM NET IS COMPUTED FROM MAXIMUM GROSS BY MULTIPLYING WITH A FACTOR, AND THIS FACTOR IS 98.42% FOR THE GROUP BDSL.
COLUMNS 4 AND 7 SHOW HI AND LI COMPUTATIONS.

Table 17-8
Heavy Vacuum Gas Oil (HVGO) Group Tanks

TANK NO.	MAXIMUM GROSS (1) O	MAXIMUM NET (2) $P = O \times 0.9577$	WORKING ULLAGE (3) Q	MAX. NET WORKING INVENTORY (4) $HI = P - Q$	MINIMUM HEEL (5) R	AVAILABLE WORKING STOCK (6) S	MINIMUM NET WORKING LLAGE (7) $LI = R + S$
111	90	86.193			4.000		
705	569	544.931			8.000		
TOTAL	659	631.124	20.000	611.124	12.000	30.000	42.000

NOTES: *LI IS MINIMUM NET WORKING ULLAGE.*
HI IS MAXIMUM NET WORKING INVENTORY.
CALCULATION:
CALCULATE THE TOTAL HI AND LI FOR EACH PRODUCT GROUP.
THE CALCULATIONS OF HI AND LI PRESENTED HERE ARE SELF-EXPLANATORY.
ALL FIGURES ARE IN THOUSAND BARRELS.
ONLY TANKS IN SERVICE ARE CONSIDERED FOR COMPUTATION OF HI AND LI.
MAXIMUM NET IS COMPUTED FROM MAXIMUM GROSS BY MULTIPLYING WITH A FACTOR,
AND THIS FACTOR IS 95.77% FOR THE GROUP HVGO.
COLUMNS 4 AND 7 SHOW HI AND LI COMPUTATIONS.

Table 17-9
Fuel Oil Group Tanks

TANK NO. COL NO.	MAXIMUM GROSS (1) O	MAXIMUM NET (2) $P = O \times 0.9725$	WORKING ULLAGE (3) Q	MAX. NET WORKING INVENTORY (4) $HI = P - Q$	MINIMUM HEEL (5) R	AVAILABLE WORKING STOCK (6) S	MINIMUM NET WORKING ULLAGE (7) $LI = R + S$
703	122	118.645			3.000		
713	123	119.618			3.000		
714	84	81.690			2.000		
715	333	323.843			6.000		
723	121	117.673			3.000		
724	123	119.618			3.000		
734	208	202.280			5.000		
743	329	319.953			6.000		
904	89	86.553			2.000		
914	127	123.508			3.000		
TOTAL	1659	1613.378	350.000	1263.378	36.000	270.000	306.000

NOTES: *LI IS MINIMUM NET WORKING ULLAGE.*
HI IS MAXIMUM NET WORKING INVENTORY.
CALCULATION:
CALCULATE THE TOTAL HI AND LI FOR EACH PRODUCT GROUP.
THE CALCULATIONS OF HI AND LI PRESENTED HERE ARE SELF-EXPLANATORY.
ALL FIGURES ARE IN THOUSAND BARRELS.
ONLY TANKS IN SERVICE ARE CONSIDERED FOR COMPUTATION OF HI AND LI.
MAXIMUM NET IS COMPUTED FROM MAXIMUM GROSS BY MULTIPLYING WITH A FACTOR, AND THIS FACTOR IS 97.25% FOR THE GROUP FO.
COLUMNS 4 AND 7 SHOW HI AND LI COMPUTATIONS.

Table 17-10
Asphalt Group Tanks

TANK NO.	MAXIMUM GROSS (1) O	MAXIMUM NET (2) $P = O \times 0.9523$	WORKING ULLAGE (3) Q	MAX. NET WORKING INVENTORY (4) $HI = P - Q$	MINIMUM HEEL (5) R	AVAILABLE WORKING STOCK (6) S	MINIMUM NET WORKING ULLAGE (7) $LI = R + S$
180	14	13.332			0.400		
171	6	5.714			0.200		
172	6	5.714			0.200		
181	13	12.380			0.300		
183	14	13.332			0.400		
170	6	5.714			0.200		
907	15	14.285			0.300		
908	12	11.428			0.300		
TOTAL	86	81.898	20.000	61.898	2.300	12.000	14.300

NOTES: *LI IS MINIMUM NET WORKING ULLAGE.*
HI IS MAXIMUM NET WORKING INVENTORY.
CALCULATION:
CALCULATE THE TOTAL HI AND LI FOR EACH PRODUCT GROUP.
THE CALCULATIONS OF HI AND LI PRESENTED HERE ARE SELF-EXPLANATORY.
ALL FIGURES ARE IN THOUSAND BARRELS.
ONLY TANKS IN SERVICE ARE CONSIDERED FOR COMPUTATION OF HI AND LI.
MAXIMUM NET IS COMPUTED FROM MAXIMUM GROSS BY MULTIPLYING WITH A FACTOR,
AND THIS FACTOR IS 95.23% FOR THE GROUP ASPH.
COLUMNS 4 AND 7 SHOW HI AND LI COMPUTATIONS.

each participant is the difference between HI and LI for every product group. These calculations are shown in Table 17-11.

ULLAGE

Ullage is the difference between HI and the actual inventory volume in a tank. Such data are important for scheduling the product offloading and the size of individual shipments.

TANK OWNED BY ONE PARTICIPANT

In joint-ownership refineries, a participant may build a tank exclusively for its own use. The capacity allocation procedure, in that case, is modified as follows:

1. Calculate the HI and LI for the group, including the exclusive tank as a part of the group.
2. Calculate the HI and LI for the group, assuming the tank is excluded.
3. The delta between the two HI and LI gives HI and LI for the exclusive tank.
4. Next HI and LI corresponding to calculations without the exclusive tank are used to allocate tankage capacity.
5. HI and LI of the exclusive tank are added to HI and LI of the participant for whose use the exclusive tank was provided.

EXAMPLE 17-2

Consider Tank 965 (Table 17-3), a WSR (naphtha) group tank built for the exclusive use of participant BOC. The calculation of LI and HI for the two participants follows:

	LI, mb	HI, mb
TOTAL EXCLUDING TANK 965	81	568
TOTAL INCLUDING TANK 965	142	1058
ESTIMATE FOR TANK 965	61	490
AOC ALLOCATION	$=0.6 \times 81$	$=568 \times 0.6$
	$=48.6$	$=340.8$
BOC ALLOCATION	$=81 \times 0.4+61$	$=568 \times 0.4+490$
	$=93.4$	$=717.2$

Table 17-11
Allocation of Available Tankage Capacity (All Figures in Thousand Barrels)

PRODUCT GROUP	TOTAL LI (1)	TOTAL HI (2)	REFINERY AVAILABLE CAPACITY (3=2−1)	AVAILABLE CAPACITY ALLOCATION		ALLOCATION OF LI		ALLOCATION OF HI	
				AOC (4=3×0.6)	BOC (5=3×0.4)	AOC (6=1×0.6)	BOC (7=1×0.4)	AOC (8=2×0.6)	BOC (9=2×0.4)
LPG	4	15	11	7	4	2	2	9	6
LSR	123	555	432	259	173	74	49	333	222
WSR	142	1058	916	550	366	85	57	635	423
GASOLINE	299	777	478	287	191	179	120	466	311
KEROSENE	355	904	549	329	220	213	142	542	362
DIESEL	210	1439	1229	737	492	126	84	863	576
BLACK DIESEL	6	23	17	10	7	4	2	14	9
HVGO	42	611	569	341	228	25	17	367	244
FUEL OIL	306	1263	957	574	383	184	122	758	505
ASPHALT	14	62	48	29	19	8	6	37	25

NOTE: AVAILABLE CAPACITY FOR ANY PRODUCT GROUP IS THE DIFFERENCE BETWEEN ITS HI AND LI WHICH IS ALOCATED TO PARTICIPANTS IN THE RATIO OF THEIR EQUITY; IN THIS EXAMPLE, 60/40.

CEDING OF REFINERY CAPACITY

Ceding is a situation in which one participant in the joint refinery chooses not to utilize its full share of refinery capacity for a prolonged period. In such a situation, it has the option to offer the unutilized capacity to the other participant. If the other participant agrees to accept the unutilized capacity, the receiving participant has to pay agreed-on charges to the leasing participant to compensate it for the depreciation of its assets and amortization of its investment in the refinery. If a participant leases part of unutilized processing unit capacity, the tankage capacity is also ceded to the receiving participant in that proportion.

Consider, for example, a refinery with crude running capacity at 260 mbpcd. The capacity rights of AOC and BOC are 60/40. The crude distillation unit capacity is split as follows:

$$\text{AOC capacity share, CDU} = 156\,\text{mbpcd}$$
$$\text{BOC capacity share, CDU} = 104\,\text{mbpcd}$$

If BOC wishes to run only, say, 40 mbpcd for a prolonged period, BOC can cede $(104 - 40) = 64$ mbpcd of its unutilized refining capacity. If AOC accepts this ceded capacity, its capacity would be $(156 + 64) = 220$ mbpcd and BOC's refining capacity would be 40 mbpcd during the period of ceding. The capacity of the downstream units ceded under the ceding agreement would equal (64/260) or 0.2461 times the downstream unit capacity, as shown in Table 17-12.

CEDING OF TANKAGE CAPACITY

The procedure for ceding tankage capacity follows the principle of ceding unit capacities. In the preceding example, the ceded tankage capacity in every product group is (64/260) of available tankage capacity. After ceding, the split of available tankage capacity is as follows:

PARTICIPANT	BEFORE CEDING	AFTER CEDING
AOC	0.6000	0.8461
BOC	0.4000	0.1539
TOTAL	1.0000	1.0000

Table 17-12
Ceded Downstream Unit Capacities (in mbpcd)

UNIT	TOTAL	AOC BEFORE CEDING	BOC BEFORE CEDING	CEDED	AOC AFTER CEDING	BOC AFTER CEDING
CRUDE DISTILLATION	260.0	156	104	64	220	40
FCCU	36.0	21.6	14.4	8.9	30.5	5.5
HYDROCRACKER	50.0	30.0	20.0	12.3	42.3	7.7
VISBREAKER	20.0	12.0	8.0	4.9	16.9	3.1
CAT REFORMER	15.0	9.0	6.0	3.7	12.7	2.3
DIESEL HDS	20.0	12.0	8.0	4.9	16.9	3.1

NOTES: AOC CRUDE DISTILLATION UNIT (CDU) CAPACITY = 60% AND BOC CDU CAPACITY = 40% TOTAL REFINERY CDU CAPACITY WITHOUT CEDING. CEDING OF DOWNSTREAM UNITS = (CDU CEDED CAPACITY)/(TOTAL CDU CAPACITY) × (UNIT TOTAL CAPACITY).

The split of total HI and LI for every group tankage is done in this ratio. Tankage capacity available to participants after ceding is shown in Table 17-13.

It is important to analyze the effect of tankage ceding on the operability of participant BOC's part of the refinery. Depending on the size and frequency of lifting, minimum ullage equal to about 17 days production is required. Referring to Table 17-14, the available storage capacity

Table 17-13
Tankage Situation under Ceding (in mb)

GROUP	TOTAL LI	AOC LI	BOC LI	TOTAL HI	AOC HI	BOC HI	TOTAL CAPACITY	AOC CAPACITY	BOC CAPACITY
LPG	4	3	1	15	13	2	11	9	2
LSR	123	104	19	555	470	85	432	366	66
WSR	142	120	22	1058	895	163	916	775	141
GASO	299	253	46	777	657	120	478	404	74
KERO	355	300	55	904	765	139	549	465	84
DSL	210	178	32	1439	1218	221	1229	1040	189
BDSL	6	5	1	23	19	4	17	14	3
HVGO	42	36	6	611	517	94	569	481	88
FUEL	306	259	47	1263	1069	194	957	810	147
ASPH	14	12	2	62	52	10	48	41	7
TOTAL	1501	1270	231	6707	5675	1032	5206	4405	801

Table 17-14
Effect of Tankage Ceding on Refinery Operability

PRODUCT GROUP	BOC AVAILS, mb (1)	BOC PROD. RATE mb (2)	AVAILABLE ULLAGE days (3=1/2)	REQUIRED ULLAGE, mb (4=2×17)	ULLAGE SHORTFALL, mb (5=4−1)
LPG	1.7	0		0	−1.7
LSR	66.5	5.4	12.3	91.8	25.3
WSR	140.9	5.8	24.3	98.6	−42.3
GASO	73.5	3.5	21.0	59.5	−14.0
KERO	84.5	6	14.1	102	17.5
DSL	189.1	13.3	14.2	226.1	37.0
BDSL	2.6	0.2	13.1	3.4	0.8
HVGO	87.5	0		0	−87.5
FUEL	147.2	7.9	18.6	134.3	−12.9
ASPH	7.4	1.5	4.9	25.5	18.1
TOTAL	800.9	43.6	18.4	741	

NOTES: (1) AVAILABLE CAPACITY (HI − LI) IS FROM THIS TABLE.
(2) APPROXIMATE YIELD PATTERN FROM PARTICIPANT BOC UNIT CAPACITIES AFTER CEDING.
(3) NUMBER OF DAYS OF ULLAGE FROM RATED PRODUCTION.
(4) VOLUME REQUIRED TO CONTAIN 17 DAYS OF RATED PRODUCTION.
(5) POSITIVE FIGURES INDICATE SHORTFALLS AND NEGATIVE FIGURES DENOTE SURPLUS IN ULLAGE, COL 5 = COL 4 − COL 1.

after ceding, in certain product groups is less than the minimum required, while in some others, the available tankage capacity is more than that required for operational reasons. Therefore, tankage capacity would have to be rationalized by switching tanks from one product group to another in such a manner that every product group has a minimum ullage equal to 17 days of its estimated production rate after ceding.

CHAPTER EIGHTEEN

Shipping Inventory Forecasts

To plan shipment of products and to schedule such shipments, the marketing department needs to know estimates of each product group's production rate and the inventory available for shipping. This information is required several weeks in advance of the actual shipment, for product sale and chartering vessels to transport products. Also, it is important to maintain the inventory level of each product group within the ullage available to the group in the refinery tankage. This is done by timely transport of products from the refinery tanks. Failure of timely off-take can cause a crisis situation in the refinery, forcing cutdown of crude throughput, shutdown or feed reduction in some downstream units, or dumping distillates into fuel oil.

To provide the marketing department with an accurate inventory forecast of all product groups, every week, the refinery estimates the production rate of each product group and uses these to revise the inventory forecasts for the remaining period of the month

WEEKLY PRODUCTION ESTIMATES

These estimates are done with a spreadsheet refinery model. Here various parameters, such as unit operating modes and blend components, can be controlled more closely than in a linear programming (LP) model, which may require a major modeling effort to reflect the actual refinery situations. The information on the feed rates, operation modes of the units, product blending, and tank inventory levels can be constantly updated, and thus these hand estimates are more realistic than the LP solutions, which are price driven.

Every week the product requirements are revised and a new estimate prepared for the remaining period of the month, incorporating the latest refinery information, such as unit throughputs, product qualities, product shipping schedule, and inventory. Also, product requirements or the specifications of a particular parcel may change, a ship may be delayed, or an important refinery unit have an unscheduled shutdown requiring revision in production estimates. If there is a major change in the normal running of refinery, such as shutting down of a key conversion unit, cut in crude rate, or any other cause that forces a significant change in the normal production rates of various product groups, a new estimate must be done to reflect these changes.

When a new estimate is made, the production rate estimate is revised. The revised production rate estimate is used for inventory forecasts for the remaining period. The production rate and inventory forecasts are updated every week.

The production rates and inventory forecasts are made for the following product groups, in a fuel refinery.

Liquefied petroleum gas (LPG).
Light straight run naphtha (LSR).
Whole straight run naphtha (WSR).
Gasolines (GASO).
Kerosene and jet fuels (KERO).
Automotive diesels (DSL).
Fuel oil (FO).
Asphalts (ASPH).

Production rates for the intermediate stocks are not generally required, as these are not exported.

REFINERY ESTIMATE SCHEDULE

The refinery estimates are revised every week for the remaining period of the month. For example, in a month, four estimates may be prepared as per the following schedule:

First estimate: From day 1 to day 30, 30 days.
Second estimate: From day 11 to day 30, 20 days.
Third estimate: From day 18 to day 30, 13 days.
Fourth estimate: From day 25 to 30, 6 days.

For the first estimate, the crude run and product grades to be made are as per the combined definitive operating plans (DOPs) of the participants. For the second estimate, the crude run is equal to the total DOP crude run minus the crude processed during the first 10 days of the month. The product target for the second estimate is the combined DOPs, plus any DOP revisions by the participants minus the product already made during the first period. Similarly, the targets for the third and fourth estimates are the parts of products still to be made.

PROCEDURE

SINGLE-OWNERSHIP REFINERIES

The procedure for establishing production rates in single-ownership refineries is straightforward. This consists of updating the refinery spreadsheet model with the latest data on available unit capacities, crude run rate, product blending, and the like. The model is run to establish refinery input and output balance. Refinery production rate estimate is used as a basis for inventory forecasts of various product groups.

JOINT-OWNERSHIP REFINERIES

The procedure for establishing production rates in joint-ownership refineries is more complex and described in detail. Here, not only the refinery production rate is established but also the production rate for every product group is split between the participants to establish separate inventory forecasts for the two participants. Here again, the refinery spreadsheet model is used to estimate the refinery production rate.

Participants' submitted DOPs form the basis for splitting refinery production in a given period. This rough allocation is used to prepare estimates of inventory available to each participant for shipping its product.

At the start of every month, the refinery receives DOPs from both the participants, which list information about the crude throughput and the product slate of each participant. Based on this information, the refinery makes its own estimates of production of various products during the month.

ALLOCATION OF DELTAS

The delta between the combined DOPs of the participants and the refinery estimates, as per spreadsheet program, for each product group is examined to determine the cause of that delta and delta assigned to the participant that caused the delta. If no particular reason can be identified, the delta is allocated to the participants in the ratio of their crude run. Production of a participant is the sum of its DOP plus the allocated delta as follows:

Participant production = participant DOP + delta

The production rate is worked out by dividing the participant's production by the number of days for which the estimates are made. In this way, the production and thus production rate estimate of the refinery during the week is split between each participant's production rate for every product group.

For the allocation of deltas for various product grades, the concept of pseudoequivalencies is used, based on how the product is blended in the refinery. Pseudoequivalency is in terms of other product grades and process stocks. Important differences from the equivalencies used for product allocation are as follows:

- Pseudoequivalency is used here for the sole purpose of estimating the production rate in terms of balancing grades and process stocks instead of only in terms of balancing grades for equivalencies used in product allocation.
- As the sole purpose of these equivalencies is to investigate the effect on product grades inventories any effect on process stocks is ignored.

The following examples should clarify the use of these pseudoequivalencies.

WSR (NAPHTHA)

The production of WSR in an estimate was 85 mb more than that listed in the combined DOPs. As no cause could be found for this additional production, this is allocated to the participants in the ratio of each one's crude run as follows:

PARTICIPANT	DOP, mb	ESTIMATES, mb	DELTA, mb	CRUDE RATIO
AOC	839		71	0.8354
BOC	96		14	0.1646
TOTAL	935	1020	85	1.0000

GASOLINE GROUP DELTAS

After an estimate, the total gasoline group delta between estimates and the DOPs was estimated as follows:

GRADE	DELTA, mb
I-387	25
I-390	24
I-395	0
I-397L	−105
I-397LL	30
TOTAL GROUP	−16

The cause of delta for certain grades could be identified, while for other grades, it could not be ascertained. In this case, delta is first allocated for grades for which the cause is known; the remaining delta is allocated to the participants in the ratio of their crude run.

Gasoline Grade 397L

The delta between I-397L production and the combined DOPs is −105 mb. This delta is split between the participants in the ratio of their crude run:

PARTICIPANT	CRUDE RATIO	DELTA, mb
AOC	0.8354	−88
BOC	0.1646	−17
TOTAL	1.0000	−105

The I-397L blend contains 25% reformate (RON 95). Therefore, a –88 production change releases 22 mb reformate for AOC. Hence, AOC's gasoline allotment decreases by 22 mb while the reformate pool increases by the same amount. Similarly, BOC's gasoline pool decreases by 4 mb and the reformate pool increases by the same amount, as follows:

STOCK	AOC, mb	BOC, mb
I-397L	–22	–4
REFORMATE (RON 95)	22	4
TOTAL	0	0

Gasoline Grade I-397LL

This grade is blended with 47% reformate and is made only for BOC. Extra production of 30 mb contains approximately 14 mb reformate, increasing the gasoline pool by that amount:

STOCK	AOC, mb	BOC, mb
I-397LL	0	14
REFORMATE (95RON)	0	–14
TOTAL	0	0

The remaining delta $= -16 - (-22 - 4 + 14)$ or -4 mb, which is allocated to the participants in their crude run ratio:

$$\text{AOC} = -3\text{mb}$$
$$\text{BOC} = 1\text{mb}$$
$$\text{AOC total delta} = (-22 - 3) = -25\,\text{mb}$$
$$\text{BOC total delta} = (-4 + 14 - 1) = 9\,\text{mb}$$

FUEL OIL

Estimates show that production of I-961 (80 cst) for AOC was 32 mb more than the DOP estimates. The effect of this delta on fuel and diesel pools was determined as follows. The refinery blends this low viscosity fuel oil by cutting normal I-961 grade with I-888 diesel as follows:

GRADE	PRODUCT	VOLUME, mb	VISCOSITY BLEND INDEX
I-961	FUEL OIL	27.8	460
I-888	DIESEL	4.2	−30
I-961 (80 cst)	FUEL OIL	32.0	398

Production of an additional 32 mb I-961 (80 cst) for AOC increases its fuel oil by (32 – 27.8) or 4.2 mb and decreases diesel by 4.2 mb as shown next:

GRADE	GROUP	PRODUCTION, mb
I-961 (80 cst)	FUEL OIL	32.0
I-961	FUEL OIL	−27.8
I-888	DIESEL	−4.2
TOTAL		0

DIESEL GRADES

In an estimate, the AOC production of diesel grade I-876 was increased by 79 mb to cover its lifting. Estimate its effect on AOC inventory.

The I-876 was blended in the refinery with I-888 diesel and I-440 to match the pour point of I-876 (21°F) as follows. Here, the I-876 delta is 79 mb.

GRADE	PRODUCT	VOLUME, mb	POUR INDEX
I-888	DIESEL	72.3	365
I-440	KEROSENE	6.7	46
I-876	DIESEL	79.0	338

The effect on the diesel and kerosene pool is as follows:

GRADE	GROUP	PRODUCTION, mb
I-876	DIESEL	79.0
I-888	DIESEL	−72.3
I-440	KEROSENE	−6.7
TOTAL		0

Here,

$$\text{AOC diesel pool} = (79.0 - 72.3)$$
$$= 6.7\,\text{mb}$$
$$\text{AOC kerosene pool} = -6.7\,\text{mb}$$

ASPHALT

In an estimate, production of asphalt was 10,000 barrels less for BOC due to lack of ullage. To estimate the effect of this change on other balancing-grade products the following the procedure was employed.

The delta of $-10,000\,\text{bbl}$ on asphalt production for BOC could be considered a blend of fuel oil and diesel, as per the asphalt equivalency, shown next:

GRADE	GROUP	VOLUME, mb	VISCOSITY BLEND INDEX
I-961	FUEL OIL	−17.6	460
I-888	DIESEL	7.6	−30
I-1138	ASPHALT	−10.0	832

As a result of this change, BOC gets $+17.6\,\text{mb}$ fuel oil and $-7.6\,\text{mb}$ diesel, as follows:

GRADE	GROUP	DELTA, mb
I-961	FUEL OIL	17.6
I-888	DIESEL	−7.6
I-1138	ASPHALT	−10.0
TOTAL		0

Examples of joint-ownership refinery weekly estimates, allocation of deltas, and determination of production rates during a month are shown in Tables 18-1 to 18-16. Table 18-1 shows the first refinery estimate and a comparison with combined participant DOPs and calculation of deltas. Table 18-2 show deltas by product groups. Table 18-3 shows the allocation of these deltas by participant, and Table 18-4 shows calculation of the production rate for all product groups based on the first estimate. Similar calculations are done for the second, third, and fourth estimates. Thus, production rates of various product groups are worked out for both the participants. In fact, the actual refinery production rate of various product groups has been split into the two participant's production rates. Whenever a new estimate is done, the new production rates replace the earlier ones in the refinery's inventory estimating program. Weekly estimates also ensure that crude run and product produced during the month is according to the refinery's DOP for that month.

INVENTORY AND ULLAGE FORCASTING SYSTEM (JOINT OWNERSHIP REFINERIES)

The inventory and ullage forcasting system (IUFS) is designed to provide information to the participants on their inventory and ullage situation up to 90 days ahead, to enable them to plan their product lifting schedule. Data on each product group are presented in a separate report. This is an on-line system in which participants can constantly update their shipping data, such as placing a new shipment, canceling an already-placed shipment, updating the estimated arrival time of expected shipments. The rest of the data are updated only by the refinery (see Tables 18-17 to 18-23).

The system presents estimates on the following, from present day to that 30–90 days ahead:

- The physical inventory of each saleable product group of the participants.
- The available inventory of each product group for shipment.
- The ullage available to each participant, for each product group, to further build their inventories for shipments.
- The estimated refinery production rate for every product group of the participants.

Table 18-1
Determination of Deltas between Refinary Estimates and Participants' DOPs, First Estimate (mb)

	AOC DOP	BOC DOP	COMBINED DOP	REFINERY ESTIMATE	DELTA
GRADE					
I-150	20	0	20	20	0
I-201	0	0	0	0	0
I-210	839	96	935	1020	85
I-220	0	165	165	165	0
I-387	0	0	0	25	25
I-390	66	0	66	90	24
I-395	120	15	135	135	0
I-397LL	0	60	60	90	30
I-397L	266	11	277	172	−105
I-397R	54	15	69	69	0
I-398	36	0	36	46	10
I-400	200	0	200	200	0
I-411	8	0	8	8	0
I-419	137	0	137	238	101
I-440	371	171	542	561	19
I-876	510	0	510	510	0
I-876zp	60	0	60	60	0
I-888	1656	347	2003	1983	−20
I-892	0	0	0	0	0
I-961	1766	260	2026	1876	−150
I-961S	0	0	0	32	32
I-1138	45	45	90	90	0
PROCESS STOCKS					
REF 90R	0	18	18	−35	−53
REF 95R	0	9	9	−26	−35
L.C.N	0	0	0	15	15
M.C.N	0	0	0	3	3
POLY	0	−3	−3	−4	−1
HSR	0	0	0	41	41
KERO BS	0	0	0	−5	−5
DIESEL BS	0	0	0	−30	−30
DES DSL	−171	−15	−186	−243	−57
LVGO	38	51	89	63	−26
L.ISOMATE	−75	−15	−90	−111	−21
HVGO	33	41	74	259	185
M. ISOMATE	−25	0	−25	−41	−16
H. ISOMATE	0	0	0	71	71
FCC CUTTER	75	15	90	0	−90
ATM. RESID	0	0	0	4	4
PRODUCTS	6154	1185	7339	7390	51
PROCESS STOCK	−125	101	−24	−39	−15
TOTAL	6029	1286	7315	7351	36

Table 18-2
Delta by Product Group, First Estimate (mb)

PRODUCT GROUPS	TOTAL DELTA	AOC DELTA ALLOCATION	BOC DELTA ALLOCATION
LPG	0	0	0
LSR	0	0	0
WSR	85	71	14
GASO	−16	−25	9
KERO	120	100	20
DSL	−20	−17	−3
BDSL	0	0	0
FUEL OIL	−118	−98	−20
ASPHALT	0	0	0
TOTAL	51	31	20

- The shipment schedule of participants: ship ID, dwt, parcel to be loaded (product grade and the quantity in tons or barrels), and the estimated time of arrival (ETA). The ETA is constantly updated by the participants.

This information is updated every time that

- A new refinery production estimate is done, typically, once a week. The refinery's new production rates, of all product groups based on the new estimate, are entered in the IUFS.
- The actual closing inventories of the participants from the previous month's allocation are available.
- The tankage capacity allocation of a participant is revised (due to releasing a storage tank for maintenance or adding a tank, which is put back in service after maintenance).
- There is a change in the ETA of a ship for the participant.

REFERENCE POINT

The reference point for any product group inventory is the closing inventories of the participants from the final allocation of the previous month.

To these figures are added, the actual past production of each product group of the participant. The basis of future production rates of a

Table 18-3
Estimation of Production Change Allocation for Inventory Forecasts, First Estimate (mb)

		AOC		BOC						
PRODUCT GROUP	TOTAL DELTA	I-397L −88	I-961(80) 32	I-397L −17	I-397LL 30	BALANCE DELTA	AOC DELTA	BOC DELTA	AOC TOTAL DELTA	BOC TOTAL DELTA
LPG	0					0	0	0	0	0
LSR	0					0	0	0	0	0
WSR	85					85	71	14	71	14
GASOLINE	−16	−22		−4	14	−4	−3	−1	−25	9
KEROSENE	120					120	100	20	100	20
DIESEL	−20		−4			−16	−13	−3	−17	−3
FUEL OIL	−118		4			−122	−102	−20	−98	−20
ASPHALT	0					0	0	0	0	0
TOTAL	51	−22	0	−4	14	63	53	10	31	20

NOTES:
CRUDE RATIO:
AOC 0.8354
BOC 0.1646

Table 18-4
Estimation of Production Rates for Inventory Forecast, First Estimate (mb)

PRODUCT GROUPS	DOP (1)	ALLOCATED DELTA (2)	PRODUCTION INVENTORY (3)	LOCAL SALES (4)	SHIPPING (5)	PRODUCTION RATE MBPCD (6)
AOC						
LPG	20	0	20	20	0	0.0
LSR	0	0	0	0	0	0.0
WSR	839	71	910	0	910	30.3
GASOLINE	542	−25	517	136	381	12.7
KERO	716	100	816	165	651	21.7
DSL	2226	−17	2209	54	2155	71.8
BDSL	0	0	0	0	0	0.0
HVGO	0	0	0	0	0	0.0
FUEL	1766	−98	1668	0	1668	55.6
ASPHALT	46	0	46	9	37	1.2
TOTAL	6155	31	6186	384	5802	193.4
BOC						
LPG	0	0	0	0	0	0
LSR	165	0	165	0	165	5.5
WSR	96	14	110	0	110	3.7
GASOLINE	101	9	110	0	110	3.7
KERO	171	20	191	107	84	2.8
DSL	347	−3	344	0	344	11.5
BDSL	0	0	0	0	0	0.0
HVGO	0	0	0	0	0	0.0
FUEL	260	−20	240	0	240	8.0
ASPHALT	45	0	45	0	45	1.5
TOTAL	1185	20	1205	107	1098	36.6

NOTES:
FIRST ESTGIMATE PERIOD = 10 days.

participant during the current month is the weekly allocation of the estimated production of the refinery between the participants. Estimates of production beyond current month are based on the advance DOPs of the participants.

For example, suppose today the date is June 10, from which date the IUFS data is to be entered.

Table 18-5
Determination of Deltas between Refinery Estimates and Combined DOP, Second Estimate (mb)

GRADE	COMBINED DOP	REFINERY ESTIMATE	DELTA
I-150	12	12	0
I-201	0	0	0
I-210	693	613	−80
I-220	161	161	0
I-383	−5	−5	0
I-387	31	31	0
I-390	42	60	18
I-395	73	73	0
I-397LL	16	16	0
I-397L	181	159	−22
I-397R	69	69	0
I-398	32	32	0
I-400	200	200	0
I-411	5	5	0
I-419	8	280	272
I-440	351	154	−197
I-800	−14	−12	2
I-876	214	214	0
I-876zp	0	0	0
I-885SP	72	72	0
I-888	1080	1047	−33
I-888 52DI	348	348	0
I-892	0	0	0
I-928	65	0	−65
I-961	1251	1158	−93
I-961S	−11	0	11
I-971	0	76	76
I-1138	68	68	0
INTERMEDIATE STOCKS			
REF 90R	−54	−27	27
REF 95R	−32	−13	19
LCN	1	1	0
MCN	−12	−12	0
POLY	−2	−2	0
HSR	36	36	0
KERO BS	39	39	0
DIESEL BS	−23	−23	0
DES DSL	−129	−129	0
LVGO	33	33	0
L. ISOMATE	−60	−60	0
HVGO	97	85	−12
M. ISOMATE	−30	−20	10
H. ISOMATE	70	85	15
FCC CUTTER	−28	0	28
ATM. RESID	−17	0	17
PRODUCTS	4942	4831	−111
PROCESS STOCKS	−111	−7	104
TOTAL	4831	4824	−7

Table 18-6
Delta by Product Group, Second Estimate (mb)

PRODUCT GROUPS	TOTAL DELTA	AOC DELTA ALLOCATION	BOC DELTA ALLOCATION
LPG	0	0	0
LSR	0	0	0
WSR	−80	−59	−21
GASO	−13	−11	−2
KERO	61	51	10
DSL	−28	−25	−3
BDSL	0	0	0
FUEL OIL	−71	−63	−8
ASPHALT	1	1	0
TOTAL	−130	−106	−24

1. The starting or reference point for the IUFS would be the closing inventories for all products as of April 30 (as per the final allocation report for the month of April of that year). The inventories of individual product grades are lumped together into product groups.
2. For the past period, May 1 to June 9, the actual refinery production rates for all product groups are available, and these are split into participant production rates on the basis of the weekly allocation of production for the IUFS for that period.
3. For the current date, June 10, to the end of current month, June 30, the production rates as established by latest weekly refinery estimates are used.
4. For the next 2 months, July and August, the participants' DOP's for July and August (advance DOPs) are used. The advance DOPs of the participants may not be very accurate but, nevertheless, contain useful data on the planned crude run, refinery units onstream factors, and projected product lifting schedule, if known beforehand.

Thus, participants have a useful tool to determine the inventory availability during the next 3 months and can schedule their vessels for product lifting accordingly.

The IUFS helps sense potential problems associated with the scheduling of ships and product loading, such as bunching ships, with the associated problems of berth availability, and delays or constraints on product loading system.

Table 18-7
Production Change Allocation For Inventory Forecasts, Second Estimate (mb)

		AOC				BOC							
PRODUCT GROUP	TOTAL DELTA	I-397L −18	I-961(80) 32	I-928 −65	I-971 76	I-397L −4	LSR ADJUST	CONDENSATE −9	BALANCE DELTA	AOC DELTA	BOC DELTA	AOC TOTAL DELTA	BOC TOTAL DELTA
LPG	0								0	0	0	0	0
LSR	0						9	−9	0	0	0	0	0
WSR	−80						−9		−71	−59	−12	−59	−21
GASOLINE	−13	−5				−1			−7	−6	−1	−11	−2
KEROSENE	61								61	51	10	51	10
DIESEL	−28		−2	−12	7				−21	−18	−3	−25	−3
FUEL OIL	−71		2	−16	−7				−50	−42	−8	−63	−8
ASPHALT	1								1	1	0	1	0
TOTAL	−130	−5	0	−28	0	−1	0	−9	−87	−73	−14	−106	−24

NOTES:
CRUDE RATIO:
AOC 0.835
BOC 0.165

Table 18-8
Estimation of Production Rates for Inventory Forecast

PRODUCT GROUPS	DOP (1)	ALLOCATED DELTA (2)	PRODUCTION INVENTORY (3)	LOCAL SALES (4)	SHIPPING INVENTORY (5)	PRODUCTION RATE, mbcd (6)
AOC						
LPG	10	0	10	10	0	0.0
LSR	134	0	134	0	134	6.7
WSR	579	−59	520	0	520	26.0
GASOLINE	367	−11	356	91	265	13.3
KERO	471	51	522	113	409	20.5
DSL	1420	−25	1395	36	1359	68.0
BDSL	0	0	0	0	0	0.0
HVGO	0	0	0	0	0	0.0
FUEL	1090	−63	1027	0	1027	51.4
ASPHALT	57	1	58	6	52	2.6
TOTAL	4128	−106	4022	256	3766	188.3
BOC						
LPG	2	0	2	0	2	0.1
LSR	27	0	27	0	27	1.4
WSR	114	−21	93	0	93	4.7
GASOLINE	72	−2	70	0	70	3.5
KERO	93	10	103	68	35	1.8
DSL	280	−3	277	0	277	13.9
BDSL	0	0	0	0	0	0.0
HVGO	0	0	0	0	0	0.0
FUEL	215	−8	207	0	207	10.4
ASPHALT	11	0	11	0	11	0.6
TOTAL	814	−24	790	68	722	36.1

ESTIMATE PERIOD = 7 DAYS

Table 18-9
Determination of Deltas between Refinery Estimates and Participants, DOP, Third Estimate (mb)

	TARGET DOP	REFINERY ESTIMATE	DELTA
GRADE			
I-150	6	6	0
I-201	0	0	0
I-210	533	451	−82
I-220	39	39	0
I-383	−2	−2	0
I-387	11	11	0
I-390	20	39	19
I-395	32	32	0
I-397LL	−21	0	21
I-397L	112	107	−5
I-397R	68	68	0
I-397C	25	0	−25
I-398	21	21	0
I-400	3	3	0
I-411	4	4	0
I-419	280	280	0
I-440	138	146	8
I-800	0	0	0
I-876	111	111	0
I-876zp	0	0	0
I-885SP	72	72	0
I-888	535	496	−39
I-888 52DI	348	348	0
I-892	0	0	0
I-928	4	0	−4
I-961	909	862	−47
I-961S	−3	0	3
I-971	−12	12	24
I-1138	53	30	−23
PROCESS			
STOCK			
REF 90R	−1	12	13
REF 95R	−39	−18	21
LCN	−1	−1	0
MCN	−3	−3	0
POLY	−7	−7	0
HSR	42	42	0
KERO BS	42	42	0
DIESEL BS	0	0	0
DES DSL	−61	−61	0
LVGO	51	51	0
L. ISOMATE	−20	−20	0
HVGO	54	54	0
M. ISOMATE	−18	44	62
H. ISOMATE	86	92	6
FCC CUTTER	−66	−66	0
ATM. RESID	−84	−84	0
PRODUCTS	3286	3136	−150
TOTAL PROCESS	−25	77	102
TOTAL	3261	3213	−48

Table 18-10
Delta by Product Groups, Third Estimate (mb)

PRODUCT GROUPS	TOTAL DELTA	AOC DELTA ALLOCATION	BOC DELTA ALLOCATION
LPG	0	0	0
LSR	0	0	0
WSR	−82	−59	−23
GASO	11	−11	22
KERO	10	51	−41
DSL	−32	−25	−7
BDSL	0	0	0
FUEL OIL	−24	−63	39
ASPHALT	−23	1	−24
TOTAL	−140	−106	−34

For example, referring to the IUFS system for whole straight run naphtha (WSR; Table 18-18),

- Column 1 refers to the calendar date.
- Column 2 refers to the physical inventory of WSR for participant AOC.
- Column 3 refers to the physical inventory of WSR for participant BOC.
- Column 4 is the total refinery inventory on the referred day, the sum of columns 2 and 3.
- Column 5 is available inventory of participant AOC, obtained by subtracting from AOC's physical inventory its allocated LI (Chapter 17).
- Column 6 is similarly the available inventory of participant BOC, obtained by subtracting from BOC's physical inventory its allocated LI.
- Column 7 is the total available inventory, the sum of columns 5 and 6.
- Column 8 is the ullage available to participant AOC for storing its product, the difference between the total storage capacity for WSR available to AOC (Chapter 17) and its present inventory.
- Column 9 is the ullage available to participant BOC for storing its product, the difference between the total storage capacity for WSR available to BOC (Chapter 17) and its present inventory.
- Column 10 is the total available ullage in the WSR group of tanks at the present inventory level. Thus,

$$\text{Inventory} + \text{ullage} = (\text{HI} - \text{LI}) \text{ for the refinery as well as for participants}$$

Table 18-11
Production Change Allocation for Inventory Forecasts, Third Estimate (mb)

		AOC					BOC									
GROUP	TOTAL DELTA	I-397L −5	I-961(80) 3	I-928 −4	I-971 24	ASPHALT −10	I-397C −25	I-397 21	ASPHALT −13	LSR ADJUST	CONDENSATE −6	BALANCE DELTA	AOC DELTA	BOC DELTA	AOC TOTAL DELTA	BOC TOTAL DELTA
LPG	0											0	0	0	0	0
LSR	0									6	−6	0	0	0	0	0
WSR	−82									−6		−76	−63	−13	−63	−19
GASOLINE	11	−1					−18	10				20	17	3	16	−5
KEROSENE	10											10	8	2	8	2
DIESEL	−32		−1	−1	3	−8			−10			−15	−13	−2	−20	−12
FUEL OIL	−24		1	−1	−2	18			23			−63	−53	−10	−37	13
ASPHALT	−23					−10			−13			0	0	0	−10	−13
TOTAL	−140	−1	0	−2	1	0	−18	10	0	0		−124	−104	−20	−106	−34

NOTES:
CRUDE RATIO:
AOC 0.835
BOC 0.165

Table 18-12
Estimation of Production Rates for Inventory Forecast, Third Estimate (bbl)

PRODUCT GROUP	DOP (1)	ALLOCATED DELTA (2)	ALLOCATION INVENTORY (3)	LOCAL SALES (4)	SHIPPING (5)	PRODUCTION RATE, mbpcd (6)
AOC						
LPG	5	0	5	8	−3	−0.2
LSR	33	0	33	0	33	2.5
WSR	445	−63	382	0	382	29.4
GASOLINE	222	16	238	59	179	13.8
KERO	355	8	363	119	244	18.8
DSL	891	−20	871	23	848	65.2
BDSL	0	0	0	0	0	0.0
HVGO	0	0	0	0	0	0.0
FUEL	750	−37	713	0	713	54.8
ASPHALT	44	−10	34	4	30	2.3
TOTAL	2745	−106	2639	213	2426	186.6
BOC						
LPG	1	0	1	0	1	0.1
LSR	6	0	6	0	6	0.5
WSR	88	−19	69	0	69	5.3
GASOLINE	44	−5	39	0	39	3.0
KERO	70	2	72	68	4	0.3
DSL	175	−12	163	0	163	12.5
BDSL	0	0	0	0	0	0.0
HVGO	0	0	0	0	0	0.0
FUEL	148	13	161	0	161	12.4
ASPHALT	9	−13	−4	0	−4	−0.3
TOTAL	541	−34	507	68	439	33.8

ESTIMATE PERIOD = 7 DAYS.

Table 18-13
Determination of Deltas between Refinery Estimates and Participants' DOP, Fourth Estimate (mb)

	TARGET DOP	REFINERY ESTIMATE	DELTA
GRADE			
I-150	−3	−3	0
I-201	0	0	0
I-210	257	262	5
I-220	−12	−33	−21
I-383	0	0	0
I-387	−5	−5	0
I-390	29	29	0
I-395	−5	0	5
I-397LL	7	7	0
I-397L	112	81	−31
I-397R	47	47	0
I-397C	0	0	0
I-398	−11	9	20
I-400	3	3	0
I-411	2	2	0
I-419	280	280	0
I-440	−94	−150	−56
I-800	0	0	0
I-876	55	55	0
I-876zp	0	0	0
I-885SP	72	72	0
I-888	−56	337	393
I-888 52DI	348	0	−348
I-892	1	0	−1
I-928	0	0	0
I-961	408	390	−18
I-961S	0	0	0
I-971	6	6	0
I-1138	15	12	−3
PROCESS STOCKS			
REF 90R	−13	−12	1
REF 95R	−16	−14	2
LCN	−18	−18	0
MCN	−9	−9	0
POLY	−8	−8	0
HSR	30	30	0
KERO BS	40	40	0
DIESEL BS	−16	−16	0
DES DSL	3	3	0
LVGO	39	39	0
L. ISOMATE	11	11	0
HVGO	−87	−20	67
M. ISOMATE	56	17	−39
H. ISOMATE	134	57	−77
FCC CUTTER	−102	−33	69
ATM. RESID	−31	−31	0
PRODUCTS	1456	1401	−55
PROCESS STOCKS	13	36	23
TOTAL	1469	1437	−32

Table 18-14
Delta by Product Groups, Fourth Estimate (mb)

PRODUCT GROUPS	TOTAL DELTA	AOC DELTA ALLOCATION	BOC DELTA ALLOCATION
LPG	0	0	0
LSR	–21	0	–21
WSR	5	–23	28
GASO	3	4	–1
KERO	–67	–56	–11
DSL	58	48	10
BDSL	–1	–1	0
FUEL OIL	–18	–14	–4
ASPHALT	–2	–2	0
TOTAL	–43	–44	1

that is, column 5 + column 8 = HI – LI, or allocated tankage capacity, for participant AOC (Chapter 17); and column 6 + column 9 = HI – LI, or allocated tankage capacity, for participant BOC.

- Columns 11 and 12 show the volume of projected lifting of products by the participants in thousand barrels on the date shown. Column 13 shows the ship ID, or identification number. Column 14 lists the product grade to be lifted by the ship. Column 15 shows the name of the ship and the likely date when the product is to be lifted. The data in columns 11 to 15 is constantly updated by the refinery, on receipt of information on ship ETA from the participants.
- Columns 16 and 17 indicate the current production rate of participants AOC and BOC. The production rate and product shipments are reflected as follows. The actual inventories of the participants (columns 1 and 2) increase at the rate of their production rates. For example, referring to the IUFS for WSR naphtha date June 19, the physical inventory of participant AOC is 796 mb. On June 20, the inventory increases to (796 + 30) or 827 mb, the production rate being 30.3 mb per day. The inventory on June 21 is (827 + 30.3 – 450) or 407 mb. Here, 450 mb is the WSR product loaded on ship Nordflex (column 11).

Table 18-15
Production Change Allocation for Inventory Forecasts, Fourth Estimate (mb)

		AOC			BOC							
GROUP	TOTAL DELTA	I-397L −26	REFORMER FEED −127	ASPHALT −2	REFORMER FEED −42	LSR ADJUST	CONDENSATE −7	BALANCE DELTA	AOC DELTA	BOC DELTA	AOC TOTAL DELTA	BOC TOTAL DELTA
LPG	0							0	0	0	0	0
LSR	−21					−14	−7	0	0	0	0	−21
WSR	5		25		24	14		−58	−48	−10	−23	28
GASOLINE	3	−7	102		18			−110	−92	−18	3	0
KEROSENE	−67							−67	−56	−11	−56	−11
DIESEL	58			−2				60	50	10	48	10
MARINE DSL	−1							−1	−1	0	−1	0
FUEL OIL	−18			4				−22	−18	−4	−14	−4
ASPHALT	−2			−2				0	0	0	−2	0
TOTAL	−43	−7	127	0	42	0		−198	−165	−33	−45	2

CRUDE RATIO:
AOC 0.8354
BOC 0.1646

Table 18-16
Estimation of Production Rates for Inventory Forecast, Fourth Estimate (mb)

PRODUCT GROUP	DOP (1)	ALLOCATED DELTA (2)	ALLOCATION INVENTORY (3)	LOCAL SALES (4)	SHIPPING (5)	PRODUCTION RATE,mbpcd (6)
AOC						
LPG	−3	0	−3	4	−7	−1.2
LSR	−10	0	−10	0	−10	−1.7
WSR	215	−23	192	0	192	32.0
GASOLINE	145	4	149	27	122	20.3
KERO	160	−56	104	55	49	8.2
DSL	351	48	399	11	388	64.7
BDSL	0	−1	−1	0	−1	−0.2
HVGO	0	0	0	0	0	0.0
FUEL	346	−14	332	0	332	55.3
ASPHALT	13	−2	11	2	9	1.5
TOTAL	1217	−44	1173	99	1074	179.0
BOC						
LPG	0	0	0	0	0	0.0
LSR	−2	−21	−23	0	−23	−3.8
WSR	42	28	70	0	70	11.7
GASOLINE	29	−1	28	0	28	4.7
KERO	31	−11	20	0	20	3.3
DSL	69	10	79	0	79	13.2
BDSL	0	0	0	0	0	0.0
HVGO	0	0	0	0	0	0.0
FUEL	68	−4	64	0	64	10.7
ASPHALT	2	0	2	0	2	0.3
TOTAL	239	1	240	0	240	40.0

ESTIMATE PERIOD = 6 DAYS.

Table 18-17
Lifting for Product Group LSR, Light Straight-Run Naphtha (mb)

	ACTUAL INVENTORY			AVAILABLE INVENTORY			ULLAGE			SHIPMENT M. BARRIES					PRODUCTION RATE, MBPCD	
DATE (1)	AOC (2)	BOC (3)	TOTAL (4)	AOC (5)	BOC (6)	TOTAL (7)	AOC (8)	BOC (9)	TOTAL (10)	AOC (11)	BOC (12)	SHIP ID (13)	GRADE CODE (14)	SHIP NAME (15)	AOC (16)	BOC (17)
1-Jun	132	76	208	80	42	122	103	80	183						0.0	5.5
2-Jun	132	81	213	80	47	127	103	75	178							
3-Jun	132	87	219	80	53	133	103	69	172							
4-Jun	132	92	224	80	58	138	103	64	167							
5-Jun	132	98	230	80	64	144	103	58	161							
6-Jun	132	103	235	80	69	149	103	53	156							
7-Jun	132	109	241	80	75	155	103	47	150							
8-Jun	132	114	246	80	80	160	103	42	145							
9-Jun	132	120	252	80	86	166	103	36	139							
10-Jun	132	125	257	80	91	171	103	31	134							
11-Jun	132	131	263	80	97	177	103	25	128							
12-Jun	132	136	268	80	102	182	103	20	123							
13-Jun	132	142	274	80	108	188	103	14	117							
14-Jun	132	147	279	80	113	193	103	9	112							
15-Jun	132	153	285	80	119	199	103	3	106							
16-Jun	132	158	290	80	124	204	103	−2	101							
17-Jun	132	164	296	80	130	210	103	−8	95							
18-Jun	132	169	301	80	135	215	103	−13	90							
19-Jun	132	175	307	80	141	221	103	−19	84							
20-Jun	132	180	312	80	146	226	103	−24	79							
21-Jun	132	186	318	80	152	232	103	−30	73							
22-Jun	132	191	323	80	157	237	103	−35	68							
23-Jun	132	197	329	80	163	243	103	−41	62							
24-Jun	132	202	334	80	168	248	103	−46	57		246	C60426	I-220	SIENNA 23–25		
25-Jun	132	−38	94	80	−72	8	103	194H	297							
26-Jun	132	−33	99	80	−67	13	103	189H	292							
27-Jun	132	−27	105	80	−61	19	103	183H	286							
28-Jun	132	−22	110	80	−56	24	103	178H	275							
29-Jun	132	−11	121	80	−50	30	103	172H	270							
30-Jun	132	313	445	80	−45	35	103	167H	264							

Table 18-18
Lifting for Product Group WSR, Whole Straight-Run Naphtha (mb)

	ACTUAL INVENTORY			AVAILABLE INVENTORY			ULLAGE			SHIPMENT					PRODUCTION RATE, Mbpcd	
DATE (1)	AOC (2)	BOC (3)	TOTAL INVENTORY (4)	AOC (5)	BOC (6)	TOTAL (7)	AOC (8)	BOC (9)	TOTAL (10)	AOC (11)	BOC (12)	SHIP ID (13)	GRADE CODE (14)	SHIP NAME (15)	AOC (16)	BOC (17)
1-Jun	476	296	772	373	173	546	106	607	713						30.3	3.7
2-Jun	506	299	805	403	176	579	76	604	680							
3-Jun	536	303	839	433	180	613	46	600	646							
4-Jun	567	307	874	464	184	648	15	596	611							
5-Jun	597	310	907	494	187	681	−15	593	578							
6-Jun	627	314	941	524	191	715	−45	589	544	225		A91460	I-210	ZENATIA		
7-Jun	433	318	751	330	195	525	149	585	734							
8-Jun	463	321	784	360	198	558	119	582	701							
9-Jun	493	325	818	390	202	592	89	578	667							
10-Jun	524	329	853	421	206	627	58	574	632							
11-Jun	554	333	887	451	210	661	28	570	598							
12-Jun	584	336	920	481	213	694	−2	567	565							
13-Jun	614	340	954	511	217	728	−32	563	531							
14-Jun	645	344	989	542	221	763	−63	559	496							
15-Jun	675	347	1022	572	224	796	−93	556	463							
16-Jun	705	351	1056	602	228	830	−123	552	429							
17-Jun	736	355	1091	633	232	865	−154	548	394							
18-Jun	766	358	1124	663	235	898	−184	545	361							
19-Jun	796	362	1158	693	239	932	−214	541	327							
20-Jun	827	366	1193	724	243	967	−245	537	292	450		A91100	I-210	NORDFLEX		
21-Jun	407	370	777	304	247	551	175	533	708							
22-Jun	437	373	810	334	250	584	145	530	675							
23-Jun	467	377	844	364	254	618	115	526	641							
24-Jun	498	381	879	395	258	653	84	522	606		236	C90426	I-210	SIENNA 23–25		
25-Jun	382	294	676	279	171	450	200	609	809							
26-Jun	412	298	710	309	175	484	170	605	775							
27-Jun	443	302	745	340	179	519	139	601	740							
28-Jun	473	305	778	370	182	552	109	598	707							
29-Jun	503	309	812	400	186	586	79	594	673							
30-Jun	534	313	847	431	190	621	48	590	638							

Table 18-19
Lifting for Product Group Gasoline (mb)

DATE	ACTUAL INVENTORY		TOTAL INVENTORY	AVAILABLE INVENTORY			ULLAGE			SHIPMENT		SHIP ID	GRADE CODE	SHIP NAME	PRODUCTION RATE, mbpcd	
(1)	AOC (2)	BOC (3)	(4)	AOC (5)	BOC (6)	TOTAL (7)	AOC (8)	BOC (9)	TOTAL (10)	AOC (11)	BOC (12)	(13)	(14)	(15)	AOC (16)	BOC (17)
1-Jun	230	327	557	51	207	258	234	−17	217						12.7	3.7
2-Jun	243	330	573	64	210	274	221	−20	201							
3-Jun	256	334	590	77	214	291	208	−24	184							
4-Jun	268	338	606	89	218	307	196	−28	168	19		A91790	I-397R	MASCARIN		
5-Jun	196	327	523	17	207	224	268	−17	251							
6-Jun	209	330	539	30	210	240	255	−20	235							
7-Jun	221	334	555	42	214	256	243	−24	219	26		A91820	I-395L	ASTROLOBE		
8-Jun	208	303	511	29	183	212	256	7	263							
9-Jun	221	306	527	42	186	228	243	4	247							
10-Jun	233	310	543	54	190	244	231	0	231							
11-Jun	246	314	560	67	194	261	218	−4	214							
12-Jun	259	317	576	80	197	277	205	−7	198							
13-Jun	272	321	593	93	201	294	192	−11	181	35		A91500	I-395L	LOTTE DANICA		
14-Jun	249	325	574	70	205	275	215	−15	200							
15-Jun	262	329	591	83	209	292	202	−19	183		35	C90421	I-395L	TBN 10–20		
16-Jun	275	297	572	96	177	273	189	13	202							
17-Jun	287	301	588	108	181	289	177	9	186							
18-Jun	300	305	605	121	185	306	164	5	169	18		A91840	I-395L	TBN 15–20		
19-Jun	295	308	603	116	188	304	169	2	171							
20-Jun	307	312	619	128	192	320	157	−2	155							
21-Jun	320	316	636	141	196	337	144	−6	138							
22-Jun	333	319	652	154	199	353	131	−9	122							
23-Jun	346	323	669	167	203	370	118	−13	105	44		A91780	I-397L	TBN 15–23		
24-Jun	314	327	641	135	207	342	150	−17	133	39		A92130	I-387	TBN 20–28		
25-Jun	288	331	619	109	211	320	176	−21	155	120		A91760	I-397L	TBN 20–30		
26-Jun	76	177	253	−103	57	−46	388	133	521							
27-Jun	88	181	269	−91	61	−30	376	129	505							
28-Jun	101	185	286	−78	65	−13	363	125	488							
29-Jun	114	188	302	−65	68	3	350	122	472							
30-Jun	126	192	318	−53	72	19	338	118	456							

Table 18-20
Lifting for Product Group Kerosene (mb)

	ACTUAL INVENTORY			AVAILABLE INVENTORY			ULLAGE			SHIPMENT					PRODUCTION RATE, Mbpcd	
DATE (1)	AOC (2)	BOC (3)	TOTAL (4)	AOC (5)	BOC (6)	TOTAL (7)	AOC (8)	BOC (9)	TOTAL (10)	AOC (11)	BOC (12)	SHIP ID (13)	GRADE CODE (14)	SHIP NAME (15)	AOC (16)	BOC (17)
1-Jun	534	333	867	286	167	453	146	120	266						21.6	2.8
2-Jun	556	336	892	308	170	478	124	117	241							
3-Jun	578	339	917	330	173	503	102	114	216							
4-Jun	599	341	940	351	175	526	81	112	193	29		A9210	I-440	MASCARIN		
										1		A9211	I-440	MASCARIN		
										22		A9212	I-440	MASCARIN		
5-Jun	569	343	912	321	177	498	111	110	221		1	C90535	I-440	MASCARIN		
6-Jun	590	346	936	342	180	522	90	107	197							
7-Jun	612	349	961	364	183	547	68	104	172	44		A9183	I-440	ASTROLOBE		
											238	C90428	I-419	ATHENIAN CHARM		
8-Jun	453	251	704	205	85	290	227	202	429							
9-Jun	474	253	727	226	87	313	206	200	406							
10-Jun	496	256	752	248	90	338	184	197	381							
11-Jun	517	259	776	269	93	362	163	194	357							
12-Jun	539	262	801	291	96	387	141	191	332							
13-Jun	561	265	826	313	99	412	119	188	307	86		A9150	I-440	LOTTE DANICA 10–1		
14-Jun	496	267	763	248	101	349	184	186	370	199		A9160	I-400	TBN 13–15		
15-Jun	319	270	589	71	104	175	361	183	544							
16-Jun	340	273	613	92	107	199	340	180	520							
17-Jun	362	276	638	114	110	224	318	177	495							
18-Jun	384	279	663	136	113	249	296	174	470	64		A9184	I-440	TBN 15–20		
19-Jun	341	281	622	93	115	208	339	172	511							
20-Jun	363	284	647	115	118	233	317	169	486							
21-Jun	384	287	671	136	121	257	296	166	462							
22-Jun	406	290	696	158	282	440	274	163	437							
23-Jun	428	293	721	180	127	307	252	160	412							
24-Jun	449	295	744	201	129	330	231	158	389							
25-Jun	471	298	769	223	132	355	209	155	364							
26-Jun	492	301	793	244	135	379	188	152	340		80	C90634	I-419	TBN 21–30		
27-Jun	514	224	738	266	58	324	166	229	395							
28-Jun	536	227	763	288	61	349	144	226	370							
29-Jun	557	229	786	309	63	372	123	224	347							
30-Jun	579	232	811	331	66	397	101	221	322							

Table 18-21
Lifting for Product Group Diesel (mb)

	ACTUAL INVENTORY			AVAILABLE INVENTORY			ULLAGE			SHIPMENT		SHIP ID	GRADE CODE	SHIP NAME	PRODUCTION RATE, Mbpcd	
DATE (1)	AOC (2)	BOC (3)	TOTAL (4)	AOC (5)	BOC (6)	TOTAL (7)	AOC (8)	BOC (9)	TOTAL (10)	AOC (11)	BOC (12)	(13)	(14)	(15)	AOC (16)	BOC (17)
1-Jun	699	174	873	541	68	609	88	350	438	1		AC2878	I-888		71.8	11.5
											1	AJ2133	I-888	AL JABER III		
											45	C20425	I-888	NEWTYNE		
										45S	45P	C20425				
2-Jun	725	185	910	567	79	646	62	339	401	1		GB2263	I-888	STRIKER		
3-Jun	795	196	991	637	90	727	−8	328	320		1	AJ2134	I-888	ALJABER IV		
4-Jun	776	73	849	618	−33	585	11	451	462	12		A29179	I-885	MASCARIN		
										9		A29210	I-885	MASCARIN		
										14		A29211	I-885	MASCARIN		
										12		A29212	I-885	MASCARIN		
										0		GB2269	I-888	SAFANIA		
											13	C290535	I-885	MASCARIN		
5-Jun	801	71	872	643	−35	608	−14	453	439		1	AJ2135	I-888	ALJABER III		
										0		GB2270	I-888	CHELSEA		
6-Jun	873	82	955	715	−24	691	−86	442	356	1		GB2268	I-888	OCEAN TRAMP		
7-Jun	936	93	1029	778	−13	765	−149	431	282	184		A29177	I-800	TEAM FROSTA		
										41		A29183	I-888	ASTROLOBE		
											37	C20538	I-888	ASTROLOBE		
8-Jun	782	68	850	624	−38	586	5	456	461	3		AC2873	I-888	USS PUGET SOUND		
9-Jun	851	79	930	693	−27	666	−64	445	381	203		A29197	I-888	OSCO SAILOR		

Table 18-21
Continued

	ACTUAL INVENTORY			AVAILABLE INVENTORY			ULLAGE			SHIPMENT					PRODUCTION RATE, Mbpcd	
DATE (1)	AOC (2)	BOC (3)	TOTAL (4)	AOC (5)	BOC (6)	TOTAL (7)	AOC (8)	BOC (9)	TOTAL (10)	AOC (11)	BOC (12)	SHIP ID (13)	GRADE CODE (14)	SHIP NAME (15)	AOC (16)	BOC (17)
											45	C20433	I-888	NEWTYNE 8–10		
10-Jun	720	46	766	566	−60	502	67	478	545	4		AC2882	I-888	USS SPICA		
11-Jun	788	57	845	630	−49	581	−1	467	466							
12-Jun	860	69	929	702	−37	665	−73	455	382	255		A29149	I-876	PARRIOT		
13-Jun	676	80	756	518	−26	492	111	444	555	66		A29150	I-888	LOTTE DANICA 10–1		
14-Jun	682	47	729	524	−59	465	105	477	582	37		A29214	I-888	FALL VII 12–15		
15-Jun	717	58	775	559	−48	511	70	466	536							
16-Jun	789	70	859	631	−36	595	−2	454	452							
17-Jun	861	81	942	703	−25	678	−74	443	369							
18-Jun	932	93	1025	774	−13	761	−145	431	286	222		A29117	I-888	GEORGIA		
19-Jun	412	104	516	254	−2	252	375	420	795	3		AC2874	I-888	USS PUGET SOUND		
20-Jun	481	116	597	323	10	333	306	408	714	5		AC2883	I-888	USS LA SALLE		
									255			A29172	I-876	COURIER		
											45	C20536	I-888	NEW TYNE 19–21		
										45S	45P	C20427				
21-Jun	248	127	375	90	21	111	539	397	936							
22-Jun	320	139	459	162	33	195	467	385	852							
23-Jun	391	150	541	233	44	277	396	374	770							
24-Jun	463	162	625	305	56	361	324	362	686	3		AC2875	I-888	USS PUGET SOUND		
25-Jun	532	173	705	374	67	441	255	351	606	185		A29115	I-888	TBN 20–30		
											148	C20427	I-888	TBN 20–30		
											45	C20537	I-888	NEW TYNE 24–26		
										95S	95P	C20427				
26-Jun	324	87	411	166	−19	147	463	437	900		80	C20634	I-419	TBN 21–30		
27-Jun	396	98	494	238	−8	230	391	426	817							
28-Jun	467	110	577	309	4	313	320	414	734	222		A29147	I-888	TBN 25–30		
29-Jun	317	121	438	159	15	174	470	403	873		45	C20635	I-888	TBN 28–30		
30-Jun	389	88	477	231	−18	213	398	436	834							

Table 18-22
Lifting for Product Group Fuel Oil (mb)

	ACTUAL INVENTORY			AVAILABLE INVENTORY			ULLAGE			SHIPMENT					PRODUCTION RATE, Mbpcd	
DATE (1)	AOC (2)	BOC (3)	TOTAL (4)	AOC (5)	BOC (6)	TOTAL (7)	AOC (8)	BOC (9)	TOTAL (10)	AOC (11)	BOC (12)	SHIP ID (13)	GRADE CODE (14)	SHIP NAME (15)	AOC (16)	BOC (17)
1-Jun	458	144	602	260	12	272	623	577	1200						55.6	8.0
2-Jun	514	152	666	316	20	336	567	569	1136							
3-Jun	569	160	729	371	28	399	512	561	1073							
4-Jun	625	168	793	427	36	463	456	553	1009	6		GA265	I-961	MASCARIN		
5-Jun	675	176	851	477	44	521	406	545	951							
6-Jun	730	184	914	532	52	584	351	537	888							
7-Jun	786	192	978	588	60	648	295	529	824	130		A9216	I-961	AL HAMIRA 6–8		
8-Jun	457	125	582	259	−7	252	624	596	1220							
9-Jun	513	133	646	315	1	316	568	588	1156	518		A9144	I-961	SEA JEWEL		
10-Jun	43	141	184	−155	9	−146	1038	580	1618							
11-Jun	98	149	247	−100	17	−83	983	572	1555							
12-Jun	154	157	311	−44	25	−19	927	564	1491							
13-Jun	209	165	374	11	33	44	872	556	1428	32		A9150	I-961	LOTTE DANICA 10–1		
14-Jun	233	173	406	35	41	76	848	548	1396							
15-Jun	289	181	470	91	49	140	792	540	1332							
16-Jun	344	189	533	146	57	203	737	532	1269							
17-Jun	400	197	597	202	65	267	681	524	1205	130		A9154	I-961	HAMIRA 15–18		
18-Jun	325	205	530	127	73	200	756	516	1272							
19-Jun	381	213	594	183	81	264	700	508	1208							
20-Jun	437	221	658	239	89	328	644	500	1144							
21-Jun	492	229	721	294	97	391	589	492	1081							
22-Jun	548	237	785	350	105	455	533	484	1017							
23-Jun	603	245	848	405	113	518	478	476	954							
24-Jun	659	253	912	461	121	582	422	468	890							
25-Jun	715	261	976	517	129	646	366	460	826							
26-Jun	770	269	1039	572	137	709	311	452	763	65		A9056	I-961	HAMIRA 25–27		
27-Jun	761	131	892	563	−1	562	320	590	910							
28-Jun	816	139	955	618	7	625	265	582	847	518		A9083	I-961	TBN 25–30		
29-Jun	354	147	501	156	15	171	727	574	1301							
30-Jun	410	155	565	212	23	235	671	566	1237							

Table 18-23
Lifting for Product Group Asphalt (mb)

	ACTUAL INVENTORY			AVAILABLE INVENTORY			ULLAGE			SHIPMENT					PRODUCTION RATE, Mbpcd	
DATE (1)	AOC (2)	BOC (3)	TOTAL (4)	AOC (5)	BOC (6)	TOTAL (7)	AOC (8)	BOC (9)	TOTAL (10)	AOC (11)	BOC (12)	SHIP ID (13)	GRADE CODE (14)	SHIP NAME (15)	AOC (16)	BOC (17)
1-Jun	37	22	59	29	16	45	−8	−3	−11		5	BA3101	I-1138		1.2	1.5
2-Jun	38	18	56	30	12	42	−9	1	−8							
3-Jun	40	20	60	32	14	46	−11	−1	−12	7		B39215	I-1138	AL JABER 4		
4-Jun	34	21	55	26	15	41	−5	−2	−7	6		B39218	I-1138	NEWQUAY		
5-Jun	29	23	52	21	17	38	0	−4	−4		5	BS3116	I-1149			
6-Jun	30	19	49	22	13	35	−1	0	−1							
7-Jun	31	21	52	23	15	38	−2	−2	−4							
8-Jun	33	22	55	25	16	41	−4	−3	−7							
9-Jun	34	24	58	26	18	44	−5	−5	−10		7	BA3102	I-1138	SEA JEWEL		
10-Jun	35	18	53	27	12	39	−6	1	−5							
11-Jun	36	20	56	28	14	42	−7	−1	−8							
12-Jun	37	21	58	29	15	44	−8	−2	−10							
13-Jun	39	23	62	31	17	48	−10	−4	−14							
14-Jun	40	24	64	32	18	50	−11	−5	−16							
15-Jun	41	26	67	33	20	53	−12	−7	−19							
16-Jun	42	27	69	34	21	55	−13	−8	−21		7	BA3103	I-1138	ALJABER 3		
17-Jun	43	15	58	35	9	44	−14	4	−10							
18-Jun	45	16	61	37	10	47	−16	3	−13	6		B39219	I-1138	NEWQUAY 17–19		
19-Jun	40	18	58	32	12	44	−11	1	−10							
20-Jun	41	19	60	33	13	46	−12	0	−12							
21-Jun	42	21	63	34	15	49	−13	−2	−15							
22-Jun	43	17	60	35	11	46	−14	2	−12							
23-Jun	45	19	64	37	13	50	−16	0	−16							
24-Jun	46	20	66	38	14	52	−17	−1	−18							
25-Jun	47	22	69	39	16	55	−18	−3	−21							
26-Jun	48	23	71	40	17	57	−19	−4	−23		6	BS3118	I-1149	AL JABER 2		
27-Jun	49	19	68	41	13	54	−20	0	−20	6		B39220	I-1138	NEWQUAY 26–28		
28-Jun	45	20	65	37	14	51	−16	−1	−17							
29-Jun	46	22	68	38	16	54	−17	−3	−20		5	BA3105	I-1138	AL JABER 2		
30-Jun	47	18	65	39	12	51	−18	1	−17							

MAXIMUM PARCEL SIZE

The maximum size of the product parcel of a participant should be less than or equal to the volume of storage space (HI – LI) available to it for a product group.

For example, consider the IUFS for WSR (whole range naphtha) for June 20. The maximum storage space for AOC for naphtha is (HI – LI) or 479 mb. Therefore, the single product lifting must be less than this volume. For a shipment larger than this volume, the participant is obliged to use the other participant's inventory.

Referring to the IUFS, on June 20, the available WSR inventory for AOC is 724 mb and has a negative ullage of 245 mb. This implies that participant AOC is using participant BOC's storage space, equal to its negative ullage of 245 mb, which can be done only with the prior permission of the BOC.

On the basis of IUFS-projected estimates, the refinery is ensured that

1. It does not develop overall negative ullage in any product group tanks at any time on a future date. This is done by directing participants to place shipments and relieve product storage tanks before such a situation actually develops.
2. Bunching of ships is avoided. This is a situation where, say, two or more large ships arrive within a small time interval, causing product availability as well as logistics problems for the terminal. For example, a ship may have to wait for an available berth or loading sea line, causing unnecessary demurrage payments to the ships.

The IUFS helps participants plan orderly transport of their product; foresee the effects on product inventory of situations, such as cancellation of a large shipment, temporary closure of the port due to weather conditions, or other emergencies; and plan remedial action with minimum cost to refinery.

SINGLE-OWNERSHIP REFINERIES

The IUFS for a single-ownership refinery is shown in Tables 18-24 to 18-30. The production rates used are the ones estimated in the weekly estimates for the whole refinery. The physical inventory for past period is the actual refinery inventory. Such data, along with production rates of

Table 18-24
Lifting for Product Group LSR, Light Straight-Run Naphtha (mb)

DATE (1)	PHYSICAL INVENTORY (2)	AVAILABLE INVENTORY (3)	ULLAGE (4)	SHIPMENT (5)	SHIP ID (6)	GRADE CODE (7)	SHIP NAME (8)	PRODUCTION RATE Mbpcd (9)
1-Jun	208	122	183					5.5
2-Jun	213	127	178					
3-Jun	219	133	172					
4-Jun	224	138	167					
5-Jun	230	144	161					
6-Jun	235	149	156					
7-Jun	241	155	150					
8-Jun	246	160	145					
9-Jun	252	166	139					
10-Jun	257	171	134					
11-Jun	263	177	128					
12-Jun	268	182	123					
13-Jun	274	188	117					
14-Jun	279	193	112					
15-Jun	285	199	106					
16-Jun	290	204	101					
17-Jun	296	210	95					
18-Jun	301	215	90					
19-Jun	307	221	84					
20-Jun	312	226	79					
21-Jun	318	232	73					
22-Jun	323	237	68					
23-Jun	329	243	62					
24-Jun	334	248	57	246	C60426	I-220	SIENNA 23–25	
25-Jun	94	8	297					
26-Jun	99	13	292					
27-Jun	105	19	286					
28-Jun	110	24	275					
29-Jun	121	30	270					
30-Jun	445	35	264					

Table 18-25
Lifting for Product Group WER, Whole Straight-Run Naphtha (mb)

DATE (1)	PHYSICAL INVENTORY (2)	AVAILABLE INVENTORY (3)	ULLAGE (4)	SHIPMENT (5)	SHIP ID (6)	GRADE CODE (7)	SHIP NAME (8)	PRODUCTION RATE Mbpcd (9)
1-Jun	772	546	713					34.0
2-Jun	805	579	680					
3-Jun	839	613	646					
4-Jun	874	648	611					
5-Jun	907	681	578					
6-Jun	941	715	544	225	A91460	I-210	ZENATIA	
7-Jun	751	525	734					
8-Jun	784	558	701					
9-Jun	818	592	667					
10-Jun	853	627	632					
11-Jun	887	661	598					
12-Jun	920	694	565					
13-Jun	954	728	531					
14-Jun	989	763	496					
15-Jun	1022	796	463					
16-Jun	1056	830	429					
17-Jun	1091	865	394					
18-Jun	1124	898	361					
19-Jun	1158	932	327					
20-Jun	1193	967	292	450	A91100	I-210	NORDFLEX	
21-Jun	777	551	708					
22-Jun	810	584	675					
23-Jun	844	618	641					
24-Jun	879	653	606	236	C90426	I-210	SIENNA 23–25	
25-Jun	676	450	809					
26-Jun	710	484	775					
27-Jun	745	519	740					
28-Jun	778	552	707					
29-Jun	812	586	673					
30-Jun	847	621	638					

Table 18-26
Lifting for Product Group Gasoline (mb)

DATE (1)	PHYSICAL INVENTORY (2)	AVAILABLE INVENTORY (3)	ULLAGE (4)	SHIPMENT (5)	SHIP ID (6)	GRADE CODE (7)	SHIP NAME (8)	PRODUCTION RATE Mbpcd (9)
1-Jun	557	258	217					16.4
2-Jun	573	274	201					
3-Jun	590	291	184					
4-Jun	606	307	168	19	A91790	I-397R	MASCARIN	
5-Jun	523	224	251					
6-Jun	539	240	235					
7-Jun	555	256	219	26	A91820	I-395L	ASTROLOBE	
8-Jun	511	212	263					
9-Jun	527	228	247					
10-Jun	543	244	231					
11-Jun	560	261	214					
12-Jun	576	277	198					
13-Jun	593	294	181	35	A91500	I-395L	LOTTE DANICA	
14-Jun	574	275	200					
15-Jun	591	292	183	35	C90421	I-395L	TBN 10–20	
16-Jun	572	273	202					
17-Jun	588	289	186					
18-Jun	605	306	169	18	A91840	I-395L	TBN 15–20	
19-Jun	603	304	171					
20-Jun	619	320	155					
21-Jun	636	337	138					
22-Jun	652	353	122					
23-Jun	669	370	105	44	A91780	I-397L	TBN 15–23	
24-Jun	641	342	133	39	A92130	I-387	TBN 20–28	
25-Jun	619	320	155	120	A91760	I-397L	TBN 20–30	
26-Jun	253	−46	521					
27-Jun	269	−30	505					
28-Jun	286	−13	488					
29-Jun	302	3	472					
30-Jun	318	19	456					

Table 18-27
Lifting for Product Group Kerosene (mb)

DATE (1)	PHYSICAL INVENTORY (2)	AVAILABLE INVENTORY (3)	ULLAGE (4)	SHIPMENT (5)	SHIP ID (6)	GRADE CODE (7)	SHIP NAME (8)	PRODUCTION RATE Mbpcd (9)
1-Jun	867	453	266					24.4
2-Jun	892	478	241					
3-Jun	917	503	216					
4-Jun	940	526	193	29	A9210	I-440	MASCARIN	
				1	A9211	I-440	MASCARIN	
				22	A9212	I-440	MASCARIN	
5-Jun	912	498	221	1	C90535	I-440	MASCARIN	
6-Jun	936	522	197					
7-Jun	961	547	172	44	A9183	I-440	ASTROLOBE	
				238	C90428	I-419	ATHENIAN CHARM	
8-Jun	704	290	429					
9-Jun	727	313	406					
10-Jun	752	338	381					
11-Jun	776	362	357					
12-Jun	801	387	332					
13-Jun	826	412	307	86	A9150	I-440	LOTTE DANICA 10–1	
14-Jun	763	349	370	199	A9160	I-400	TBN 13–15	
15-Jun	589	175	544					
16-Jun	613	199	520					
17-Jun	638	224	495					
18-Jun	663	249	470	64	A9184	I-440	TBN 15–20	
19-Jun	622	208	511					
20-Jun	647	233	486					
21-Jun	671	257	462					
22-Jun	696	440	437					
23-Jun	721	307	412					
24-Jun	744	330	389					
25-Jun	769	355	364					
26-Jun	793	379	340	80	C90634	I-419	TBN 21–30	
27-Jun	738	324	395					
28-Jun	763	349	370					
29-Jun	786	372	347					
30-Jun	811	397	322					

Table 18-28
Lifting for Product Group Diesel (mb)

DATE (1)	PHYSICAL INVENTORY (2)	AVAILABLE INVENTORY (3)	ULLAGE (4)	SHIPMENT (5)	SHIP ID (6)	GRADE CODE (7)	SHIP NAME (8)	PRODUCTION RATE Mbpcd (9)
1-Jun	873	609	438	1	AC2878	I-888		83.3
				1	AJ2133	I-888	AL JABER III	
				90	C20425	I-888	NEWTYNE	
				45	C20425			
2-Jun	910	646	401	1	GB2263	I-888	STRIKER	
3-Jun	991	727	320	2	AJ2134	I-888	ALJABER IV	
4-Jun	849	585	462	12	A29179	I-885	MASCARIN	
				9	A29210	I-885	MASCARIN	
				14	A29211	I-885	MASCARIN	
				12	A29212	I-885	MASCARIN	
				0	GB2269	I-888	SAFANIA	
				13	C290535	I-885	MASCARIN	
5-Jun	872	608	439	1	AJ2135	I-888	ALJABER III	
					GB2270	I-888	CHELSEA	
6-Jun	955	691	356	1	GB2268	I-888	OCEAN TRAMP	
7-Jun	1029	765	282	184	A29177	I-800	TEAM FROSTA	
				41	A29183	I-888	ASTROLOBE	
				37	C20538	I-888	ASTROLOBE	
8-Jun	850	586	461	3	AC2873	I-888	USS PUGET SOUND	
9-Jun	930	666	381	203	A29197	I-888	OSCO SAILOR	
				45	C20433	I-888	NEWTYNE	

10-Jun	766	502	545	4	AC2882	I-888	USS SPICA
11-Jun	845	581	466				
12-Jun	929	665	382	255	A29149	I-876	PARRIOT
13-Jun	756	492	555	66	A29150	I-888	LOTTE DANICA 10–1
14-Jun	729	465	582	37	A29214	I-888	FALL VII 12–15
15-Jun	775	511	536				
16-Jun	859	595	452				
17-Jun	942	678	369				
18-Jun	1025	761	286	222	A29117	I-888	GEORGIA
19-Jun	516	252	795	3	AC2874	I-888	USS PUGET SOUND
20-Jun	597	333	714	5	AC2883	I-888	USS LA SALLE
				255	A29172	I-876	COURIER
				45	C20536	I-888	NEW TYNE 19–21
				90	C20427		
21-Jun	375	111	936				
22-Jun	459	195	852				
23-Jun	541	277	770				
24-Jun	625	361	686	3	AC2875	I-888	USS PUGET SOUND
25-Jun	705	441	606	185	A 29115	I-888	TBN 20–30
				148	C20427	I-888	TBN 20–30
				45	C20537	I-888	NEW TYNE 24–26
				190	C20427		
26-Jun	411	147	900	80	C20634	I-419	TBN 21–30
27-Jun	494	230	817				
28-Jun	577	313	734	222	A29147	I-888	TBN 25–30
29-Jun	438	174	873	45	C20635	I-888	TBN 28–30
30-Jun	477	213	834				

Table 18-29
Lifting for Product Group Fuel Oil (mb)

DATE (1)	PHYSICAL INVENTORY (2)	AVAILABLE INVENTORY (3)	ULLAGE (4)	SHIPMENT (5)	SHIP ID (6)	GRADE CODE (7)	SHIP NAME (8)	PRODUCTION RATE Mbpcd (9)
1-Jun	602	272	1200					63.6
2-Jun	666	336	1136					
3-Jun	729	399	1073					
4-Jun	793	463	1009	6	GA265	I-961	MASCARIN	
5-Jun	851	521	951					
6-Jun	914	584	888					
7-Jun	978	648	824	130	A9216	I-961	AL HAMIRA 6–8	
8-Jun	582	252	1220					
9-Jun	646	316	1156	518	A9144	I-961	SEA JEWEL	
10-Jun	184	−146	1618					
11-Jun	247	−83	1555					
12-Jun	311	−19	1491					
13-Jun	374	44	1428	32	A9150	I-961	LOTTE DANICA 10–1	
14-Jun	406	76	1396					
15-Jun	470	140	1332					
16-Jun	533	203	1269					
17-Jun	597	267	1205	130	A9154	I-961	HAMIRA 15–18	
18-Jun	530	200	1272					
19-Jun	594	264	1208					
20-Jun	658	328	1144					
21-Jun	721	391	1081					
22-Jun	785	455	1017					
23-Jun	848	518	954					
24-Jun	912	582	890					
25-Jun	976	646	826					
26-Jun	1039	709	763	65	A9056	I-961	HAMIRA 25–27	
27-Jun	892	562	910					
28-Jun	955	625	847	518	A9083	I-961	TBN 25–30	
29-Jun	501	171	1301					
30-Jun	565	235	1237					

Table 18-30
Lifting for Product Group Asphalt (mb)

DATE (1)	PHYSICAL INVENTORY (2)	AVAILABLE INVENTORY (3)	ULLAGE (4)	SHIPMENT (5)	SHIP ID (6)	GRADE CODE (7)	SHIP NAME (8)	PRODUCTION RATE Mbpcd (9)
1-Jun	59	45	−11	5	BA3101	I-1138		2.7
2-Jun	56	42	−8					
3-Jun	60	46	−12	7	B39215	I-1138	AL JABER 4	
4-Jun	55	41	−7	6	B39218	I-1138	NEWQUAY	
5-Jun	52	38	−4	5	BS3116	I-1149		
6-Jun	49	35	−1					
7-Jun	52	38	−4					
8-Jun	55	41	−7					
9-Jun	58	44	−10	7	BA3102	I-1138	SEA JEWEL	
10-Jun	53	39	−5					
11-Jun	56	42	−8					
12-Jun	58	44	−10					
13-Jun	62	48	−14					
14-Jun	64	50	−16					
15-Jun	67	53	−19					
16-Jun	69	55	−21	7	BA3103	I-1138	ALJABER 3	
17-Jun	58	44	−10					
18-Jun	61	47	−13	6	B39219	I-1138	NEWQUAY 17–19	
19-Jun	58	44	−10					
20-Jun	60	46	−12					
21-Jun	63	49	−15					
22-Jun	60	46	−12					
23-Jun	64	50	−16					
24-Jun	66	52	−18					
25-Jun	69	55	−21					
26-Jun	71	57	−23	6	BS3118	I-1149	AL JABER 2	
27-Jun	68	54	−20	6	B39220	I-1138	NEWQUAY 26–28	
28-Jun	65	51	−17					
29-Jun	68	54	−20	5	BA3105	I-1138	AL JABER 2	
30-Jun	65	51	−17					

each product group, are used for future physical inventory forecasts. LI and HI data for every product group tankage (Chapter 17) is used to estimate the available inventory:

$$\text{Available inventory} = \text{physical inventory} - \text{LI}.$$

$$\text{Total volume available for storing a product group} = (\text{HI} - \text{LI})$$

$$(\text{HI} - \text{LI}) = \text{actual inventory} + \text{ullage}$$

where ullage is estimated from the preceding relationship. The rest of format for a single-ownership IUFS system is almost identical to that for joint-ownership refineries discussed earlier.

The IUFS is updated every time that

- A new refinery production estimate is done, typically, once a week. Refinery new production rates, for all product groups, based on the new estimate, are entered in IUFS.
- The tankage capacity in a product group service is revised (due to releasing a storage tank for maintenance or adding a tank, put back in service after maintenance).
- There is a change in the ETA of a ship, a new shipment is inserted, or data on an existing projected shipment is revised.

CHAPTER NINETEEN

Refinery Operating Cost

Petroleum refining is a capital-intensive business. A grassroots refinery of average complexity processing 100 mb crude per day may cost a billion dollars to build. For a refinery to be economically viable, its operating cost must be minimized. Joint-ownership refineries are built and operated with these objectives in view. Large throughput refineries can be built with initial investment spread over the resources of two companies instead of one. As many operating cost elements, such as depreciation, insurance, and personnel, cost remain constant with refinery throughput, operating cost per barrel crude processed is reduced, thereby increasing refinery profit for the participants. This chapter discusses the equitable sharing of operating costs of the refinery (allocation of operating costs) between the participants.

A refinery operating cost can be classified under the following heads:

Personnel cost. This includes salaries and wages of regular employees, employee benefits, contract maintenance labor, and other contracted services.

Maintenance cost. This includes maintenance materials, contract maintenance labor, and equipment rental.

Insurance. Insurance is needed for the fixed assets of the refinery and its hydrocarbon inventory.

Depreciation. Depreciation must be assessed on refinery assets: plant machinery, storage tanks, marine terminal, and the like.

General and administrative costs. This includes all office and other administrative expenses.

Chemicals and additives. These are the compounds used in processing petroleum and final blending, such as antioxidants, antistatic additives and anti-icing agents, pour point depressants, anticorrosion agents, dyes, water treatment chemicals, and so forth.

Catalysts. Proprietary catalysts used in various process units.

Royalties. Royalties are paid either in a lump sum or running royalty purchased for know-how.

Purchased utilities. This may include electric power, steam, water, and so on.

Purchased refinery fuel. This may include natural gas purchased by the refinery for use as refinery fuel and feedstock for hydrogen production.

ALLOCATION OF OPERATING COST

In a jointly-operated refinery, the individual operating expenses under different cost headings, as just described, can be allocated to the participants by one of the following methods: the system costing method, the theoretical sales realization valuation (TSRV) method, or on an actual usage basis.

SYSTEM COSTING METHOD

In the system costing method, costs are allocated to a participant in the ratio of its equity in the refinery All operating expenses involved in processing feedstocks in the refinery and related general and administrative services are allocated in this manner. The portion of current operating expenses related to major maintenance and repair items, such as unit shutdowns, emergency repairs, and large expenditure on replacements and renewals, which do not extend the life of fixed assets, are segregated. These are spread over a 12-month period by including in the operating expenses for each operating period, a monthly amount equal to 1/12th the estimated amount of such expenses, ensuring a 12-month period to recover the actual expenditure.

THEORETICAL SALES REALIZATION VALUATION METHOD

In the TSRV method, the total expense is allocated to the participants in the ratio of TSRV of its product. The following example illustrates the methodology involved.

EXAMPLE 19-1

The total operating expenses of a marine terminal of a refinery during a month were $1.3 million. We want to allocate these expenses to the participants using the TSRV method. The product shipments during the month from the terminal were as follows:

PRODUCT	AOC'S SHIPMENTS, bbl	BOC'S SHIPMENTS, bbl
NAPHTHA	817,149	511,711
GASOLINE	412,477	78,417
KEROSENE	632,858	101,675
DIESEL	1,900,245	460,552
FUEL OIL	1,706,555	376,461
ASPHALT	29,221	50,832
TOTAL	5,498,505	1,579,648

The first step in allocating the cost by the TSRV method is to estimate the value of the product shipped by both the participants. This is done by multiplying the shipment volumes by per unit cost of the product as follows. The cost used here is the average mean of Platts (MOP) published prices of the products during the month:

PRODUCT	MOP PRICE, $/bbl	AOC SHIPMENTS, $ millions	BOC SHIPMENTS, $ millions
NAPHTHA	18.681	15.265	6.373
GASOLINE	25.761	10.626	1.347
KEROSENE	27.252	17.247	1.847
DIESEL	23.234	44.149	7.133
FUEL OIL	13.422	22.905	3.369
ASPHALT	15.000	0.438	0.508
TOTAL		110.631	20.577

Total value of the product shipped over marine terminal = $131.208 million

Value of the product shipped by participant AOC = $110.631 million

Value of the product shipped by participant BOC = $20.577 million

Total value of the product shipped = \$131.208 million
Participant AOC's product share = 84.0%
Participant BOC's product share = 16.0%
Total operating expenses of the marine terminal = \$1.31 million
Participant AOC's share of operating cost (84%) = \$1.092 million
Participant BOC's share of operating cost (16.0%) = \$0.121 million

COST ALLOCATION FOR ACTUAL USAGE

The following cost items are allocated to the participants as per their actual usage:

1. The cost of the chemicals and additives, such as antiknock compounds, pour point depressants, and antistatic dissipaters. It is possible to accurately estimate the quantity of antiknock compound, pour point depressants, and other additives used in the final blending of their products from shipment and quality data records.
2. All operating expenses involved in receiving crude oil and other feedstocks in each operating period are segregated and allocated to the participants on the basis of that received by each in the period. For example, if a participant brings in a crude or another feedstock for processing in its share of refining capacity, all expenses related to receiving the crude is allocated to that participant. If a crude is brought in by pipeline for processing by both the participants, the pipeline-related expense is allocated to the participants in the ratio of the crude received.
3. All operating expenses involved in the manufacture and shipping of solid products, such as asphalt and sulfur, in each operating period are segregated and allocated to the participants on the basis of their respective shares of shipment of such products.

UNUSED CAPACITY CHARGE

After each operating period, the refinery establishes the amount of total crude distillation capacity available to each participant during the operating period but not used by that participant. The per-barrel charge to be applied to payable unused capacity is calculated as follows. If we let

$$\text{Total operating expenses during a month} = A \text{ (million \$)}$$
$$\text{Total available refinery crude distillation capacity during month} = B \text{ (mb)}$$
$$\text{Per barrel capacity charge} = A/B$$

Suppose the participants' equity in refinery is 60/40. The capacity available to participants during the month is

$$\text{AOC} = 0.6B \text{ mb}$$
$$\text{BOC} = 0.4B \text{ mb}$$

If one participant, say, BOC, utilizes only 95% of its available capacity,

$$\text{BOC's unused capacity} = 0.4 \times 0.05B \text{ mb}$$
$$= 0.02\text{B mb}$$
$$\text{Unused capacity charge payable by BOC} = \$[(A/B) \times (0.02B)] \text{ million}$$

The unused capacity charges are deducted from the total operating expenses of the refinery before allocating these expenses to the participants. An example of monthly allocation of various cost elements in an actual refinery is shown in Table 19-1. In the table the total monthly operating expenses are shown in column 4. The basis of allocation of each operating expense is indicated in column 1.

We see that

- Personnel, maintenance, insurance for the refinery plant and machinery, depreciation, royalties, catalyst costs, and so forth are allocated on the basis of participant equity in the refinery.
- Insurance for the hydrocarbon inventory in the refinery's tanks is on the basis of average inventory held by each participant.
- Natural gas import costs are allocated to the participants in the ratio of their usage in the allocation LPs.

Table 19-1
Monthly Allocation of Various Cost Elements

EXPENSE	BASIS OF ALLOCATION (1)	AOC ALLOCATION, $ (2)	BOC ALLOCATION, $ (3)	TOTAL EXPENSE, $ (4)
PERSONNEL	EQUITY	1,710,000	1,140,000	2,850,000
MAINTENANCE	EQUITY	958,500	639,000	1,597,500
INSURANCE				
ASSETS	EQUITY	112,500	75,000	187,500
INVENTORY	INV. RATIO	12,375	10,951	23,326
TAXES/LICENCES	EQUITY	4,926	3,284	8,211
DEPRECIATION	EQUITY	143,100	95,400	238,500
NATURAL GAS	USAGE	751,511	499,185	1,250,696
CHEMICALS/ADDITIVES	USAGE	69,438	103,273	172711
CATALYSTS	EQUITY	180,000	120,000	300,000
ROYALTIES	EQUITY	682	454	1,136
UTILITIES	CRUDE RUN	225,000	150,000	375,000
CRUDE RECEIVING	CRUDE RECEIPTS	2,932	3,058	5,990
PIPELINE				
MARINE TERMINAL	TSRV	624,268	633,639	1,257,907
ASPHALT PRODUCTION	PRODUCTION	756	504	1,260
ASPHALT DRUM FILLING	PRODUCTION	1,260	840	2,100
SULFUR PLANT	TOTAL CRUDE RUN	4,927	3,284	8,211
TOTAL		4,802,175	3,477,873	8,280,048

- Chemical and additive costs is allocated to the participants on the basis of actual usage. For example, the cost of anti-icing additives used for blending jet fuel by one participant can be estimated from the volume of jet fuel blended and the additive dosage rate. For additives and chemicals whose use by an individual participant cannot be identified is allocated in the ratio of the crude run of the participant.
- Utilities costs are allocated on the basis of the participant's crude run.
- If crude is received by way of a pipeline for use of more than one participant, all pipeline expenses are allocated to the participants in the ratio of their crude run in the refinery. If crude or feedstock is received by refinery for exclusive use of one participant, all expenses relating to that import are allocated to the receiving participant.
- All expenses relating to product export from the refinery marine terminal are allocated by the TSRV method. In this method, the value of total product exported by each participant is estimated and the total operating expenses of the terminal is allocated in that ratio. An example of the methodology involved is shown in Example 19-1.
- Asphalt production and drumming is a noncontinuous activity in many refineries. Special operating costs, such as drum filling, are segregated from other unit operating costs and allocated to participants in the ratio of the product shipped. Similarly, sulfur plant special costs, such as prilling plant costs, are allocated to participants in the ratio of their sulfur shipped.

REVENUES FROM LEASING EXCESS TANKAGE CAPACITY

A situation can arise in which the refinery has excess storage capacity available over that required for operation of the refinery at its maximum crude throughput. In such situation, the refinery can lease its surplus tankage capacity on a long-term lease to other companies for storage of its products and earn some extra profit. Any profit from such an operation is split between the participants in the ratio of their equity in the refinery.

EXAMPLE 19-2

A jointly owned company is awarded a tender from the Defense Department to store 1 million bbl petroleum products for a period of 5 years. The products comprise the following:

PRODUCTS	VOLUME, bbl
KEROSENE/JET FUEL	350,000
AUTOMOTIVE DIESEL	300,000
FUEL OIL	350,000
TOTAL STORED	1,000,000

The business is offered to the refinery, at $5.00/bbl/year. The refinery, however, has to use one storage tank of 500 mb capacity built exclusively for participant BOC, in addition to its own excess tankage capacity, to meet the storage capacity required for this business. The equity of the participants in the refinery is 60 and 40%. The participants agree to share revenues from this venture 50:50. Estimate net revenue of each participant from this business.

Total revenue from leasing tankage capacity = 1 million × 5 = $5 million/year. The revenue sharing of $5 million per year will be 50:50, as follows, for the 5-year period of the contract.

YEAR	AOC REVENUE, $ million	BOC REVENUE, $ million
1	2.5	2.5
2	2.5	2.5
3	2.5	2.5
4	2.5	2.5
5	2.5	2.5
TOTAL	12.5	12.5

To provide the required storage capacity, participant BOC leases its exclusively owned tank of 500 mb capacity to the refinery for this business at $5.0 bbl/year for a period of 5 years. As the equity of participants in the refinery is 60:40 for AOC and BOC, the revenue of the participants will be as follows:

Capacity of participant BOC tank leased to the refinery = 500 mb

Annual lease charge payable to the participant BOC = $500,000 × 5
= $2.5 million

This lease cost is shared between the participants in their equity ratio (60/40 here), as follows:

YEAR	LEASE COST, $ million		
	TO AOC	TO BOC	TOTAL
1	1.5	1.0	2.5
2	1.5	1.0	2.5
3	1.5	1.0	2.5
4	1.5	1.0	2.5
5	1.5	1.0	2.5
TOTAL	7.5	5.0	12.5

Revenue to BOC from leasing the tank to refinery, because of its ownership of the leased tank, is as follows:

YEAR	REVENUE TO BOC, $ million
1	2.5
2	2.5
3	2.5
4	2.5
5	2.5
TOTAL	12.5

The net profit to each participant from storage tender and participant BOC leasing its tank can be determined as follows:

$$\text{Participant AOC} = \$2.5 - \$1.5 = \$1.0\,\text{million/year}$$

$$\text{Participant BOC} = \$2.5 - \$1.0 + \$2.5 = \$4.0\,\text{million/year.}$$

	CUMULATIVE REVENUE, $ million	
YEAR	AOC	BOC
1	1.0	4.0
2	1.0	4.0
3	1.0	4.0
4	1.0	4.0
5	1.0	4.0
TOTAL	5.0	20.0

We see that the total revenue of the participants remains constant at $25 million, after 5 years of business. However, the split of this revenue between the participants has changed from 50:50 without BOC tank leasing to the refinery, to 20/80 in participant BOC favor as a result of tank leasing.

APPENDIX

Processing Agreement for Joint-Ownership Refinery

Operation of a joint-ownership refinery requires that the companies enter into a proper agreement on the procedures to be employed for establishing the ownership of stock and sharing the operating costs of the refinery. Also, the processing agreement defines the procedure for day-to-day running of the refinery, the split of processing unit and storage tanks capacity between the participants for processing their crude and storing inventory. It must, however, be emphasized that there is no physical split of processing unit or tankage capacities in the refinery. The split exits only in the refinery linear programming (LP) models and accounting procedures.

For proper operation of the refinery, a joint operating company (JOC) is formed. The JOC is responsible for the day-to-day operation of the refinery. Coordination between the refinery and the participants is through a Process Coordination Committee (PCC). The PCC has at least one member representing each participant and a member from the JOC. The structure of JOC is shown in Figure A-1.

Most refineries export their products from a marine terminal, the operation of which is closely controlled by the refinery. As the participating companies load their product from the marine terminal, close coordination between the participants and between the refinery and the participants is required. Processing agreement lays down the procedure for coordination and smooth operation of the marine terminal.

Presented here is an outline of an agreement used successfully for operation of a joint-ownership refinery which can serve as a model for other joint-ownership refining ventures, elsewhere.

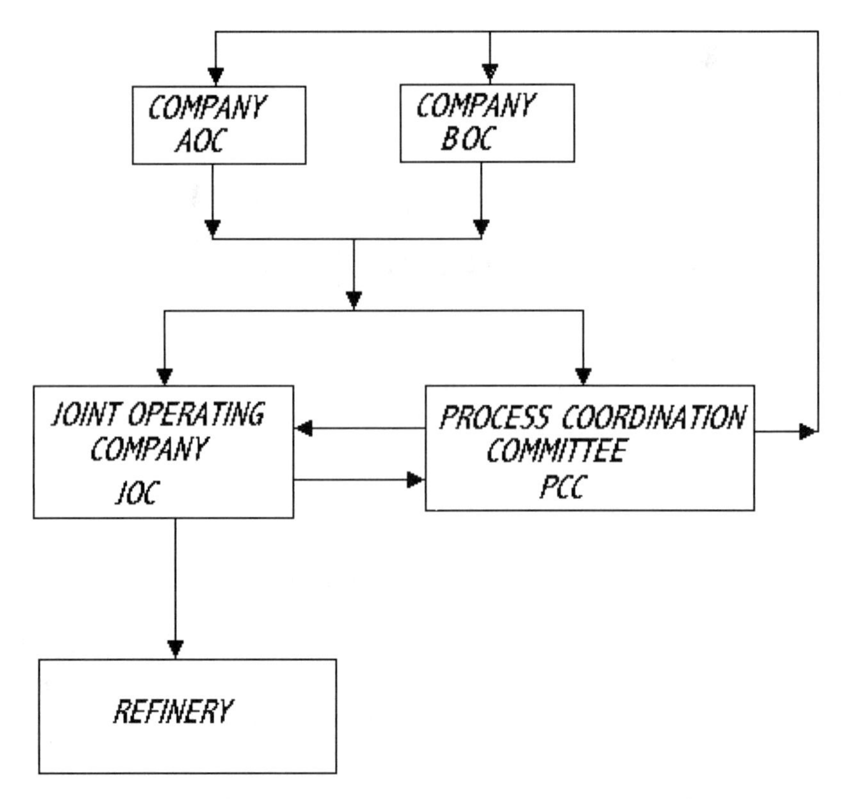

Figure A-1. Organizational structure of joint-ownership refinery.

PROCESSING AGREEMENT (PRO FORMA) BETWEEN COMPANIES AOC AND BOC FOR OPERATION OF A JOINT-OWNERSHIP REFINERY

Whereas the companies AOC and BOC have entered into a participation and operating agreement, setting out the basis of their relationship with respect to their jointly owned refinery and;

Whereas this is intended to be a participation and operating agreement, referred to as the Processing Agreement and executed by AOC and BOC on (DD/MM/YY).

Therefore, in consideration of the premises and mutual agreement hereinafter set forth, the parties agree as follows:

1. EFFECTIVE DATE

This agreement shall be effective from (DD/MM/YY) unless modified or terminated in accordance with Paragraph 11 and shall continue until (DD/MM/YY).

2. CAPACITY RIGHTS

2.1. During any operating period, each participant shall have the right to process feedstocks to a natural yield through any or all of the processing systems that constitute the total refining facilities ("Systems") available during each Operating Period in proportion to its equity share in the refinery ("Basic Capacity Right"). If either Participant elects not to use its share in one or more of the Systems available during an Operating Period, such spare capacity shall be made available to the other participant and, if used or finally accepted by the other participant, shall be treated as part of that participant's Basic Capacity Right, for the period of time agreed on.

2.2. Subject to the rights of the other participant and ability of the refinery to process it, each participant may deliver to the refinery for processing within its Basic Capacity Right, any feedstock approved in accordance with Paragraph 2.3.

2.3. At any time, either participant may give to the JOC, a detailed analysis of any feedstock not until then approved for processing in the refinery and request data on the yield and refining characteristics of that feedstock. The test run shall be made by the JOC to obtain such data, provided an allocation of the related cost is made to the participant requesting that test run and such test does not impair the JOC's ability to meet the Definitive Operating Program (DOP) of the other participant. The JOC shall provide the results of the test run to the Process Coordination Committee (PCC) established pursuant to Paragraph 9 of this agreement. If approved by both participants, the feedstock in question may be delivered by either participant for processing in the refinery and the yield and any capacity limitations encountered shall be included in the Basic Data Book of the refinery.

2.4. If either participant wishes to store, blend, or process stocks in refinery facilities, without requiring use of its Basic Capacity Rights, it may do so on the following terms:

a. The PCC first determines that
   i. JOC's ability to meet the DOP of the other participant will not be impaired by this action.
   ii. That such stock and any of its derivatives is compatible with other stocks with which they may comingle.
b. JOC shall ensure that
   i. An allocation of related cost is made to the participant supplying such stock to cover all expenses directly attributable to processing of this stock.
   ii. Such participant shall cause such stock or its derivatives to be lifted in accordance with a preagreed-on schedule between the participants and the JOC.

2.5. Slop from external sources shall be processed at the JOC's discretion and credited to a reduction in overall refinery losses.

3. BASIC DATA BOOK

The basic data shall be agreed on by the participants and contain the information, procedures, and other parameters that govern the operation of the refinery. This includes the following:

a. Refinery stock balance data sheets.
b. Feedstock assays.
c. Product equivalencies and conversion factors.
d. Product specifications.
e. Refinery and shipping terminal tankage summary.
f. Schedule of shipping facilities.
g. Port information.

All revisions, updates, and addition of new data shall require the approval of both participants through the PCC.

4. REFINERY PROGRAMMING PROCEDURE

4.1. The Process Coordination Committee shall develop the Definitive Operating Programs for each operating period in a manner established in the Refinery Programming Procedure. Under such procedure, each participant shall have the right to process feedstocks and receive back products according to latest Definitive Operating

Program. The objective shall be for the JOC to operate the refinery as close as possible to the combined DOPs, which incorporates each participant's DOP.

4.2. This procedure shall allow each participant to exercise its primary right to obtain a natural yield of products from that share of the refinery system which corresponds to its Basic Capacity Right during a given operating period.

4.3. This procedure shall also provide for a programming cycle during which all relevant information, including spare unit capacity, offtake requirements, feedstock availability, and overall refinery operating constraints, is provided by the JOC to the PCC so that the PCC can make alterations to the operating program to benefit the participants.

4.4. The manner in which the PCC will exchange information and evaluate shipping slates to establish an agreed-on definitive offtake program for each participant shall also be included in the procedure. The product offtake program shall meet, as closely as possible, the requested shipping slate for each participant. A combined definitive product offtake program for both participants shall be developed, which will allow performance of each of participant's agreed-on DOP, taking into account the required minimum and maximum inventory levels in the refinery.

4.5. In practice, it will be desirable from time to time to update the operating program, to utilize spare capacity or take advantage of unforeseen developments. Either participant may request the PCC to consider modifications at any time, and to the extent practical, the operating program shall be modified.

4.6. If any unforeseen event, not caused by one or both the participants, occurs (such as but not limited to an unscheduled shutdown) and in the JOC's opinion will prevent it from meeting a participant's DOP and product offtake program, the JOC shall promptly inform the participants through the PCC and take necessary action to minimize the disruption to planned refinery operations. The JOC shall also propose the necessary revisions to such programs and any further action that needs to be taken, and such programs shall be modified by the the PCC to the extent necessary.

4.7. If, during an operating period, a participant fails to deliver the feedstock required to meet its Definitive Operating Program or remove products in accordance with its definitive product offtake program and if such failure prejudices the right of the other participant under this agreement or prevents the JOC from operating the refinery to meet the agreed-on combined DOP, the JOC shall immediately inform both participants through the PCC and take necessary action to minimize disruption of the nondefaulting participant's program. The JOC shall also propose necessary provisions to the defaulting participant's DOP and its product offtake program shall be modified to the extent necessary. A revision to the nondefaulting participant's program may be made only if absolutely necessary for operational reasons or if such participant agrees. In any case, all additional cost incurred by the JOC as a result of the default shall be allocated to defaulting participant to the exclusion of any further claim by one participant against the other arising from such failure.

4.8. All communication related to the refinery programming procedures between the JOC and the participants shall be through the PCC.

5. ALLOCATION OF FEEDSTOCK AND PRODUCTION

5.1. When the volume of feedstocks actually processed during an operating period and or the actual volumes of product manufactured differs from the volumes estimated in the DOP, the actual volumes shall be allocated to the participants in the following manner, after taking into account all adjustments under Paragraphs 4.5, 4.6, and 4.7.

5.2. Actual feedstock processed for only one participant shall be allocated entirely to that participant. Actual feedstock processed for both participants shall be allocated so as to achieve in total the same overall ratio of the total feedstock as utilized in the final DOP.

5.3. The final DOP for each participant shall be modified in such a manner that the production of fixed grades and process stocks in the combined DOP is made equal to the actual production of fixed grades and process stocks. Also, the crude run in the combined DOP should equal the actual refinery crude run during the month. Such reworked program shall be called the "retro DOP," or retrospective operating program for the participants.

5.4. The retrospective DOP is compared with actual refinery production during the month. The deltas remain only in the balancing grade products and losses. These are allocated to the participants in the ratio of their retro DOP productions expressed in balancing grades.

The procedure for allocating the balancing grades is described in Attachment 3.

5.5. Sulfur produced shall be allocated to the participants in proportion to their respective quantities of feedstock actually processed during the operating period.

5.6. Stocks handled or processed for participants in accordance with Paragraph 2.4 shall be allocated entirely to the participant supplying such stocks.

6. QUALITY OF PRODUCTS

The products manufactured and delivered by the JOC to each participant shall conform to the Product Specifications set out in the product specifications book.

7. PRODUCT MOVEMENT AND MEASUREMENT CONDITIONS

The terms and conditions that shall apply to vessels, loading or discharging products or crude oil at the refinery marine terminal, as well as measurement, testing, and related responsibilities of the participants are contained in the Attachment 2 of this agreement.

8. OWNERSHIP, RISK, AND LOSSES

8.1. Ownership of the feedstocks and other stocks delivered to the refinery by each participant or such entity or entities for whom such participants are processing ("Designee") and of process stocks and finished products derived from them shall remain vested in participants or entities as the case may be.

8.2. The JOC shall submit monthly account statements to each participant showing quantities of its various feedstocks received, products manufactured, and products shipped during the month and quantities in process and in inventory at the end of each month. Such statement shall also show products transferred by agreement from one participant to the other.

8.3. Normal operating losses or gains of feedstocks and other delivered stocks, process stocks, and finished products shall be apportioned to participants or their designees as a part of allocations made in accordance with Paragraph 5.

8.4. Other losses shall be separated from normal operating losses and allocated to participants as follows:

a. For all stocks that can be identified as property of one participant or its designee, the losses shall be allocated entirely to that participant or its designee.
b. For all stocks that cannot be so identified, such as comingled stocks, the losses shall be allocated in proportion to the respective ownership of such stocks determined in an equitable manner at the time of the incident that caused the loss.

8.5. All stocks delivered by each participant, process stocks, and finished products derived from them shall be in the JOC's care and custody from the time

a. Crude and other stocks obtained locally pass the line of refinery fence.
b. Pipeline crude passes the refinery fence.
c. Feedstocks and other stocks pass the flange connecting the ocean tanker loading lines to the shore hose or unloading arm at the wharf until the time the products derived from them passes through the flange connecting the receiving line of the ocean tankers.

9. PROCESS COORDINATION COMMITTEE

The Process Coordination Committee (PCC) shall coordinate the implementation of this agreement between the participants and the JOC. The PCC shall be established as follows:

a. Each participant shall appoint one representative and up to two alternatives.
b. The refinery shall appoint the third representative, who shall be fully qualified in scheduling of the refinery operations.
c. The JOC shall appoint the secretary for all meetings and shall act as secretariat for the PCC, supplying all secretarial services the PCC may require.
d. The representatives shall meet in the refinery as often as required. The representatives of the JOC shall act in an advisory capacity. In all matters, decisions shall be made only by agreement between the representatives of both participants.

10. ALLOCATION OF REFINERY OPERATING COSTS

Allocation of refinery operating costs between the participants shall be calculated in accordance with the procedure described in Attachment 4.

The essence of this procedure is to achieve a fair and equitable allocation of refinery costs. This procedure shall continue until a new or revised procedure of allocating refinery operating costs is available and put in use by the agreement between the participants.

11. TERMINATION OF AGREEMENT

This agreement can be modified or terminated, in whole or in part, by agreement in writing, duly executed by participants.

(signatures)

For the Company AOC

For the Company BOC

ATTACHMENT 1. DEFINITIONS

Feedstocks. Feedstock includes any feedstock to the refinery's crude distillation facilities excluding slops.

Natural yield. The natural yield is any balanced yield of finished products that the JOC is able to obtain from feedstock processed, taking into account the capability of total refining facilities at its disposal.

Operating period. The operating period consists of one calendar month according to Gregorian calendar.

Operating program. The operating program is the JOC's best estimate of the quantity of finished products that will be produced for a participant when specified quantities of feedstocks and other stocks are processed through the refinery in a manner consistent with the product pattern desired by each participant. The JOC shall report the operating conditions for major refinery units and other relevant conditions in such programs.

Product offtake program. The product offtake program is a schedule of the volume quantities of each product grade to be lifted by a participant during an operating period. It shall indicate the scheduled vessel arrival at the refinery's loading terminal, cargo sizes, makeup of each cargo parcel, and lifting dates. It shall also include product delivery schedule via pipeline.

ATTACHMENT 2. LOADING AND DISCHARGE CONDITIONS

1. It is the responsibility of each participant to ensure that regulations and instructions issued from time to time by port authorities or the JOC contained in this attachment relating to the use of the port or its approaches are advised to the masters of vessels or their representatives nominated to offtake or discharge oil cargoes at the port.
2. Each participant, with respect to the vessel nominated for offtake or discharge of products or crude oil, shall arrange with vessel's masters or their representative to give three notices by radio to the JOC via independent channels, of the estimated time of arrival (ETA) at the loading port: the first notice approximately 72 hours, the second notice approximately 48 hours, and the third notice approximately 24 in advance of the ETA. Failure to give any notice, at least 24 hours in advance of arrival will increase the laytime allowed by the JOC by an amount equal to the difference between 24 hours and the number of hours prior to the arrival of the vessel that notice of ETA is received.
3. Vessels arriving on the dates or within the period established by definitive product offtake programs to lift or discharge cargoes shall be accepted by the JOC for loading or discharging in their order of arrival. Exceptions can be made by the JOC if

   a. In the JOC's opinion a suitable berth is not available.
   b. Safety would be jeopardized.
   c. The product or products are not available.
   d. Ullage in receiving tanks is not available.

   If a participant requests that one of its nominated vessels be given priority over another of its nominated vessels, the JOC, will decide whether such priority could be given, taking into account the other participant's rights. Vessels lifting or discharging cargo shall take priority over vessels lifting bunkers unless, in its sole discretion, the

JOC decides otherwise. The JOC shall not be held liable for deviation from the preceding provisions whenever, in its opinion, circumstances warrant such deviation.

4. The vessels shall be assigned to suitable berths by the JOC. However, if, in the opinion of the JOC, a vessel is not suitably sized and equipped so that it can be safely handled, moored, loaded, or discharged and unmoored or it is unsafe in any other respect, the vessel shall not be accepted and the nominating participant shall be advised accordingly.
5. Determination of the suitability of vessel's cargo tanks to load the nominated cargo shall be the responsibility of vessel's master or representative. Participants may request the services of an independent inspector to assess the suitability of the vessel's tanks. All costs or fees so incurred shall be on the account of participant requesting such inspection services.
6. Each participant shall arrange with the master or representative to give the JOC notice that the vessel is ready to receive or discharge products (hereinafter referred to as vessel's "notice of readiness," or NOR). The NOR shall be given to the JOC on arrival at the port by any available means followed by written confirmation as soon as practical. Except as provided in Paragraph 2, laytime for loading or discharge shall commence after a berth is available upon:

   a. Expiration of 6 hours after giving NOR.
   b. Commencement of loading or discharging, whichever occurs first.

   However period of delay in the commencement or actual loading/discharging of vessel caused by or due to action or deficiencies of the vessel shall be deducted from the calculation of lay-hours used.
7. In the event of a participant's vessel arriving outside the period or on a date other than the one scheduled by Definitive Operating Plan and product offtake program, such vessel, notwithstanding the order of arrival, shall await the berthing of other vessels scheduled for and arriving during that period or date. The JOC shall decide when such vessel may be berthed, taking into account the product availability and the other participant's rights. In such a case, the laytime shall not commence until the subject vessel commences loading or discharging.

8. Subject to Paragraphs 6 and 7, the JOC shall allow as laytime that amount of time which could be required for loading or discharging the cargo (including ballast, slops, and water) at such rates as established by the JOC from time to time. The JOC shall be allowed a minimum of thirty-six hours laytime for discharging any fully loaded vessel or loading a full cargo load. Adjustments for partly loaded vessel shall be done accordingly. Used laytime shall cease when loading/discharging line has been disconnected.
9. The JOC shall have the right of shifting vessels from one berth to another, in and out of berth, and shall pay all charges and expenses incurred in connection with shifting except when it is done for reasons of safety or *force majeure*. Each vessel shall vacate her berth as soon as practicable after loading or discharging is complete. Any loss or damage incurred by the JOC as a result of a vessel's failure to vacate the berth promptly, including any demurrage incurred due to the resulting delay to other vessels awaiting turn to load or discharge, shall be paid to JOC by the participant nominating such vessel.
10. If, in any case whatsoever reasonably beyond its control, the JOC is delayed, hindered, or prevented from furnishing the product required or any part thereof or from loading or discharging the vessel, the time so lost shall not be counted as laytime.
11. When the time is used in loading or discharging and the time counted as laytime exceeds the allowed laytime pursuant to the preceding paragraphs, JOC shall, if satisfied by the validity of the claim, pay the participant demurrage at the rate applicable to the size of the vessel used for the cargo concerned and based on the Worldwide Tanker Nominal Freight Scale (Worldscale) as amended from time to time by the London Tanker Broker Panel Average Freight Rate Assessment (AFRA) applicable to which shall be added any premium that may be applicable from time to time and in effect at the date the vessel tenders its notice of readiness to load or discharge or, if applicable, at the demurrage rate being paid under the charter for the vessel in question, whichever is less. Should Worldscale be cancelled and or AFRA cease to be promulgated, the demurrage rates shall be determined by reference to such voyage rate or scale as is generally deemed to be substituted by the trade from the time Worldscale or AFRA cease to exist as the industry standard. In no event, however, shall the JOC pay demurrage if not incurred by the participant.

12. If the actions or deficiencies of a participant's nominated vessel cause delay to a following vessel or vessels (not to exceed three) resulting in payment of demurrage to a third party by the JOC, such participant shall reimburse to JOC the amount paid in demurrage.

1. MEASUREMENT, SAMPLING, AND TESTING

a. The JOC shall measure all feedstocks either by means of shore tank gauges or by metering devices as necessary and appropriate. Such quantities and, in particular, API gravity and density shall be calculated net of sediment and water at 60°F or at such temperature as otherwise agreed by the participants from time to time. Regular daily samples of feedstocks and other stocks shall be taken by JOC from incoming supply line or from receiving tank or both for quality checks. Significant variation in quality shall be advised to participants by the JOC. When recommended by the JOC and approved by the PCC, representative samples of feedstocks shall be evaluated by competent outside agency to verify or recommend the correctness of assay data of such crudes shown in the basic data book.
b. The JOC shall measure all other stock quantities by means of shore tank gauges, metering devices, or weigh bridges as necessary and appropriate. Such quantities shall be calculated at 60°F or at such temperature as shall be agreed on by participants from time to time. Exceptions can be made, however, in case of certain products, such as LPG, sulfur, or bitumen, where vessels measurements can be accepted.
c. The quality of product shipped shall be determined by testing of shore tank samples prior to loading. The JOC's obligation to test shall be limited to those tests that guarantee product quality.
d. The JOC's responsibility for the quality of the product shall cease after the product passes the flange connecting the shore hose or loading arm to the receiving vessel, pipeline, or tank truck.
e. Each participant shall have the right, at its own expense, to have an independent inspector verify the quality and quantities of feedstocks, other stocks, and products.
f. The JOC shall retain the ship's composite samples of each discrete product shipment for a period of 6 months together with a spot sample of products from sea line taken during loading. Samples of products such as LPG and asphalt need not be retained.

g. All measurement, sampling, and testing shall be in accordance with pertinent ASTM or API standards. Modifications, if necessary to meet locally established practices, may be made with the participant's approval.
h. The JOC, on completion of loading each vessel, shall undertake the following on behalf of each participant or its designee:

    i. Prepare and sign for each vessel the loading certificate stating the quality and quantity of each product delivered on board or discharged from the vessel and analysis of each product.
    ii. Prepare and sign the bill of lading for each cargo.
    iii. Prepare and sign each vessel's time sheet.
    iv. If required by the participants, provide a certificate of origin.
    v. Furnish to each participant the number of copies of the certificates relevant to its shipment as it may require.
    vi. Advise participants promptly of the quantities loaded.

ATTACHMENT 3. ALLOCATION OF DIFFERENCES BETWEEN ACTUAL PRODUCTION AND THAT IN THE RETROSPECTIVE DOPS

1. The actual amount of each product manufactured in an operating period will be different from planned production of those products according to retrospective operating program (retro DOP) because of estimating inaccuracies and changes to operating conditions in actual practice. This appendix describes how these differences, which will normally be small, are to be allocated to the participants.
2. As a first step, the total delta for each product is allocated to the participants in proportion to their respective shares of production in the combined retro DOP for the operating period in question. Deltas for products that only one participant planned to make is allocated 100% to that participant. Deltas for products that neither participant planned to make are allocated to the participants in proportion to their respective share of total feedstock input in the combined retro DOP for the period in question.
3. By the nature of calculations described in Paragraph 2, the sum of each participant's delta allocation will be imbalanced between participants but equal in amount and of opposite plus and minus signs, since the sum of such imbalances shall equal zero.

Accordingly, a further step is required to eliminate these imbalances. This is done by "reverse allocation" as follows:

a. The total delta, for each product that both participants had planned to make are numerically added, without regard to its plus or minus sign, resulting in absolute values of combined total delta.
b. The ratio of total delta for each product, without regard to its plus or minus sign, divided by the absolute value of combined total delta is multiplied by the total imbalance (sum of delta allocations) with negative sign to obtain the reverse allocation for each participant.

4. The total product available to each participant for the operating period in question (composite allocation) is the amount of that product available to each participant from its share of the combined retrospective DOP plus the sum of allocations for such products made to it according to Paragraphs 2 and 3.
5. A pro forma example of the calculation procedure described is enclosed (Table A-1).

ATTACHMENT 4. ALLOCATION OF REFINERY OPERATING COSTS

1. UNUSED CAPACITY CHARGE

1.1. After each operating period, the JOC shall establish the amount of total Basic Capacity Right to crude distillation capacity that had been available to each participant during such operating period (after adjusting any such capacity offered to the other participant and accepted or finally used) but that was unused in practice by the participant or other participant.

1.2. After each operating period, the JOC shall determine per barrel charge to be applied to payable unused capacity. This charge shall be the total current operating expenses (with some adjustments described later) divided by the total crude distillation capacity that had been available during such period.

1.3. After each operating period, the JOC shall calculate each participant's payable unused capacity charge for such operating period by

Table A-1
Pro Forma Example of Product Allocation (ml)

BALANCING GRADE PRODUCTS	RETROSPECTIVE DOP			ACTUAL PROD.	TOTAL DELTA	*FIRST STEP*, SPLITTING TOT. DELTA		REVERSE ALLOCATION		SUM OF ALLOCATION DELTAS		COMPOSITE FINAL ALLOCATION		
	AOC (1)	BOC (2)	TOTAL (3)	(4)	(5)	AOC DELTA (6)	BOC DELTA (7)	AOC (8)	BOC (9)	AOC (10)	BOC (11)	AOC (12)	BOC (13)	TOTAL (14)
I-180 (LPG)	20	6	26	20	−6	−5	−1	1	−1	−4	−2	16	4	20
I-210 (NAPHTHA)	1045	260	1305	899	−406	−325	−81	36	−36	−289	−117	756	143	899
I-397 (MOGAS)	310	242	552	851	299	168	131	27	−27	194	105	504	347	851
I-440 (KERO)	733	223	956	1035	79	61	18	7	−7	68	11	801	234	1035
I-888 (DIESEL)	1650	870	2520	2540	20	13	7	2	−2	15	5	1665	875	2540
I-981 (FUEL OIL)	2028	1020	3048	2974	−74	−49	−25	7	−7	−43	−31	1985	989	2974
LOSS	54	39	93	181	88	51	37	8	−8	59	29	113	68	181
TOTAL	5840	2660	8500	8500	0	−86	86	86	−86	0	0	5840	2660	8500

NOTES:

ABSOLUTE VALUE OF COMBINED TOTAL DELTA IS 972

(1) RETROSPECTIVE AND ACTUAL TOTAL OPERATING PROGRAMS MUST HAVE SAME CRUDE RUN AND TOTAL PRODUCTION.

(2) TO MAKE THE SUM OF COLUMNS 6 AND 7 EQUAL TO ZERO, THE TOTAL IMBALANCE (86) IS SPREAD OVER ALL PRODUCT GRADES IN THE RATIO OF THEIR PRODUCTION AND DEDUCTED FROM THEM.

(3) AVCTD IS THE ABSOLUTE VALUE OF COMBINED TOTAL DELTAS IN COLUMN 5.

(4) COLUMN12 = COLUMN 1 + COLUMN 10.

(5) COLUMN13 = COLUMN 2 + COLUMN 11.

multiplying the barrels of payable unused capacity for such participant by the per-barrel charge for such participant as described and charge such participant accordingly. The sum of these amounts shall be deducted for the operating period in question prior to allocation of current operating expenses in Paragraph 2.

2. ALLOCATION OF OPERATING COSTS

2.1. The total operating expenses shall be computed by the JOC for each operating period and allocated to the participants as outlined here.

2.2. The portion of current operating expenses relating to major maintenance and repair items, such as unit shutdowns, emergency repairs, and large expenditure on replacements and renewals that do not extend the life of fixed assets, will vary in amount from month to month depending on the nature of the work. Such expenses shall be segregated and spread over a 12-month period by including in the operating expenses for each operating period, an agreed monthly amount based on 1/12th the estimated amount of such expenses for the ensuing 12-month period adjusted from time to time to recover actual expenditure incurred.

2.3. All current operating expenses adjusted in accordance with Paragraph 2.2 involved in the processing of feedstocks in the refinery and relating to general and administrative services, excluding the expenses defined in Paragraphs 1.3, 2.4, 2.5, and 2.6 shall be allocated to the participants in the ratio of their equity in the refinery.

2.4. All operating expenses involved in the shipping of products over the marine terminal in each operating period shall be segregated and allocated to all products so shipped during such period in the same ratio that the TSRV of each product so shipped bears to the total TSRV of all such products. The unit TSRV of a product shall be the simple arithmetic average of the prices per barrel in effect for that product, as posted or published for export products during the period in question by AOC and BOC. The total cost thus allocated to each product shall be to the account of each of the participants on the basis of their respective shares of actual product shipped.

2.5. All operating expenses involved in receiving feedstocks and other stocks in each operating period shall be segregated and allocated to the participants on the basis of the particular feedstock and other stock received by each in such period.

2.6. All operating expenses involved in the manufacture and shipping of solid products, such as asphalt sulfur, in each operating period shall be segregated and allocated to the participants on the basis of their respective shares of the shipment of such products in that period.

3. CAPITAL CHARGE FOR CEDED CAPACITY

Should one participant agree to cede a portion of its Basic Capacity Rights for a period in excess of 2 months, the participant utilizing such ceded capacity shall pay to the other participant an agreed-on charge to compensate it appropriately for depreciation and amortization of its investment in the refinery.

Index